Regional Geology Reviews

The Geology of—series seeks to systematically present the geology of each country, region and continent on Earth. Each book aims to provide the reader with the state-of-the-art understanding of a regions geology with subsequent updated editions appearing every 5 to 10 years and accompanied by an online "must read" reference list, which will be updated each year. The books should form the basis of understanding that students, researchers and professional geologists require when beginning investigations in a particular area and are encouraged to include as much information as possible such as: Maps and Cross-sections, Past and current models, Geophysical investigations, Geochemical Datasets, Economic Geology, Geotourism (Geoparks etc.), Geo-environmental/ecological concerns, etc.

More information about this series at http://www.springer.com/series/8643

Martin Meschede • Laurence N. Warr

The Geology of Germany

A Process-Oriented Approach

Springer

Martin Meschede
Institute of Geography and Geology
University of Greifswald
Greifswald, Germany

Laurence N. Warr
Institute of Geography and Geology
University of Greifswald
Greifswald, Germany

ISSN 2364-6438 ISSN 2364-6446 (electronic)
Regional Geology Reviews
ISBN 978-3-319-76101-5 ISBN 978-3-319-76102-2 (eBook)
https://doi.org/10.1007/978-3-319-76102-2

Library of Congress Control Number: 2018949344

This Springer imprint is published by the registered company Springer Nature Switzerland AG
The registered company address is: Gewerbestrasse 11, 6330 Cham, Switzerland

Preface

Our scientific perception of the world has radically changed over the last 200 years and a significant part of this originated from the valuable insight into past Earth processes gained by studying the regional and global geological record. The oldest geological maps of Germany date back to the eighteenth and nineteenth centuries, when the first systematic surveying of the land was undertaken. During the discovery and unraveling of the rock record, Germany played center stage to a number of important geoscientific discussions of global importance. Probably, the earliest relevant debate concerned the origin of basaltic rock (hence known as the basalt conflict) between two 18th century prominent science schools: the Neptunists and the Plutonists.

Since the founding of modern geological science in the middle of the 19th century, many generations of geologists have continued to unravel the details of Earth's history. This work has been anchored by systematic mapping in the form of national geological surveys, and in-depth research and scientific debate conducted by universities and other research centers. Paleontologists have contributed not only to the understanding of the origin of life but have also established a valuable biostratigraphy that enables age classification of sedimentary rocks. The practice of geological science has become an interdisciplinary part of the Earth's Sciences curriculum, based on the interaction of geochemists, geophysicists, hydrogeologists, mineralogists, paleontologists, petrologists, physical geographers, sedimentologists, structural geologists, and volcanologists. All of these disciplines are encompassed today by the unifying concept of plate tectonics that was established in the 1960s, which evolved from the findings of Alfred Wegener with his groundbreaking 1912 publication on the continental drift theory. It took more than 50 years for the theory of plate tectonics to be widely accepted, following the discovery of seafloor spreading. Viewing geological history in the light of current modern scientific principles, based on the concept of plate tectonics and the regional geology of Germany and Central Europe, is the foundation of this book.

Since geology is a science based on a combination of both observational and analytical methods, outcrops that once provided excellent insights into the geological record should not be lost: such as backfilled, flooded or overgrown quarries or sealed, inaccessible roadside cuts and mines. Several renowned and well-documented outcrops important to the geological history of Germany, as those found at Hagen-Vorhalle, are now found deteriorated or even completely closed. Therefore, this book intends to broaden the awareness of the need to preserve significant geological sites by taking into account the importance of the conservation of geological monuments and newly formed Geoparks. Geotope protection is to geologists just as important as biotope conservation is to ecologists. For this reason, geoscientific issues should also have a chance to be at the forefront of conservation needs, especially when it comes to accessing and preserving rock exposures in quarries or road cuts, which are fundamental to progressing geoscientific knowledge.

This book is primarily intended for students of geosciences, other science-oriented university degree programs, and a broad readership interested in the natural sciences. Originating from an introductory lecture series on the regional geology of Germany and Central Europe, its text is therefore written at a level that does not require prior geological expertise. In order to make it easier for all interested readers to understand the geology, a large number of supplementary "Boxes" are inserted to cover individual topics and terms in a concise, summarized form. Some general and introductory chapters on the themes of time, rocks, age dating, and plate tectonics precede the regional geological chapters. The regional geology starts with a consideration of earthquake activity in Germany in the context of large-scale plate tectonics. All subsequent chapters are ordered according to their stratigraphic development, starting with the oldest rocks and minerals found in Germany and finishing with the most recent ones. This particular book structure places the various successive geological events into the context of the Earth´s complex evolving history. Following a detailed description of the Variscan Orogeny, which occurred at the end of the Paleozoic and brought together the geological units of Germany, the following Mesozic history of the country is explained as part of Central Europe.

The unique development of the younger, Alpine region is of fundamental importance for understanding the geological development of Central Europe. The geology of the Alpine region is addressed in a separate chapter, regardless of country borders. The following Tertiary chapter covers the Cenozoic evolution of Germany at the foothills of the Alps with its special basin development and volcanic events. The asteroid impacts of the Nördlinger Ries and the Steinheim Basin are also described in a separate chapter devoted to these events. As a large part of the German landscape was shaped by the special climatic conditions of the geologically recent glaciation, the Quaternary deposits and erosional features caused by the glaciations are dealt with in the final chapter of the book.

The compilation of such a book is not possible without critical review. We would like to thank all those who helped with valuable hints, comments, and additions to the text, as well as those who supplied their photographic material in addition to drawings and designs. Our special thanks go to Horst Hann (Stuttgart), Helmut Weller, Jürgen Eidam†, and Heiko Hüneke (all Greifswald) for their help on the subject of the Paleozoic and older rocks. Werner Stackebrandt (Potsdam) corrected and supplemented the Mesozoic chapter and provided valuable input. Wolfgang Frisch (Vienna/Austria) and Hans-Jürgen Gawlick (Leoben/Austria) made helpful contributions to the Alpine chapter by suggesting corrections and additions. Kurt Goth (Dresden) helped by providing comments and corrections on the Tertiary chapter and Horst Kämpf (Potsdam) provided input on earthquakes and volcanism. Finally, our thanks go to Henrik Rother (Halle) for his revision and correction of the chapter on the Ice Ages and to Matthias Meschede (Paris/France) for his comments on the introductory chapters as well as Manfred Menning (Potsdam) for his help on stratigraphic tables and terminology. The former coworkers in our graphics lab, Heike Sengpiehl and Dagmar Lau, created the basis for many of the drawings, often by the tedious digitizing of templates. We also thank Stefan Meng (Greifswald) and Claus Meschede (Aachen) for providing useful photographs. Jörg Hartleib and André Deutschmann (Greifswald) helped with the satellite imagery and digital elevation models. Finally, yet importantly, our thanks go to our families for their understanding of the many hours spent clicking and typing on the computer during the compilation of this book.

This textbook presents an updated English version of the German "Geologie Deutschlands" by Martin Meschede, which was first published in 2015 with its second edition appearing in 2018. Such a work can only attempt to convey the current state of science and, as a result, future revisions and changes are therefore inevitable. We are most grateful to colleagues who have helped to make improvements to the first edition of the German version of the book, which have been incorporated into this new English version. We would like to particularly thank Wolfgang Franke (Frankfurt a. M.), who has provided extensive feedback on the older units covered in the book (Paleozoic and older) by providing a number of valuable references and constructive suggestions concerning the controversial nature of Late Paleozoic plate tectonic model. Furthermore, we thank Peter Suhr and Kurt Goth, both from the State Office for the Environment and Geology (Saxony) as well as Eckhard Groll (Eberswalde), Thorsten

Bickert, (Bremen), and Wolfgang Frisch (Vienna) for their corrections and clarifications of the text. Erica Alio-Warr is also thanked for editing the language and providing corrections.

We are always open to constructive comments and would be pleased to receive any feedback or suggestions that may improve the content of this book book (email: meschede@uni-greifswald.de, warr@uni-greifswald.de).

Greifswald, Germany
December 2018

Martin Meschede

Laurence N. Warr

Contents

About the Authors

Martin Meschede (born 1957) is Professor in regional and structural geology at the University of Greifswald since 2001. From 1986 to 2001, he was a research fellow at the University of Tübingen. He works largely in themes dealing with plate tectonics, marine geology, geodynamics, structural geology, and the regional geology of Central America, the east Pacific area, and northern central Europe. In addition to other scientific publications, he coauthored together with Wolfgang Frisch (Tübingen/Vienna) the textbook *Plate tectonics* and acted together with Jan Harff (Sczcecin/Poland), Sven Petersen (Kiel), and Jörn Thiede (Kiel/St. Petersburg/Russia) as editor for the *Encyclopedia of Marine Geosciences*.

Laurence N. Warr (born 1965) is Professor in economic geology and geochemistry at the University of Greifswald since 2007. From 1990 to 2002, he was a research fellow at the University of Heidelberg and between 2003 and 2007 researched for the CNRS at the University of Strasbourg. He works largely in the field of clay mineralogy, diagenesis, metamorphism, and deformation processes in fault zones. His regional geology interests lie in SW England, central Europe, California and New Zealand. In addition to other scientific publications, he is a contributing author to the textbook *Geological History of Britain and Ireland (2009; Wiley, 2nd edition).*

List of Figures

List of Tables

List of Boxes

1 Introduction

Wilt thou wander on forever? See the good that lies so near.

Johann Wolfgang von Goethe wrote these words in his poem *Erinnerung—Memory (1789)*—but they could be equally used to describe the diverse geology of Germany. Situated in the middle of Europe, Germany offers a rich and diverse geological history which, compared to other countries, occurs in quite a small area. We do not have to travel to distant places to visit special sites of geological interest; here we find them virtually on our doorstep. The reason for the geological diversity is its special location in the middle of a structural crossroads during the last 500–600 million years. Here, the breakup and collision of continents and smaller continental plates took place. Oceans and open seas spread and subsequently disappeared leaving nothing but a few crustal remnants, and finally, mountain ranges formed that were leveled to the base we can observe today.

Despite Germany's diverse geology, a complete insight into its geological history is only possible when we include neighboring countries. In northern Germany, this is mainly due to the fact that the ancient basement is covered by thick sedimentary deposits and is therefore only accessible at the surface in the adjacent regions of Scandinavia and Great Britain or at specific points when penetrated by deep boreholes. Fortunately, due to the intensive exploration activities of the former GDR (German Democratic Republic), there exist now a large number of boreholes in the northeastern part of Germany.

In the south, only a very small part of the Alps belongs to the German nation. To understand the evolution of the Alps, it is therefore necessary to look at the development of the entire mountain range. This is similar in the west and east of Germany when studying the ancient mountain belts of the Rheinisches Schiefergebirge, the Erzgebirge, Riesengebirge, and the Bohemian Massif. They pass through Germany from Belgium and France and continue to Poland, Czech, and Slovakia. Germany therefore represents just one piece of an entire European puzzle. Therefore, political borders today are irrelevant when it comes to studying geological processes. Historically, however, the diverse number of languages and cultural approaches used has never made European geology an easy subject. International political divisions are here mentioned only for orientation purposes and to set limits for the study area. To understand the structure of the basement and the geological evolution of Germany and its surrounding areas, a division into subordinate tectonic units is necessary to define regions formed during a distinct epoch of geological time or involved in profound changes common to that time.

This textbook illustrates the geological history of Central Europe written in a comprehensible language. It is a diverse history, which involves the formation and disappearance of oceans and mountain ranges over the course of millions of years. Central Europe, most specifically, Germany, has always been located at the heart of this structural evolution. Here, the oceans of the Variscan Orogeny were closed and mountain ranges formed only to be subsequently eroded immediately after. The formation of the Alps also induced significant tectonic movements, which can still be observed today in the form of low mountain ranges lying adjacent to the older Variscan structures or more ancient tectonic lines.

M. Meschede and L. N. Warr, *The Geology of Germany*, Regional Geology Reviews,
https://doi.org/10.1007/978-3-319-76102-2_1

2 Time and Geological Periods

Major global changes that have occurred over time are generally grouped into geological periods. For human beings, these periods represent quite unimaginably long intervals and they extend back to equally unconceivable ages that are difficult to grasp.

Because humans, in their contemporary environment, have a life expectancy of about 80 years, it is difficult to understand and relate to a period lasting for millions or billions of years. Geologists typically have to learn to handle such time periods that do not correspond to our known horizon. With the help of our parents and grandparents, we can properly comprehend a period of about 100 years. We may gain a better impression of the past by speaking with contemporary witnesses or by reading history books. We also achieve time recognition by taking into our hands and feeling objects of the present and the recent past that are taken from our surroundings. Yet, already it becomes more difficult to picture life before World War I. In fact, the further we try to look back in time, the more difficult it becomes to form an accurate mental picture. Our history changes dynamically in terms of both social circumstances and technology, whereby the things we take for granted today may not be around much longer in the future. Changes can be so dramatic that reports from the past sometimes appear like tales from a different world, even when considering the last 200–300 years. The motor vehicle, for example, did not exist much more than 125 years ago, whereas the first programmable computer appeared within the last 100 years. Many of us find it difficult to perceive life without modern computers or even mobile telephones, both of which have become normal parts of our everyday lives only since the Millennium. The internet, which is today the most important communication medium, started with its triumphal advance only a little more than 10 years ago (Figs. 2.1 and 2.2).

Geological periods, however, encompass not only hundreds of years but assume vast proportions of time. We count in tens of thousand years (10's Kyr), hundreds of thousand years (100's Kyr), and millions of years (Myr) depending on which phase of the Earth's evolution is being considered. To understand the enormous length of time encompassed in the periods of Earth's history, we need to visualize what, for instance, a period of 1 Myr actually represents. 1 Myr is a completely different period of time than that used in history or **archeology**. A period of 1 Myr is 200 times older than the beginning of modern civilization and the building of the Egyptian pyramids, the oldest of them being about 4600 years old. Another analogy that helps comprehend the length of 1 Myr is to imagine counting from 1 to 1,000,000 loud pronouncing every number. Let us assume that we need exactly one second for each number, then our counting experiment requires exactly 1,000,000 s. This is equivalent to a daily count of 11½ days with uninterrupted counting. Taking a 40 h working week with corresponding breaks as a basis, we would need a total of 7 weeks counting (it is though questionable if we are able to pronounce a series of digits like, for example, 637,859 within one second).

Even 1 Myr in a geological context is a short period of Earth's history. Geological processes are often so slow that they extend over many millions of years. The entire history of the Earth occurred over a formidable 4.51 billion years (Gyr), which is equivalent to a duration of 4510 Myr. The history of the universe itself started even before 14 Gyr ago. If attempting to visualize these incredible long periods, we soon realize the great difficulty in trying to perceive such vastness of time. However, despite problems of perception, it is evident that ample time was available for key processes such as the evolution of the species in the sense of Darwin or the eventual dawn of our more familiar world.

M. Meschede and L. N. Warr, *The Geology of Germany*, Regional Geology Reviews,
https://doi.org/10.1007/978-3-319-76102-2_2

1,000,000 years | 100,000 years | 10,000 years

Homo erectus | hand axe | megalithic tomb

Fig. 2.1 A selection of changes in the development of humankind over time. From left: skull of *Homo erectus* (© INTERFOTO/Mary Evans/Natural History Museum); hand axe from Sarstedt, Lower Saxony (courtesy of the Lower Saxony State Office of cultural heritage

To illustrate the historical development of the Earth and its temporal dimension, it is useful to equate Earth's history from its very beginning 4.51 billion years ago (Ga) until today with the duration of one year (365 days; Fig. 2.3). In this analogy, one day counts for a period of 12½ Myr. The birth of the **Earth–Moon system** took place on January 1, which formed as a consequence of the collision of **proto-Earth** (a somewhat smaller ancestor of the present-day Earth) with a planet of about the size of Mars, named **Theia**. The major portion of the unified mass (more than 98%) was maintained in the Earth in its present form, whereas the remaining material formed the Moon that still orbits the Earth since its collision.

As a result of their common origin, both terrestrial bodies of the Earth and Moon have the same age. The oldest minerals (zircons) detected following the initial formation of Earth originated in the analogy of a year at about mid-January (4.3 Ga), and the oldest rocks appeared in the first half of February (such as the Acasta Gneiss found in Northwest Canada, dated at 4.03 Ga by Bowring and Williams 1999). After nearly 3 months (3.5 Ga), the first signs of life appeared on planet Earth in the form of unicellular organisms. Plate tectonic movement started before about 3 Ga and from that same time also the oldest sedimentary rocks are known.

Around the middle to the end of May, the first life forms (such as the blue-green algae) started photosynthesis (2.7 Ga) to produce the first free oxygen dissolved in the ocean waters. Some weeks later, around mid-June, by the boundary between the Archean and the Proterozoic (2.5 Ga), about two-thirds to three-quarters of the continental crust already had formed and the first, more complex, multicellular organisms evolved. The transition from the Archean to the Proterozoic was also marked by the first proven glaciation. In the Early Proterozoic (at the end of July in Fig. 2.3), free oxygen could be detected for the first time, which was, however, below 0.3%. In the Late Proterozoic, life was still restricted to the oceans because the oxygen content of the Earth's atmosphere lay at a maximum of only 3%, far below its present value of 21%. Only in November, at the end of the Proterozoic, did the oxygen concentration of the Earth's atmosphere suddenly increase, with the multicellular organisms undergoing explosive diversification (about 620–600 Ma). This stage of development led to the well-known **Ediacara fauna**, which appeared as an exotic assemblage of soft-bodied animals. The end of the Proterozoic was reached on November 18 (542 Ma), and during the Early Cambrian on November 19 (at about 530 Ma) the first vertebrates evolved. This occurred together with the appearance of the first primitive plants that colonized the land by the end of November (about 420 Ma in the Silurian). The so-called Cambrian explosion provided the basis for the evolution of the more highly developed creatures that exist on the Earth today. Around December 13, the era of the dinosaurs began and lasted until December 27 (from 250 to 65 Ma, Triassic to Cretaceous). Only on December 29, did the Alpine Mountain Range of Europe we know today start to form, which before 20 Ma ago had been a relatively flat area with some parts submerged under marine water (Frisch et al. 2000).

Let us now consider separately the last day, and subsequently, the last hour of our one-year geological analogy to illustrate the number of important events that occurred in this short time span (Fig. 2.3). On December 31 at 12:00 O'clock, 12½ Myr of recent Earth history remains. At this time (Early Miocene), it was still tropically warm and brown coal seams formed in the areas of the Lower Rhine, Leipzig, and Helmstedt. Around noon, the Mediterranean Sea was closed off from the Atlantic and the Indian Ocean and dried up completely over a period of 600,000 years to leave a basin that lay more than 4000 m below sea-level. In the middle of

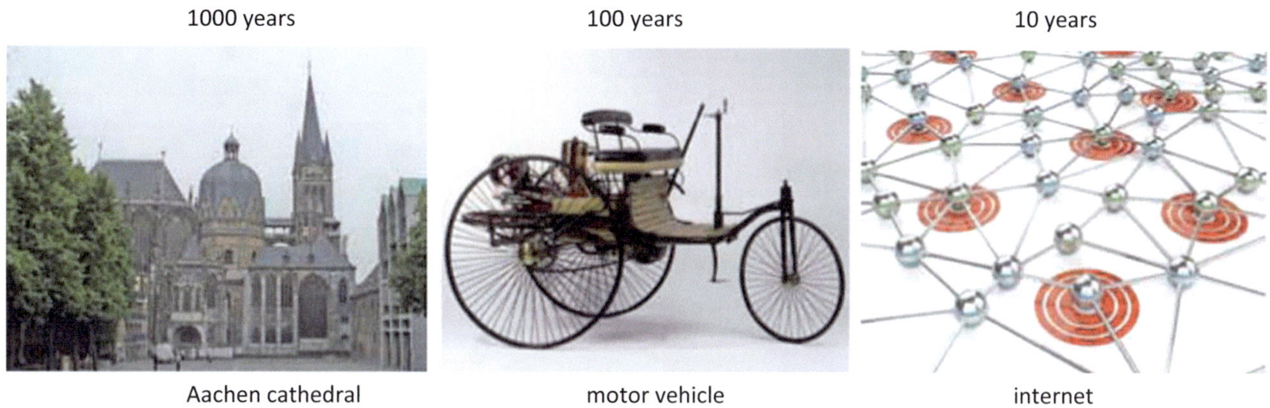

preservation); megalith tomb, near Sassnitz, Rügen Island; Aachen cathedral; Benz patented motor vehicle from 1886 (by courtesy of the Daimler AG, Mercedes-Benz Classic, Stuttgart); networking of the internet (© Dreaming Andy—Fotolia.com)

this depression, salt crystallized from the evaporated seawater to form thick layers of up to 100 m thickness (between 5.9 and 5.3 Ma). The first ancestors of the humans (genus *Homininae*) arrived between 10 and 11 O'clock (about 7 Ma). The direct **ancestors of the modern humans** (*Homo erectus*) appeared just 3½ h before the end of the year (about 1.8 Ma) and only shortly after the beginning of the last **glaciation** that disappeared only 1½ min before the end of the year (about 13.000 years b.p.). Shortly after, at 23:58:29:19 O'clock, the **Laacher See volcano** erupted in Rheinland-Pfalz (10.965 B.C.). Our recorded history (with the Nativity of Christ more than 2000 years ago) started in this analogy just a little more than 14 s before the stroke of midnight. Just to illustrate the vastness of time, the fall of the Berlin Wall (November 9, 1989 A.D.) happened only an instant before the years close: exactly 17 hundredths of a second before the end of our "geological" year.

Let us now turn towards the recent past witnessed and documented by the human race. It becomes immediately clear how short-lived our communications are and how insufficient our knowledge is about the past epochs of time beyond this short window of information. We can ask contemporary witnesses about the last 100 years and may have access to cinematic, photographic, or written documentation from that time. However, when looking further back, the number of preserved records decreases dramatically. The oldest photographs are no more than 200 years old and documentary movies extend back to a little more than 100 years ago. Paper books are a bit older, but they do not document more than the last 500 years. Moreover, paper is quite unstable and is only preserved over long periods of time under fortunate circumstances. In Europe, paper was not used until medieval times. Prior to this, books were printed on parchment with only a few copies made. Therefore, already for the medieval times, about 1000 years ago, we have to be satisfied with comparatively scarce remnants of information to help reconstruct events of that time.

It becomes even more difficult when we try to reconstruct the times of the Bronze Age or Stone Age. There are very few remnants of these former cultures, typically found in the form of isolated remains of massive or durable materials. In extremely rare cases, we can obtain a more complete historical record as, for instance, with the 5000-year-old glacial mummy of "Ötzi", discovered in 2001 in the Hintertux Alps. Numerous items of equipment and materials, which under normal conditions are not preserved, were deeply frozen in the ice and thus survived. Such circumstances are true luck, but such cases provide an invaluable insight into an otherwise unclear past. Easily degradable materials commonly used today, such as paper and wood, have a limited lifespan. In 1000 years, remnants of these materials will survive only under favorable conditions. Everybody can observe the destructiveness of nature in their own garden when following the annual cycle of plant growth or that of a compost pile. It is, therefore, all the more difficult to find and correctly interpret remnants of the geological past. Under oxygenated conditions, organic material decays within a few years and the solid components are frequently degraded, dissolved, and removed. Little of such material is preserved and, the further we go back in time, the sparser these remnants become and the less information obtained (Fig. 2.2).

Geological remnants are, in contrast to the materials of archeology or the historical sciences, often many times overprinted and modified. Archeologists normally deal with original objects although commonly rare in nature. Geologists, however, obtain their information from strongly modified or fossilized materials, in particular when these remnants are many millions of years old. Nevertheless, over a number of decades of modern scientific practice and with the particular help of radiogenic isotopes and fossilized

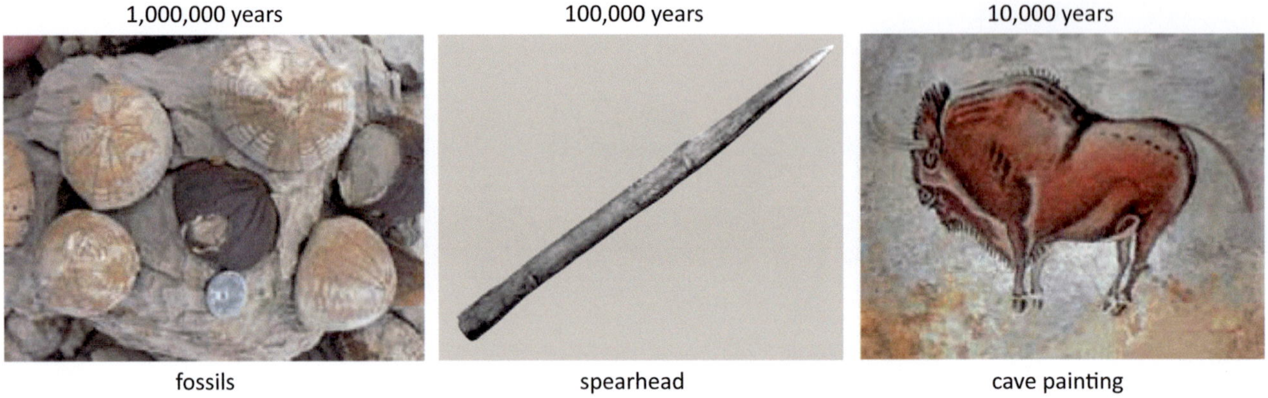

Fig. 2.2 What remains? Preservation over time. From the left: Eocene sea urchins; the lance of Lehringen (courtesy of the Historical Museum of Verden); Cave painting from the Altamira Cave, northern Spain (© picture-alliance/akg-images); mosaic from the 10th century of the Hagia

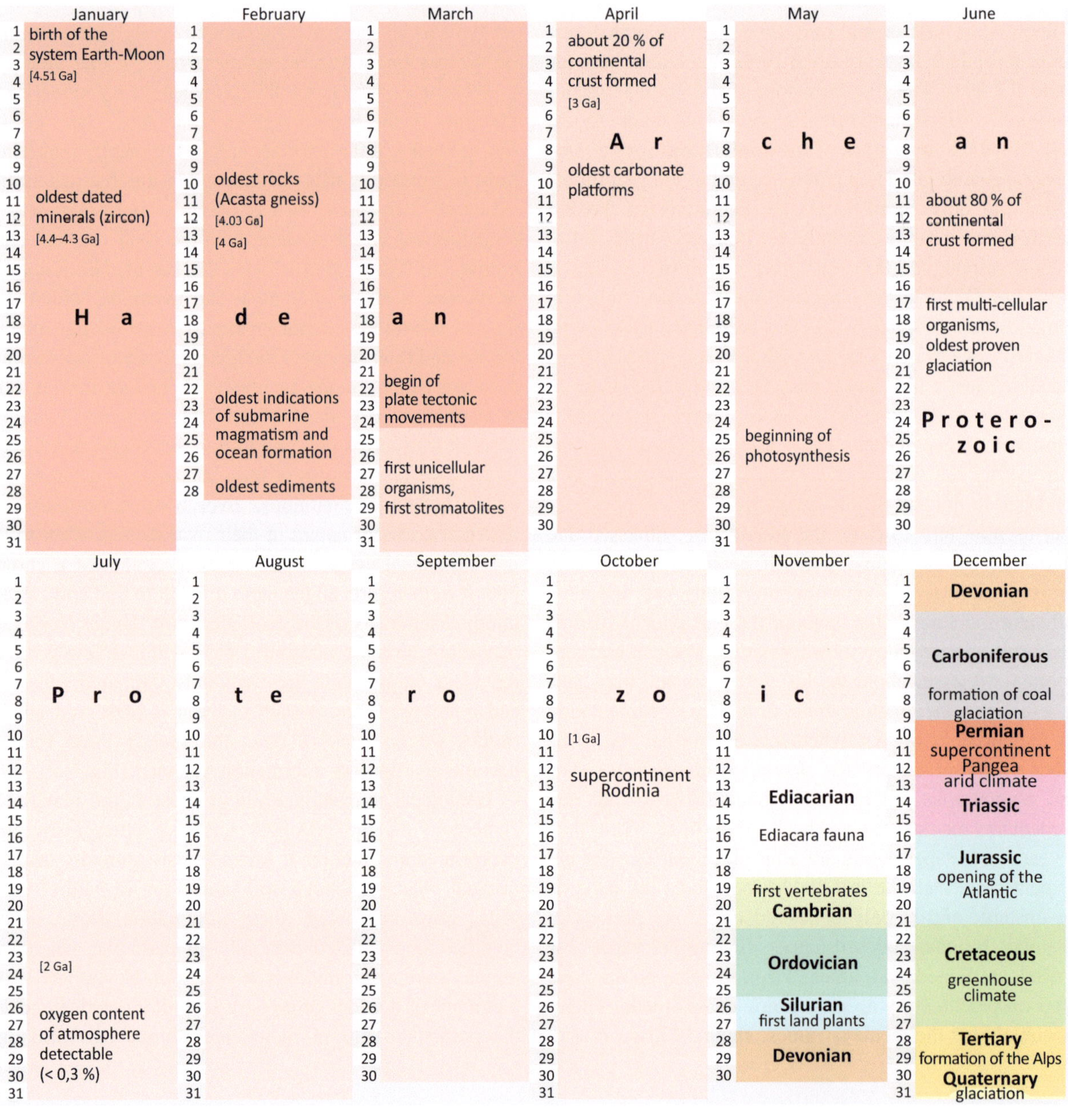

Fig. 2.3 The evolution of the Earth extrapolated to a calendar year; b.p. before present; B.C. Before Christ; A.D. in the year of the Lord

Sophia, Istanbul (© thinkstock/Getty Images International); hand colored photograph (status around 1900) and digital photograph (status of 2008) of the Rhône glacier (© J. Alean/M. Hambrey)

31. December, in the mornings		31. December, in the afternoon	
1	formation of brown coal	13	
2	in the Bay of Cologne	14	drying-up of the Mediterranean
3		15	
4		16	
5		17	
6		18	
7		19	
8		20	onset of the ice age
9		21	*Homo erectus*
10	first hominines	22	
11		23	
12		24	*Homo sapiens*

31. December, 23:00–24:00 hrs

23:37	*Homo sapiens sapiens* and *Homo neanderthalensis* (200,000 yrs. b.p.)
23:51:37	eruption of the Toba volcano (74,000 yrs. b.p.)
23:56	extinction of *Homo neanderthalensis* (28,000 yrs. b.p.)
23:58:29	end of last glaciation (13,000 yrs. b.p.)
23:58:29:19	eruption of the Laacher See volcano (10,965 B.C.)
23:59:29:41	building of the Egyptian pyramides (2500 B.C.)
23:59:45:91	begin of the modern calculation of time (the Nativity)
23:59:45:97	battle of the Teutoburger Wald (battle of Varus, 9 A.D.)
23:59:52:59	coronation of Charlemagne (800 A.D. [748 - 814 A.D.])
23:59:56:35	discovery of America by Christopher Columbus (1492 A.D.)
23:59:59:69	moon landing (21st July 1969 A.D.)
23:59:59:82	fall of the Berlin Wall (9th November 1989 A.D.)

Fig. 2.3 (continued)

organisms, it has become possible to reconstruct the evolution of the Earth from the lithified remnants of diverse geological processes. Using this approach, geologists have deduced a story of changing continents, landscapes, and climatic conditions that extends back to the very origin of our planet. Earth's story is, however, certainly incomplete in many details but the general picture now provides a clear view of global change and life's evolution, which has diversified and developed over many millions of years. Comparing this to a film recording, the history of the Earth would be like watching an incomplete series of frozen images and, although ongoing research provides us with an ever-increasing coverage, we will always have to infer most of the picture by putting together the pieces.

3 Rocks and the Geological Record

For geologists, rocks are the key to the past. They also offer the possibility to find out something about the inside of the Earth, how it is built up, and to which temperatures and pressure the rocks have been subjected to at different depths.

Though we will never be able to physically probe much deeper than the deepest borehole, which is currently a little more than 12 km deep on the Kola peninsula of Russia (north of Scandinavia), we know from the careful study of rocks and associated geophysical information rather well how the Earth's interior is built up. With the help of earthquakes that produce **seismic waves** (Greek *seismos*: vibration, earthquake), we are able to "listen into the Earth" and use the signals that bounce back to deduce its internal structure. This method resembles many medical procedures, for example, ultrasound computer tomography (USCT) that permits a physician to look into our body without the use of a scalpel but by utilizing the different penetrability or absorption behavior of our bones and soft tissues. In a similar way, seismic waves propagate at different velocities through different layers of the Earth's body that vary in composition, density, and temperature. This methodology allows the detailed structure of the interior of the Earth to be determined after data processing and analyses of an enormous number of wave measurements and calculations of the differences in their runtimes through the Earth (Fig. 3.1). This procedure is called **seismic tomography**.

Within the Earth and in particular the Earth's crust, three major rock groups can be recognized according to their origin: **magmatic**, **sedimentary**, and **metamorphic**, all of which are closely related to the **geological rock cycle** (Fig. 3.2).

Large volumes of **magmatic rock** formed at the beginning of the Earth´s history. Such rocks are generated by cooling and crystallization from a hot liquid, semi-molten-rock material known as **magma** (Greek *mágma*: kneaded mass). Such a mixture of melt-mineral-rock and dissolved gas originates where the temperature rises to levels of at least 700 °C. At the beginning of its history, 4.5 billion years ago, the Earth was significantly hotter than today and the rocks were generally softer and more ductile. As a result, the thinner outermost shell of the Earth was subjected to more frequent cataclysmic events, such as intense volcanism and meteorite bombardment.

Accordingly, the first rocks that contributed to the construction of the solid Earth's crust formed over a long time at greater depths. Even today, minor remnants of those original crustal rocks still exist. Rocks that form at depth by

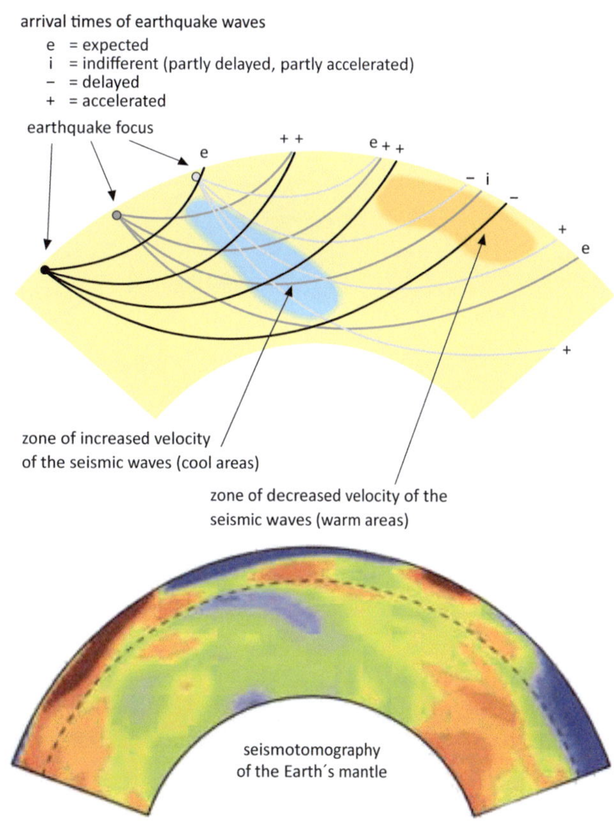

Fig. 3.1 How we can "look" into the earth: the principle of seismotomography (modified after Frisch and Meschede 2013)

M. Meschede and L. N. Warr, *The Geology of Germany*, Regional Geology Reviews,
https://doi.org/10.1007/978-3-319-76102-2_3

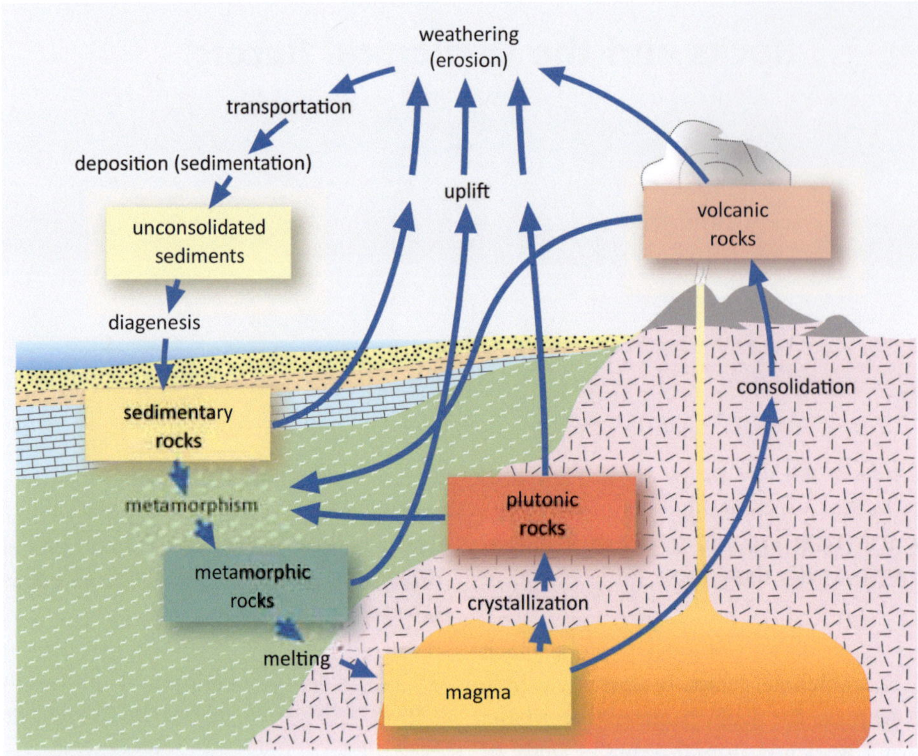

Fig. 3.2 Earth's rock cycle

Fig. 3.3 Granite (location: Weißenstein ruin, north of Erbendorf, in Upper Palatinate). Black—biotite, gray—quartz, white—feldspar (plagioclase, alkali feldspar) [N 49°54′50″/E 12°04′58″]

Fig. 3.4 Coarse-grained gabbro (location: Coverack, Great Britain). Black—pyroxene, little brown minerals—weathered olivine, white—plagioclase [N 50°00′14″/W 05°06′44″]

crystallization directly from magma are known as **plutonic igneous rocks** (after *Pluto*, the God of the underworld) or intrusive rocks. As intrusive magma rises and penetrates into the cooler and solid wall rock at higher crustal levels, the molten magma phase itself cools down and solidifies by crystallization. At a depth of several kilometers, the cooling of magma is typically very slow (taking millions of years) and ample time is available to grow large and visible interlocking crystals typical of the common varieties of coarse-grained igneous rocks such as granite or gabbro (Figs. 3.3 and 3.4; Box 1: Granite).

Box 1: Granite

Granite is an igneous rock composed largely of the rock-forming minerals quartz, alkali feldspar, plagioclase, and mica. It differs significantly from gabbro (the coarse-grained intrusive rock equivalent of basalt), which does not contain quartz, alkali feldspar and mica, but instead pyroxene and commonly olivine. Both rocks, however, originate from a magma that crystallizes in the Earth's crust. Why does granite form in one case and in another case gabbro? A gabbroic melt, such as that which develops at a mid-ocean ridge, is a product of the asthenosphere, i.e., the upper mantle (see Fig. 5.4). It represents the beginning of a magmatic series and a more primitive state of a magma that may evolve through different stages to a granitic melt by a process known as fractional crystallization. As magma cools, the various rock-forming minerals crystallize from the melt-phase and sink down to the bottom of the magma chamber. The minerals, however, have different melting and crystallization temperatures. Some minerals crystallize earlier during cooling and are extracted from the melt as a solid fraction, whereas other minerals require lower temperatures to crystallize. As a result, the composition of the melt changes continuously. All the main rock-forming minerals belong to the **silicate group**, i.e., they contain different proportions of silicon dioxide (SiO_2), also known as silicic acid. Therefore, gabbroic (or basaltic) melts are defined as basic (little SiO_2) and granitic (or rhyolitic) melts as acid (lots of SiO_2). For gabbroic-basaltic melts, olivine and pyroxene crystallize first along with plagioclase. Olivine and pyroxene contain relatively high amounts of Fe and Mg, whereas the SiO_2 content is lower. As these minerals are extracted from the melt by crystallization, they deprive the remaining melt of elements necessary for the further formation of mafic minerals. The composition of the melt changes as larger amounts of Si, Al, and alkali metals, namely Na and K, remain in the melt and are passively enriched by this process. As a result, the composition of the melt becomes more "acid". With a

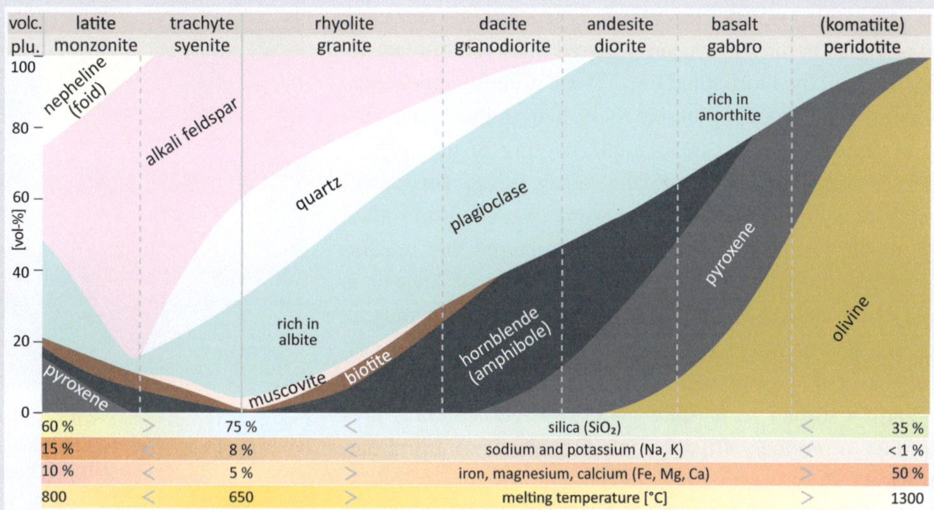

Fig. 3.5 Mineral composition of important igneous rock types ranging from ultramafic (right side of image) to granitic and quartz-free varieties with feldspathoid minerals (left side of image). V. volcanic rocks, P. plutonic rocks (modified after Frisch and Meschede 2013)

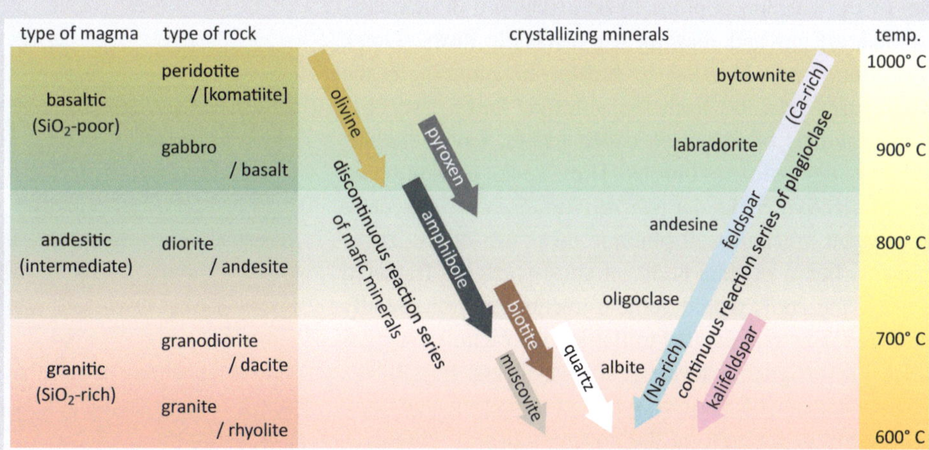

Fig. 3.6 Bowen's reaction series: discontinuous series of mafic minerals, continuous series of plagioclase

further decrease in temperature, other minerals with different compositions continue to crystallize from the melt-phase until completely solidified. This process of differentiation results in a series of minerals arranged in a predefined sequence and makes up the basis of a simple classification scheme for magmatic rocks (Fig. 3.5). The end of the process is marked by the sole crystallization of quartz (SiO_2) that forms when all other elements have been consumed, as in the case of granite. Overall, this process of magma composition changing where minerals are extracted from the melt in a specific sequence is known as **magmatic differentiation** (Fig. 3.6).

The granite that forms from the differentiation of gabbroic magma is known as **plagiogranite**, named after its most abundant feldspar mineral that is almost exclusively composed of plagioclase. These rocks form at mid-ocean ridges, although they are not of common occurrence. In this setting, there is often insufficient time to pass through the entire differentiation process and fractionation does not approach completion before the magma solidifies. These granites are, compared to those intruded into continental crust, characterized by a relatively low SiO_2 content. Although not common, the differentiation mechanism of granite formation does represent the original process by which granitic rocks evolved during Earth's early history. In contrast, continental granites typically form by the palingenesis of rocks with a chemical composition similar to that derived from the melting of metamorphic gneiss or clastic sedimentary rocks. Significant volumes of granites also develop at convergent boundaries above subduction zones whereby a combination of magma differentiation and palingenetic processes operate. The major portion of granite, however, forms within substantially thickened continental crust where magma has more opportunity to crystallize slowly into typical coarse-grained granite rock.

With time, the primitive Earth's crust became solid and firm as it cooled down. Henceforth, igneous rocks that were also extruded at the surface, are referred to as **volcanic rocks** (after *Vulcan*, the God of the fire). These differ in appearance from the slowly cooled plutonic rocks both in their grain size and in texture. In the case of rapid cooling of the melt as observed today in active volcanic areas, either very small crystals develop or the molten phase is quenched into glass without the formation of crystals. The magma that reaches the surface commonly flows out in the form of lava or is explosively ejected as ash. The difference between magma and lava is that magma contains different amounts of gaseous phases that are expelled and degassed as it emerges at the Earth's surface to produce lava. In its cooled state, lava solidifies to produce **effusive rock** (lat. *effusio*: pouring out, extrusion). If the magma has a very high dissolved volatile (gaseous) content, the **eruption** (lat. *eruptio*: outfall or outburst) of the volcano is prone to be explosive. In this case, hot clouds of gas and rock form **pyroclastic flows** (Greek *pyros*: fire; Greek *klastein*: to break). In addition to the volatile gases released from the volcano, flows carry significant volumes of **ash** (also called tephra; Greek *tephra*: ash) and larger **volcanic bombs**. These rocks constitute the group of **ejective rocks** (lat. *eiectare*: to discard). A generic link between volcanic and plutonic rocks are **dikes**. These originate when a magma body intrudes overlying fractured wall rock and cools within vertical fractures or horizontally as **sills** that penetrate parallel to flat-lying structures, such as sedimentary or igneous layers (intrusive sheets; Fig. 3.7).

Rocks consist of various types of minerals that vary considerably depending on the concentration of chemical

Fig. 3.7 Basaltic dike with chilled margin, intruded into granite (till boulder of the Weichselian glaciation; location: Vierow, Baltic Sea shoreline, Mecklenburg-West Pomerania). White minerals in the basalt are calcite-filled vesicular cavities. [N 54°07′55″/E 13°34′44″]

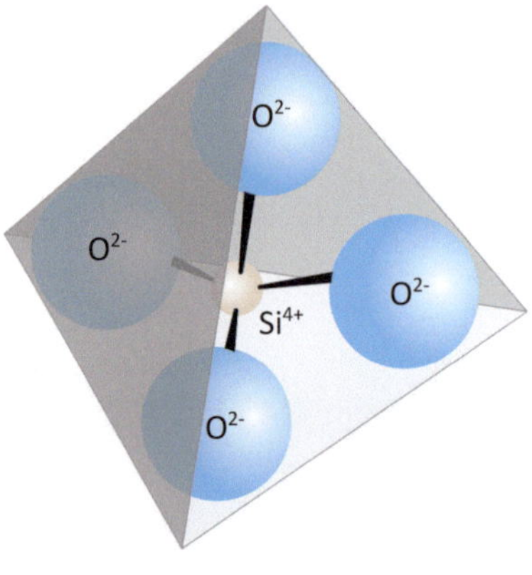

Fig. 3.8 SiO_4 tetrahedra with four bonding possibilities at the corners of the tetrahedron

Box 2: Basalt

Basalts represent the most common rock type occurring in the upper part of the Earth's crust. It crystallizes from a molten rock of basaltic composition when cooled rapidly at or close to the Earth's surface. Its composition is defined as a rock that consists mainly of the minerals pyroxene, plagioclase and/or olivine, although its precise mineralogy varies dependent on the tectonic conditions of melt generation. Basalts formed at mid-ocean ridges are often significantly different from those erupted within a continental plate of normal lithospheric thickness. Such differences are evident in chemical composition and, in particular, the concentration of the alkali metals Na and K, as well as a number of trace elements. Basalts formed at mid-ocean ridges contain only small amounts of alkali metals, whereas those formed within a lithospheric plate are often enriched in Na and K and therefore referred to as alkali basalts. The content of SiO_2 is also important in the normative mineral content (i.e., the idealized hypothetical composition of rocks calculated from their elemental content according to a norm), whereby alkali basalts are undersaturated in SiO_2 and the so-called subalkaline basalts are saturated or even supersaturated in SiO_2.

The various types of basalt are often very similar in mineralogical composition and cannot be assigned to a tectonic environment based only on mineral content. The basaltic rock type known as **tholeiite**, which accounts for the largest portion of all basalts of the Earth, forms mostly at mid-ocean ridges and is named after its type locality,

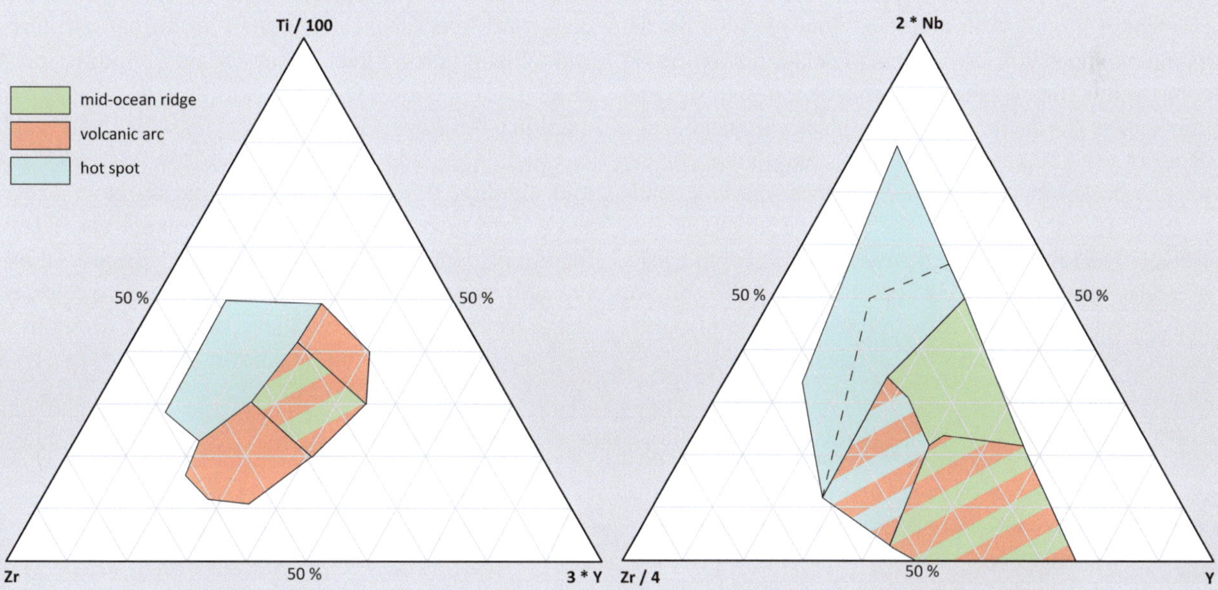

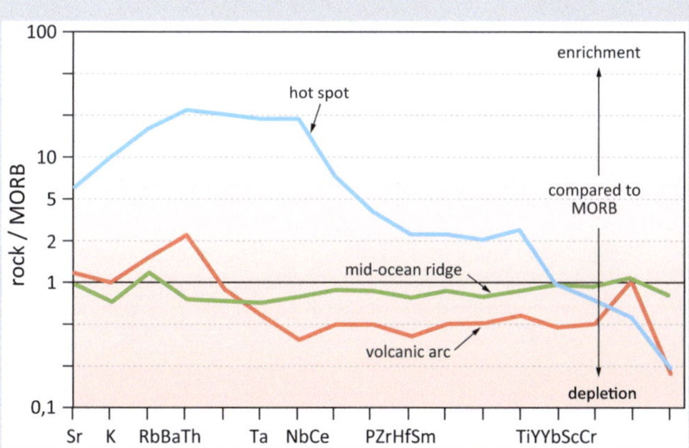

Fig. 3.9 Diagrams used to distinguish between different types of basalts occurring at mid-ocean ridges, volcanic arcs or above subduction zones and hot spots. The triangular diagram requires a recalculation of element contents normalized to a 100% total. To make fields more visible, some elements are scaled up or down by a given factor. The so-called spider diagram (bottom left) shows the concentration of elements normalized to average MORB (mid-ocean ridge basalt) values. The scale on the y-axis of the spider diagram is logarithmic. Ti/100–Zr–3–Y diagram after Pearce and Cann (1973), 2Nb–Zr/4–Y diagram after Meschede (1986), spider diagram patterns from Pearce (1983)

Tholey, a small town in the Saarland. However, the tholeiite from the Schaumberg near Tholey is undoubtedly an intracontinental basalt (see Sect. 11.2.1, Saar–Nahe basin) that formed during the Rotliegend (Early Permian). Its trace element distribution, i.e., the quantitative distribution of those elements which occur in very low concentrations in a rock (typically measured in parts per million or billion), and in this case particularly the rare earth element content, do differ slightly from classical **mid-ocean ridge basalts (MORB)** and are classified as subalkaline MORB or commonly, **tholeiitic basalts**.

The basalt formed at a mid-ocean ridge is fed by a magma chamber within the oceanic crust that develops at only a few kilometers depth. Within continental crust, magma chambers are usually located much deeper. The magma chambers beneath oceanic crust contain melt derived from the underlying, partially molten, asthenosphere, i.e., the uppermost layer of the mantle. Only the molten fraction of the asthenosphere reaches the magma chamber at a mid-ocean ridge and is responsible for the basaltic composition of the melt. For the differences in composition, the degree of partial melting is of vital importance. Normally, the asthenosphere commences at a depth of 90–120 km underneath the lithosphere. Below the mid-ocean ridge, however, the asthenosphere extends upwards to only a few kilometers beneath the Earth's surface. The low overburden pressure at this depth results in a relatively high degree of partial melting (20–30% of molten mantle material). The "normal" degree of partial melting of the asthenosphere at greater depths is only about 2–5%.

Many individual elements (typically cations) are incorporated to varying degrees within the crystal lattice of the basaltic rock-forming minerals that crystallize under mantle or crustal conditions. The degree of incorporation depends largely on their ionic radii and electrostatic charge. Well-fitting elements (e.g., Si, Al, Mg, Ti, Zr, P) are referred to as **compatible** and the less-fitting elements as **incompatible** (e.g., alkali elements like Na, K, Sr, Rb). Incompatible elements generally remain longer in the melt and are passively enriched as the solidified portion of the asthenosphere becomes correspondingly depleted. At low degrees of partial melting, the concentration of incompatible elements in the melt phase is high and decreases with increasing melting due to a higher degree of mixing. As a result, basalts at mid-ocean ridges have a significantly lower portion of incompatible elements than those formed in areas of normal lithospheric thickness. The third type of basalt also petrographically designated as a tholeiite is known to form above subduction zones. At a depth of 110–120 km, basaltic melts are generated and mix with subducted rocks usually of sedimentary origin to form hybrid magmas. Since the admixed rocks usually contain only very small proportions of both compatible and incompatible trace elements, the amount of incompatible elements in the basaltic melt is reduced and results in a characteristic depletion. Based on trace element distribution patterns or so-called discrimination diagrams (lat. *discriminare*: subdivide), it is possible to distinguish routinely the different types of basalt (Fig. 3.9).

As basaltic melt extrudes out at the surface, it often cools down too quickly to crystallize minerals. The still molten rock material then solidifies as amorphous material (without crystalline form) in the form of volcanic glass. In young basalts, volcanic glass is commonly observed in thin sections of the rock. In the course of time, however, the glass devitrifies and turns very slowly into crystals. When the melt solidifies deep in the Earth's crust, the minerals form large crystals and a **gabbro** develops. However, many basaltic melts reach the Earth's surface because of their low viscosity and high fluidity. Basaltic melts also typically have a relatively low proportion of dissolved volatile gases and, therefore, explosive eruptions are rare. Basaltic flows typically pour out at the surface as lava over large areas. If the viscosity increases, the cooled lava develops a very rough surface and is known as **Aa lava**. If it

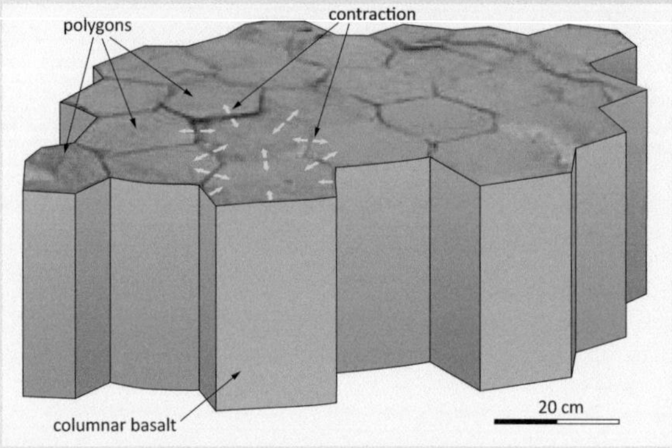

Fig. 3.10 Columnar basalt with polygonal structures visible at the surface. The hexagonally arranged fractures are caused by shrinkage during the cooling of the basaltic rock that originated from melt

maintains its very low viscosity, a **pahoehoe lava** forms characterized by smooth or ropy shaped surfaces. However, if basaltic melt intrudes at shallow crustal levels and encounters groundwater, it may result in phreatomagmatic eruptions and the formation of maars (Box 24: Maars).

Magmatic melts tend to shrink slightly during cooling, with a volume reduction of about 1–2%. Many lava flows develop an insulating solidified skin and a network of shrinkage cracks that form at the margin of the flow. Such a network is similar to the desiccation cracks developed in the upper clay layers of dried out mud puddles. With continuous cooling, the cracks propagate perpendicular to the cooling surface into the solidifying basaltic lava. This results in characteristic **columnar basalts**, which in the past have been considered by some scientists to represent crystallization structures, but in fact form as cooling and shrinkage cracks (or joints) in a solidified lava flow. The column axes of these joints typically lie perpendicular to the cooling front (Fig. 3.10). In lava flows, the joints of columnar basalts are therefore upright but in volcanic chimneys, they are frequently inclined, horizontal or arcuate. Ideally, the shapes of the columns are hexagonal, which represents the smallest surface area to volume ratio. However, the columns commonly depart from their ideal shape by developing fewer or more corners. Such columnar structures are not limited to basaltic lavas but also occur in lavas of different composition. In addition, they are much more frequently developed in basaltic lavas than in SiO_2 enriched andesitic or rhyolitic melts due to their low viscosities that are maintained as they erupt at the Earth's surface.

elements within them. The type and texture of the minerals present are responsible for the chemical and physical properties of the rock. Most minerals belong to the **silicate** group, i.e., minerals where the two most abundant elements of the Earth´s crust, silica (Si) and oxygen (O) occur in the form of tetrahedrons of SiO_4 and are most commonly ionically bonded with elements such as iron (Fe), magnesium (Mg) or aluminum (Al) (Fig. 3.8). In many cases, the color of a rock is already a good indicator of its composition. Dark minerals typically contain abundant Fe, Mg, and calcium (Ca). They constitute the group of mafic minerals (*mafic* comes from the composition of *Mg* and *Fe*), which includes pyroxene, amphibole, olivine, and biotite. A typical **mafic rock** is a **basalt** (Box 2: Basalt), which occurs in abundance around Europe. Young basalt occurs from the lava flows of recent volcanic activity, such as the well-known active volcanoes of Vesuvius and Stromboli or the older complex of the Eifel Mountains. Significantly older basalts occur in the geological record that are frequently several hundred million years old, such as those of the Rhenohercynian zone of the Variscan Orogeny where the igneous rock variety is that of a diabase (Box 8: Diabase and Schalstein).

In contrast to mafic minerals, abundant sodium (Na), potassium (K) and Al characterize light-colored silicate minerals. These are termed felsic minerals and this group is dominated by feldspar and quartz. A typical **felsic rock** is a granite with a high content of quartz and feldspar. Granite is one of the major components of the continental crust and is present in the internal, central part of most mountain belts around the world. Numerous examples of granite plutons occur in the young mountain range of the European Alps, as well as in the remnants of the more ancient, eroded orogenic belts that evolved many hundreds of millions or billions of years ago. These are now found within the stable continental **cratons**, such as the Baltic shield or that of Greenland.

To subdivide the different types of igneous rock, the so-called TAS diagram is particularly useful. TAS stands for total alkali versus silica and describes a diagram that plots the weight components of alkali metals ($Na_2O + K_2O$) against silica (SiO_2) (Fig. 3.11). This discrimination diagram allows the various rock suites that may crystallize from original magmas of different compositions to be subdivided. Large differences in the concentration of alkali metals occur, which appear in rocks most commonly together with SiO_2 incorporated in **feldspar** (plagioclase, alkali feldspar) and **feldspathoids** (e.g., nepheline, leucite, among others). Feldspathoids, more simply termed **foids** (abbreviation for feldspathoids), develop when a crystallizing magma is depleted of SiO_2 so that feldspar cannot be formed together with alkali metals. Foids, on the other hand, may still form since they require smaller amounts of SiO_2. In particular, if the magma contains a high amount of alkali metals, mainly rocks with feldspathoid minerals will form. Figure 3.11 presents rocks of the subalkaline series (blue colored) and partly of the transitional series (green colored) that contain free quartz. Foids and quartz cannot coexist since alkali metals are required to build feldspar in the case of SiO_2 excess. For this reason, feldspathoid minerals do not occur in granites that, by definition, contain significant quantities of quartz.

Water did not exist on the Earth´s surface during the formation of the planets, but during gradual cooling of the crust, it rapidly condensed to form a water cover of variable thickness. Since then, the presence of surface water has played a major role in the evolution and weathering of rocks during much of the history of the

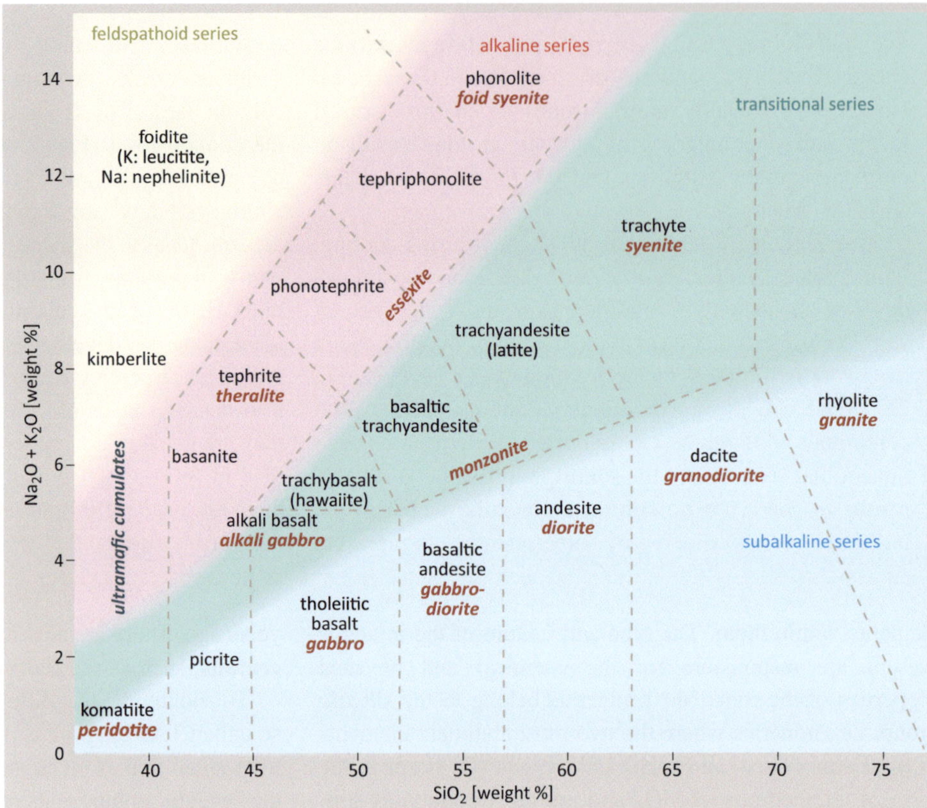

Fig. 3.11 TAS diagram (total alkali vs. silica) used to subdivide magmatic rocks and their evolution according to their content of alkali metals and SiO_2. Normal letters—volcanic rocks; italic/red—intrusive plutonic rocks equivalent to the extrusive volcanic varieties (modified after Le Maitre 2005)

planet. Rain, wind, and waves, which during the Earth´s early evolution were probably more intense than seen today, weathered the rocks, and transported the eroded material away as sediment to be deposited in the first sedimentary basins. Weathering is a surface process where rocks are disintegrated by physical and chemical mechanisms to gravel, sand, silt, or clay-sized particles. These erosional products are transported by wind, water, or ice before been deposited on land, in lakes, rivers, or in the open sea where continuous subsidence eventually leads to the accumulation of sedimentary sequences several kilometers thick. The longer the duration of the process, the thicker the sedimentary layers and the stronger the compaction and solidification of the deepest layers at the bottom of the sequence. In addition, circulating waters transport soluble elements that may lead to the precipitation of new minerals that fill available pore space in the sediments.

It is, therefore, an important part of the geological rock cycle that new consolidated rock develops from unconsolidated sediment to form lithified sedimentary lithologies, which if once again uplifted are subject to renewed weathering and erosion (Fig. 3.2). As a result, sand, regardless of whether it accumulations in a dune, on the bank of a river or on the beach, consists as all other clastic sediments, of crushed material derived from older rocks.

Sedimentary rocks include not only clastic rocks, i.e., those deposited from the erosional products of other rocks, but also those that are chemically precipitated, such as salt or gypsum, or biogenically formed by the accumulation of various types of organic remains. Chemical sedimentary rocks form, for example, by the evaporation of seawater containing dissolved substances that precipitate as rock salt. The average seawater of today contains 3.5–3.8 wt% of dissolved salt, with table salt (NaCl) representing the major portion with an abundance of 85%. However, only in the course of geological history, have all these salts been added to the increasing volume of water. The necessary anions and cations originate from the dissolved constituents of igneous rocks that end up in the rock cycle by chemical weathering. In contrast, biogenic sedimentary rocks are formed by the deposition of organic remains, with calcareous species being by far the most abundant source. Almost all limestones contain the remains of calcareous organisms, in particular, coccoliths and foraminifera. The great bulk of these remains, however, are so small that they cannot be identified properly even with a good optical microscope. Only by using an electron microscope can we observe their detailed structure and diversity of form. Also of importance are the noncalcareous biogenic sedimentary rocks that include the radiolarites and diatomites. These are constructed from the remains of silica-shelled organisms (i.e., radiolarians and

Fig. 3.12 Stages of metamorphism and the types of meta-basaltic rocks that develop. (From top left) non-metamorphic: basalt; low-grade metamorphic: greenschist (greenschist facies); medium-grade metamorphic: amphibolite (amphibolite facies); high-grade metamorphic: pyroxene granulite (granulite facies)

diatoms). Other types of noncalcareous, organically derived deposits are the economically important coals, oil, and gas shales, as well as their petroleum derivatives.

The third large group of rocks encompasses those of metamorphic origin (Greek *metamórphosis*: transformation). As rocks of all types are subjected to increasing pressure and/or temperature, they are slowly transformed in the solid state into various types of metamorphic rocks. An analogy for the formation of a **metamorphic rock** is a construction kit consisting of variably colored building stone. Inside the kit, the number of building stones and their colors do not change, however, we are able to construct very different things using the blocks. This situation is similar in rocks where the available chemical elements can be considered as building stones that according to the prevalent pressure and temperature can be used to construct different minerals. In the case of changing conditions, some minerals become unstable and rebuild themselves to form other more stable minerals. In this way, although the bulk chemical composition of the rock is largely retained, the mineralogical assemblage and its texture will adjust to the applied pressure–temperature conditions. It is by studying the assemblages and textures of rocks (including foliations, folding and mineral elongations) that geologists are able to reconstruct the metamorphic conditions to which they were subjected and relate them to key geological processes, such as past tectono-thermal events.

In the geological record, rock sequences occur that reflect the varying **degrees of metamorphism** characterized by different pressure and temperature ranges. Let us consider basalt as the parent rock material (Fig. 3.12), the so-called **protolith** (prefix *proto*: first; Greek *lithos*: rock): After low-grade metamorphism at elevated temperatures and pressures (320–500 °C, burial depth 10–30 km), basalt is transformed to a metamorphic rock known as a **greenschist**. Under these conditions, the originally preexisting minerals such as pyroxene, plagioclase and if present olivine are transformed into epidote, actinolite (a mineral of the amphibole group) and chlorite. With increasing metamorphic grade and increasing pressure and temperature (550–700 °C, burial depth 15–40 km), a greenschist rock transforms into an **amphibolite**. Here, again the mineral composition changes to become dominated by amphibole (hornblende) and plagioclase with subordinate garnet, biotite, quartz, or epidote.

The highest grade of metamorphism occurs under exceptional conditions where rocks begin to partially melt in the presence of water at temperatures above 700 °C. Metamorphism of a basaltic protolith under these conditions (>700 °C, burial depth > 40 km) results in a **pyroxene granulite**, which consists essentially of plagioclase, pyroxene, and garnet. Other diverse conditions of metamorphism result in completely different rocks that form from the same starting material. If the same basalt is transported down a subduction zone and passes to the mantle, it is subjected to high-pressure, low-temperature metamorphism. As the oceanic crust is subducted to depths at high rates over a geologically short time period, the pressure increases rapidly but the temperature lags behind the normal geothermal depth

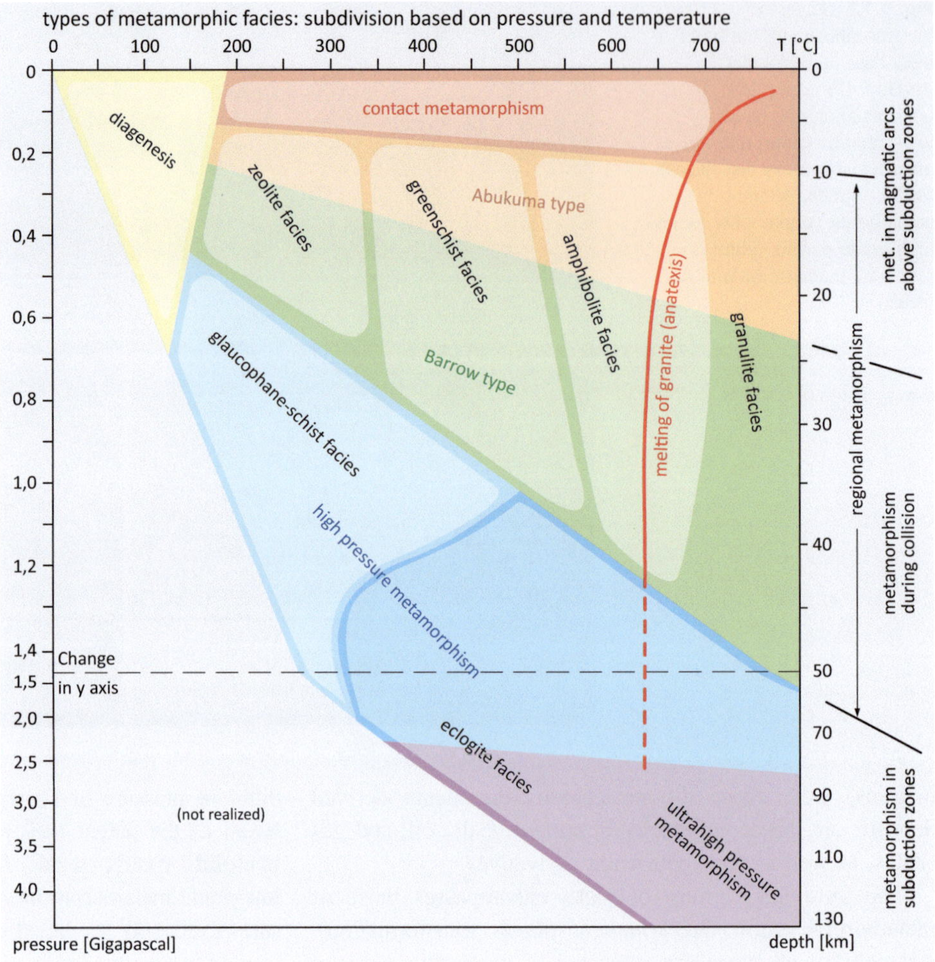

Fig. 3.13 Diagram showing the types of metamorphic facies that develop under different conditions of pressure and temperature (after Frisch and Meschede 2013)

gradient. The reason for this is that rock is a poor conductor of heat, which under these circumstances results in a negative temperature anomaly. The mineral transformations that occur under these anomalous conditions produce special rock types, among them the **blueschist** that contains glaucophane, a blue-colored amphibole. Under more extreme conditions **eclogites** form, consisting of omphacite, a pyroxene typical for high-pressure facies, and garnet. If these rock types are later encountered in orogenic belts, it can be assumed that they were initially metamorphosed in a subduction zone before been transported (obducted) back to the surface during the mountain building process. Numerous examples of both blueschists and eclogites can be found across Europe and these rocks are of importance in paleogeographic reconstructions when locating the sites of ancient ocean margins.

As illustrated, basaltic rocks are of major significance for recognizing the different metamorphic zones and can be subdivided according to the rock types and minerals encountered. These zones are referred to as **metamorphic facies** and, in terms of geological conditions, they occur in three main types of settings (Fig. 3.13). (1) A **regional metamorphic setting** characterized typically by burial metamorphism during orogenic crustal thickening where both temperature and pressure continuously increase in a normal to low geothermal gradient. In this setting, metamorphic rocks belonging to the **greenschist**, **amphibolite** and **granulite facies** are encountered. (2) A high-pressure, low-temperature metamorphic setting typical of subduction zones that result in **blueschist** and **eclogite facies** rocks. (3) A high-temperature, low-pressure metamorphic setting, referred to as contact metamorphism. This occurs when rocks are strongly heated at relatively shallow crustal depths. This type of metamorphism is found at the margin of intruding igneous plutons, or when volcanic magma is extruded at the surface. At such locations the country rock is subjected to low-pressure, very high-temperature conditions (**contact metamorphism**) that approach the point of melting, which in turn creates characteristic mineral assemblages and textures (e.g., hornfels).

The Age of the Rocks

4

Knowing the age of rocks for geoscientific investigations is of utmost importance. Very many methods that can provide age information are available to the geologist. Nevertheless, in many cases, it is not possible to determine an absolute rock age.

Best known and widely used is the **biostratigraphic classification** of sediments and sedimentary rocks based on the occurrence of index fossils. These specimens are referred to as **index fossils** for, in the course of Earth's history, they only occur during relatively short periods in a temporal sequence. When an appropriate index fossil is found, if it is clear that the lithology in which it occurs was not redeposited at a later stage, an age determination can be made. The age of an index fossil can be determined by locating it in detailed stratigraphic tables. In addition to macrofossils, such as the well-known ammonites (Box 3: Ammonites), various kinds of microfossils also make useful index fossils and can be used for determining ages when only small amounts of rock material are available. Microfossils are of vital importance in dating the age of sedimentary rocks recovered during the drilling of boreholes.

Box 3: Ammonites

Ammonites are an extinct group of marine **cephalopod** animals (Greek *kephale*: head; Greek *podós*: foot). They are molluscs whose modern representatives include the living pelagic cephalopod, *Nautilus*. Some genera of cephalopods built a shell (ammonites, nautiloids) that was in most cases between 1 and 30 cm in size (e.g., Figure 4.1), although larger specimens also existed. The largest known ammonite fossil (*Parapuzosia seppenradensis*) was found near Münster in Westphalia with a diameter of 1.8 m. Large ammonites of the genus *Arietites* from the Early Jurassic are also well known and frequently exceeded half a meter in size.

Fig. 4.1 Ammonite (*Schloenbachia varians*) from the Cenomanian, Late Cretaceous (location: Rüthen/Möhne, Mildental). [N 51°29′59″/ E 08°25′13″]

The name ammonite was initially used in the ancient world when Pliny the Elder described fossils as so-called Ammon horns (after the God *Ammon*, who was often depicted with ram horns). In 1884, the German paleontologist Karl Alfred von Zittel (1839–1904) was the first to use it as a collective term to describe the ammonite group (Zittel 1884), although Friedrich August Quenstedt (1846–1849) had previously published extensive studies on the systematics of these cephalopods.

Due to the extraordinary diversity of shell forms and their global distribution, the ammonites are highly suited for the stratigraphic classification of sediment and sedimentary rock in which they were embedded. Especially in geological successions of Jurassic and Cretaceous age, many stratigraphic zones are delineated by fossil ammonites and, as a result, they represent the most important index fossils for determining the relative age of Mesozoic strata. The ideal **index fossils** are represented by species that lasted for very short periods of geological time (up to 1 Myr) but were globally widespread. By finding the appropriate index fossils in sedimentary rocks of unknown age, the strata can be easily dated by comparison with stratigraphic charts.

M. Meschede and L. N. Warr, *The Geology of Germany*, Regional Geology Reviews,
https://doi.org/10.1007/978-3-319-76102-2_4

The method of **magnetostratigraphy** uses the property of many rocks to preserve the orientation of the Earth's magnetic field at the time of formation. Since in the course of the Earth's history the **polarity of the geomagnetic field** has been repeatedly reversed in an irregular but continuous way, the retained magnetic information can be used to determine rock ages. This works in a similar way as the bar code found today on nearly every product that we buy. The information is enclosed in lines of different thickness, which are arranged in irregular but precisely defined intervals. Magnetic information can be read in the same way as special readers at the checkout counter can determine the price of a product. Magnetic age information is characterized by sequences of normal or inverse magnetism in rocks and age determination can be made by matching these reversal patterns to the Global Magnetic Polarity Time Scale (GMPTS). Typically, **reversals** vary between 10,000 and 100,000 years and in some cases, single reversals extended over many millions of years, whereby during the Cretaceous, one magnetic quiet period actually lasted almost 40 Myr.

In young geological deposits, seasonal changes in sedimentation as well as the seasonal growth of some organisms, may be used to determine the age of rocks. Such an approach is applicable to **varved clays** occurring in lake sediments deposited along the marginal areas of glaciers, but also to corals and tree rings. Fluctuating weather conditions have an effect on the thickness of the sediment layers and the growth rings of trees or corals. With the help of detailed correlations, such fluctuation can be traced back several thousand years. These methods have been successfully used to determine the age of a number of recent events, such as the last major volcanic eruption in the Eifel that formed the Laacher See.

Paleomagnetic-based methods are not applicable to rocks that solidified at depth in the Earth's crust or mantle (plutonic rocks) or those that have been subjected to intense metamorphism. Magnetostratigraphy is most suitable for dating the younger geological periods (<160 Ma) where detailed knowledge of the polarity reversals preserved in the oceanic crust is available. Moreover, the magnetic information is preserved only up to certain temperatures before been lost and reset when heated above 580 °C, as in the case of magnetite-bearing rocks.

For very old rocks that contain no fossils and there is little possibility of paleomagnetic dating, the only reliable method of determining a rock age is by means of a special property of certain chemical elements that undergo radioactive decay. Such elements are **unstable isotopes** that spontaneously decay to other isotopes whereby their decay products remain as traces in the rock. Measuring the amount of the parent isotope and resulting **decay products** may be used to calculate an absolute age. This method is called **radiometric age dating** (Box 4: Radiometric age dating).

Box 4: Radiometric age dating

Minerals are composed of a range of different elements. An element is defined as a group of atoms, all of which have the same chemical properties. The number of existing protons in an atom defines the chemical element to which it belongs. Every positively charged mass particle has a corresponding number of negatively charged electrons. The number of protons gives the atomic number, which determines the position of the element in the periodic table. Moreover, each atom contains a number of neutrons (also mass particles), which together with the protons define the atomic mass number. Atoms with different mass numbers but the same number of protons belong to the same chemical element and represent different isotopes of that element. Together they constitute the isotope series of an element as, for example, the naturally occurring isotopes of the element uranium [$_{92}$U, atomic number 92] that consists exclusively of the radioactive isotopes ^{238}U [99.27%], ^{235}U [0.72%], ^{234}U [0.0055%], and ^{236}U [traces]. The individual isotopes of a series, therefore, differ in both mass number and atomic weight.

However, some isotopes are not stable and after a certain time, they decay spontaneously to return to a more energetically favorable and stable state. Since the decay of isotopes involves the release of energy in the form of α-, β- and/or γ-radiation, they are referred to as **radioactive isotopes**. Their decay can occur directly from the **mother isotope** to a stable **daughter isotope**, but much more commonly it takes place through a number of transitional radioactive nuclei referred to as a series, which form and decay in a defined sequence before the stable daughter isotope is reached. In such a radioactive series, the daughter isotopes may be unstable and form further daughter isotopes. Some of these isotopes are extremely short-lived (sometimes less than a second) and can only be detected during specially designed

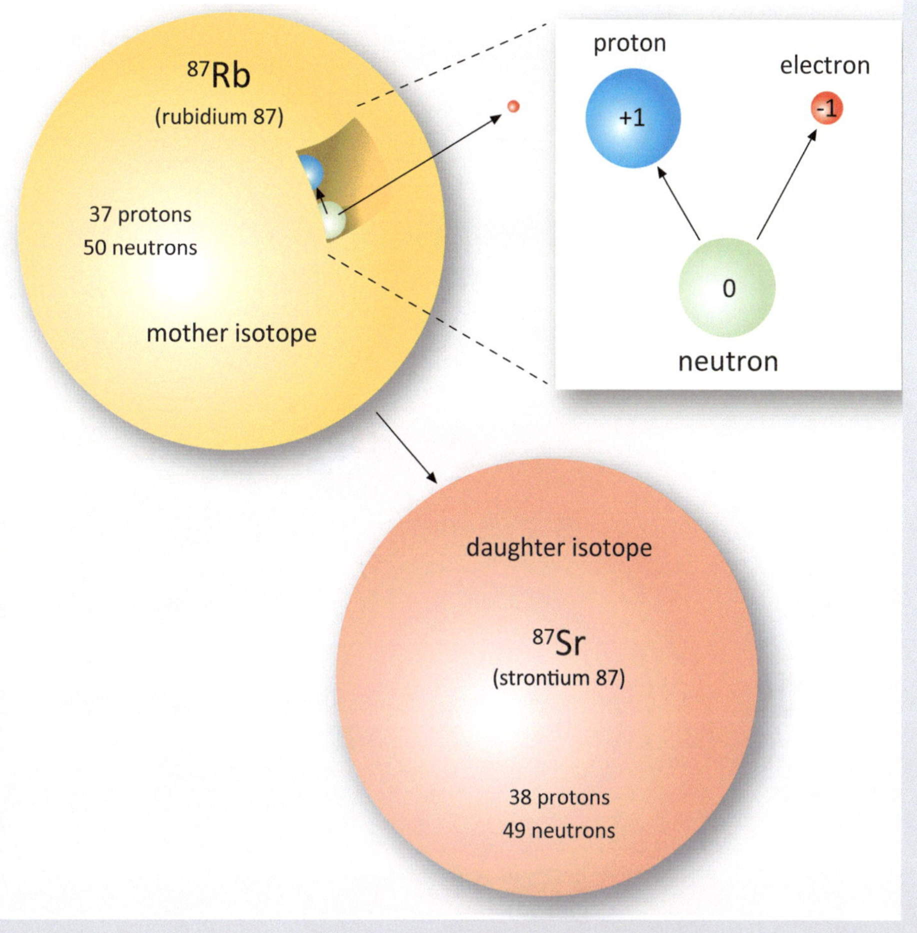

Fig. 4.2 Decay of the radioactive Rb isotope (^{87}Rb) to the stable Sr isotope (^{87}Sr) by releasing an electron and transforming a neutron into a proton

experiments. Mother and daughter isotopes of a radioactive series always belong to different elements because the number of both neutrons and protons changes (Fig. 4.2).

Each radioactive isotope has its own characteristic **half-life**. This is the time required for exactly half of the unstable parent isotope to decay and be replaced by a corresponding daughter isotope (Fig. 4.3.) This specific characteristic of unstable isotopes is used for **radiometric age dating** by measuring the number of remaining parent isotopes and newly formed daughter isotopes in a rock or mineral at the time of measurement. Based on the measured concentrations of parent and daughter isotopes, it is possible to calculate the absolute age of a sample if the radioactive constant of the decaying isotope is known and the number of daughter isotopes is zero or is known at the beginning of the decay process (i.e., the date of origin). Moreover, the "geological clock" should not be disturbed or reset to zero during heating above a certain point, which is known as the closure temperature because below this point the system becomes isotopically "closed". However, subsequent chemical exchange processes during fluid–rock interaction may lead to the subsequent loss or gain of either parent or daughter isotopes that can significantly distort measured ages. Table 4.1 lists some of the most geologically important radioactive isotopes, their half-lives, and some datable minerals.

A mass spectrometer is used to measure the concentration of the different radiogenic isotopes and to determine the ratio of the decaying and newly formed ones. It is possible to directly measure the current amount of a daughter isotope in any given rock sample, whereas information concerning the initial isotope ratios at the time of resetting can be worked out indirectly by applying the **isochron method**.

An example is the case of the Rb–Sr (rubidium–strontium) dating method (decay of ^{87}Rb [proton number 37] to ^{87}Sr [proton number 38]; Fig. 4.2) where the Rb and Sr contents are measured in a broad range of samples from a selected rock unit. The concentration of the daughter isotope (^{87}Sr), which forms by the decay of ^{87}Rb, has to be different in each fraction and the data points should arrange along a line in the diagram (Fig. 4.4), provided that no heating has occurred during a

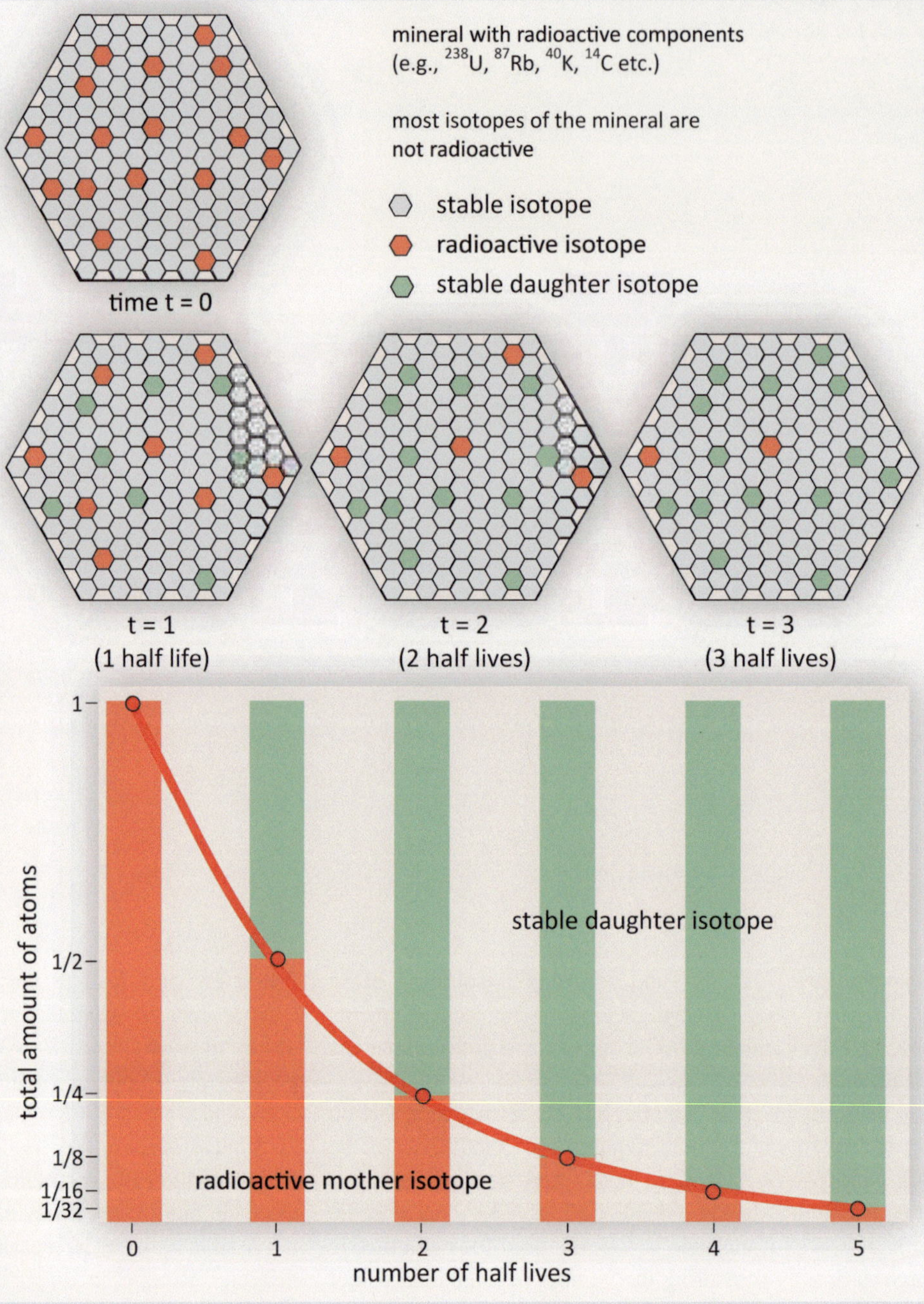

Fig. 4.3 Modification of the proportion of radioactive mother isotopes to stable daughter isotopes with an increasing number of half-lives

Table. 4.1 Half-life times (in years) of the most important radioactive isotopes used for geological age dating

Parent	Daughter	Half-life	Dateable minerals
^{87}Rb	^{87}Sr	48.80 Gyr	Muscovite, biotite, K-feldspar, plagioclase, hornblende
^{232}Th	^{208}Pb	14.05 Gyr	Monazite
^{238}U	^{206}Pb	4.47 Gyr	Zircon, uraninite, monazite, titanite, calcite
^{40}K	^{40}Ar	1.25 Gyr	Muscovite, biotite, K-feldspar, hornblende, illite, glauconite
^{235}U	^{207}Pb	704 Myr	Zircon, uraninite, monazite, titanite, calcite
^{14}C	^{14}N	5730 yrs	Organic matter

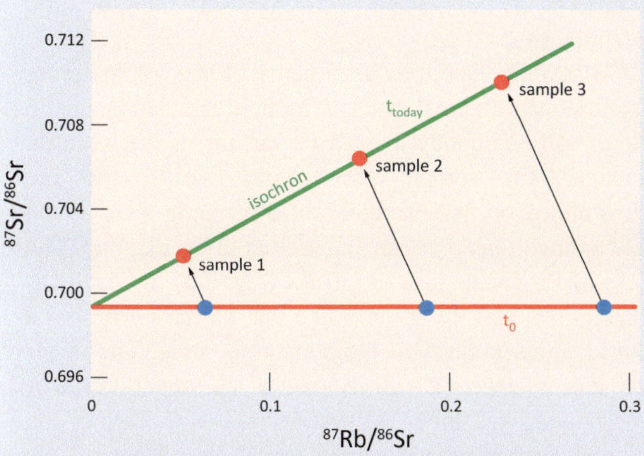

Fig. 4.4 Isochron diagram of the Rb–Sr age-dating method

later thermal event. In such a case, the system can be regarded as closed and the y-intercept represents the original isotopic ratio of the parent isotope compared to a non-decaying relative (e.g., ^{87}Sr). The slope of the isochron line (equivalent to the evolving radiogenic isotopic ratio) represents the age since the last reset of the system, for example, the cooling age following the intrusion of a magmatic rock or a metamorphic event. This is referred to as the radiogenic isochron age and is geologically relevant if the system was closed during the complete period of decay.

The **K–Ar method** is based on the decay of ^{40}K to ^{40}Ar. Argon is a noble gas that is not strongly incorporated in the crystal lattice of minerals and degasses from a rock above a certain temperature. A rock that forms by crystallization of a magma or suffers metamorphism does not normally contain any radiogenic Ar at the stage of formation since the "radiometric clock" is set to zero at high temperatures. Below the closure temperature, the newly generated radiogenic ^{40}Ar commonly remains in the system and when measured reflects either the cooling age of the magmatic melt or the last metamorphic event. Minerals rich in K, such as muscovite (closing temperature 350 °C), biotite (300 °C) or hornblende (500 °C) are particularly suitable for dating by the K–Ar dating method. However, this method is rather sensitive to disturbances, such as the loss of radiogenic Ar, which may not always be detectable. Therefore, today the enhanced **^{40}Ar–^{39}Ar method** is generally preferred where more information can be gained on the history of the sample by examining isochron plots and the Ar degassing behavior of the sample.

In the ^{40}Ar–^{39}Ar method, the amount of the parent isotope ^{40}K is determined indirectly by neutron irradiation of a sample and the measurement of the ^{39}Ar arising by this procedure. The advantage is that both parent and daughter isotopes (^{39}Ar and ^{40}Ar) can be measured on the same sample portion at the same time, as opposed to ^{40}K that has to be measured separately. In addition, detailed information of the isotopic system is obtained using stepwise heating and monitoring of the degassed Ar released at various temperatures. At low temperatures, Ar initially degasses from the outer regions of mineral grains, whereas at high temperatures the inner parts of crystals degas, making it possible to detect any zoned distribution of Ar. If the mineral grains are only slightly isotopically disturbed at the margins, a constant age value will characterize most of the degassed spectrum and form a plateau, which is used to calculate a geologically relevant age. The more uniform the individual age data, the better the resulting plateau.

Somewhat different is the **U–Pb dating method** using the concordia diagram. In a U-bearing mineral, two independent **radioactive series** occur where both parent isotopes and daughter isotopes belong to the same respective elements and thus have an identical closure time and chemical behavior. ^{235}U decays with a half-life of 704 Myr through a complicated series of radioactive elements to ^{207}Pb and ^{238}U decays with a half-life of 4.468 Gyr to ^{206}Pb.

The actual ratio of the two U isotopes ^{238}U and ^{235}U is identical in all rocks around the world ($^{238}U/^{235}U = 137.88$), with only a few exceptions in the case of natural or man-induced fission reactions. It follows that this ratio documents the decay of U that existed at the beginning of Earth's history assuming that there has been no significant U supply from the outside since that time. It is possible to calculate the original portion of the two isotopes using their known half-life time. About half of the originally existing ^{238}U, but only about 1% of the much faster-decaying ^{235}U is left today. The isotope ratios of mother and daughter isotopes of the radioactive series are plotted in the concordia diagram against each other, $^{207}Pb/^{235}U$ on the x-axis and $^{206}Pb/^{238}U$ on the y-axis (Fig. 4.5). The result is a curve, the **concordia**, with a curvature that results from the different half-lives of ^{238}U and ^{235}U.

The use of this diagram is illustrated by plotting the measured isotopic data of a natural sample. If it has a concordant age, the isotopic ratio of the sample lies exactly on the concordia line, i.e., both radioactive series result in the same age. In this case, the system was always closed since the last isotopic closure of the

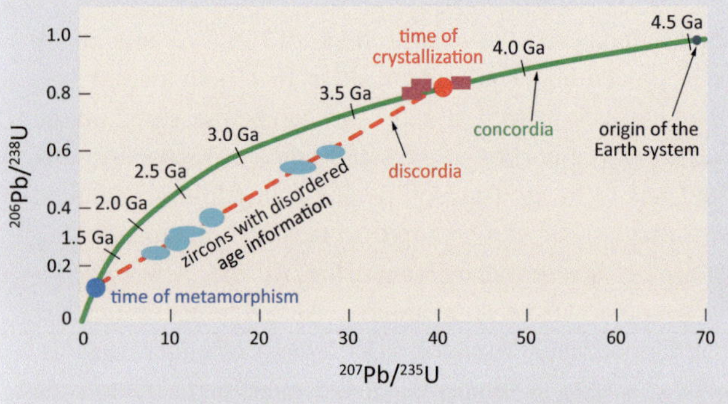

Fig. 4.5 U–Pb age dating of zircons from a gneiss sample using the concordia–discordia diagram

system and no loss of the daughter isotope Pb has occurred. In the case of an igneous origin, this age dates the crystallization of the rock and thus the time of intrusion.

Loss of Pb caused by disturbance of the system during a subsequent thermal event or by fluid circulation can be detected by adopting a similar approach as the isochron method. The isotopic ratios of the two U series are determined on, for example, different grain sizes of the U-bearing mineral zircon and plotted in the diagram. The loss of Pb results in a **discordia** (Fig. 4.5) in which the data points usually lie on a line that has an upper and a lower intersection with the concordia curve. The upper point of intersection is interpreted as the formation age of the mineral and the lower intersection as the age of the geological event that disturbed the U–Pb system.

So far, the oldest known rocks formed on Earth that have been dated using radiometric methods give an age of 4.28 Ga (located in the Nuwuagittuq greenstone belt in Canada; O'Neil et al. 2008). Even ages of up to 4.4 Ga have been determined in the center of individual detrital zircon minerals in Western Australia (Wilde et al. 2001). The oldest dated rock material are meteorites, with ages typically between 4.5 and 4.6 Ga that correspond to the formation of our planetary system.

The ages given in stratigraphic tables today are usually calibrated, radiometrically, i.e., contain time markers that are constrained by radiometric age data. A well-known series of calibration profiles is located, for example, in Italy in the territory of Gubbio. At this location, an almost uninterrupted succession of sediment of the Cretaceous and Tertiary periods is exposed. This sedimentary record has been closely examined down to the centimeter range using radiometric, biostratigraphy, and magnetostratigraphy so that today even brief reversals of the geomagnetic field with a duration of only a few 10,000 years are distinguishable.

5 Plate Tectonics, the Unifying Theory

What are the driving forces that subject rocks to high pressures and temperatures, and lead to a never-ending cycle that turns sediment into rock and rock back into sediment? It is caused by the constantly active transformation processes that occur in Earth's system, which, since its origin 4.5 billion years ago has undergone continuous dynamic change. Nothing is permanent, even if it is sometimes presented to us in the form of an ancient rock.

The entire Earth's system is subject to constant modification caused by heat that wants to escape to the outside. The heat stored inside the Earth originated from the early period of its development and is constantly regenerated by energy produced from the decay of radioactive elements. As the **thermal conductivity of rock** is generally very poor, the release of heat is achieved with the aid of **convection currents** (Fig. 5.1). These currents ensure that hot viscous material from the deeper regions of the mantle rises to the top because it is slightly lighter, while denser cooler material from the outer parts of the Earth's mantle descents. This creates a kind of conveyor belt system that removes heat from the Earth. As heat passes upwards from the core and is released to the Earth's exterior in zones of structural weakness, colder material descends into the mantle and cools down the deeper regions that keep the flow of heat transfer in motion. This mechanism of heat flow avoids overheating of the Earth's crust that without convective cooling would eventually begin to glow.

The transfer of heat within the Earth takes place over long periods of geological time. This is coupled with the flow of mass in the plastic mantle and in the partially liquid core, which circulates continuously by convection. In contrast, the more rigid outer shell of the Earth, known as the **lithosphere** (Greek *lithos*: solid), is relatively fixed and responds to the underlying mass flow by the build up of stress. As a result, **lithospheric plates** virtually "swim" on the uppermost layer of the mantle, known as the **asthenosphere** (Greek *asthenos*: soft), and are shifted horizontally by the forces of convection. The motion of plates was first proposed by **Alfred Wegener (1880–1930)**, a German meteorologist and the pioneer of plate tectonics (Wegener 1912, 1915). His **continental drift theory** made a major advance towards understanding how the distribution of today's continents resulted from the drift of tectonic plates. However, Wegener could not explain its mechanism. For this reason, the theory received strong resistance among the experts of that time. Despite plausible evidence for continental drift based on identical fossils found on both sides of the Atlantic Ocean or the remains of an ancient glaciation in the today's Sahara Desert, it took another 50 years until the **theory of plate tectonics** reached scientific acceptance. Today, the concept of plate tectonics provides an important foundation for understanding geological processes and is well documented in modern-day textbooks (e.g., Frisch et al. 2011; Frisch and Meschede 2013).

Movements resulting from the convection currents in the mantle leave their trace on the Earth's surface. We can recognize the dynamics of the Earth's interior by studying devastating **earthquakes** or spectacular **volcanic eruptions**. These events happen when the fixed and rigid (stiff, non-plastic) lithospheric plates build up stresses that are quickly released during fracturing and failure of the rocks in the upper crust. Connected to this activity is the buildup of magma, which, when overpressured by heating or additional melt production, moves upward along fracture systems and may even reach the Earth's surface as volcanic eruptions.

Rigid lithospheric plates vary significantly in size and have thicknesses that normally range from 70 to 150 km or more. Beneath the continents, the lithosphere is significantly thicker than under the oceans, and near high mountain ranges reaches more than 200 km. The lithosphere contains a crust with a thickness between 5 and more than 70 km, depending on whether it is oceanic or continental in nature. The rest of the lithosphere is comprised of an underlying part of the upper mantle (Fig. 5.2). Both the **lithospheric mantle** and, beneath it, the **asthenosphere** do not differ in composition. However, the asthenosphere is partially molten to a

M. Meschede and L. N. Warr, *The Geology of Germany*, Regional Geology Reviews,
https://doi.org/10.1007/978-3-319-76102-2_5

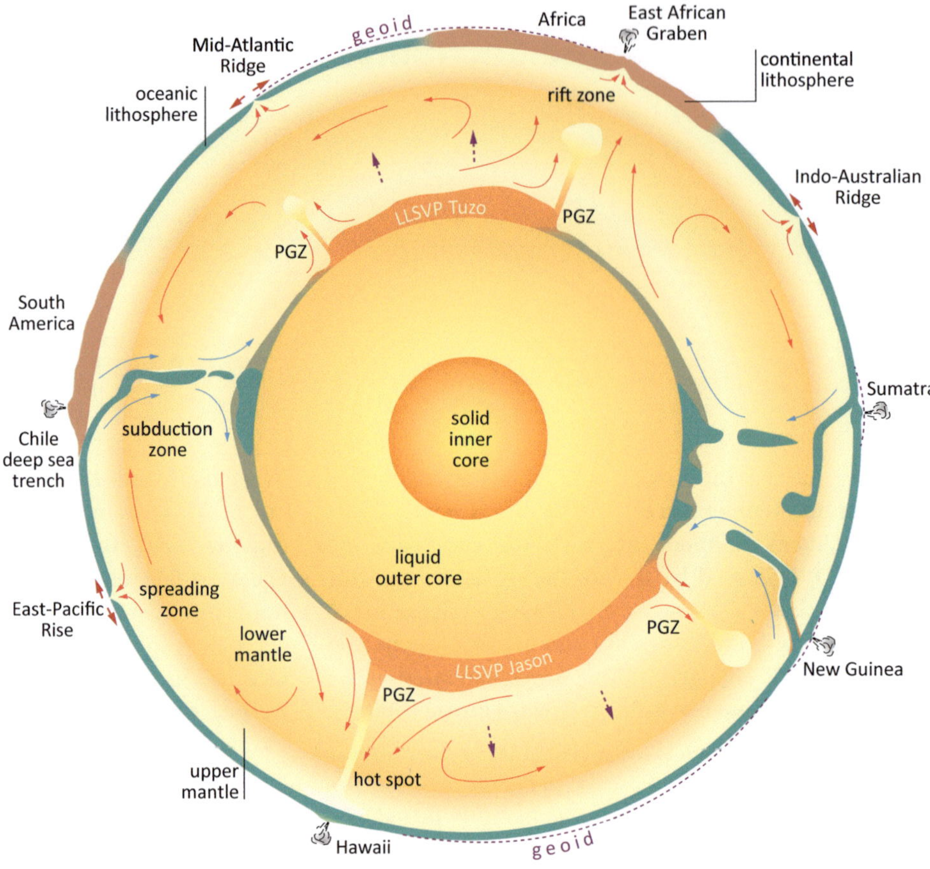

Fig. 5.1 Schematic cross-section of the Earth as seen from the South Pole. To illustrate the differences between the oceanic and continental lithosphere, the thickness of these units is exaggerated. With a realistic scaling, the lithosphere would be thinner than 1 mm in the illustration (comp. Fig. 5.2). Modified after Condie (1997), Schubert et al. (2001), Tackley (2008), Torsvik et al. (2014). LLSVP large low shear wave velocity provinces beneath Africa (Tuzo) and the Pacific (Jason), PGZ plume generation zones, thin arrows above Tuzo and Jason indicate buoyant upwelling resulting in the deflection of the geoid (dashed lines)

small extent, gaining the ability to move by **plastic flow**. Although the amount of partial melting is only a few percent of the total rock volume, this small percentage is enough for the lithospheric plates to move on its ductile, nonliquid underlay. The velocity of plate movement is measured in the range of centimeters per year. This is comparable with the rate of growth of our fingernails. We do not notice such rates as a visible movement but, after some time, the effects do become clear. In a period of 1 Myr, an unimaginably large time span for humans to conceive, but still a short time span in terms of geological processes, a plate drifting at a slow rate of 1 cm per year will move 10 km and 1000 km in 100 Myr. Throughout the Earth's history, plate movements of rates up to 20 cm/year have been known to occur. Therefore, it can be deduced that the surface of the Earth has completely changed several times throughout its history. Today's distribution of continents and oceans (Fig. 5.3) therefore provides only a snapshot of the interplay of tectonic forces. In the following chapter (Chap. 6), the tectonic evolution of Central Europe is presented over the past 750 Myr and a framework given for the changing distribution of the tectonic units, which ends in the familiar pattern of lithospheric plates and continents we know today.

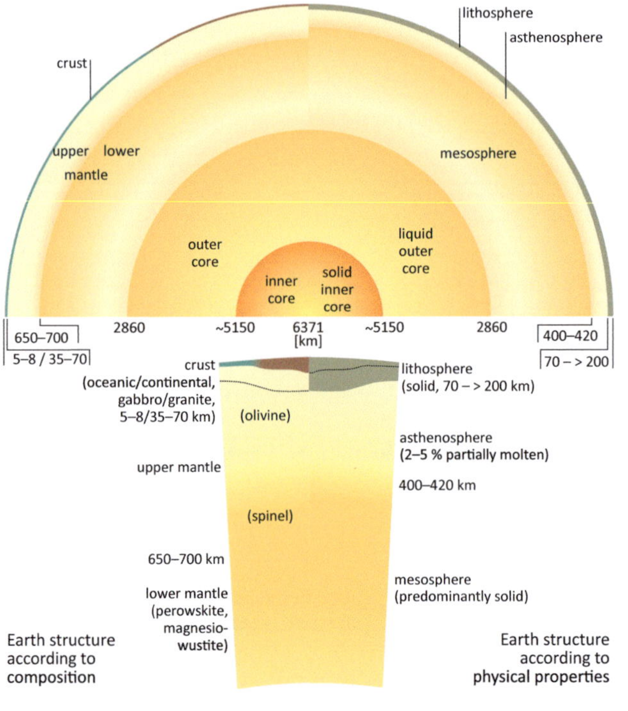

Fig. 5.2 Earth structure according to its mineralogical, petrological, and geophysical properties. Modified after Stüwe (2000)

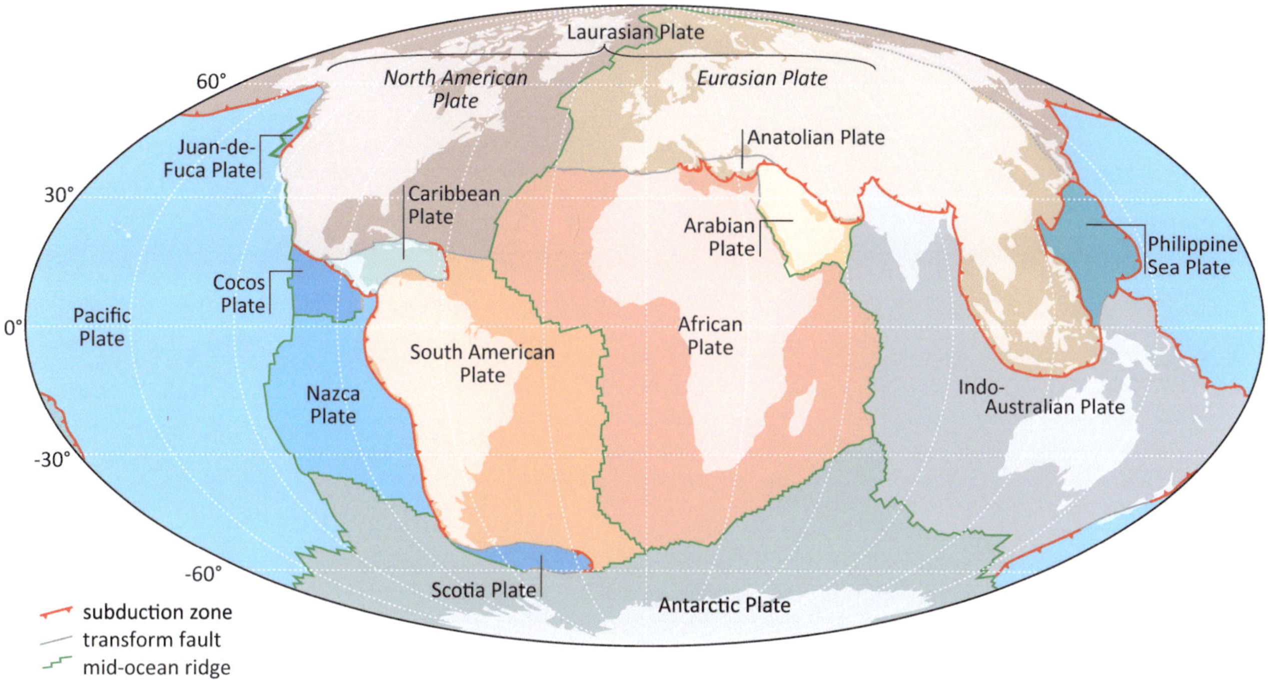

Fig. 5.3 Earth's current distribution of lithospheric plates

At present, the lithospheric plates move at different rates and in different directions, but since these plates are all arranged together without gaps to form the Earth's spherical crust, the interaction between the individual plates is unavoidable. If two plates diverge from each other at any location, new crustal material must originate in between to fill the gap. Such plate boundaries are called **constructive plate boundaries** because of the generation of new crustal material (Fig. 5.4). However, the addition between the two plates must be removed at another location in order to maintain the spherical shape of the Earth. If not, the Earth would continuously expand. This removal takes place where two plates collide with each other to form a **subduction zone**. At this location, a lithospheric plate is thrust beneath the other and is recycled into the mantle. Accretionary prisms often form at the surface along plate boundaries where sediment is typically scraped off from a subducting oceanic slab and piled up on the edge of the adjacent continent. Such plate margins are referred to as **destructive plate boundaries**. These two types of margins are connected by plate boundaries where the crust is neither formed nor recycled into the mantle. This type of **conservative plate boundary** is characterized by large **transform faults** along which two lithospheric plates slide laterally past each other.

Wegener (1912, 1915) had recognized the following different types of crust (Fig. 5.4): (1) The **oceanic crust** with an average thickness of 5–8 km and (2) The **continental crust** with an average thickness of 30–35 km. Beneath mountain ranges, the thickness of continental crust even exceeds 70 km. Oceanic crust consists almost exclusively of basalt and at depth its coarse-grained equivalent, gabbro (Fig. 3.12). These rocks have an average density of 3.0 g/cm^3 and are therefore considerably heavier than the continental crust, composed of a variety of less dense rocks with an average density of 2.6–2.8 g/cm^3.

Oceanic crust is continuously formed at **mid-ocean ridges** located along constructive plate boundaries. These ridges are elongated submarine lineaments positioned in the middle of an ocean and are partly elevated more than 3000 m above the **abyssal plains**. For example, the Mid-Atlantic Ridge is, at its shallowest water depth, about 2500 m below sea-level. In contrast, the adjacent deep abyssal plains have water depths >5000 m. In the middle of the ridge lies the fracture zone between the two plates that are moving away from each other. This is called the **spreading center**. At this location, oceanic crust constantly forms and produces by far the greatest proportion of volcanic material that reaches the Earth's surface every year. To be more precise, mid-ocean ridges represent about 75% of the total global production of volcanic rock (Schmincke 2013). As these volcanoes lie mostly far below the water surface, their activity goes largely unobserved. In contrast, long chains of subsurface and active volcanoes are frequently located above subduction zones, including almost 500 of the ca. 550 active volcanoes known on Earth.

Continental crust is much more complicated in both its composition and structure compared to oceanic crust. Whereas oceanic crust is dense enough to be returned into the mantle along subduction zones and recycled by melting,

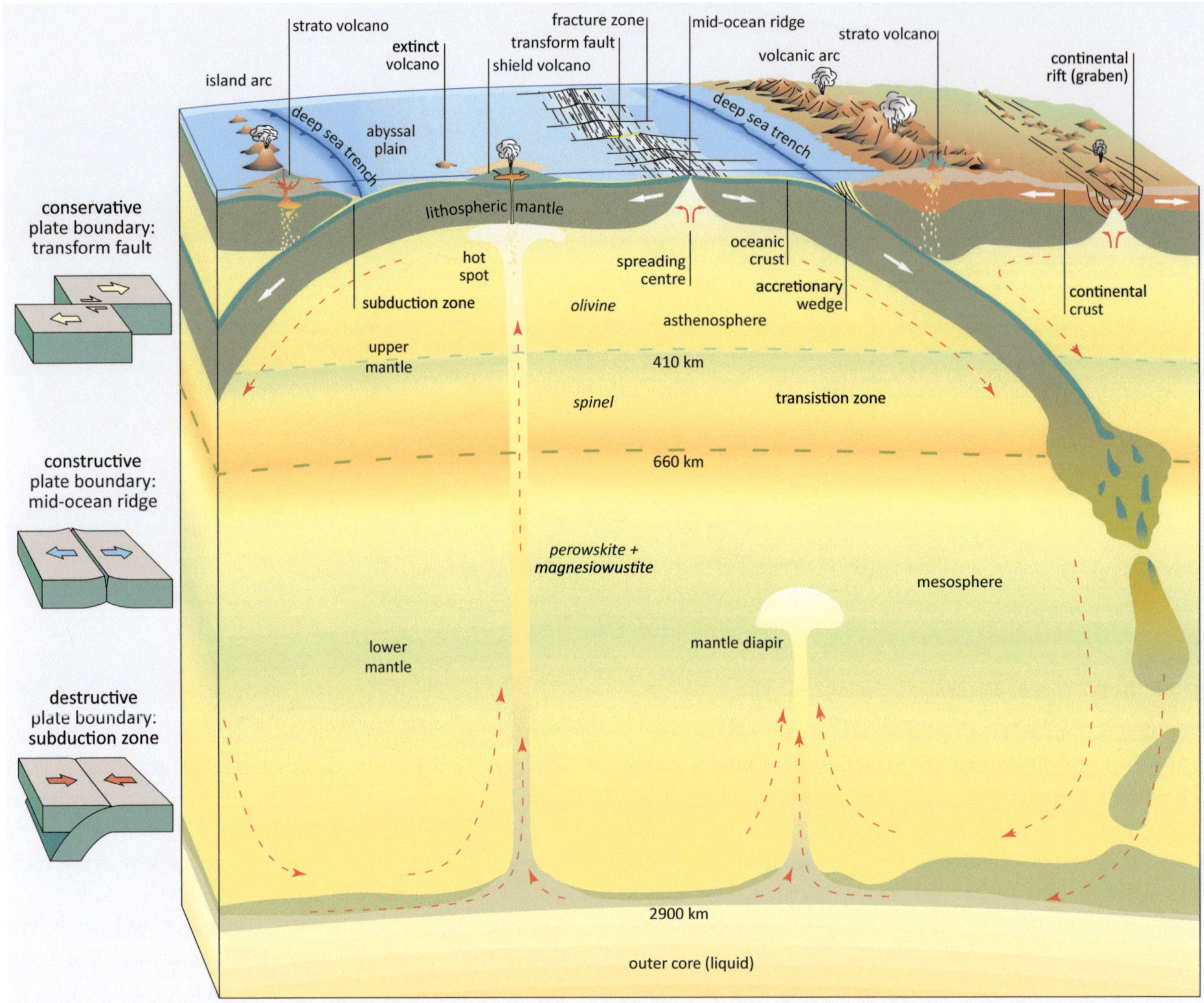

Fig. 5.4 Schematic representation of plate tectonic processes acting on the Earth's surface and mantle (various sources)

continental crust material is too light to be subducted. Therefore, over many hundreds of millions of years, a light "skin" of rock has accumulated, modified by displacement, deformation, and partial melting, which today represents our continental crust. In these rocks, most of the Earth's history is preserved, although sometimes changed beyond recognition. The oldest oceanic crust occurring today originated nearly 200 million years ago, as part of the western Pacific, but the rocks of the continental crust are considerably older. Some rock remnants have been identified with ages of around 4.1 Ga, 20 times older than the oldest oceanic crust.

Subduction zones are characterized by strong earthquakes and intense volcanic activity. At this location, lithospheric plates collide with each other resulting in the buildup of significant stress that is released in the form of potentially devastating earthquakes. These are the zones where the dynamics of the Earth are most evident. However, even in the midst of relatively stable lithospheric plates, some strong earthquakes may occasionally occur, such as the 1999 quake of Gujarabat in India or the violent volcanic eruptions of Mauna Loa in Hawaii. Although subduction zones are typically located at the margins of oceans, not all oceans are surrounded by subduction zones. A distinction is made between **active continental margins**, where oceanic lithosphere is subducted and recycled back to the mantle, and **passive continental margins** where both continental and oceanic crust is firmly connected.

At the sites of subduction, oceanic lithosphere is pulled into the mantle, where it is melted and transformed. If the descending slab sinks down to the outer edge of the Earth's core, the oceanic material may again be heated and continuously transformed before reascending to the surface after many millions of years at a site referred to as a **hot spot**. During the slow drift of an oceanic plate, sediments are deposited on its surface. These sediments are, in contrast to the oceanic crust, not able to be recycled into the mantle under normal circumstances,

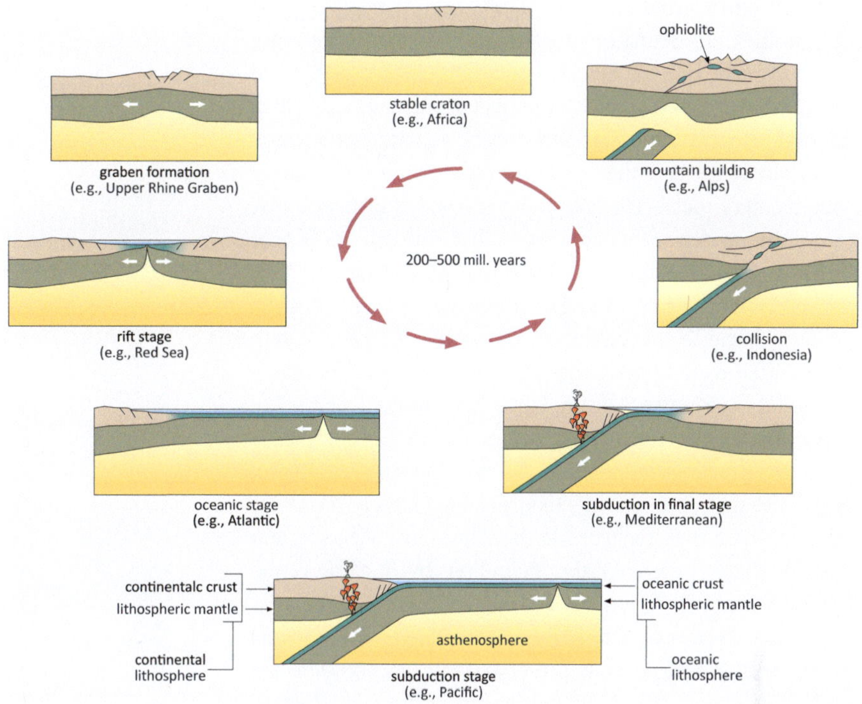

Fig. 5.5 Cyclic succession of plate tectonic processes according to the Wilson Cycle

because their density is too low to be transported downward. Instead, these sediments are either scraped off from the surface and deposited in an **accretionary prism** along the subduction zone or melted at a relatively shallow depth of about 110–120 km before migrating upward again as magma. Such magma either is extruded at the surface through volcanoes to form effusive rocks or crystallizes at some kilometers depth to form plutonic rocks, most commonly as granite.

Volcanic eruptions occurring in the midst of a lithospheric plate have other causes. These are mostly induced by **hot spots** that protrude like chimneys from very deep areas of the mantle and transport hot asthenosphere upwards to the base of the lithosphere (Fig. 5.4). Such chimneys are called **mantle diapirs**, which contribute significantly to the heat balance of the Earth. Magma chambers beneath the solid lithosphere are fed by ascending hot flows that lead to **bulging of the lithosphere**. Gigantic volcanic eruptions with global catastrophic consequences sometimes eject from these bulges at the surface with explosive intensities only matched by an asteroid impact. In volcanology, they are referred to as **supervolcanoes** that in contrast to the well-known conical volcanoes do not form volcanic mountains but leave huge cauldron-like craters, known as **calderas**. About 74,000 years ago, the Toba supervolcano erupted in Indonesia and produced about 2800 km^3 of volcanic rock material that was partly ejected up into the upper atmosphere. As a comparison, this amount of ejected material would cover the entire area of the Federal Republic of Germany with an 8-m-thick ash layer. The eruption caused global cooling by several °C over a period of several years and this time coincides, according to recent findings, with the coldest years of the Vistula and Würm glaciations. A highly controversial theory assumes that the ancestors of today's human race were drastically reduced in population by the eruption of the Toba volcano, with only a small group of a few thousand individuals surviving. All people living today are therefore considered to have descended from this limited population. Such super eruptions happen at irregular intervals of ca. every 500,000 to 1 Myr and they will happen again. However, nobody knows precisely how long it will take until the next supervolcanic eruption, which is likely to occur in either Indonesia, the Yellowstone area of the USA, or Chile.

Plate tectonic processes have operated throughout most of the Earth's history and occur as very long-term "Wilson" cycles: named after J. Tuzo Wilson (1908– 1993), one of the founding fathers of the modern theory of plate tectonics. The so-called **Wilson Cycle** begins with the breakup of continental plates in a graben system (Fig. 5.5). This is caused by stretching of the continental lithosphere whereby partially molten asthenosphere rises upward towards the surface beneath a rift graben system. Continuous extension leads to widening of the continental rift system and eventually to the formation of large volumes of oceanic crust along a

Box 5: Ophiolites

Ophiolites are defined as rock sequences that represent a section of the oceanic lithosphere that has been uplifted and commonly emplaced onto continental crust. Ideally, they consist of portions of the lithospheric mantle at the base, overlain by rocks of the oceanic crust and topped by marine sedimentary rocks. The bottom layer consists of **peridotite**, i.e., olivine-rich ultramafic rocks that form in the upper mantle (e.g., harzburgite, lherzolite). Under near-surface conditions, they are mostly converted to serpentinites, mainly because the olivine is not stable in contact with water and alters to hydrous serpentine minerals. The name ophiolite originates from the sometimes snake-like appearance of serpentinitic rocks in these complexes (Greek *ophis*: snake; Greek *lithos*: stone) and was first used by the French naturalist Alexandre Brongniart (1770–1847) to describe a rock sequence in the Alps. Gustav Steinmann (1856–1929) expanded the definition by including the serpentinites, pillow basalts, and radiolarites within the term ophiolite, known as the "Steinmann trinity". The igneous rocks of the oceanic crust preserved with an ophiolite are typically characterized by **gabbro**, **dolerite,** and **basalt**. Cumulate rocks, formed by the accumulation of heavy minerals at the bottom of the magma chamber by settling and differential segregation, are often found in the transitional zone to the peridotite. The boundary between the gabbro and peridotite that corresponds to the crust–mantle boundary is named the Mohorovičić discontinuity or Moho for short after the Croatian geophysicist Andrija Mohorovičić (1857–1936). However, the petrological Moho is different from the Moho recognized on deep seismic profiles (Fig. 5.6). Whereas the petrological Moho is characterized by a change in rock composition, the seismic Moho is based on the runtime differences of seismic velocities that only arise when entering the peridotitic mantle, with the zone of cumulates exhibiting similar seismic velocities as the gabbro.

Above the gabbro in an ophiolite sequence is the **sheeted dike complex.** These are comprised of parallel sets of dike intrusions that penetrate vertically into existing oceanic crust at the spreading zone of a mid-ocean ridge. Due to their parallel orientation, the dikes appear as "layered sheets", which is why they are called sheeted dike complexes. Above the sheeted dike complex occur **pillow basalts** (or pillow lava) that make up the upper layer of the oceanic crust. Pillow basalt develops when lava pours out at the seafloor and encounters water. The sudden cooling leads to the formation of rounded and tubular structures that can reach up to 1 m diameter. In the interstices of the pillows, **hyaloclastites** accumulate composed of quenched and broken volcanic glass. The uppermost layer in an ophiolite sequence, lying above the pillow basalt, is typically marked by **deep marine sediments** deposited following the extrusion of the pillow lava on the seafloor.

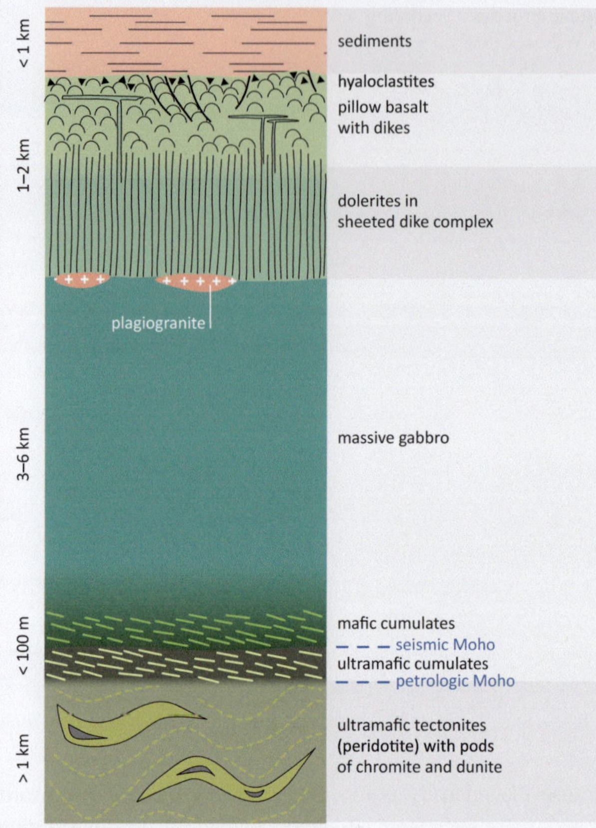

Fig. 5.6 Schematic structure of a typical ophiolite sequence

spreading center that finally leads to the opening of an ocean. Following the closure of the ocean by subduction and subsequent slab breakoff, a mountain range develops in a zone of compression. Subsequent erosion and gravitational collapse of the mountain belt finally leads back to a normal crustal thickness as at the beginning of the Wilson Cycle.

In 100 Myr, a rift in a continental plate may expand to form an open ocean of a width of several 1000 km's. While oceanic lithosphere is continually generated and recycled back into the mantle in subduction zones, the continental lithosphere remains closer to the surface, since it is not easily subducted because of its much lower density. The oldest oceanic crust known in its original location of formation is just a little less than 200 Ma old. All older oceanic crust has been consumed along convergent plate margins or small volumes scraped off and obducted into continental crust, such as in the Mediterranean area (the Troodos ophiolite of Cyprus). This mechanism is associated with mountain building processes whereby fragments of oceanic crust are sliced into the accretionary prism along active continental margins and avoid subduction by forming an **ophiolite** (Box 5: Ophiolites). Continental crust also preserves large volumes of ancient rock, which may be repeatedly involved in mountain building processes and modified by multiple metamorphic events.

Tectonics Units of Europe

6

Europe represents a mosaic of different continental crustal blocks that have been assembled over a prolonged period of geological time. The continent can be roughly divided into two major regions, a stable Precambrian craton in the northeast, known as the East European Platform or Baltica, and a group of crustal blocks successively attached to the old cratonic core. The boundary between these two units is the Trans-European Suture Zone (TESZ), which extends for more than 2000 km from the Black Sea to the North Sea where it is also called the Tornquist–Teisseyre zone (Fig. 6.1). Based on the main phases of mountain building processes in Europe, four main tectonic units of different age can be recognized:

6.1 Ancestral Europe

This is the oldest unit comprised predominantly of Precambrian rocks of the East European Platform and contains the **Baltic Shield**, the **Russian Platform**, and the **Ukrainian Shield** (Fig. 6.1). Paleogeographically these areas belonged to Baltica, which over a long period of time existed as a separate tectonic plate. A shield is referred to an area of the crust composed of ancient continental crust, typically several billion years old.

The East European Platform consists largely of high-grade metamorphic or magmatic rocks mostly of Precambrian and Paleozoic age. The platform is bordered by three younger areas that were attached during Paleozoic mountain building processes after the end of the Precambrian, 542 Ma ago. Germany is located in the midst of these units and was strongly influenced by several mountain building "orogenic" phases. Its location makes it extremely well suited for studying the geological history and in particular the last two orogenic stages of the continental development of Europe. The three major tectonic units attached to the East European Platform are each of younger age when passing from north to south. These units are described as follows:

6.2 Paleo-Europe

Paleo-Europe lies in the northwest, directly adjacent to the East European Platform and contains the **Caledonian Mountain Range**. It extends from Norway to Scotland and Ireland, and borders the **North American Greenlandic Platform** in the far northwest of Scotland. The Caledonian Mountain Belt formed along the margin of the **Laurentia** continental plate, named after the St. Lawrence River in North America. A small remnant of the Laurentia craton is today located in the northwest corner of Scotland (Fig. 6.1). The vast majority of the craton is found today on the other side of the Atlantic as a result of the opening of the Atlantic Ocean during the Mesozoic and Cenozoic.

Caused by the collision of Baltica and Laurentia, the Caledonian Mountain Range (lat. *Caledonia*: Scotland) was formed during the Ordovician and Silurian by the closure of the **Iapetus Ocean** (named after the Greek titan *Iapetos*). In the south, a smaller continental block designated as **Avalonia** (named after the Avalon Peninsula on the Canadian island of Newfoundland) was attached as a terrane to Baltica by the closure of a smaller branch of the Iapetus Ocean, known as the **Tornquist Ocean**.

Originally Avalonia was part of **Gondwana** (named by Eduard Suess, 1831–1914, after *gondavana*, the "land of Gonds" [Sanskrit] in India), but was separated from this continent in the Early Paleozoic, prior to moving northwards. A terrane is an independent tectonic block that migrates passively on a larger lithospheric plate and is attached to a continent when colliding with it. The Avalonian terrane (Fig. 6.1) includes the southern parts of Britain and Ireland, the basement rocks of northern Germany, and the Variscan units of the Rhenohercynian (Rhine-Harz zone). Rocks of this terrane are also located in Newfoundland on the American side of the Atlantic. Avalonia was

M. Meschede and L. N. Warr, *The Geology of Germany*, Regional Geology Reviews,
https://doi.org/10.1007/978-3-319-76102-2_6

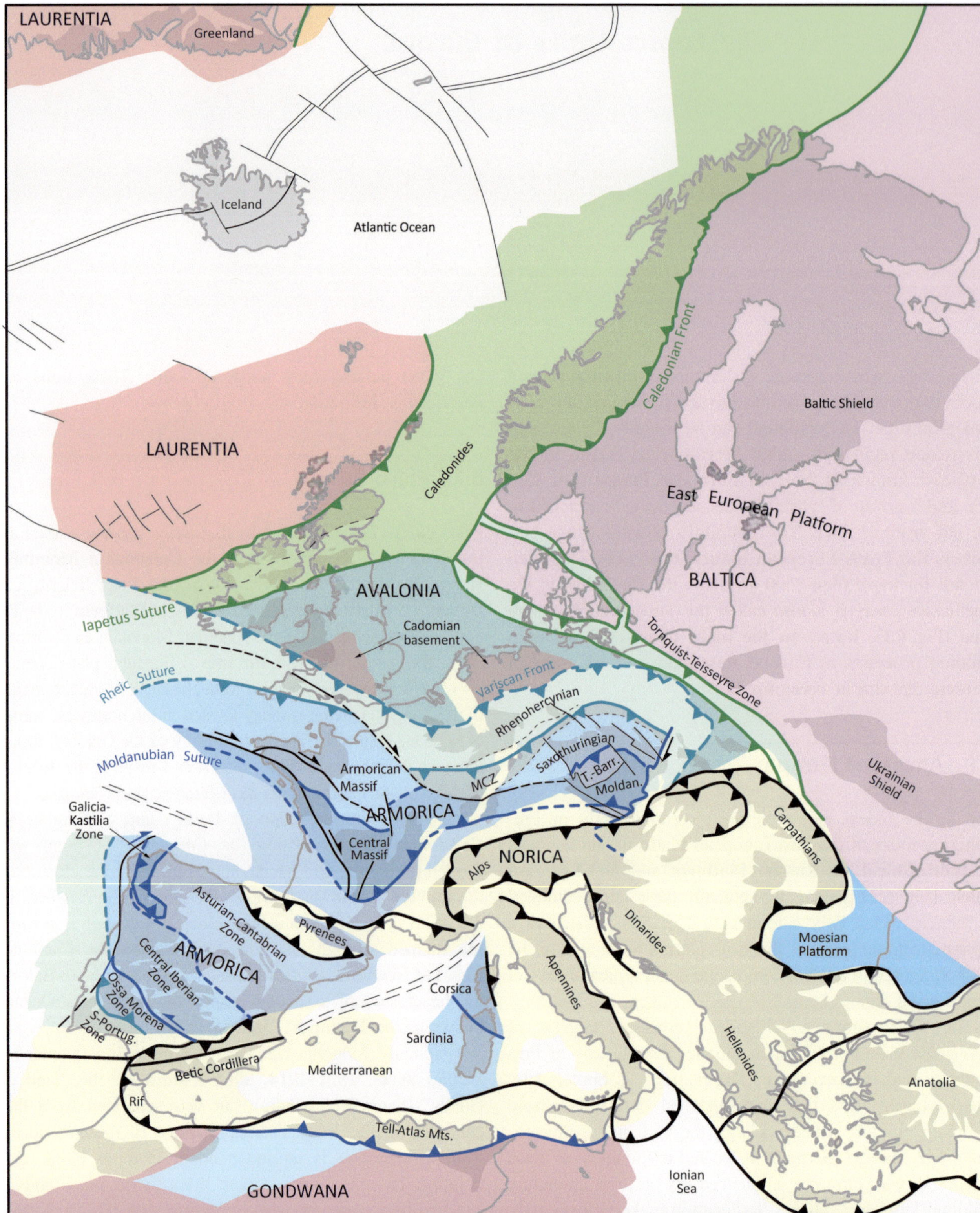

Fig. 6.1 The tectonic subdivision of Europe showing the three major cratonic units Baltica, Laurentia, and Gondwana and assignment to the microplates of Avalonia, Armorica, and Norica. Modified after Kossmat (1927), Franke (1989), Martínez et al. (2007) and others sources

connected to the East European Platform along the Tornquist–Teisseyre zone (named after the German and Polish geologists Alexander Tornquist, 1868–1944, and Wawrzyniec Teisseyre, 1860–1939). This Trans-European Suture Zone, along which the Tornquist Ocean was closed, is one of the most important tectonic sutures in Europe. A **suture** (lat. *sutura*: seam) marks the contact boundary between two tectonic units connected during a collision.

6.3 Meso-Europe

Meso-Europe is made up of the eroded remnants of the **Variscan Mountains** that were named by Eduard Suess (1831–1914) after *Curia Variscorum*, the Latin name for the city of Hof in the Fichtelgebirge (Suess 1883). The name refers to the Celtic tribe of Variscans that lived in the region at the time of the Romans. This unit is comprised of the southern part of **Avalonia** and large parts of **Armorica** (named after the Armorican Massif of northwestern France). The latter is an elongated terrane composed of multiple, partly quite different tectonic units more appropriately termed the **Armorican Terrane Complex**. This complex includes, in addition to Iberia, the Armorican Massif, the Central Massif, most of the Vosges, the Black Forest, the Bohemian Massif, and the Saxothuringian (Saxonian–Thuringian zone) in the sense of Kossmat (1927; see below) (Fig. 6.1).

The **Variscides** formed a mountain belt with a number of recognizable internal and external tectonostratigraphic zones assembled during Devonian and Carboniferous times. Only a part of this mountain belt is located in Europe, with a longer section continuing in the **Appalachian Mountains** on the other side of the Atlantic. Similar to Avalonia, Armorica also once belonged to Gondwana, but later split off and then drifted independently to the north. During the Variscan Orogeny, Armorica became attached to Avalonia following the closure of the intervening **Rheic Ocean** (named after the Greek goddess *Rhea*). At this time, Avalonia already belonged to the northern continent of **Laurussia** since its collision during the Caledonian Orogeny. The sediments of the Rhenohercynian zone were deposited on the northern edge of the Rheic Ocean, an area which represented a passive margin undergoing a high degree of crustal extension and thinning of the continental crust.

South of the Rheic Ocean, the Saxothuringian developed with a similar continental shelf facies as part of the Armorican Terrane Complex and the **Moldanubian** (Vltava–Danube zone). It is separated from the adjacent units to the south belonging to the Norica terrane (see below) by the **Moldanubian suture**. The subdivision of the Variscan Mountain Range into the Rhenohercynian, Saxothuringian and Moldanubian zones goes back to Franz Kossmat (1871–1938), who introduced this concept in 1927 to the geological literature (Kossmat 1927). Some units in the south and southwest were probably not completely separated from Gondwana prior to the Variscan Orogeny. These include, for example, the **Norica** terrane (named after the Roman term *Alpes Noricae* for part of the Eastern Alps), which formed during the Variscan Orogeny and is today located in the Alps (Fig. 6.1), as well as other Paleozoic basement complexes in the Alpine chain. The formation of the Variscan Mountain Range, therefore, reflects the amalgamation of a number of continental plate fragments by the end of the Carboniferous to form the Permian supercontinent of **Pangea**.

6.4 Neo-Europe

Neo-Europe is represented by the most recent attachment to the Eurasian continent in the form of the **Alpine Mountain Belt** together with its southerly located microplates. In addition to the Alps, the Alpine Mountain Belt includes the Carpathians, the Dinarides, the Hellenides, the Apennines, the Atlas Mountains, the Betic Cordillera, and the Pyrenees. There are, however, different opinions as to the number and size of the plates involved in this collision zone. In this book, the **Adriatic–Apulian Plate** and the **Anatolian Plate** are recognized and described. An early (Eoalpine) collision stage in the Jurassic was followed by the main **Alpine Orogeny** at the end of the Cretaceous. The latter orogenic event is still active to this day, with the continued uplift of the Alpine chain. The end of the orogeny will be achieved when Africa and Eurasia complete its collision by the closure of the **Mediterranean Sea,** which in its eastern part still has residual parts of oceanic crust from the **Neotethys Ocean**. At this stage, in a few tens of million years, the Alpine Mountain Range is likely to have increased its lateral extend to include this area.

7 Overview of the Plate Tectonic History of Europe

Figure 7.1 illustrates the drift of continents over the globe following the breakup of the supercontinent of Rodinia (Russian *rodit*: to give birth) about 750 million years ago. It also shows how the present land—sea distribution has changed significantly over time during constant plate tectonic motion. The old Rodinia supercontinent, the starting point of the breakup, was assembled during the Grenville Orogeny about 1.1–1.0 Ga (Fig. 7.1a). The few crustal blocks of Rodinia whose remnants can be identified today in Germany were originally located at that time on the southern edge of the supercontinent, close to West Africa and Amazonia.

The breakup of Rodinia initiated before ca. 750 million years formed several smaller continents, among them Baltica, Laurentia, Siberia, and Northern China (Fig. 7.1b). The larger southern continental mass of **Gondwana**, which persisted throughout the entire Paleozoic, formed somewhat later during the **Pan-African Orogeny** ca. 650–500 million years ago from a number of fragments of Rodinia that included Amazonia, Rio de la Plata, Australia, Antarctica, India, Congo, and Kalahari. Baltica formed as an independent plate and underwent a significant counterclockwise rotation during the late Proterozoic. At this time, Avalonia remained firmly attached to Gondwana.

During the Paleozoic, several oceans formed that were subsequently closed a short time later. Important for the European development during the Cambrian and Ordovician was the opening of the **Iapetus Ocean** between Baltica and Laurentia, and the **Tornquist Ocean** between Baltica and Gondwana (Fig. 7.1c). Obducted remnants of these oceans were preserved as **ophiolites** (Box 5: Ophiolites) in a number of locations across Europe, such as the well-developed Ballentrae complex of Scotland. Both of these oceans closed during the Ordovician and Silurian, simultaneous with the opening of a new Rheic Ocean, which then separated the microplate of Avalonia from Gondwana. The collision of Laurentia with Baltica occurred in the Silurian with the development of the Caledonian Mountain Belt, remnants of which are well exposed in parts of Norway and Scotland.

During the Cambrian and Ordovician, the Central European microplates of **Avalonia** and **Armorica** were still located close to the South Pole, where remnants of glacial sediments indicate it was glaciated. At this time and later in the Silurian and Devonian (Fig. 7.1d) Avalonia and Armorica drifted as separate microplates or terranes northward. Avalonia collided first with Laurentia in the Early Devonian and then Armorica with Avalonia and Baltica during the Early Carboniferous (Fig. 7.1e). In the Devonian, sediments were deposited in the **Rheic Ocean** between Avalonia and Armorica, which can be found today in the Variscan complexes of Central Europe. Devonian sediments also accumulated in the **Moldanubian Ocean** between Armorica and **Norica** that lay to the north of Gondwana. These were later incorporated into the Alpine Mountain Belt. The area including Germany was located along the paleo-suture zone between the Avalonian and Armorican microplates, a line across which the Rheic Ocean was consumed.

During the Carboniferous (Fig. 7.1f) and Permian, almost all continental plates rejoined together to form the supercontinent of **Pangea** (Greek *pan*: all-embracing; Greek *gaia*: Earth, land), which was originally named by Alfred Wegener (1912, 1915). This amalgamation followed the closure of the Rheic and Moldanubian Oceans and was accompanied by the Variscan Orogeny. The deep orogenic roots of the Variscan Mountains are preserved as the crystalline basement of Central Europe. These rocks extend laterally to eastern North America and run along the Appalachian Mountains. At the time of the Variscan Orogeny, Central Europe was located near the equator. At this latitude, the warm tropical climatic conditions were ideal for widespread forest development and the formation of abundant coal deposits characteristic of the Carboniferous period. The poles experienced a distinct period of glaciation during

M. Meschede and L. N. Warr, *The Geology of Germany*, Regional Geology Reviews,
https://doi.org/10.1007/978-3-319-76102-2_7

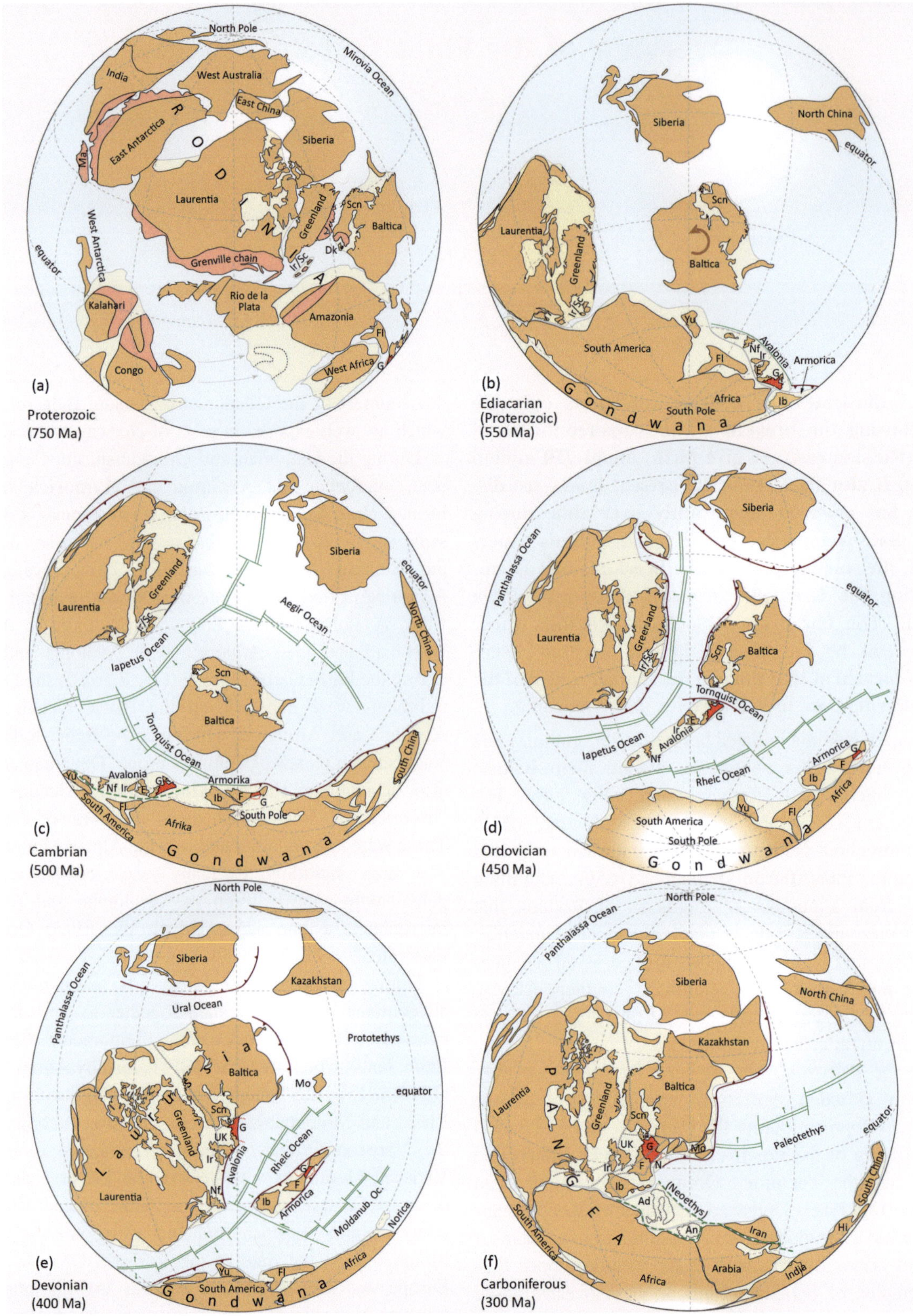

Fig. 7.1 Reconstruction of the global plate tectonic evolution showing events of Central Europe following the breakup of the supercontinent of Rodinia. *Ad* Adriatic-Apulian Plate, *Al* Alps, *An* Anatolian Plate, *Ca* Carpathians, *Dk* Denmark, *E* England, *F* France, *Fl* Florida, *G* Germany, *GB* Great Britain, *Ib* Iberia, *Ir* Ireland, *Pe* Penninic Ocean, *Ma* Madagascar, *Mo* Moesia (presently Romania), *N* Norica, *Nf* Newfoundland, *Py* Pyrenees Ocean, Sc Scotland, Scn Scandinavia, Yu Yucatán. Modified after Tait et al. (1997), Hunter and Easterbrook (2004), Lawver et al. (2009), Frisch et al. (2011), Frisch and Meschede (2013) and other sources

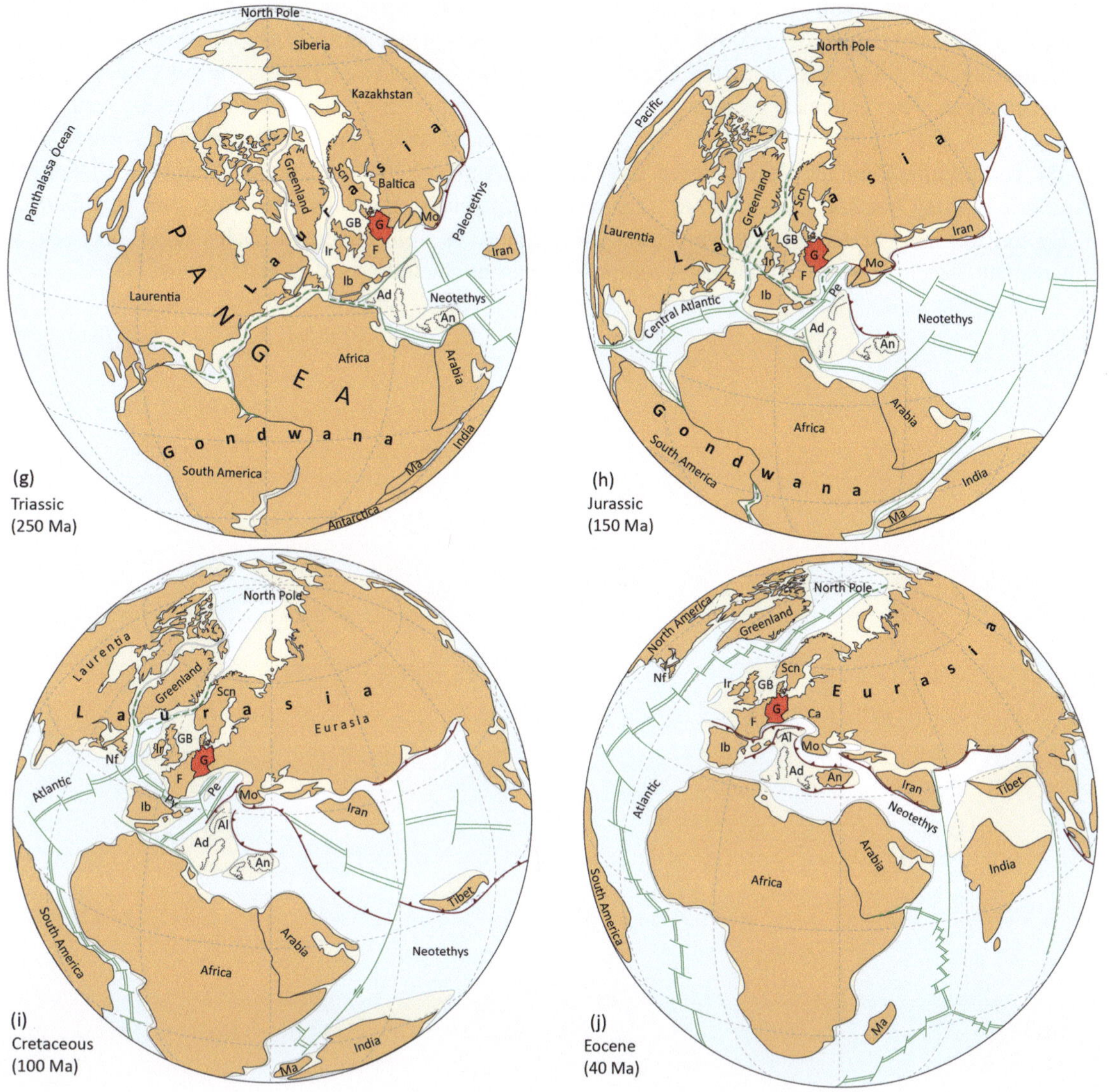

Fig. 7.1 (continued)

Carboniferous and Permian times, while during the Permian widespread arid climatic conditions extended across large parts of Pangea. This lead to the development of red beds and huge salt deposits that accumulated in restricted basins. At this stage, the Atlantic did not exist, which today separates Europe and North America by several thousand kilometers.

The breakup of the Pangean supercontinent started in the Triassic (Fig. 7.1g) and still continues to this day. Central Europe slowly drifted northward in the Mesozoic but remained in very warm regions. The ice caps at the poles disappeared and the arid climatic conditions that continued into the Triassic produced typical red beds and salt deposits within basins formed by the initial breakup of Pangea. During the Jurassic, Central Europe was located on the passive margins of newly formed oceans that cut through Pangea from east to west (Fig. 7.1h): namely the **Neotethys** in the east, passing through the **Penninic Ocean** to the **Central Atlantic**, which in turn was connected to the Pacific via the Caribbean partly by transform fault systems. At this time the climate became notably more humid.

In the following Cretaceous to Early Tertiary period, the spreading direction shifted and the opening of the oceans switched to an E-W direction, resulting in N-S oriented spreading axes. This period saw the opening of the Atlantic over its entire length from north to south, separating Africa from South America and in the north Laurentia from Eurasia (Fig. 7.1i). In the area of the Neotethys, the strongest phase of crustal shortening began. The South Atlantic opened in the Middle Cretaceous, followed by the North Atlantic in the Paleocene/Eocene (Fig. 7.1j). The climate in the Cretaceous period was characterized by extremely warm greenhouse conditions.

In the Early Tertiary, the Penninic Ocean closed and the **Adriatic-Apulian Plate** became attached to Eurasia (Fig. 7.1j). By the Eocene, narrowing between Africa and Eurasia intensified leading to the almost complete subduction of the Neotethys, apart from a few remnants that still exist in the form of the Mediterranean, the Black Sea, and the Caspian Sea. During the most recent Tertiary period, N-S compression across Central Europe resulted in the formation of the Alpine mountain chain, which encompasses the Alps, the Pyrenees, the Apennines, the Dinarides and the Carpathians. Further to the east, this mountain chain continues as fold-and-thrust belts lying between India and Europe, including the Himalayas.

8 The Dynamic Earth, Earthquakes in Germany

Although Germany is not currently located along an active plate boundary characterized by large earthquakes, some areas of the county are regularly affected by notable seismicity. This activity, which is the subject of this chapter, is related to the buildup and release of tectonic stresses in the German upper crust caused largely by extensional movements of the European Cenozoic Rift System.

In the early morning of April 13, 1992, the residents of the German-Dutch border area between Roermond and Düren, and as far as Cologne and Bonn, were awakened from their sleep by tremors that caused bookshelves to tumble, cups to clank in the cupboards and tiles to fall from the roofs. An **earthquake** had occurred, which with a magnitude of 5.9 belongs to one of the strongest earthquakes yet to be experienced in Germany. In North Rhine-Westphalia, one person died and more than 30 people were injured in the incident, mostly hurt from the falling bricks and roof tiles. The material damage on both sides of the border amounted at that time to about 125 million euros. Some houses were so badly damaged that they had to be demolished and in Cologne, at a distance of 80 km from the earthquake focus, a pinnacle from a turret fell off the roof of the city's cathedral. This was the worst earthquake in the Rhineland since the Düren 1756 earthquake. The **epicenter** of the quake was located about 4 km southwest of the Dutch town of Roermond and the earthquake focus was placed at a depth of 18 km (Fig. 8.1). The last noticeable earthquake to occur in this region took place on July 22, 2002, to the northeast of Aachen, where the tremors reached a magnitude of 5.0.

Similar earthquakes have occurred in the south of Germany. On September 3, 1978, an earthquake of magnitude 5.7 was recorded at six o'clock around the area of Onstmettingen, located in the Zollernalbkreis of the Swabian Alb (Fig. 8.1; Box 6: Earthquake magnitude). At the nearby Hohenzollern castle in Balingen, a side tower partially collapsed (Fig. 8.2). Overall, 25 people were injured, more than 6000 houses damaged and the material costs of repair were estimated to be equivalent to 100 million euros. This area is influenced by fault movements within the **Hohenzollern Graben** system and is periodically shaken by earthquakes. Records have shown that during the last hundred years, an earthquake of magnitude 5–6 has occurred on average every 30–40 years.

Another area in Germany that is regularly affected by earthquakes is the area of the **Eger Graben** in the Vogtland and Fichtel Mountains (Fig. 8.1). The last major earthquake in this area occurred in October 2008, when an earthquake swarm with a maximum magnitude of 4.1 robbed the residents of the region of a peaceful nights sleep.

But why are some regions in Germany regularly shaken by earthquakes, while other areas remain unaffected? The reason for this is the buildup of geological stresses within the structure of the deep subsurface of Europe's continental crust.

Earthquakes in Europe are actually not as frequent as in many other regions of the world, but they do occur repeatedly at irregular intervals and occasionally cause considerable damage. Historical reports tell of the widespread destruction of the Lisbon earthquake of November 1755 with up to 50,000 casualties (Fig. 8.3) or the Basel earthquake of October 1356 with more than 300 casualties. In Lisbon, the destructive effect of the earthquake was reinforced by a subsequent tsunami similar to that occurring in the Indian Ocean on December 26, 2004, and in Japan on March 11, 2011.

Usually, we cannot perceive crustal movements because they occur too slowly for our human perception of time. Earthquakes, however, are rapid events that are visible and can be felt at the Earth's surface. Slow movements of the Earth's crust occur at an average rate of only a few millimeters per year, but over the time scale of many millions of years this motion constantly changes the face of the Earth and shapes the present mountains, plains and oceans. The continents of Europe and America, for instance, are currently moving away from each other with the distance between them increasing every year by about 2–3 cm, i.e. at the rate, our fingernails grow each year. To explain this phenomenon, earth scientists apply today the modern theory of plate tectonics, which elegantly explains the formation and disappearance of crustal plates, the movements of the Earth's crust resulting in the formation of

M. Meschede and L. N. Warr, *The Geology of Germany*, Regional Geology Reviews,
https://doi.org/10.1007/978-3-319-76102-2_8

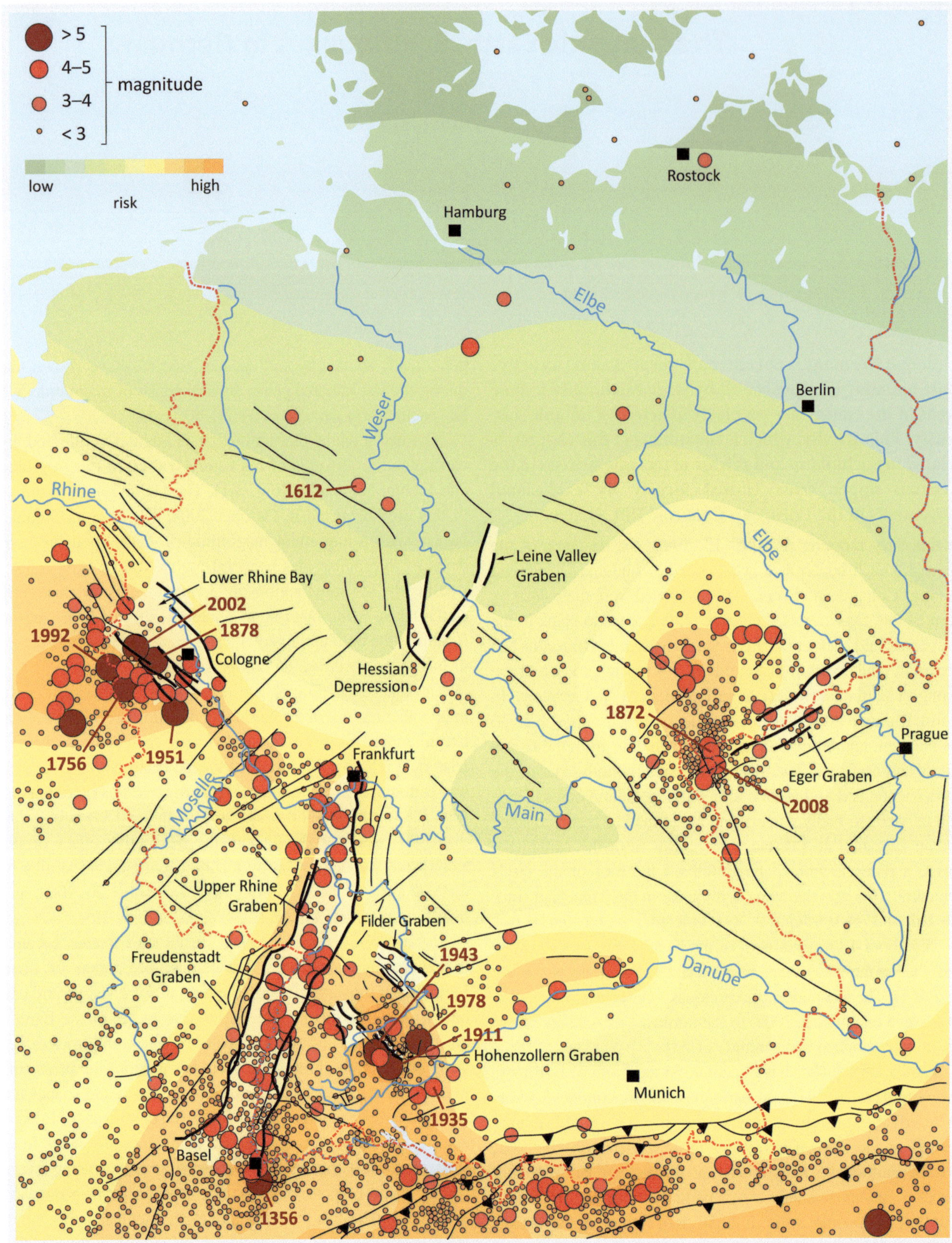

Fig. 8.1 Earthquake activity of Germany and neighboring areas showing a selection of important earthquakes that occurred over the last 100 years. The colored marking on the map indicates the probability of an earthquake reoccurring based on past records. The highest risk areas correspond with young graben structures. Modified after Grünthal and Wahlström (2003), and Leydecker (2011)

Box 6: Earthquake magnitude scales

Readers are certainly familiar with the headlines: "The earthquake had a magnitude of (…) on the open-ended Richter scale." However, what is the Richter scale? The **Richter scale**, named after the American geophysicist Charles F. Richter (1900–1985), is only one of several **magnitude scales** used to describe the strength of an earthquake (Richter 1935). Initially used as a local magnitude (M_L) scale, it was adopted worldwide in the 1930s. It was, however, initially unsuitable for measuring larger earthquakes of magnitudes above 6.5, since it was based on local amplitude measurements from seismograms that only produced reliable values in a range of 100 km from the epicenter of the earthquake. Even though the Richter scale is still used in the media, today the strength of an earthquake is measured using the **momentum magnitude** (M_W). This refers to the size of the earthquake source and represents a measure of the released energy during faulting. With earthquake magnitudes up to 6.5, the values of the logarithmic Richter and the momentum magnitude scales agree well, however, for earthquake magnitudes >6.5, only the momentum magnitude scale is valid. Although there is theoretically no upper limit to the Richter scale, the momentum magnitude does have a calculated maximum value of 10.5 (M_W), above which the Earth's crust would break apart. The largest magnitude recorded so far was the 1960 Valdivia earthquake in Chile with an impressive magnitude of 9.5 (M_W). Since the beginning of earthquake records, there have been only five earthquakes worldwide with a magnitude of 9 and higher. This includes two earthquakes that caused devastating tsunamis, one in 2004 on the edge of the Indian Ocean and the other in 2011 on the eastern side of Japan that led to the Fukushima Daiichi nuclear accident.

As earthquake magnitudes are measured on a logarithmic scale, a quake of magnitude 7 is about 3 times stronger than a quake of magnitude 6 and 10 times stronger than a quake of magnitude 5. It is not possible to calculate the magnitude of historical earthquakes because seismographs were invented just a little more than 100 years ago. For these earthquakes, the descriptive **Mercalli scale** (named after the Italian volcanologist Giuseppe Mercalli, 1850–1914) is used, which attempts to classify earthquake magnitude based on the damage caused (Table 8.1). Using this kind of assessment it is possible to compare the strength of historical earthquakes such as the Basel earthquake of 1356 or the Lisbon earthquake of 1755, with the damage caused by events that are more recent.

Table 8.1 Comparison of the momentum magnitude and the Mercalli intensity scales for describing the intensity of earthquakes

Effect	Brief description	Momentum magnitude	Intensity (Mercalli)
Only detectable with instruments	Imperceptible	2.0	I
Sporadically sensible	Very weak	2.5	II
Slight vibrations	Weak	3.0	III
Objects oscillate	Moderate	3.5–4.0	IV
Doors clatter	Rather strong	4.0–4.5	V
Small damage to buildings	Strong	4.5–5.0	VI
Moderate damage to buildings	Very strong	5.5	VII
Strong damage to buildings, pediments collapse	Disruptive	6.0–6.5	VIII
Strong damage to buildings, fissures in the ground	Devastating	6.5–7.0	IX
Strong demolition	Destructive	7.5	X
Strong destruction	Disaster	7.5–8.0	XI
Total demolition	Great disaster	8.5–9.5	XII

mountains, the occurrence of earthquakes, and the distribution of volcanic activity, all in a coherent and self-contained picture (Chap. 5). In this context, most earthquakes result from the rapid release of crustal stress along geological faults, caused directly or indirectly by plate motions.

Earthquakes that occur in Germany mostly result from extensional stresses affecting the Earth's crust. At present, the active graben structures of Hohenzollern, the Rur, the Lower Rhine Bay and the Eger Graben are responsible for such disturbances. At these locations, the continental crust is been stretched and as a result, the graben structures are subsiding. The tectonic movement associated with the release of dilatational stresses results in earthquakes. The largest and most striking geomorphological graben system

Fig. 8.2 During the earthquake of September 3, 1978, in the Swabian Alb, a side tower of Hohenzollern Castle, located in the Hohenzollern Graben, was severely damaged. (© picture-alliance/dpa). [N 48°19′23″/ E08°58′2″]

is located in southwest Germany and is known as the Upper Rhine Graben. This structure completed its extensional phase of development ca. 6–7 million years ago and since then underwent lateral movements along the boundary faults of the graben. Originally, the main extensional direction was in the WNW–ESE direction and thus perpendicular to the NNE–SSW extending graben axis, which represented the main direction of compression. The Upper Rhine Graben represents the central part of a tectonically important Cenozoic fault system that crosses a large portion of continental Europe from north to south (Fig. 8.4). Previously it was termed the Mediterranean-Mjøsa zone (Stille 1925), but today it is generally known as the **European Cenozoic Rift System** (ECRIS; Ziegler 1992). It extends from the south, starting in the Gulf of Lyon, and continues through the graben structures of the Rhone, Bresse, and Limagne, to a group of smaller parallel structures in the north. These structures are connected to the Upper Rhine Graben via the Rhine-Bresse Transform Fault Zone (Laubscher 1970; Frisch et al. 2011; Frisch and Meschede 2013). Further to the north the ECRI system branches out into the northeastern Hessian Depression and several smaller structural depressions that merge into the southern Lower Saxonian Basin and the Lower Rhine Bay. The latter branches out to the northwest into some smaller structural lows in the Middle Rhine area (Neuwied Basin). The Rhenish triple junction, where these branches meet, is located beneath the Mainz Basin as first suggested by Cloos (1938). The northwestern branch eventually joins with the graben structures of the North Sea (North Sea Graben and the Viking Graben). Contrary to previous assumptions about the Cenozoic rift system, the Leine Valley, and Oslo grabens are not to be included as they were only active prior to the Tertiary period.

The breakup of the European continental plate ceased in the Miocene with the reorientation of the European stress system. The compressional direction changed to the NW-SE direction and in doing so marked the end of subsidence in the Upper Rhine Graben. Today, the active graben and fault systems whose main extensional direction coincides with the current stress directions are represented by the shorter and narrower Hohenzollern Graben and some parallel structures (e.g., Freudenstadt Graben, Filder Graben), as well as the Rur Graben and related structures (e.g., Erft Step), and the Lower Rhine Bay (Fig. 8.5). The ascent of magmatic melt along NW-SE running fault zones in the Eifel also follows the orientation of the graben system (Figs. 14.9 and 14.10).

During historical times, earthquakes with the strongest magnitudes in Germany occurred in the Lower Rhine Bay and its surrounding area. The 1992 Roermond earthquake with a magnitude of 5.9 is one of the strongest earthquakes measured in this area (Figs. 8.1 and 8.6). The **Lower Rhine Bay** is a young, active graben structure that post-dates the formation of the Upper Rhine Graben. It is ironic that the Roermond earthquake happened just prior to the scheduled revision of the construction permit for the nuclear power plant at Mülheim-Kärlich. The reactor, which was initially planned to be built directly on a fault of the graben system, was moved several tens of meters away from the fault line. As this was done without a reapproved construction permit, the site remained controversial and the plant eventually closed partly due to the seismic risk. Future earthquakes that may even exceed magnitude 6 are to be expected in the Lower Rhine Bay area.

The **Hohenzollern Graben** of the Swabian Alb is an extensional structure that is not easy to recognize at the surface. It belongs to a branched graben system that reflects tectonic stresses in the subsurface crust. This system includes also the Freudenstadt Graben and some smaller extensional structures that lie parallel to the Hohenzollern Graben. Figure 8.7 shows the most important tectonic structures of southwest Germany in the surroundings of the

Fig. 8.3 Contemporary illustration of the Lisbon earthquake of November 1755. (© picture-alliance/akg-images)

Hohenzollern Graben. The distribution of earthquake foci detected by seismographs, including those of very small magnitude, do not occur parallel to the axis of the Hohenzollern Graben but instead align roughly in an N–S direction. This reflects **decoupling** at the upper crustal level in response to dilatational stresses that form extensional fractures similar to tension gashes in a shear zone. The deeper crustal level beneath deforms more by ductile flow with movement along the evaporate rock horizon of the Middle Muschelkalk, which is oriented along old inherited structures similar in orientation to the Upper Rhine Graben (Fig. 8.8). The earthquakes that have occurred along this structure, known as **Albstadt Shear Zone**, are not characterized by extension but instead show clear lateral displacements. Thus, structures at the surface that are clearly identified as a tectonic graben are actually caused by strike-slip faulting at deeper crustal depth (Reinecker and Schneider 2002).

The tectonic displacement of the Hohenzollern Graben is best understood by taking a hike up to the Raichberg from where there is an impressive view of Hohenzollern Castle. Although the castle is located within the graben structure, it does not occur as a graben-like depression but instead forms a classic example of **inverted relief** (Fig. 8.9). Since the castle hill is separated geomorphologically from the plateau of the Swabian Alb, it represents a **cuesta outlier**, which is an elevated area that was once part of the Alb. At this location, the graben structure has displaced erosion-resistant limestones of the lower Late Jurassic to deeper levels than in the surrounding area. The resultant thick carbonate sequence in the rift forms the cap to the mountain that continues to protect the underlying softer Middle Jurassic marl from erosion (for the nomenclature of stratigraphic units see Fig. 11.56 and Sect. 11.6).

Located outside the Hohenzollern Graben, the slightly higher Raichberg forms part of the elevated plateau area of the Swabian Alb. The summit area of the Raichberg is characterized by upper Late Jurassic rocks. If you walk along the trail known as the Hanging Rocks (*Hangender Stein*), you will cross the boundary fault of the graben that runs just below the trail. The Hanging Rocks are in fact large blocks that are about to detach from the cuesta and slide downslope. In some places, gaps of several meters occur between the Hanging Rocks and the intact layer of limestone. This regressive erosion is a common feature of the Alb Cuesta (*Albtrauf*) and is the cause of the prominent cuesta made up of light-colored limestones of Late Jurassic age (Box 7: Cuesta landscapes). An offset along the graben fault can also be observed from the viewing platform of the Hanging Rocks. The view extends to a cuesta on the opposite side of the valley that lies outside of the graben structure. At this location, older Late Jurassic rocks occur at exactly the same topographical height due to a structural displacement of the stratigraphy by about 100 m along the boundary fault of the graben.

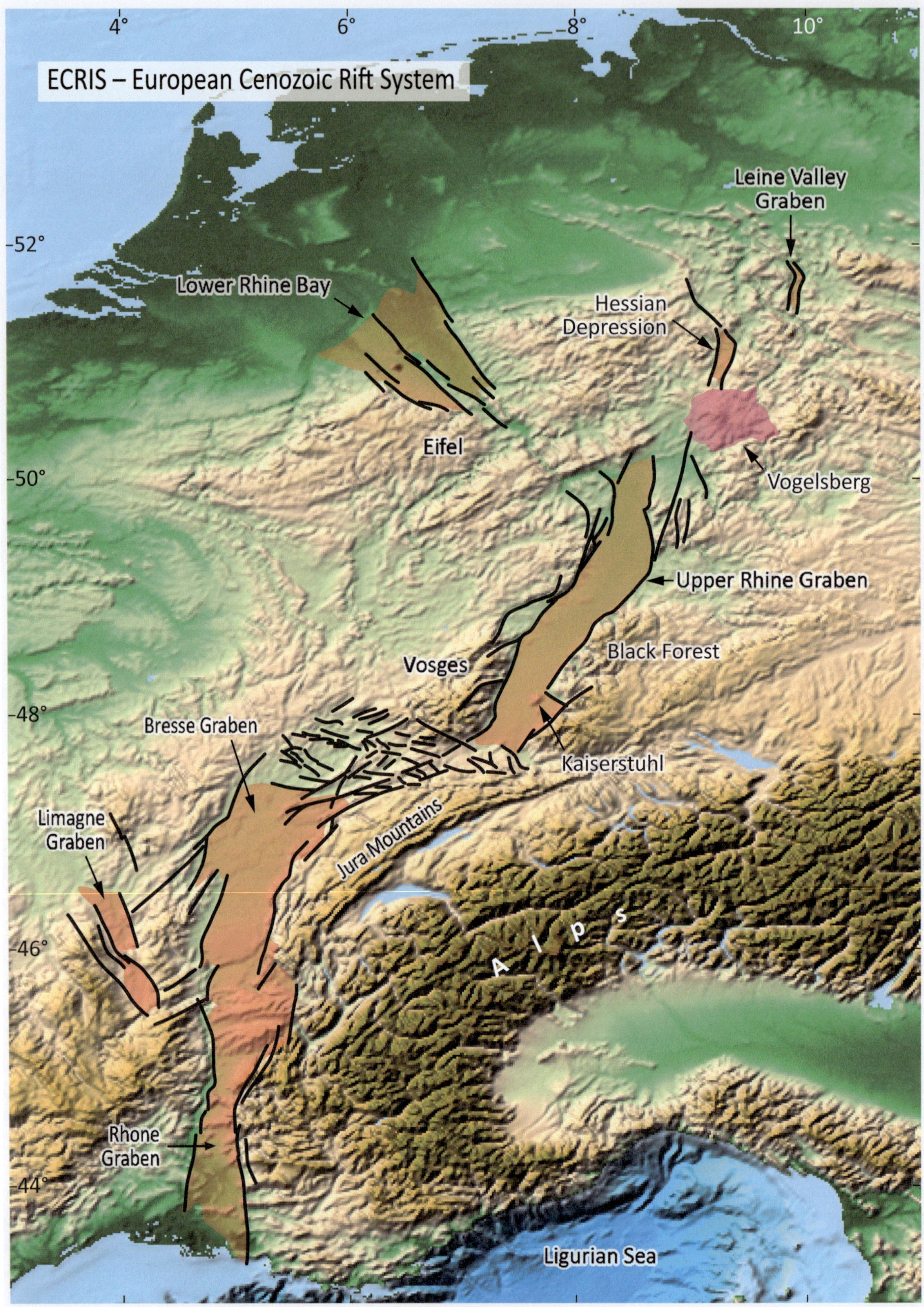

Fig. 8.4 The European Cenozoic Rift System (ECRIS), which extends from the Rhone Graben across the Bresse and the Upper Rhine Graben, to the Lower Rhine Bay. Modified after Frisch et al. (2011)

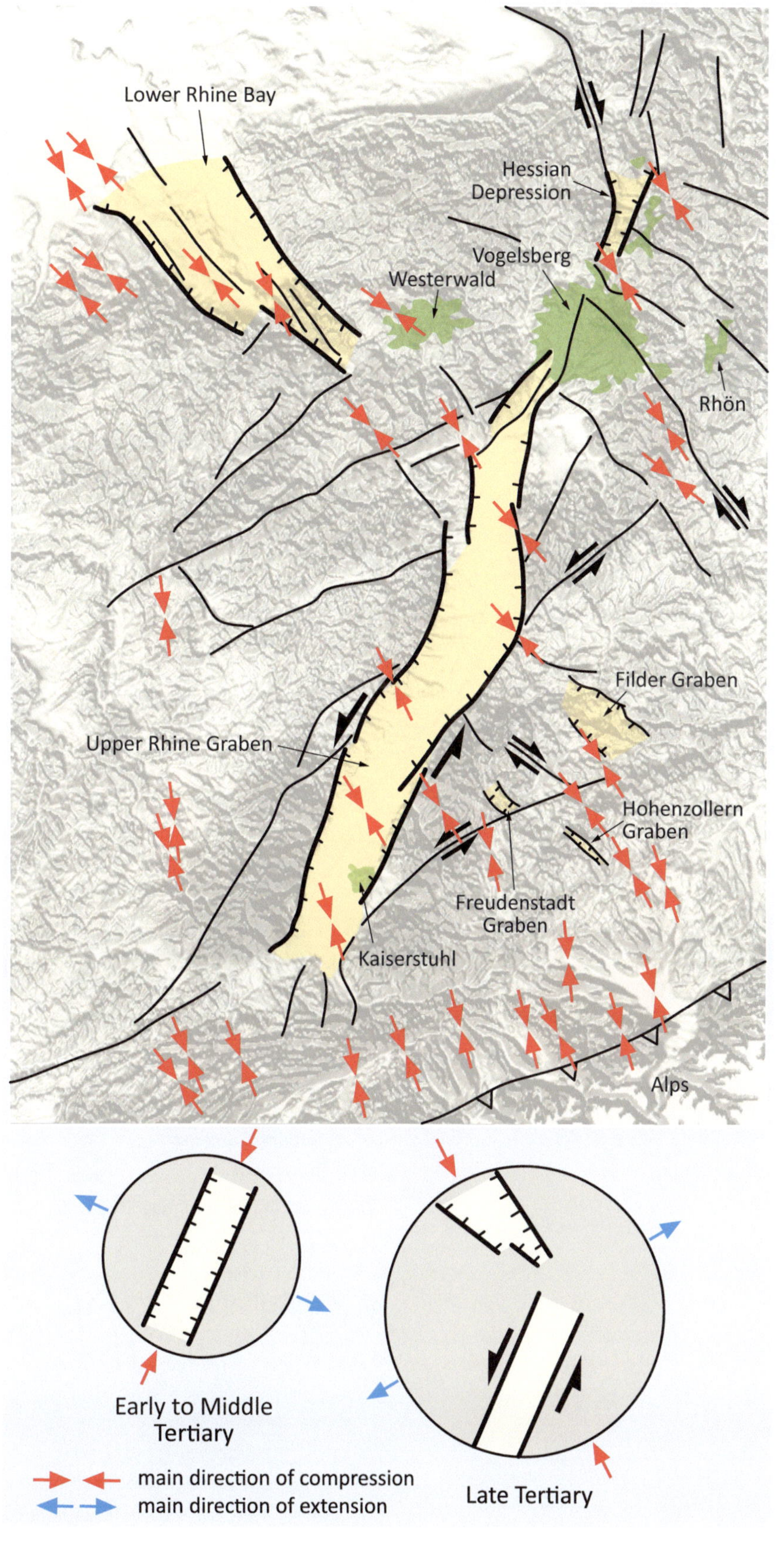

Fig. 8.5 Tertiary and Quaternary graben systems located in southwest Germany. Red arrows indicate the direction of today's dominant horizontally oriented principal compressive stresses (Blundell et al. 1992; Heidbach et al. 2008). The change of stress field that occurred between Early, Middle and Late Tertiary is shown in the two schematic figures below. The older stress field caused the formation of the Upper Rhine Graben. The younger stress field formed the Lower Rhine Bay graben, some smaller graben systems, and the left-lateral strike slip movements of the Upper Rhine Graben. Volcanoes related to the Tertiary development of the graben systems are shown in green. Modified after Frisch and Meschede (2013)

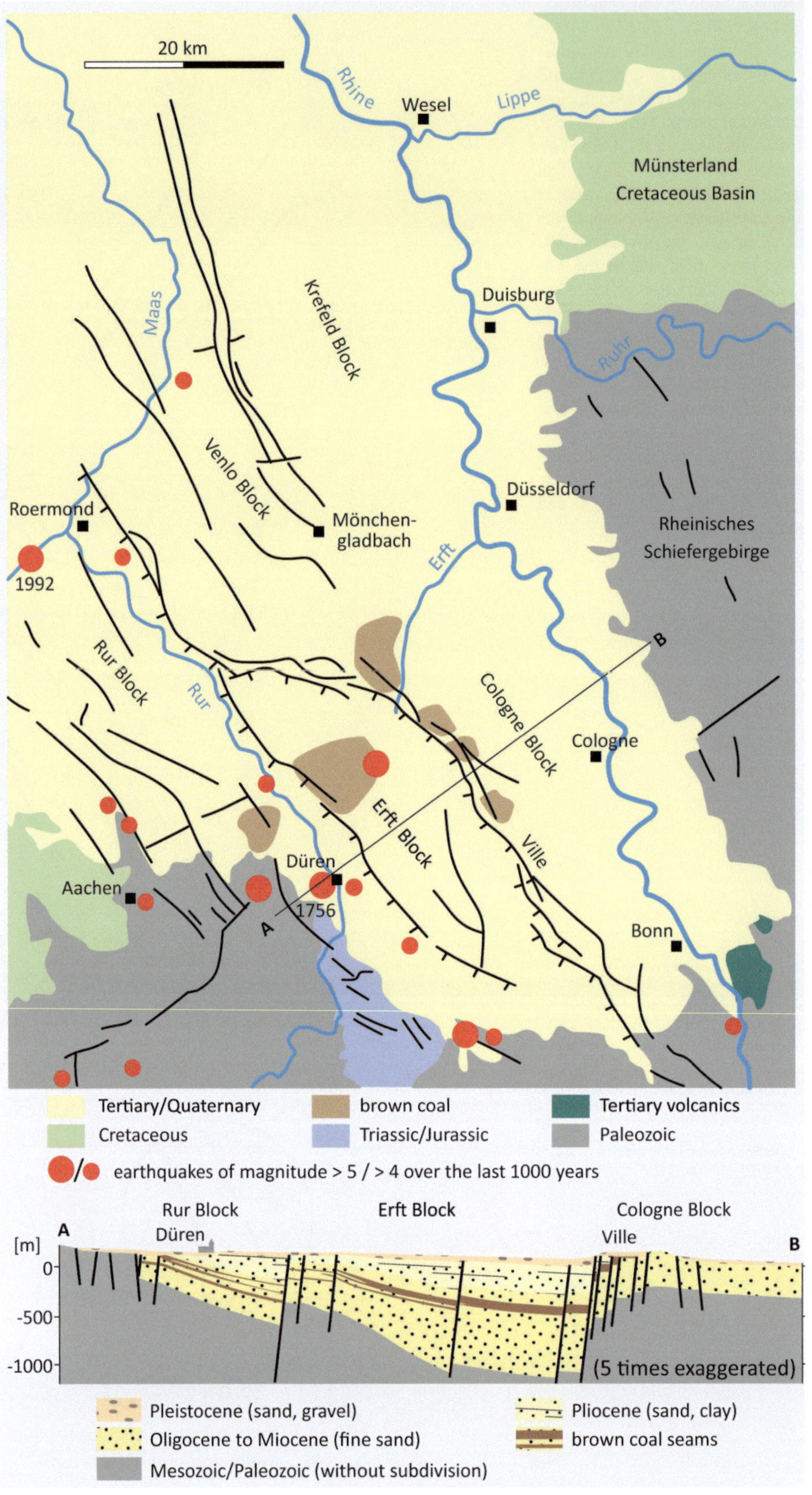

Fig. 8.6 Tectonic overview and geological cross-section (A–B) of the Lower Rhine Bay. Modified after Walter (2007), Geologisches Landesamt Nordrhein-Westfalen (1988)

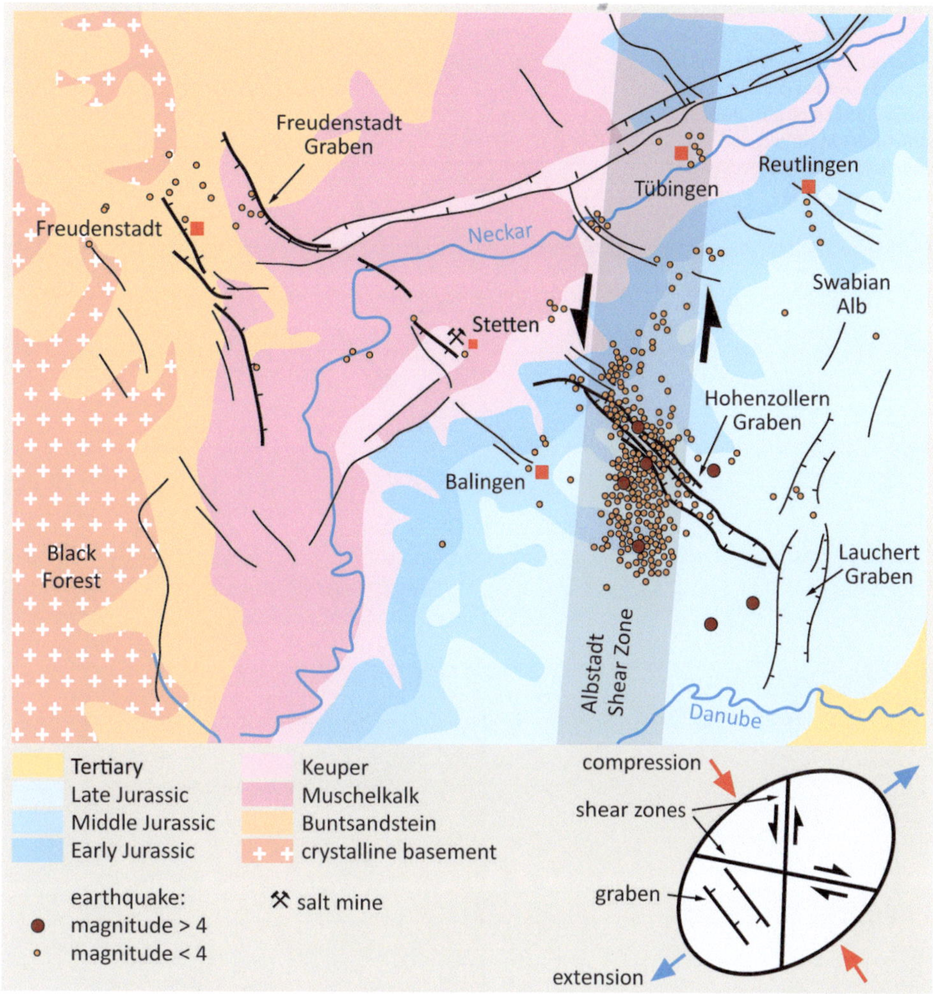

Fig. 8.7 Tectonic overview of the graben system near of the Albstadt Shear Zone. Modified after Reinecker and Schneider (2002)

The cause of graben development across Germany has a number of different origins. The Hohenzollern Graben was formed during lateral displacement, the Lower Rhine Bay by extension in the direction of the currently existing stress field, and the graben of the Vogtland area possibly by the presence of a magma chamber at depth. The N–S oriented tectonic structures of the Lower Rhine Bay area were in fact initiated during the Variscan Orogeny and reactivated as strike-slip faults during younger tectonic movements (Bankwitz et al. 2003). Reactivation of these faults is the probable cause of earthquakes recorded in this area.

The occurrence of earthquakes in Vogtland and northwest Bohemia on the Czech side (Fig. 8.1) reflect structural movements near the Eger Graben. These are mostly characterized by **earthquake swarms** distinguishable by a series of tremors at roughly the same location. Sometimes up to several thousand micro-earthquakes can occur within a few weeks. They are not particularly strong, with most tremors lying below a magnitude of 3 and represent a level of seismicity not normally felt by humans. Occasionally, a magnitude 4 earthquake is measured and in exceptional cases magnitude 5, which last occurred in October and November of 2008. This region is one of the most seismically active regions in Germany where significant earthquakes recur every 8–10 years. The cause of these earthquake swarms is not yet clear but may relate to the formation of a magma chamber at deeper crustal levels beneath the Eger Graben. This idea is based on the occurrence of gas leaks at the surface with an unusual chemical composition that may originate from **degassing magma**. The elevated ^{3}He/^{4} He ratio of the gas indicates mantle-derived helium (Chap. 14; Bräuer et al. 2009). The gas probably accumulates under a well-sealed layer, causing pressure to increase continuously until the rock fractures and triggers episodes of earthquake swarms that sometimes last for months.

Another less known area of Germany where earthquakes occur is located at the edge of the Teutoburg Forest, close to the **Osning Thrust** (Fig. 8.1). In 1612, an earthquake occurred in the Bielefeld area that damaged buildings by the collapse of chimneys and falling roof tiles (Keiter 2013). According to the intensity scale of Mercalli (Box 6: Earthquake magnitude), the earthquake can be assigned to a magnitude of VI, which corresponds to a magnitude of 4.5–5 on the moment magnitude scale. This earthquake resulted from the reactivation of a fault that was formed during the Late Cretaceous and occurred together with compressive movements along the Osning Thrust, the Northern Harz Boundary Fault, the

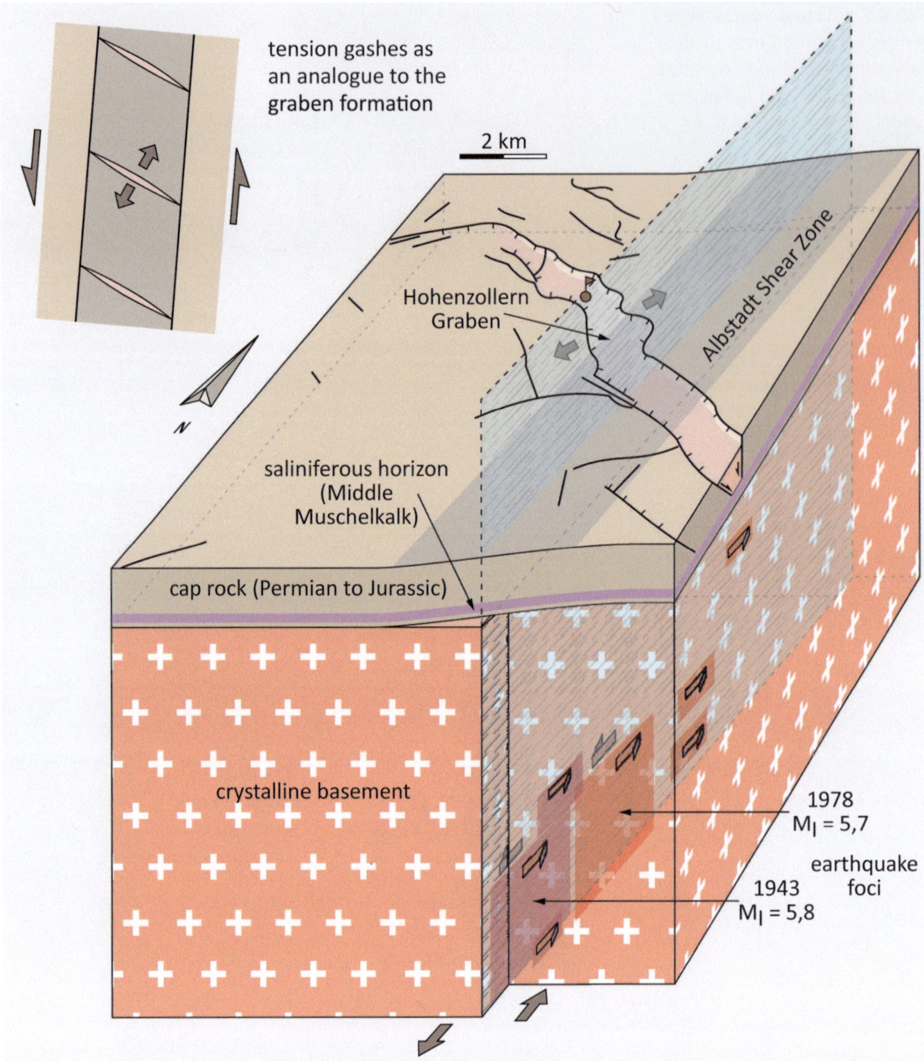

Fig. 8.8 3D-view of the Albstadt Shear Zone. The shear zone occurs as a strike-slip fault in the basement and closer to the surface transfers into graben structures. Modified and completed after Reinecker and Schneider (2002)

Lusatian Fault and numerous other structures aligned in the Hercynian direction (Sect. 11.7.2).

Earthquakes of an entirely different origin occur regularly in former and presently active mining areas. In the Ruhr area, magnitudes of up to 3.3 have been measured. Although each year more than 1000 **mining-induced earthquakes** occur, the vast majority of these events are very small with magnitudes below 2.0. Even the most intense quakes of this kind are, if noticed at all, only felt in a radius of a few hundred meters to a few kilometers from the earthquake focus. These earthquakes result from near-surface subsidence above collapsing cavities, which induce shearing movements that trigger the seismic waves. The subsidence is a consequence of extensive underground mining, where large volumes of raw material (coal, salt, ores and even natural gas) were removed during hundreds of years of exploitation. Following the collapse, the resulting cavities are filled by the overlying rock. Although there are only a few incidences of sudden and uncontrollable damage at the surface, some sizable parts of the Ruhr area have subsided by up to 20 meters as a result of mining. To avoid inundation by the surrounding rivers, water in the subsiding area has been constantly pumped out over the past 100 years. The expense of action is locally referred to as "**eternity costs**". Without these pumps, the area of subsidence would flood within a few weeks to form a large lake comparable in size to Lake Constance within the densely populated area between Duisburg and Hamm.

Among one of the world's most powerful earthquakes caused by human activity is the **rockfall** of Völkershausen in the Rhön. Caused by the collapse of an excavated chamber, combined with a gas explosion in the nearby potash mine, an earthquake of magnitude 5.6–5.8 occurred in March 1989. It ended up destroying a large area of the small town of Völkershausen. There are other examples of man-induced seismic activity in Germany. A few years ago in 2004, an earthquake with a magnitude of 4.4 (Fig. 8.1) occurred in an otherwise seismically quiet area near Rotenburg/Wümme between Bremen and Hamburg. This quake was probably related to the exploitation of natural gas in this region (Dahm et al. 2007), although a natural tectonic cause cannot be entirely ruled out.

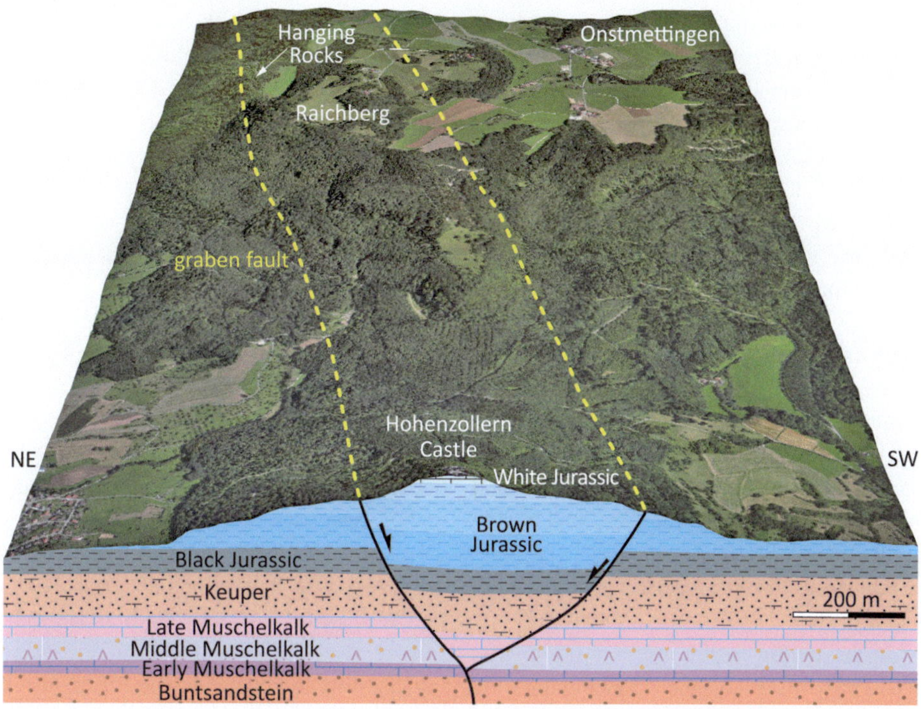

Fig. 8.9 Inverted relief of the Hohenzollern Graben in the area of the Hohenzollern Castle. Modified after Geyer and Gwinner (2011)

Box 7 Cuesta landscape

The landscape of central and southwestern Germany is in many areas characterized by well-exposed geomorphological escarpments known as cuestas. Such landscapes occur when sedimentary layers composed of rocks with different erosional resistance are shaped during weathering. Such sedimentary rock originated as sediment that was lithified following deposition and compaction. The rocks were then uplifted and slightly tilted to about 2°–7° from the horizontal plane. When a well-lithified stratum such as a hard limestone overlies weak or loosely cemented soft marl or mudstone, a **cuesta** can develop. The harder layer forms the prominent, step-like layer of the cuesta, which is often well fractured and permeable to water. When

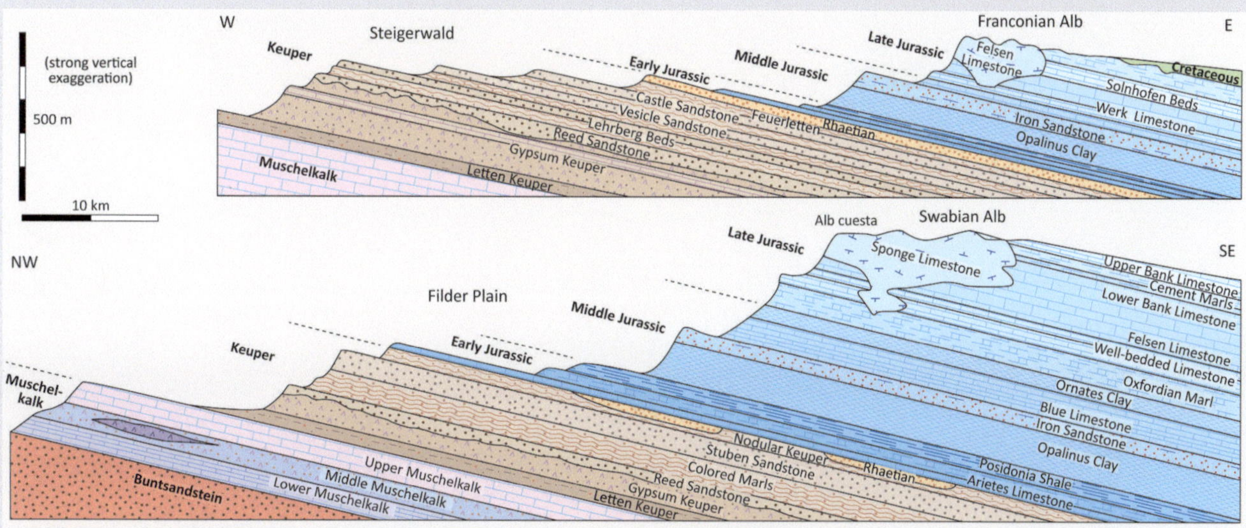

Fig. 8.10 Step-forming layers in the gently dipping Triassic and Jurassic sedimentary rocks of southern Germany (Swabian and Franconian Alb). Modified after Wagner (1960)

the underlying rocks are watertight beds, a **spring line** frequently develops between the two layers and this plays an important role in formation of the cuesta. The **step-forming layer** is eroded at a slower rate than the underlying soft layer, which leads to the eventual undermining and breakage of the hanging harder rock. The resulting edge moves backward by this continuous process, which is known as **regressive erosion**. Numerous steep-sided valleys formed by erosion, cut into the cuesta, giving rise to a heavily dissected topography. Frequently, isolated blocks form **outliers** that are capped by a protective erosion-resistant layer. Based on the distribution of such outliers, the former coverage of the cuesta can be reconstructed.

The most prominent cuestas in Germany occur in the Swabian and Franconian Alb, where they are developed in Late Jurassic strata (Fig. 8.10). These cuestas, which in the Swabian Alb are referred to as the **Alb Cuesta** (Albtrauf), extend as a visible escarpment stretching from the Black Forest in the southwest to close to Nuremberg in the northeast. The cuestas of the Franconian Alb continue almost perpendicular to it and run from SSE to NNW. In addition to the cuesta landscape, comprised of Late Jurassic rocks, three other rock formations occur as distinctive step-forming layers: The Buntsandstein (Early Triassic) lying on the Variscan basement, the Muschelkalk (Middle Triassic) above the Buntsandstein and the Keuper sandstone (Late Triassic). The latter is partly overlain by hard limestone beds of Early Jurassic age (the Arietes limestone; Fig. 8.10) that overlies softer Keuper marl and claystone.

The formation of the southwest German cuesta landscape was related to the development of the Upper Rhine Graben (Sect. 13.2; Fig. 13.14). The uplift of the mountain flanks on either side of the graben (the Black Forest, Vosges, Odenwald, and Palatinate Forest) occurred at the same time as Oligocene subside in the graben and tilting of the sedimentary layers along the elevated ridges. This, in combination with incipient erosion, gave rise to a prominent cuestas landscape. Based on the present location of outliers and other relics of the former overlying sedimentary layers, reconstructions indicate that the cuesta of the Swabian Alb has retreated by up to 60 km since the Oligocene.

Early Geological Evolution of Germany 9

The deep continental crust of Germany is comprised of ancient crystalline and sedimentary rock that is generally much older than the geological units formed during the Variscan orogenic phase. The oldest rocks originated in the Late Proterozoic, although relicts of older tectono-thermal events are recognizable from the remains of individual minerals grains, in particular, detrital zircons. Younger Proterozoic rocks were formed during the Cadomian orogeny and subsequently metamorphosed under diverse conditions. Early Paleozoic sequences are also preserved in places, which pass stratigraphically upward into syntectonic lithologies formed during the Variscan orogeny. Within the Variscan Mountains, these rocks were variably metamorphosed and today occur as very low-grade slate, medium-grade schist and amphibolite, and high-grade gneiss and eclogite.

There are only a few occurrences of **Precambrian** rocks known in Germany, despite this time period encompassing almost 90% of the Earth's history when including the Proterozoic (Greek *proteros*: early; Greek *zoon*: creatures), Archean (Greek *archaios*: old), and Hadean (named after Hades, the underworld of the Greek mythology). The scarcity of Precambrian rocks in Germany is due to the fact that they were strongly overprinted during the younger orogenic phases of the Caledonian, Variscan or Alpine events, and only bordered by older cratonic units, such as the Baltic Shield.

Germany's oldest rocks are assigned to the Late Proterozoic **Cadomian Orogeny** with an age between 650 and 535 Ma. The term Cadomian is named after *Cadomus*, the Latin name of the city of Caen in northern France, which is the type locality for the Cadomian unconformity that marks the uplift and erosion of that time. This orogenic event occurred on the northern edge of Gondwana and along the eastern edge of Baltica. It was contemporaneous with the Pan-African Orogeny that affected the entire African continent. The latter event is marked by a distinct angular unconformity at the base of the Early Cambrian. The uppermost Proterozoic (635–545 Ma) is now referred to as **Ediacarian** according to the most recent definition by the International Stratigraphic Commission. This replaces older terms such as the Neoproterozoic III or Vendian. The Ediacaran is named after the Ediacara Hills in Southern Australia and is famous for its fossil fauna containing the first highly developed and larger multicellular organisms.

In Germany, pre-Cadomian rocks, which formed the crystalline basement during the Cadomian Orogeny, are not exposed at the surface. The history of these rocks can only be derived indirectly from mineral remnants such as old **zircon** grains that were detritally derived by erosion, sedimentation, and lithification or preserved within magmatic rocks. In the Bohemian Massif (Sect. 10.3.4), some metamorphosed sedimentary rocks contain such displaced zircons derived from older source areas. The oldest detrital zircons of Central Europe dated, at 3.84 Ga, were recovered from rocks near Regensburg in Eastern Bavaria (Gebauer et al. 1989). Overall, the range of zircon ages from the Bohemian Massif falls into distinct age groups of 2.5–2.0, 1.9–1.6 Ga and 950–535 Ma. These groups indicate a close connection to the northern edge of the Gondwana continent during the Proterozoic. At isolated locations in the Bohemian Massif, Proterozoic aged rocks have also been dated: the Svetlik orthogneiss (Southern Bohemia) of 2.1 Ga and the Dobra orthogneiss of 1.36 Ga (Wendt et al. 1993). The latter is widely regarded as the oldest rock of Austria (Friedl et al. 2004). Both occurrences of orthogneiss probably formed from granite that originated above a subduction zone (Patočka et al. 2003). In some Late Proterozoic graywackes and conglomerates of the **Barrandian**, cobbles that provide information on the pre-Cadomian crystalline have been preserved. These fragmentary remnants suggest that large parts of Central Europe are underlain by very old, possibly even Archean, continental crust.

Rocks of Early Paleozoic age occur only sporadically in Germany and do not cover large areas. Usually, they arise in sequences that begin with rocks of Late Proterozoic age and continue into Cambrian, Ordovician and Silurian strata, although parts of the stratigraphy are often incomplete. All these strata were influenced by the Variscan Orogeny (Chap. 10), with varying degrees of deformation and metamorphism. Today they appear within different structural units in the Variscan Mountains.

M. Meschede and L. N. Warr, *The Geology of Germany*, Regional Geology Reviews,
https://doi.org/10.1007/978-3-319-76102-2_9

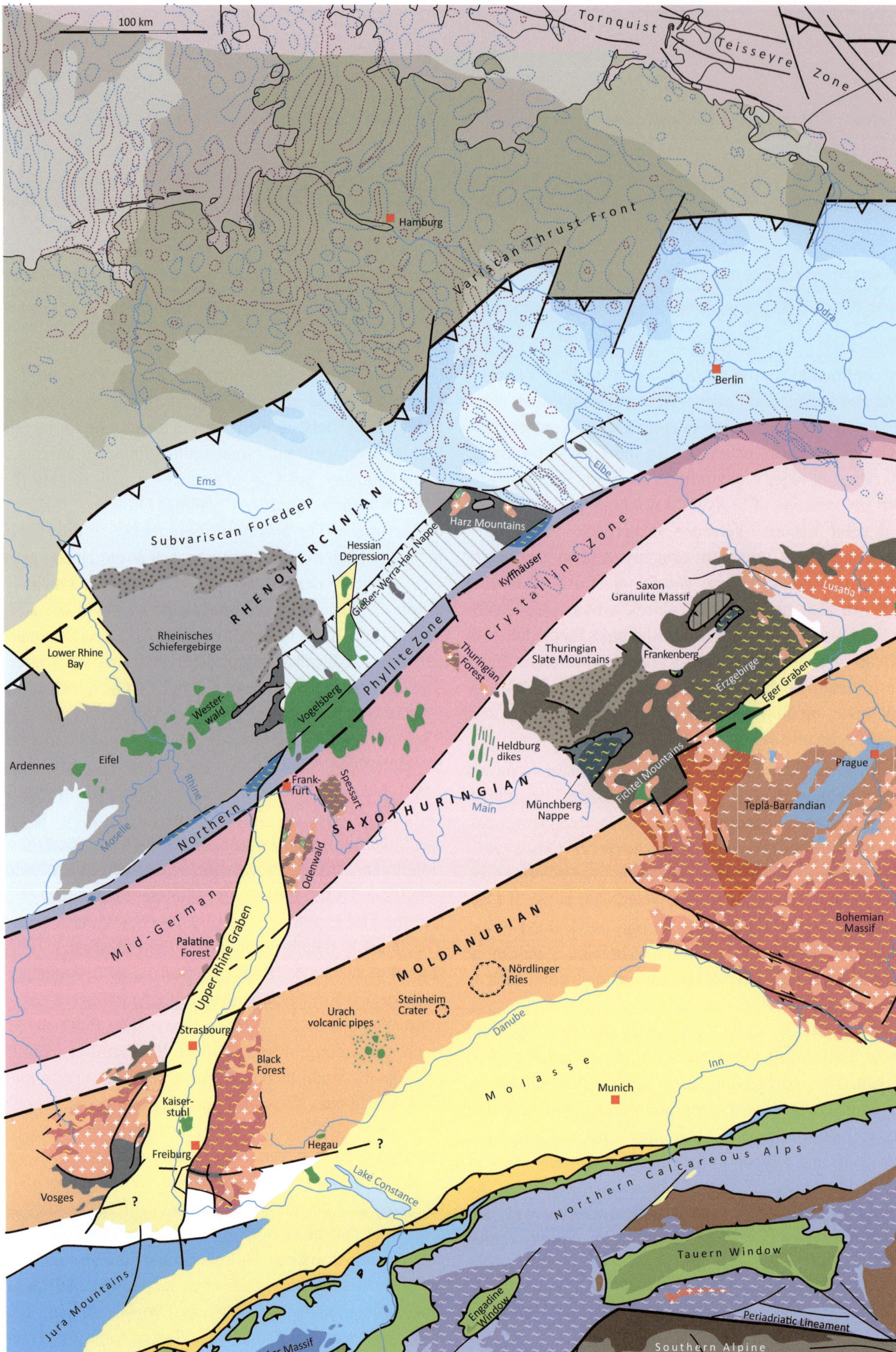
100 km
Tornquist
Teisseyre
Zone
Hamburg
Variscan Thrust Front
Odra
Berlin
Ems
Elbe
Subvariscan Foredeep
RHENOHERCYNIAN
Harz Mountains
Hessian Depression
Gießen-Werra-Harz Nappe
Kyffhäuser
Crystalline Zone
Saxon Granulite Massif
Lusatia
Lower Rhine Bay
Rheinisches Schiefergebirge
Phyllite Zone
Thuringian Forest
Thuringian Slate Mountains
Frankenberg
Erzgebirge
Eger Graben
Wester-wald
Vogelsberg
Heldburg dikes
Fichtel Mountains
Prague
Ardennes
Eifel
Rhine
Frank-furt
Spessart
SAXOTHURINGIAN
Main
Münchberg Nappe
Teplá-Barrandian
Moselle
Northern
Odenwald
Mid-German
Upper Rhine Graben
Palatine Forest
MOLDANUBIAN
Bohemian Massif
Nördlinger Ries
Steinheim Crater
Urach volcanic pipes
Danube
Strasbourg
Black Forest
Molasse
Inn
Munich
Kaiser-stuhl
Freiburg
Hegau
?
Lake Constance
Northern Calcareous Alps
Vosges
?
Tauern Window
Jura Mountains
Engadine Window
Periadriatic Lineament
Aar Massif
Southern Alpine

Fig. 9.1 Tectonic map of Germany and adjacent regions without the Permo-Mesozoic cover. In addition to the Kossmat zones of the Variscan fold-and-thrust belt, the distribution of young graben structures, selected salt structures, and the Alpine Molasse are shown. Modified from the European Geotraverse map of Burollet et al. (1992)

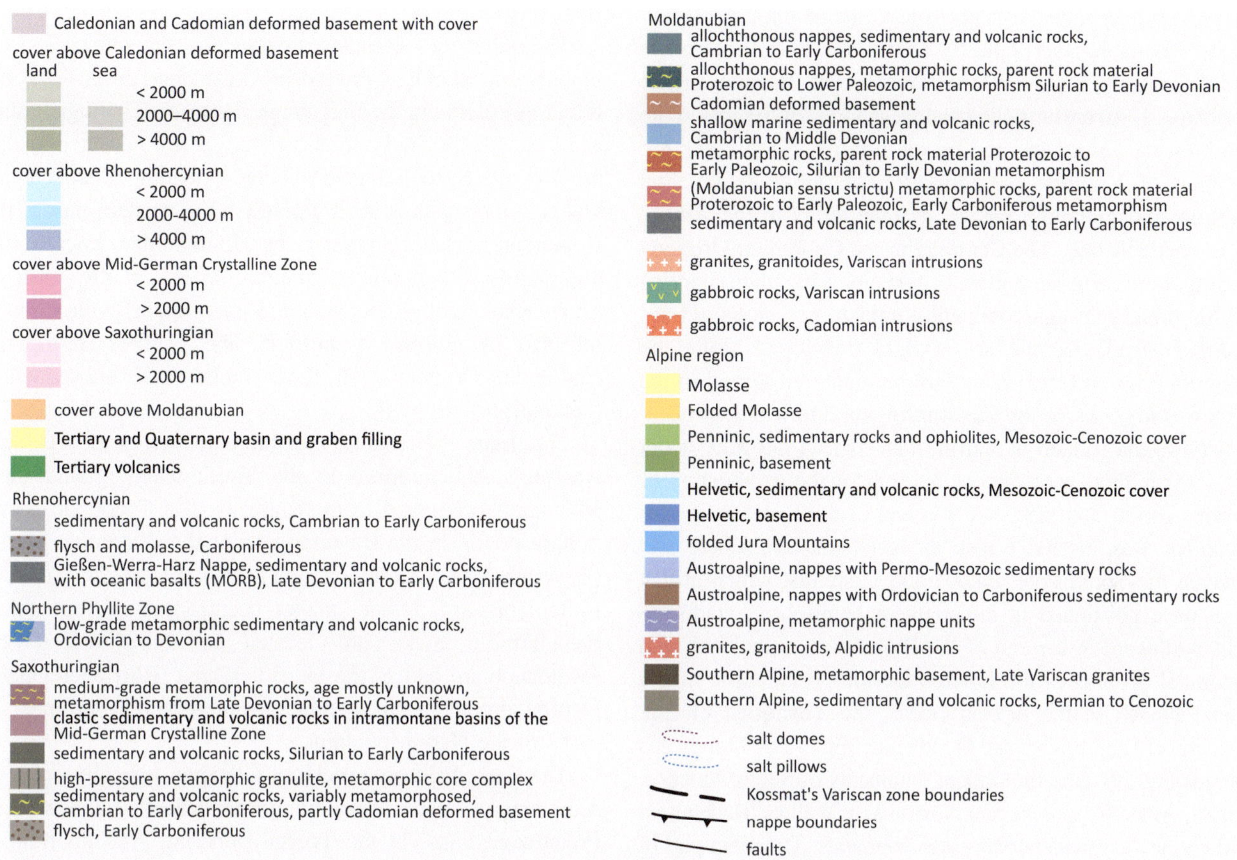

Fig. 9.1 (continued)

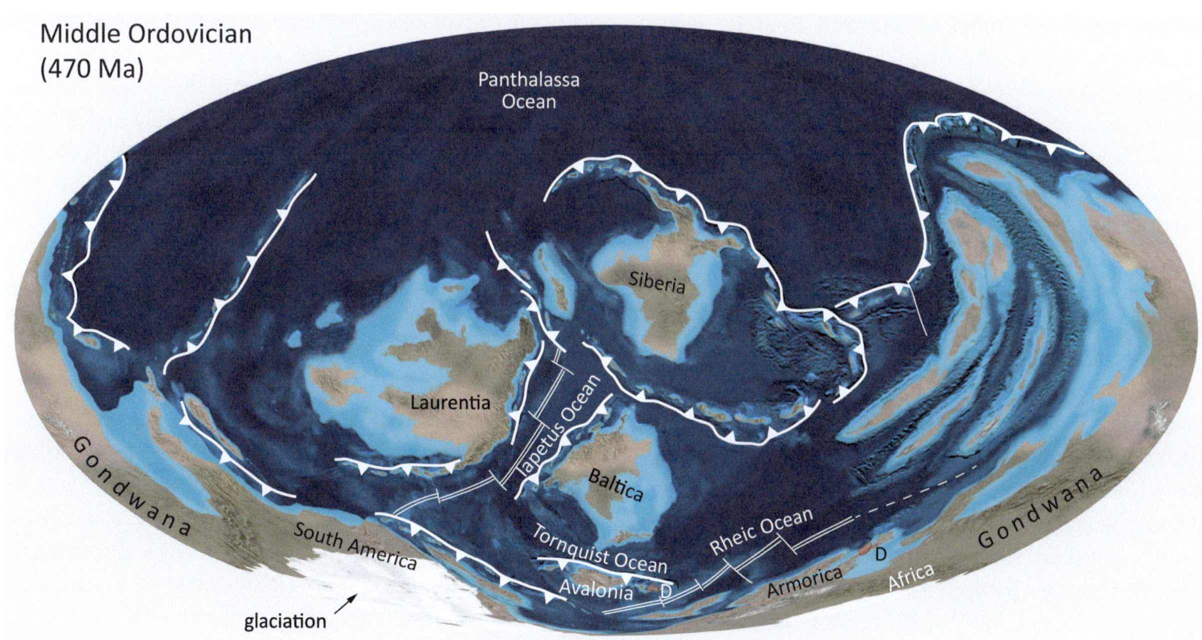

Fig. 9.2 Distribution of continents and oceans during the Middle Ordovician (470 Ma). The topographic base for the paleogeographic map was provided by Ron Blakey, Flagstaff, Arizona (Blakey 2003)

9.1 The Pre-variscan Basement

About 750 million years ago, the supercontinent of **Rodinia** broke up and a number of independent tectonic plates developed that played a major role in the tectonic evolution of the Paleozoic (Fig. 7.1a, b). During the Late Proterozoic and Paleozoic, the continental masses of **Baltica** in the northeast, **Laurentia** in the northwest and **Gondwana** in the south, were important in Central Europe. Soon after the breakup of Rodinia, the formation of active continental margins and mountain ranges, as seen today in the Andes, were reestablished. The Central Europe Cadomian Orogeny became noticeable on a global scale along the plate margins of that time, although continent-continent collisions did not occur. Instead, accretionary wedges developed along the northern edge of Gondwana with remnants preserved in the Proterozoic rocks of the Saxothuringian and Moldanubian.

Gondwana formed during the Pan-African Orogeny about 650–600 million years ago as a result of the aggregation of various continental plates. At the end of the Proterozoic, this continent was located across the southern hemisphere and covered the South Pole. Baltica and Laurentia, which at that time were positioned in the southern hemisphere, started to drift northward at the end of the Proterozoic, coincident with the initial opening of the **Iapetus Ocean**. At the same time a lateral branch of the Iapetus Ocean, the **Tornquist Ocean,** opened between Baltica and Gondwana (Fig. 7.1c). Gondwana included at this time the present continents of South America, Africa, Antarctica, India, and Australia, as well as Madagascar and Guinea as smaller continental fragments. The main part of Gondwana existed as a large continent during the entire Paleozoic before breaking up into today's parts in the Jurassic. Only a few smaller fragments, such as the Avalonia terrane and the Armorican Terrane Complex, separated from its more northern margin during the Paleozoic.

The basement rocks of Germany contain evidence of three successive orogenic phases: the Cadomian, the Caledonian, and the Variscan Orogeny. Some central parts of the crustal blocks welded together during these orogenies, originated from older Proterozoic complexes that were formed during the Late Proterozoic Cadomian Orogeny. In Germany, Cadomian rocks are exposed only in a few places, but they are more common at depth and form the basement rocks of the continental crust. Slightly younger Caledonian rocks are not directly exposed but do occur in drillings of the deeper crust in northern Germany. Caledonian rock series occur extensively in England and Scotland, where they are exposed at the surface, and extend beneath the North Sea to Scandinavia, Northern Germany, and Poland. Today many of these rocks are covered by thick layers of younger sedimentary strata deposited in the Mesozoic and Cenozoic. The boundary between the Caledonian and Variscan basement extends along a WSW-ENE-trending fault line running approximately from Muenster (Westphalia) to Szczecin (Poland). It is interrupted by several lateral structural displacements (Fig. 9.1). To the north, the Caledonian basement is limited by the **Tornquist-Teisseyre zone**. Here, the Tornquist Ocean, a lateral branch of the much larger Iapetus Ocean, closed during the collision of Avalonia and Baltica (Fig. 7.1d).

The oldest rock complexes exposed at the surface in Germany are dated as Late Proterozoic (Ediacaran) in age. They are found in Lusatia, in the Erzgebirge and in the Thuringian Slate Mountains (s.l.). They are also found in several other smaller areas of the Saxothuringian (Torgau-Doberlug, Clanzschwitz, Dohna; Linnemann et al. 2008) and in the western part of the Bohemian Massif assigned to the Moldanubian. These rocks are mainly Cadomian sedimentary rocks in the form of graywackes, conglomerates and carbonates, which were intruded by granites, granitoids, and gabbros during the Cadomian Orogeny. All these rocks were subsequently metamorphosed.

The Late Proterozoic metamorphic series of the Schwarzburg Anticlinorium in the Thuringian-Vogtland Slate Mountains originated from sediments of a Cadomian accretionary prism. In the Bavarian part of the Bohemian Massif (Bavarican), the Drosendorf unit is also at least partly of Late Proterozoic age. Other smaller occurrences of Proterozoic metamorphic rocks are located in the Black Forest (Moldanubian) and in the Hunsrück area (Rhenohercynian). Detrital zircons of this age are also preserved in sedimentary rocks of the Harz Mountains.

The Early Paleozoic rocks of Germany typically occur as sedimentary successions lying discordantly above Late Proterozoic units. In the Torgau-Doberlug Anticlinorium, a low-grade metamorphic Cambrian sedimentary series unconformably overlies rocks of the Late Proterozoic Rothstein Formation. Such Cambrian rocks are poorly exposed and usually only accessible by drilling through the thick cover of younger strata. Ordovician and Silurian rocks are in contrast more commonly exposed. They occur in the core of anticlinal structures formed during the Variscan Orogeny and in some areas were subjected to high-grade metamorphism such as in the Mid-German Crystalline Zone (Sect. 10.4.3). In these cases, stratigraphic classification is only possible using radiometric age dating. These rocks originated as sediments deposited on the passive margin of Gondwana, in related deep basins, and across-shelf areas in the vicinity of the newly forming Armorican Terrane Complex. At this time, the bulk of these continents and continental blocks were located at high latitudes in the southern hemisphere around the South Pole.

Around the time of the Proterozoic-Cambrian boundary, the microcontinent of Avalonia started to drift away towards the north from its origin at the edge of Gondwana; a movement which led to the opening of the Rheic Ocean. Somewhat later, the Armorican Terrane Complex also drifted northward away from Gondwana, resulting in the opening of additional oceans between Gondwana and its fragmented terranes. At the end of the Ordovician, Avalonia collided with Baltica following the

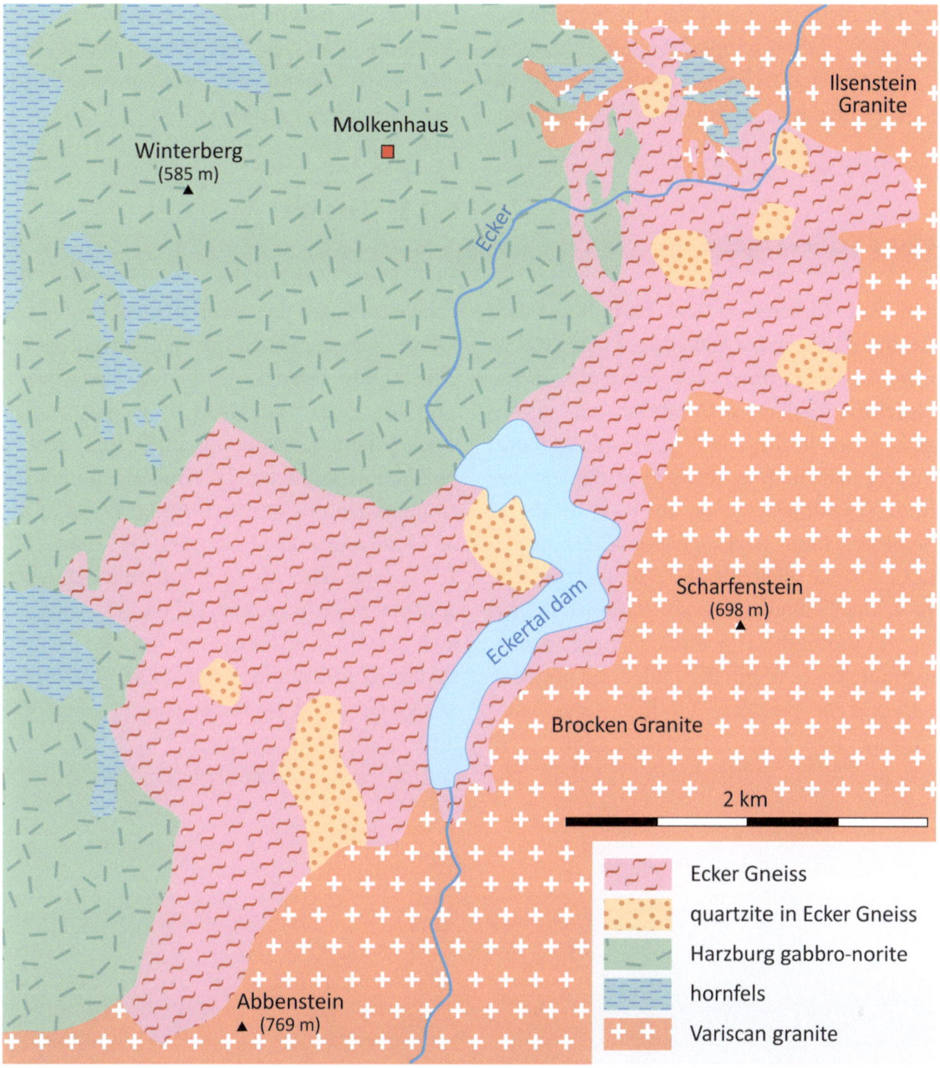

Fig. 9.3 Geological map of the Ecker Valley dam area showing the Ecker Gneiss lying between the Harzburg gabbro-norite and the Brocken Granite. Modified after Franzke (2009)

closure of the Tornquist Ocean. Subsequently, Baltica collided with Laurentia in the Silurian after closure of the Iapetus Ocean (Fig. 9.2). The amalgamation of the newly formed large continent of Laurussia (composed of Laurentia, Baltica, and Avalonia) was of particular importance in the Central European development of the Variscides. This northern continent is also known in older publications as the "Old Red Continent".

9.2 Occurrences of Proterozoic and Early Paleozoic Units

9.2.1 Harz Mountains

Detrital zircons recovered from the **Ecker Gneiss complex** in the Harz Mountains have been dated between 900 Ma and 1.8 Ga (Geisler et al. 2005). The Ecker Gneiss, which is nicely exposed at the Eckertal Dam lies between the gabbro-norite of Harzburg and the Brocken Pluton (Fig. 9.3). The gneiss is mainly comprised of cordierite-bearing gneiss and mica-schist. The detrital zircons within the gneiss were partly derived from the erosion products of the basement rocks of Baltica, which is not consistent with the view of the Ecker Gneiss being metamorphosed during the Cadomian Orogeny (Baumann et al. 1991). In addition to the Proterozoic zircon ages of the Ecker Gneiss, significantly younger Silurian and Devonian ages have been determined for the zircons of subsequent igneous intrusions. These ages indicate that both the sedimentation of the parent material and its syntectonic metamorphism to amphibolite facies occurred after the Early Devonian (Martin-Gombojav 2003; Geisler et al. 2005). Stratigraphic equivalents of the Ecker Gneiss occurring in the northwest and middle of the Harz Mountains (Early Devonian clastic sediments) were only overprinted by a weak anchimetamorphism (ca. 200–300 °C). The gneisses were later affected by metamorphism ranging up to granulite facies conditions. This process was related to the formation of the Harz nappes and the intrusion of the Brocken Pluton around the time of the Carboniferous-Permian boundary. The Ecker Gneiss is interpreted to lie in a thrust sheet of the Harz nappes that were emplaced during the Variscan Orogeny (Sect. 10.4.2).

9.2.2 Rheinisches Schiefergebirge

Few occurrences of Late Proterozoic and Early Paleozoic rocks are known in the *Rheinisches Schiefergebirge* (Rhenish Slate Mountains). In the northwest of Germany, in the **Stavelot-Venn anticline**, rocks of Cambrian and Early Ordovician age are exposed in the center of a fold structure. A Cadomian history is indicated by the age of detrital zircons in the Cambrian sandstones, which are used to assign the Rheinisches Schiefergebirge to the Avalonia terrane.

The oldest rocks of the Rheinisches Schiefergebirge occur as gneiss in a small exposure near Wartenstein and Schweppenhausen in southern Hunsrück. These rocks, which appear as tectonic lenses along the fault zone between the Rhenohercynian and the Northern Phyllite Zone, were metamorphosed during the Late Proterozoic and yield detrital zircons ages of 574 Ma (Molzahn et al. 1998). The zircons are interpreted as remnants of the old basement of Avalonia that was overprinted during the Cadomian Orogeny prior to connecting to Laurussia during the Caledonian Orogeny (Franke 2000).

9.2.3 Lusatia

The most widespread sedimentary deposits of Late Proterozoic age preserved in Germany are located in Lusatia. These were deposited during the Cadomian Orogeny and are collectively referred to as the **Lusatian Block** (Fig. 9.4). The strata are composed of weakly metamorphosed graywacke and subordinate tuff that received sedimentary material from a newly formed volcanic arc. The detritus contains zircons derived from erosion of the passive margin of Gondwana dated between 560 and 580 Ma and around 1.7–3 Gyr. The up to 3 km thick graywacke sequence is considered to represent subduction-related sediment (Kemnitz 2007), deposited in a backarc basin of an island arc system (Linnemann et al. 2000). Deformation of the graywacke occurred by folding into large open anticlinal structures and was accompanied by weak metamorphism at the end of the Late Proterozoic and beginning of the Cambrian.

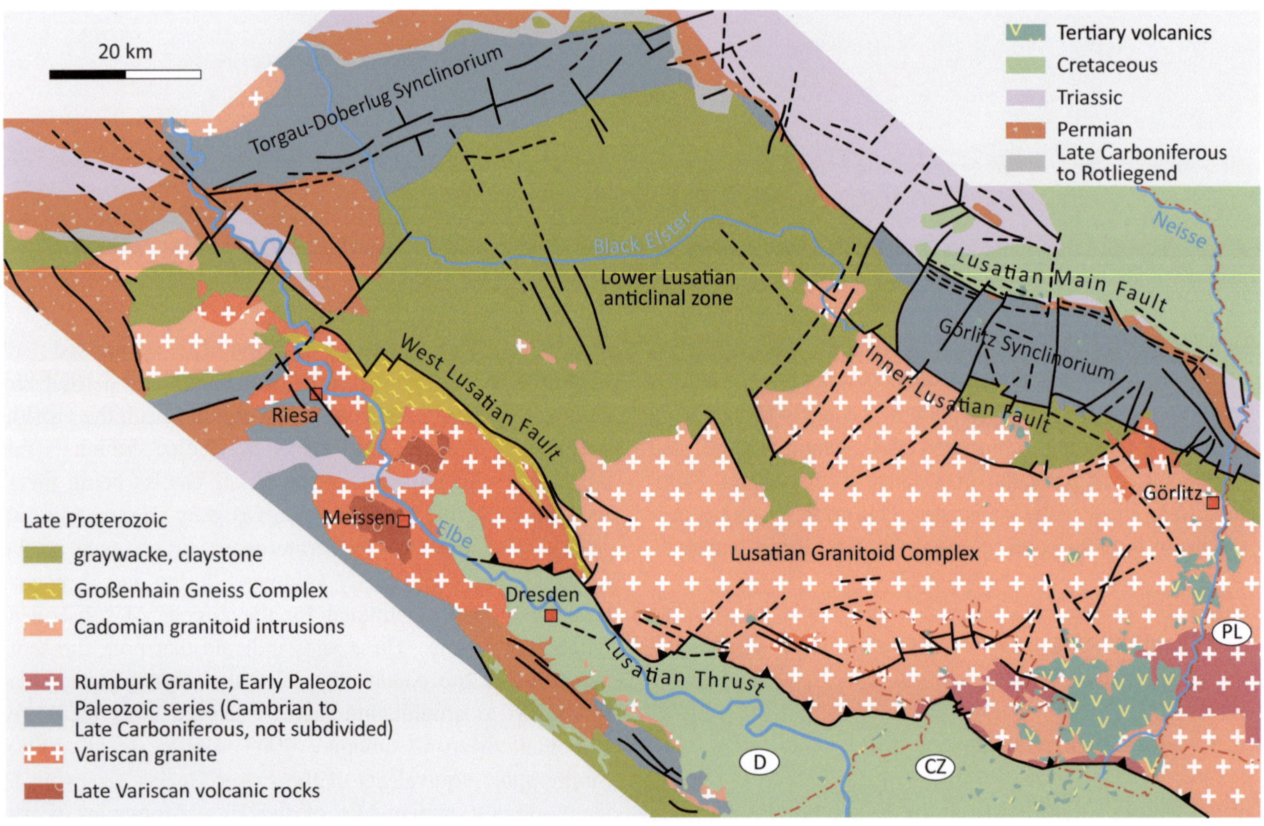

Fig. 9.4 Geological map of Lusatia and the Elbe Zone showing the occurrence of Late Proterozoic and Early Paleozoic rocks (without the Quaternary cover). Modified and simplified after Kemnitz (2007) and Berger et al. (2008)

The rocks are still mined for crushed stone in many places in Lusatia, which includes some large quarries.

In the final phase of the Cadomian Orogeny during the Early Cambrian, **granodiorite** intrusions resulted in contact metamorphism of the graywacke sequence. Four varieties of granodiorite can be recognized that reflect multiple stages of melting (Krauss et al. 1992). Extensive melting of the graywacke combined with the subsequent mixing with mafic melts initially led to the development of (i) two-mica (muscovite and biotite) granodiorite interpreted as anatectic migmatite. This was followed by (ii) mica- and biotite-bearing granodiorite, (iii) biotite granodiorites and, lastly (iv) monzogranites that represent the final fractionation of the melt. The intrusion of the Rumburk Granite in Lusatia occurred in connection with large-scale extensional movements associated with the northward drift of Avalonia, at the turn of the Cambrian and Ordovician (490 Ma) (Fig. 9.4; Hammer et al. 1999). The Variscan Orogeny is manifested in the Lusatian Block only by poorly developed shear zones and foliations and some Late Variscan granite intrusions of Late Carboniferous to Early Permian age (Eidam et al. 1995). Towards the east, the intensity of deformation increased significantly and resulted in gneissification of the Rumburk Granite (Isergneis) in the Western Sudetes.

9.2.4 Elbe Zone

Intense shearing movements occurred during the Variscan Orogeny along the margins of the Lusatian Block. In the **Elbe Zone**, this resulted in strong recrystallization of the Proterozoic graywackes and granitoids and metamorphism of the **Großenhain Gneiss** (Fig. 9.4). In contrast, Late Proterozoic graywacke and granodiorite lying southwest of Dresden, in the Elbtalschiefergebirge near Weesenstein, are of significantly lower metamorphic grade. The graywacke can be traced to Leipzig, where it has been quarry mined in the last century. The Elbe Zone was formed during the Variscan Orogeny as an NW-SE trending, right-lateral shear zone (Rauche 1991). Here, the Late Proterozoic and Paleozoic rocks exposed in Lusatia were ductily deformed and metamorphosed simultaneously to the intrusion of basic and granitic igneous rocks and the formation of volcanic rocks in the region (Meissen Pluton and volcanics of the uppermost Early Carboniferous; Fig. 9.4). The Elbe Zone separates the low-grade metamorphic autochthonous units of Lusatia from the partly high-grade metamorphosed rocks of the Erzgebirge. They are offset along shear zones where strongly deformed ductile mylonites and brittle deformed cataclasites were formed. Occurrences of granodiorite, graywacke, and mica-schist are also known from the northwest Saxon Anticlinorium and the northwestern portion of the Elbe Zone. In the area of Torgau-Doberlug (granodiorite of Schilda), however, they are usually covered by younger sedimentary rocks but can be detected beneath the surface by shallow drilling or geophysical surveys.

9.2.5 Erzgebirge

The **Erzgebirge** is subdivided into two parts: a western area composed predominantly of low- and medium-grade metamorphic Ordovician sedimentary and igneous rocks and an eastern area with abundant medium-grade metamorphic Cadomian graywacke and granitoid (Fig. 10.47). These two areas are separated by the **Flöha Zone**, which represents a Variscan shear zone. Following the Variscan Orogeny, the Erzgebirge Block was tilted and the eastern portions more uplifted and deeply eroded than to the west. In a few places, the Cadomian sedimentary rocks of Late Proterozoic to Cambrian age contain some calcareous series that today occur as marble. Near Lengefeld in the Erzgebirge, dolomite marbles were metasomatically overprinted during the Variscan Orogeny and diagenetically dolomitized. These marbles are mined underground and used as filler material and binders in industry. Most of the Cadomian rocks located in the eastern Erzgebirge are medium-grade metamorphic gneisses (gray and red gneisses of the East Erzgebirge Group). They form large dome-shaped structures such as the Freiberg dome with its core composed of granodiorite. This igneous rock was intruded during the Early Cambrian into an envelope of Late Proterozoic graywacke that lay along the active continental margin of Gondwana (Sebastian 2013). This sequence has similarities with the low-grade metamorphic rocks of Lusatia (Sect. 9.2.3). Other similar dome structures are found in Lauenstein, Reitzenhain, and Annaberg. The outer parts of these domes consist of Ordovician orthogneiss and other gneiss of unknown age. Separated by thrust fault zones, high-pressure metamorphic units are overlain by orthogneiss and paragneiss. These units contain eclogite and serpentinite as well as diamond-bearing quartzite. Also part of the Erzgebirge are the well-known garnet-bearing serpentinites of Zöblitz that lie in the Flöha Zone, near Zschopau. These rocks were mined historically over a long time for decorative purposes, though today they are used only as crushed stone.

Research investigations have shown the parent rock of the Variscan gneiss preserved in the thrust nappe piles of the eastern Erzgebirge was formed from an Ordovician protolith with an age of around 480–490 Ma (Kroner and Willner 1998). In the central and western Erzgebirge, the degree of metamorphism decreased in the tectonically higher units from west to northwest. The uppermost tectonic unit is composed of low-grade metamorphic phyllite of the Elsterberg region and West Erzgebirge. Although the age of the metamorphic rocks varied from Late Proterozoic to

Devonian, all units of the Erzgebirge were metamorphosed during the Variscan Orogeny (Sect. 10.3.3). The metamorphic climax is radiometrically dated to have occurred during the Early Carboniferous.

9.2.6 Schwarzburg Anticlinorium, Vesser Zone

Located at the northwestern edge of the Thuringian Slate Mountains, a sequence of Early Paleozoic and Late Proterozoic rocks is exposed in the center of the **Schwarzburg Anticlinorium** (Fig. 9.5). Alternating horizons of graywacke, partly siliceous shale and metavolcanic rocks occur that were only weakly deformed and metamorphosed during the Cadomian Orogeny. They are interpreted as sedimentary deposits formed at a convergent plate boundary that was active during the Cadomian collision (Heuse et al. 2001). Around the transition from Proterozoic to Cambrian, the intrusion of granitoids marked the termination of the orogenic event. These rocks were later deformed by intensive shearing and southeast-directed thrust tectonics during the Variscan Orogeny (Sommer and Katzung 2004).

Late Proterozoic rocks occur in the **Katzhütte complex** where they are overlain unconformably by the Cambro-Ordovician Rotseifen Formation. The deposition of these marine sediments marked the beginning of the **Thuringian Facies** that continued until the Early Carboniferous. These sediments include the Frauenbach and Phycodes Group deposited during Ordovician and earliest Silurian times. The Thuringian Facies, which is well developed across the Saxothuringian, encompasses a several thousand meters thick sequence of initially shallow marine and later deeper marine deposits. In the Schwarzburg Anticlinorium, the thick shallow marine shelf sediments changed to deeper marine, pelagic strata during the Silurian. This change in facies reflects progressive extension and subsidence of the passive continental margin at the edge of the Armorican Terrane Complex during the Cambrian and Early Ordovician. The resulting thinning of the lithosphere during the Ordovician also led to the generation and ascent of granitoid intrusions and related volcanic melts.

A special feature of the Thuringian Facies sequence is the Ordovician ***Lederschiefer***, which occurs in both the Schwarzburg Anticlinorium and the Berga Anticline farther

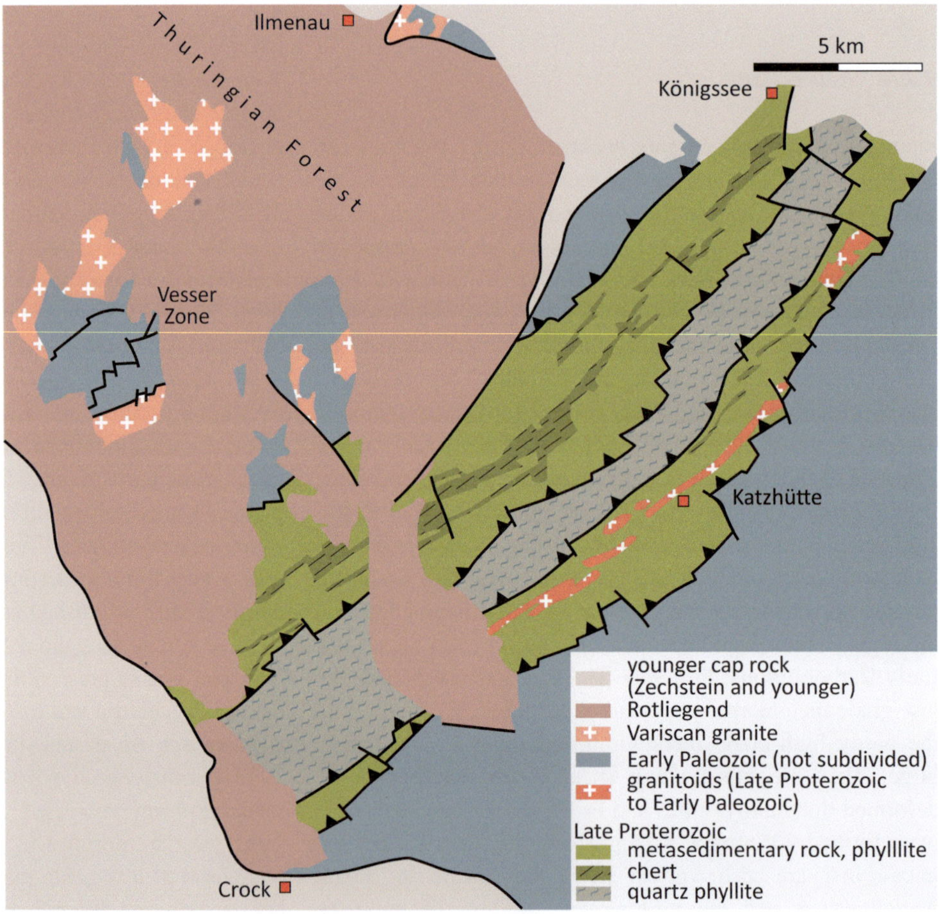

Fig. 9.5 Geological map of the central Schwarzburg Anticlinorium and the Vesser Zone showing the distribution of Late Proterozoic and Early Paleozoic rocks. Modified after Bankwitz et al. (1994) and Sommer and Katzung (2004)

Fig. 9.6 A dropstone in the Ordovician *Lederschiefer* at the northeastern limit of the village of Steinach, Thuringia. The exposure is located at the "Georgsfelsen" above the railway embankment. [N 50°26'19"/E 11° 09'19"]

to the southeast (Fig. 10.43; Katzung 1961; Hofmann and Linnemann 2013). The fine-grained sediment of the *Lederschiefer Formation* (Leather Shale Formation) contains in places numerous large clasts of up to 20 cm. These clasts are often poorly rounded and not of the same composition as the enclosing sediment. An easily accessible exposure is located at the Georgsfelsen in Steinach, Thuringia (Fig. 9.6). The exotic clasts consist of various rocks (typically quartzite) and occasionally exhibit striations. These clasts are interpreted as glacial **dropstones** that were transported as a sedimentary load by glaciers and drifted into the marine environment within icebergs. During melting of the iceberg, the transported rocks were released and sedimented as dropstones on the seabed. This way the coarse exotic components ended up incorporated in the often very fine-grained marine sediment. In the Thuringian Facies, they were deposited in medium water depths of ca. 500 m on the continental slope of the Armorican Terrane Complex, which at that time was still connected to Gondwana. These sediments, known as **glaciomarine diamictites**, indicate that a glaciation affected the southern hemisphere during the Late Ordovician. The extent of the glaciation at this time can be established from similar occurrences in other places. A notable example is known in the Sahara, which at that time was located in the midst of Gondwana. Here, Late Ordovician tillites (=sediments deposited by a glacier) are widespread and as a result, this glaciation phase is referred to as the "Sahara glaciation".

In the Vesser zone that lies to the west of the Schwarzburg Anticlinorium (Fig. 9.5), metamorphosed clastic sedimentary rock, volcanics and subordinate limestone of Late Cambrian and Ordovician age occur. This zone, which is both geochemically and petrologically distinct, was formed at the northern edge of the Saxothuringian when Avalonia separated from Gondwana. Later, it became attached to the Armorican Terrane Complex. It is considered to represent the rift zone (*Vesser-Rift*) of the Rheic Ocean (Bankwitz et al. 1994, Schäfer et al. 1998).

9.2.7 Bohemian Massif

The **Teplá-Barrandian** is located only in a small part of Germany, with most of the unit occurring in the Czech Republic (Fig. 10.49). These rocks comprise the northern part of the **Bohemian Massif** and were mainly affected by the Cadomian Orogeny. The center of the unit, the Barrandium, is composed of weakly metamorphosed to low-grade metamorphic Paleozoic sedimentary rock sequences (Sect. 10.1.3). The other units consist of low-grade metamorphic sedimentary and volcanic units composed of Late Proterozoic shales, graywackes, and cherts, together with some intermediate to acidic volcanics (keratophyre) and Cadomian granitoid intrusions. These rock sequences also occur in the Upper Palatinate Forest along the northwestern boundary of the Moldanubian, where a transition to higher metamorphic units occurs to the south. These units have been correlated to Late Proterozoic sequences in Lusatia and in the Erzgebirge (Sects. 9.2.3 and 9.2.5), and indicate a similar paleogeographic position on the northern edge of Gondwana during the Late Proterozoic. The Bavarian part of the Moldanubian (also referred to as Bavaricum) is characterized by high-grade metamorphic and partly anatectic rocks (gneiss and mica-schist) and Late Carboniferous granitoids. The parent rocks were formed by subduction-related Cadomian magmatism in the Proterozoic (Teipel et al. 2004). As in similar areas of the Moldanubian, these rocks were metamorphosed during the Variscan Orogeny (Sect. 10.3.4).

9.2.8 Black Forest

A variety of metamorphic rocks occur in the Black Forest, where protoliths were formed during Proterozoic and Early Paleozoic times. All of these units were influenced by high-grade metamorphism during the Variscan Orogeny (see Sect. 10.3.4, paragraph "Black Forest" for a detailed description). In the granitic area of the Northern Black Forest, only a few gneiss blocks occur. This contrasts to the central Black Forest where widespread gneisses that were likely formed by metamorphism of Proterozoic and Lower Paleozoic graywacke and mudrock are found. Based on radiometric dating, remnants of ophiolite and eclogite have been determined to be Cambrian to Ordovician in age. This includes an eclogite near Hinterzarten, which has been considered for a long time to be the oldest rock in Germany. However, isotopic studies have since shown these rocks to belong to an Ordovician ophiolite (age of the protolith), with only a few older relic zircons radiometrically dated at around 2.1 Ga. The ophiolite is surrounded by Early Paleozoic sedimentary rocks (Cambrian to Devonian; Kalt 1991; Chen et al. 2003) and all of these units suffered metamorphism during the Devonian.

The Early Paleozoic rocks of the Black Forest developed in a series of ocean basins during the crustal extension that began at the beginning of the Paleozoic. This extensional motion culminated later in the separation of the Armorican Terrane Complex, consisting of the Saxothuringian and Moldanubian units, from Gondwana. This separation then led to the opening of the Rheic Ocean and some associated smaller oceans (Chap. 10). The eclogites were probably derived from Ordovician basalts prior to high-grade Variscan metamorphism (Sect. 10.3.4).

Late Paleozoic of Germany

10

The Variscan orogeny is the most important mountain-building process in Central Europe that began in the Devonian, reached its climax during the Late Carboniferous times and ended during the Early Permian. The orogeny saw the close of the Rheic ocean and collision of the southern continent of Gondwana with the northern continent of Laurussia to form the supercontinent of Pangea. The collision was particularly complex due to the amalgamation of a range of Gondwana-derived, intervening terranes and terrane complexes (Avalonia, Armorica, Norica). Almost all areas in Germany were affected by the Variscan orogeny and occur in a number of structural units known as the Rhenohercynian, Saxothuringian, and Moldanubian, as first introduced by Kossmat (1927). This chapter highlights the sedimentary and tectono-thermal history of these units across Central Europe and discusses current plate tectonic models for their development.

10.1 The Variscan Orogeny

The long-lasting phase of deformation and metamorphism that occurred across Central Europe in the Devonian and Carboniferous is referred to as the **Variscan Orogeny**. The accompanying mountains that formed during this orogeny are known as the **Variscides**. This Late Paleozoic orogeny represents the final collision between the great southern continent of **Gondwana** and its northern counterpart, **Laurussia**. The collision also included a number of intervening terranes and terrane complexes. Large parts of the pre-Permian basement in Central Europe are today composed of the continental crust that was formed during this orogeny. The termination of the Variscan Orogeny also corresponded roughly with the end of the Paleozoic era.

A large section of the eroded Variscan Mountain Range is today located on the other side of the Atlantic in the **Appalachian Mountains** and extends for many thousands of kilometers along the entire eastern seaboard of the North American continent. The main orogenic phase of the Appalachians is called the **Acadian Orogeny** (named after a former French colonial area on Nova Scotia Island, eastern Canada). At the end of the Paleozoic, the European and American parts of the mountains were still connected as part of the supercontinent of Pangea (Fig. 7.1f). The different orogenic phases are sometimes grouped together as the Variscan–Acadian Orogeny, where the Acadian Orogeny started a little earlier than the Variscan during the Middle Devonian. The final closure of the **Rheic Ocean**, which is considered to have been the largest and most important ocean during the Variscan orogenic phase, did not occur until the Late Permian in the southern Appalachian Mountains and in the Mauretanides of northwest Africa. The collision following the late closure of the Rheic Ocean is known as the Alleghenian Orogeny (after the Allegheny Mountains in southern North America). At this time, extensional movements in Central Europe were already evident with the incipient breakup of Pangea. In English-speaking countries, the name Hercynian is sometimes used for the Variscan Orogeny. However, this term was originally used to describe WNW-ESE structures or facies variations in the Harz Mountains (Lat. *montes hercyniae*) of Germany (see Hercynian facies; Sect. 10.2.1). These features were related to the Late Cretaceous reactivation of Variscan structures along the Northern Harz Boundary Fault, during which the Harz was thrust onto the Harz foreland toward the north (Sect. 11.7.2).

The Variscan Orogeny resulted from the collision of various small microcontinents and terranes with the northern continent of **Laurussia** (Fig. 7.1e). The subsequent emergence of these continental plates then formed the **Pangea** supercontinent. The various oceans that lay between the continental units were closed. Remnants of these oceans are preserved today in Europe only in exceptional circumstances, such as the Lizard ophiolite on the Lizard Peninsula of Cornwall, southwest England. There, rocks of the oceanic lithosphere, probably belonging to some part of the Rheic Ocean complex, have been preserved. Large ophiolite complexes known from other mostly younger collisional zones are, however, largely absent in the European Variscides.

M. Meschede and L. N. Warr, *The Geology of Germany*, Regional Geology Reviews,
https://doi.org/10.1007/978-3-319-76102-2_10

The most important suture line of the Variscan orogenic belt is the **Rheic Suture**, which marks the zone where the Rheic Ocean closed between the Silurian and Permian (Fig. 6.1). In Central Europe, this closure probably took place mainly during Late Devonian and Early Carboniferous times. North of the suture line, which largely corresponds with the boundary between the Rhenohercynian and Saxothuringian Zones (Fig. 9.1), the continental units were attached to Baltica and Laurentia following the Caledonian Orogeny as part of the landmass of Laurussia. The most important tectonostratigraphic element in this context was the **Avalonia** Terrane, which was separated from Gondwana during the Early Paleozoic. To the south, other crustal terranes followed that were once connected to northern Gondwana and moved independently during the Paleozoic toward the north before attaching to the northern continent of Laurussia. These crustal units included the **Armorican Terrane Complex** and associated terranes that now form part of the Alpine Variscan basement (such as the Norica Terrane).

A different origin of the continental fragments located south of the Rheic suture line is evident from the occurrence of glacially formed marine sediments (glaciomarine diamictites) that developed during the Late Ordovician glaciation phase (Sahara glaciation). There is, however, no evidence of such glacial sediments in Avalonia, implying that this continental fragment had already drifted away from Gondwana a long time before the Ordovician glaciation. Another indication of a different paleogeographic history of Avalonia and Armorica is obtained from radiometric age data determined from magmatic and metamorphic mica. Caledonian mica ages are recorded from Avalonia or the Rhenohercynian but not from Armorica and rocks of the Saxothuringian. As Armorica was connected to Gondwana for a longer time than Avalonia, the absence of Caledonian ages south of the suture line is taken to indicate a wide ocean existed between Avalonia and Armorica during a large part of the Late Paleozoic.

The Variscides of Europe form an up to 2000-km-wide fold-and-thrust belt that is notably curved due to strong oroclinal bending. The **Bohemian Arc** is located at the eastern end, where the general strike of the orogenic belt changes by about 90°. An even stronger oroclinal bend occurs at the western end of the belt, where the Cantabrian Mountains that form part of the **Ibero-Armorican Arc** change strike by 180°. This structure originated prior to the Tertiary opening of the Bay of Biscay and therefore in order to reconstruct the original geometry of Iberia, a clockwise rotation of about 40° is required.

The Variscan units can be subdivided into an inner zone dominated by crystalline rocks, sometimes referred to as the **Internides**, and an external zone comprised of more weakly metamorphosed sedimentary and igneous rocks. The inner zone represents the central axis of the Variscan Mountain Range. In the north and west are two concentrically arranged **fold-and-thrust belts** of diagenetic to greenschist facies metamorphic grade. The belt located south of the Internides was strongly overprinted by the Alpine Orogeny so that only the high-grade metamorphic and highly deformed crustal parts are preserved today.

Remnants of Paleozoic rocks formed on the passive margin of Gondwana and deformed during the Variscan Orogeny can be traced across Central Europe, the Montagne Noire of France, to the southern Cantabrian Zone and Western Iberian Peninsula of Spain. The Internides zone is composed of high-grade metamorphic rocks that contain eclogites and granulites, as well as ophiolite remnants, and is considered to mark the hinterland of the Variscan Mountains. During the late stage of orogenic development, large volumes of granitic melt were generated as a result of the thickened continental crust to form the numerous granite intrusions exposed today in the central zone of the Variscan orogenic belt.

Although not well constrained, lithospheric subduction of oceanic crust is considered to have occurred during the Variscan collisional phases from Late Silurian until the Early Carboniferous times (Fig. 10.1). Initially, subduction of the Rheic Ocean beneath the northerly-located Avalonia occurred, which was already part of the northern continent of Laurussia at that time. This was followed by further subduction of the Rheic Ocean beneath the Armorican Terrane Complex and formation of a volcanic arc now represented by the **Mid-German Crystalline Zone**. In the European segment, the Rheic Ocean was probably closed by the Early Carboniferous. The subsequent collision and deformation of the passive margin of Laurussia then lasted until the Late Carboniferous and formed the external Variscan fold-and-thrust belt (Franke 2006). South of the Mid-German Crystalline Zone lay the Saxothuringian Basin and the **Saxothuringian Ocean**. These were limited by an active plate margin to the southeast (Fig. 9.1). In the Devonian, the Saxothuringian was joined together with the Moldanubian, which was also part of the Armorican Terrane Complex. To the south, lay the **Moldanubian Ocean** that closed in the late Middle to Late Devonian.

The central European Variscides are classically subdivided into a number of distinct tectonic–stratigraphic zones known as the Rhenohercynian, Saxothuringian, and Moldanubian units (Kossmat 1927). The **Kossmat zones** are still considered valid and the classification has been expanded only by the addition of the Mid-German Crystalline Zone (MCZ) as a separate tectonic element (Fig. 9.1). Interestingly, Kossmat presented this division at a time when the concept of plate tectonics was not yet known. Today, these units are interpreted as an amalgamation of independent lithospheric microplates and terranes surrounded by suture zones (Fig. 6.1). There is, however, no generally accepted conceptual model yet for the development of the Variscan Orogeny and it is not entirely clear how many suture zones lay beneath the individual units. There are also contrasting opinions concerning the number, scale, and width of the consumed oceans that lay between the units, and a range of different names have been introduced. Some models prefer a simple narrowing in the NW-SE direction (e.g., Matte 1986;

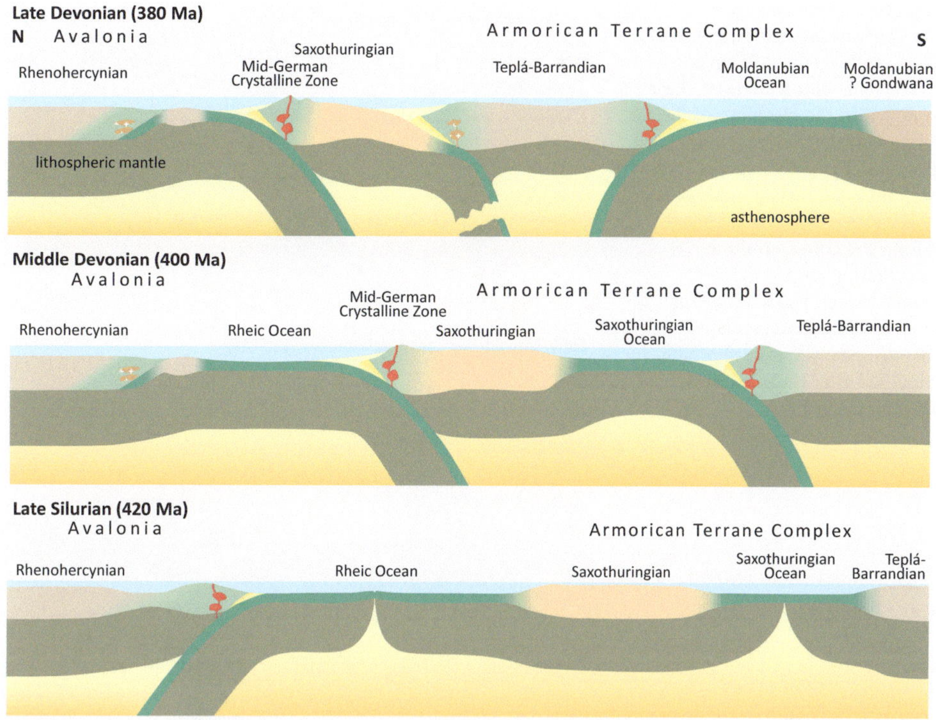

Fig. 10.1 Cross-sections showing the Late Silurian to Devonian evolution of the Variscides as a consequence of the collision between the Armorican Terrane Complex and Avalonia. Modified after Franke (2006)

Franke 1992), where the boundaries between the zones represent the respective subduction zones. There are also indications of large lateral displacements along these zones, which are particularly evident in the internal, high-grade metamorphic areas (e.g., Martinez et al. 2007). The model presented here is therefore just one of the many, often contradictory reconstructions of the Variscan Orogeny. Future research will certainly bring more insight into this topic.

10.1.1 Rhenohercynian Zone

The **Rhenohercynian Zone** is the northernmost unit of the Kossmat zones and is considered to represent part of the microcontinent of Avalonia, which was welded to Laurentia and Baltica during the Caledonian Orogeny to form Laurussia. The Rhenohercynian Zone extends from the **Moravo-Silesian belt** (Moravia and Silesia) east of the Bohemian Massif, through the **Harz** and the **Rheinisches Schiefergebirge,** the **Ardenne** of Belgium, as far as southwest England, South Wales, and Ireland (Warr 2012). It also continues around the oroclinal bend of the Iberian arc and connects with the South Portuguese Zone (Fig. 6.1). It is mainly composed of clastic rocks and volcanics of Devonian and Early Carboniferous age, which formed in a shelf area on the southern edge of Laurussia (including the already attached Avalonia). The sedimentary material was derived from a northern land mass known as the **Old-Red Continent**. On the southern edge of the shelf, a deeper basin developed with the deposition of fine-grained clay-rich sediments. This area extended out to a region of extensively thinned continental crust that developed at the Laurussian passive continental margin of the Rheic Ocean. As new oceanic crust did not form during this phase of crustal thinning, the Rhenohercynian Basin is considered to have been part of an **epicontinental sea** not underlain by oceanic crust. That the area south of this basin is sometimes referred to as the "Rhenohercynian Ocean" is somewhat misleading, with most recent paleogeographic reconstructions presenting it as a narrow oceanic basin formed as the northeastern continuation of the closing Rheic Ocean (Franke et al. 2017).

The Gießen-Werra-Harz Nappes were thrust northwestward onto the Rhenohercynian rocks during the Late Devonian stage of the Variscan Orogeny (Sect. 10.3.1; Fig. 10.2). Here, Devonian basalts of mid-ocean ridge chemistry (Box 2: Basalt) and pelagic sediments (fine-grained Middle Devonian siliceous sediments and Late Devonian flysch; Box 20: Turbidites) are preserved. These presumably originated from the Rheic Ocean or a northerly branch off the main ocean (Franke 2000). Lying to the north of the Rhenohercynian zone is the **Subvariscan Foredeep**, which evolved in advance of the northerly migrating fold-and-thrust belt with the deposition of thick clastic sequences with numerous coal seams.

The Rheic Ocean, sometimes referred to as the Rhea or Rhenanian Ocean, is considered by most authors as the largest and most important ocean that closed during the Variscan Orogeny. It initially opened in the Early Ordovician between Gondwana and Avalonia, as Avalonia began to drift northward. However, it's width and timing of closure are not well constrained. There is also controversy concerning the location of subduction zones that led to the consumption of the ocean. It is well accepted that the Mid-German Crystalline Zone represents the deep crustal level of a former volcanic arc that lay above a

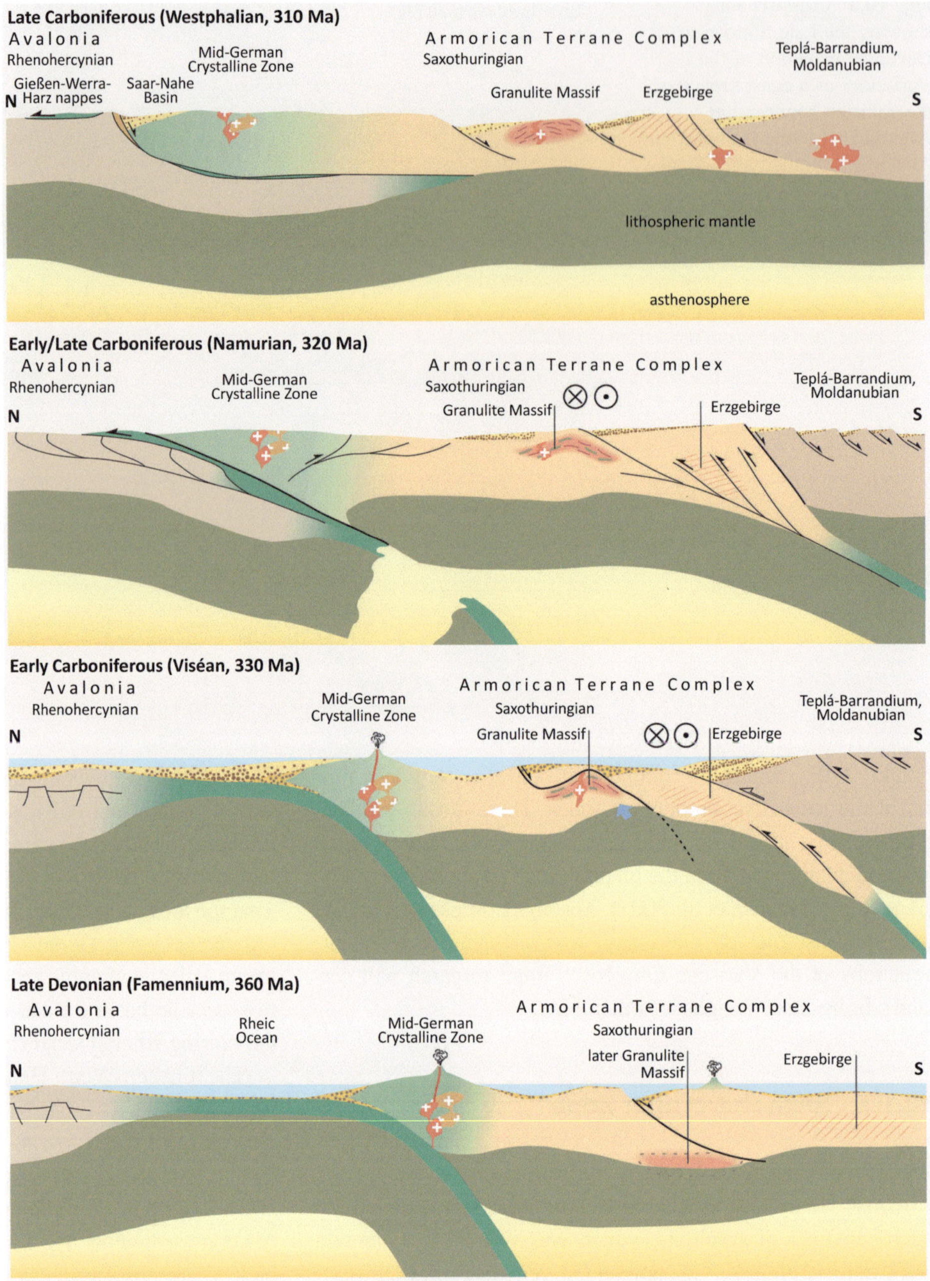

Fig. 10.2 The Variscan Orogeny and the development of the Saxothuringian. Extensional movements along a crustal shear zone led to the elevation of the Granulite Massif as a metamorphic core complex. The high-pressure metamorphic rocks of the Erzgebirge were formed during the collision of the Saxothuringian and the Moldanubian. Circles represent respective midpoints relative to the observer. x: strike-slip fault. Modified after Enderle (1998) and Oncken (1993)

subduction zone. The polarity of the subducting slab is not so clear. Some models assume Laurussia to be the overriding plate in which the subduction zone dipped to the north (e.g., Kroner et al. 2008). Other models propose a southward-dipping subduction zone at the northern edge of Armorica (e.g., Franke 2000; Oncken 1997). There are indications for a northwest-directed subduction at the southern edge of Laurussia during an early phase of the Variscan Orogeny (Early Devonian). This was later replaced by a southeast-directed subduction zone beneath the Mid-German Crystalline Zone at around 380–360 Ma (Franke 2000, Franke et al. 2017).

The southern boundary of the Rhenohercynian is marked by the **Northern Phyllite Zone** (Fig. 9.1), which extends from the southern edge of Hunsrück through the Taunus and into the Wippra Zone of southeastern Harz. The narrow zone consists of elevated pressure, low-temperature metamorphic rocks that originated partly in the Rhenohercynian and partly in the Saxothuringian. As a result, the Northern Phyllite Zone is considered to present the suture zone between Armorica and Laurussia and the site of Rheic Ocean closure. Adjoining, is the Mid-German Crystalline Zone that represents the deep crustal level of a Devonian to Carboniferous volcanic arc complex. Although there was a broad range of deformation conditions that influenced this zone, the structural and metamorphic events are more typical of the Saxothuringian.

10.1.2 Saxothuringian Zone

The **Saxothuringian Zone** is subdivided into the **Mid-German Crystalline Zone** and a southeastern area of basement rocks underlain by Cadomian-formed continental crust composed mostly of low-grade metamorphic sedimentary and volcanic rocks. The former volcanic arc rocks of the Mid-German Crystalline Zone consist of different units that formed on the northern edge of Armorica, but the orientation and sequence of these units are probably not that of the original subduction complex (Oncken 1997). Remnants of Devonian igneous rocks occur in Kyffhäuser and the western Odenwald, whereas older crystalline basement rocks are exposed in Spessart, the eastern Odenwald and the Ruhla area of the Thuringian Forest. These crystalline rocks can be traced further beneath the Mesozoic and Cenozoic sedimentary cover (Fig. 9.1). Rocks of the Saxothuringian also occur in the Thuringian Slate Mountains (s.l.), the Fichtel Mountains, and the Erzgebirge. There is an additional narrow band located at the northern edge of the Vosges and the Black Forest.

The rocks of the Saxothuringian Zone are notably heterogeneous and consist of Late Proterozoic to Early Carboniferous strongly folded schist and gneiss of varying metamorphic grade. Similar rocks to the Mid-German Crystalline Zone are also found in a narrow belt at the southern edge of the Ossa–Morena Zone (metamorphic belt of Aracena). These are located in the western part of the Iberian Peninsula (Fig. 6.1) where the main part of the zone as well as the northern and central Armorican Zone can be correlated with the Saxothuringian units of Germany.

In its plate tectonic context, the Saxothuringian Zone is considered to represent part of the Armorican Terrane Complex (Tait et al. 1997), which was composed of several different crustal fragments. Alternatively, it has been referred to as the Armorican Terrane Collage (Linnemann et al. 1999). The terrane complex consists of a number of tectono-stratigraphic units composed of sedimentary, igneous, and metamorphic rocks of Proterozoic to Early Carboniferous age (Iberia, Armorica, Bohemia, Thuringia, Franconia; after Franke et al. 2017). The individual units, the boundaries of which are often only vaguely defined, have their own geological histories and according to paleomagnetic data (Tait et al. 1997) underwent separate translations, rotations, and collisional events between them. Eventually, they, were all subject to a common drift pattern and became attached to the Laurussia continent during the Variscan Orogeny following the closure of the Rheic Ocean that separated Armorica from Laurussia. This event mainly took place in the Late Devonian and represented a "soft" collision. The former volcanic arc of the Mid-German Crystalline Zone marked the leading northern edge of Armorica. In southwest Poland and the western Sudetes, this zone contains gabbros that are regarded as direct remnants of oceanic crust from the Saxothuringian Ocean. Another narrow and short-lived ocean that was subducted during the Variscan Orogeny was located between the Saxothuringian and Moldanubian zones (Fig. 10.1).

The Saxothuringian differs from the Moldanubian, which lies to the southeast, in terms of its pressure–temperature and rock deformation history during the Variscan Orogeny. In the Saxothuringian, the continental crust was stretched and thinned during the Devonian and Carboniferous. This resulted in uplift of the Saxon Granulite Massif along a subduction channel and exhumation of deeply buried crustal and mantle rocks (Franke and Stein 2000; Sect. 10.3.3; Fig. 10.2). The Moldanubian, on the other hand is characterized by initial crustal thickening due to the Variscan collision. Once thickened and elevated, crustal thinning and collapse of the orogeny occurred to return to a state of equilibrium. This history of thickening and collapse led to the exhumation of high-grade metamorphic rocks deeper than those exposed in the Saxothuringian, such as the crystalline Moldanubian rocks. As a result, high-grade metamorphic rocks are widely exposed at the surface of the Bohemian Massif.

Between the Saxothuringian and Moldanubian lies $a < 5$ km wide intensely sheared zone that is interpreted as a thrust and strike-slip fault zone (Fig. 10.2). During the Variscan collision, nappes of the Moldanubian were thrust onto the units of the Saxothuringian. An example is the allochthonous nappe of the Münchberg Gneiss Massif, which was thrust upon the Saxothuringian (Greek *allos*: differently; Greek *chthōn*: Earth). The Münchberg Gneiss Massif represents in part a highly metamorphosed crystalline series that was structurally displaced onto weakly metamorphosed rock units and is bounded by faults on all sides (Fig. 10.3).

10.1.3 Moldanubian Zone

The **Moldanubian Zone** (in the broader sense) is located in the axial zone of the Variscan Mountains and is bounded to the north by the suture zone between the Saxothuringian and Moldanubian. To the south, it is limited by the Moldanubian suture, which represents the boundary to other Variscan terranes now incorporated into the Alpine Mountain chain (Fig. 6.1). The major part of the **Bohemian Massif** and its western continuation, the **Black Forest** and the **Vosges**

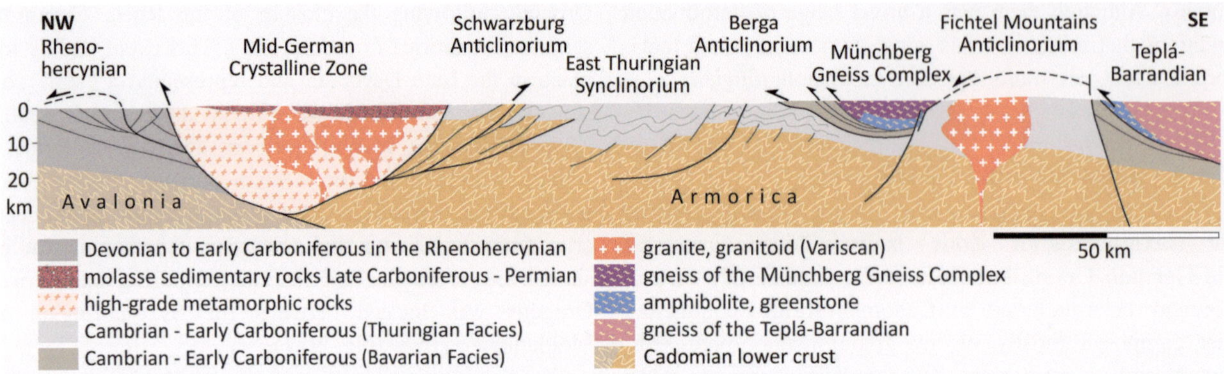

Fig. 10.3 Schematic cross-section through the Saxothuringian along a line crossing the Fichtel Mountains and the Thuringian Forest. Modified after Walter (2007)

Mountains, belong to the Moldanubian. These crystalline areas are mainly characterized by high-grade metamorphic rocks and Variscan magmatic intrusions. The **Bohemian** or **Teplá-Barrandian**, named after the village of Teplá in the northeastern Bohemian Massif (Czech Republic) and after the French geologist Joachim Barrande (1799–1883), represents one of the most important areas where Paleozoic rocks are exposed. It is characterized by sedimentary and volcanic rocks of the Late Proterozoic that were folded and weakly metamorphosed during the Cadomian Orogeny (Fig. 9.1). These rocks are overlain by an almost complete Paleozoic rock sequence ranging from Cambrian to Devonian.

The sedimentary and volcanic sequence of the **Barrandian** (s.str.) is well known for its abundant fossil assemblages that were extensively investigated in the nineteenth century. As a result, this sequence has become the type locality in this region for a number of time intervals for the Silurian and Devonian (Přídolí, Lochkovian, and Pragian). A number of very old Precambrian ages up to 3.84 Ga have been recorded from preserved detrital zircons in a range of metamorphic rocks of the Bohemian Massif, such as paragneiss, amphibolite, graphite schist, anatectic gneiss, and granulite (Gebauer et al. 1989). These rocks were metamorphosed during the young Proterozoic and Early Paleozoic events typical of the large southern continent of Gondwana with which Armorica was at that time connected. The Moldanubian rocks of the Bohemian Massif can be traced through the Central Massif and southern Armorica of France to the Galician–Castilian Zone of the Iberian Peninsula (Fig. 6.1). There, similar high-grade metamorphic rocks and age relationships occur, which indicates its Gondwana origin.

The Moldanubian (s. l.) forms part of the Armorican Terrane Complex, which in spite of having a distinct metamorphic overprint different from the Saxothuringian, probably collided with Laurussia as a terrane assemblage. The **Münchberg Gneiss Massif** represents a nappe that was thrust upon the less metamorphic rocks of the Saxothuringian during the collision between the two terranes (Sect. 10.3.4; Figs. 9.1 and 10.3). The existence of a small ocean closed during the Variscan Orogeny and known as the Saxothuringian Ocean has been proposed (Fig. 10.1). Remnants of this ocean in the form of eclogites and amphibolites occur in the Münchberg Gneiss Massif. Some authors have suggested that the Bohemian Massif existed as an independent terrane called "Perunica" which had at least in part an autonomous development (Cocks and Torsvik 2006). It was separated from Gondwana already in the Ordovician and collided with the Armorican Terrane Complex in the Devonian. Since this terrane includes the Saxothuringian, the Tepla–Barrandian, and the Moldanubian (s.str.), it is not presented as a different paleogeographic unit.

In the region of the Alps, the continuation of the Variscides is difficult to reconstruct due to the strong overprint of the Alpine Orogeny (Chap. 12). The Variscan massifs of the Alpine mountain chain, which lie in the central Eastern Alps, are attributed to the Norica Terrane. Other more isolated occurrences are the Belledonne, Pelvoux, and Mont Blanc massifs in the western Alps of France, the Argentera–Mercantour Massif in the Maritime Alps and the Aar Massif in the Central Alps of Switzerland (Fig. 12.1). It is not clear whether these complexes were separate from Gondwana and followed a different wandering path. The Chamrousse ophiolite of the Belledonne Massif that formed in the Ordovician probably represents a preserved slice of the Moldanubian Ocean.

10.2 Development of the Variscan Orogeny Through Time

Although the Rheic Ocean began to open in the Early Ordovician, with subduction along the southern edge of Laurussia from Silurian times onward (Franke 2006; Fig. 10.1), the collisional stage of the Variscan Orogeny and hence the start of mountain building did not commence until the Devonian (Fig. 10.2). The size and regional extent of the Rheic Ocean are still rather uncertain due to the lack of widespread paleomagnetic constraints. Based on fauna and flora, as well as some paleomagnetic reconstructions, an ocean with a width of over 1000 km is commonly assumed. In contrast, some models present it as a narrow ocean, comparable to about the size of today's Gulf of California. Despite the differences in opinion as to the timing and extent of the Rheic Ocean prior to the Variscan Orogeny, it is well accepted that the orogeny involved major lateral motions and strike-slip displacements during collision. The intense oroclinal bending of the Variscan orogenic belt (Fig. 6.1) has been proposed to result from major late Variscan right-lateral strike-slip faulting combined with the rotation of the Iberian Plate by more than 90° counterclockwise (e.g., Martinez et al. 2007). The strong oroclinal bending may also represent a Variscan Terrane Wreck (Warr 2012), similar to the Early Tertiary Great Alaskan Terrane Wreck of northwestern America.

10.2.1 Devonian

At the beginning of the Devonian (named after the English county of Devonshire), a passive continental margin developed on the southern edge of Laurussia, associated with thinning of the continental crust and extensional tectonics. During rifting of the upper crust, a shallow marginal sea formed with deposition of sediments in a **Rhenohercynian Basin** (Fig. 10.4). In the north, shallow marine, brackish, fluvial, and terrestrial sediments were deposited in a deltaic environment. Typically, sand and coarser-grained lithologies dominated, which originated from the Old-Red continent and belong to the "Old-Red facies" (Fig. 10.5). To the south, slightly deeper areas of sedimentation occurred on the continental shelf characterized by significantly fine-grained sediment. This type of clastic-dominated facies, which is typical of the northwestern Rheinisches Schiefergebirge and southwestern England, is referred to as the **Rhenish facies**. Toward the center of the basin, an offshore facies developed comprised of fine-grained and calcareous sediments known as the **Hercynian facies**. These occur mainly in the Harz and eastern Rheinisches Schiefergebirge.

Rocks of the Rhenish facies are characterized by graywacke, sandstones, and sandy shale, whereas the Hercynian facies is dominated by shales, pelagic limestones, and all transitions between the two (e.g., nodular limestone, nodular slate, etc.). The different facies were developed together in the Early Devonian of the Rhenohercynian, with coarser clastics developed to the north and fine basinal sediments to the south. In places, oxygen-free zones occurred on the seafloor due to undisturbed sedimentation in restricted basins (Box 17: Euxinic environment), with the preservation of organic matter and the fauna and flora. A classic location for well-preserved sea and brittle stars, sea lilies, and trilobites is located in the Bundenbach Slate in the Hunsrück area (Fig. 10.6). Beautifully preserved fossils are displayed in the small fossil museum in Bundenbach. Deposited at the same time were sediments of the 900-m-thick Wissenbach Slate, located in the Harz Mountains. Intercalations of sandy

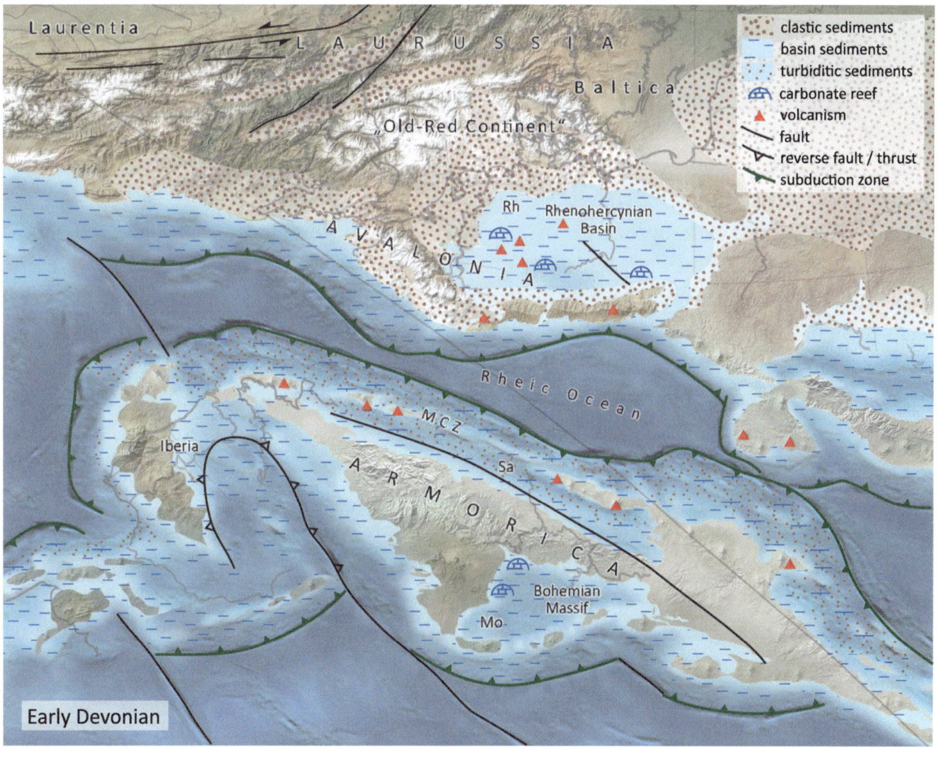

Fig. 10.4 Early Devonian paleogeographic reconstruction of Central Europe. *MCZ* Mid-German Crystalline Zone, *Mo* Moldanubian, *Rh* Rhenohercynian, *Sa* Saxothuringian. Topographic base map from Blakey (2011), modified after Ziegler (1990)

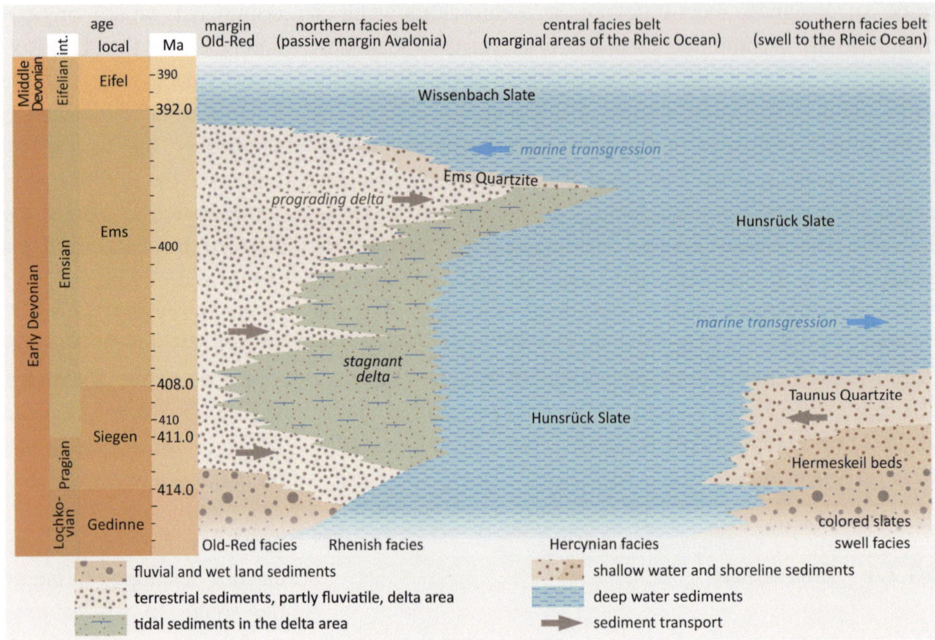

Fig. 10.5 Early Devonian facies distribution across the Rhenohercynian zone from north to south along a line running parallel to the Rhine River. Modified after Stets and Schäfer (2002), stratigraphy after STD 2016, Deutsche Stratigraphische Kommission (2016)

successions formed in the far south of the Rhenohercynian Basin (Hermeskeil Beds, Taunus Quartzite; Fig. 10.5). These developed in shallow water depths on top of an elevated swell that separated the Rhenohercynian Basin from the Rheic Ocean to the south (Fig. 10.4). At this time, the Rheic Ocean was already constricted to a narrow basin and was subsequently closed during Middle to Late Devonian times.

Marine sediments in the area of the Saxothuringian and Moldanubian were deposited during the Early Devonian in water depths similar to the deeper parts of the Rhenohercynian Basin. Devonian slate and nodular limestone that formed in shallow marine basins are also considered as "Hercynian or Bohemian facies," with examples occurring in the Thuringian Slate Mountains (s.l.) of the Saxothuringian and the Barrandian sequence of the Moldanubian. There is a distinction between the two terms "Thuringian facies" and "Bavarian facies," which in the Saxothuringian Zone are applied to most of the Paleozoic sequence ranging from the Cambrian to the Early Carboniferous (Sect. 9.2.6). During the Early Carboniferous, the Bavarian facies was comprised of coarser-grained clastic sediments that developed in a more proximal situation than the Thuringian facies, whereas during the Silurian and Devonian it was characterized by fine-clastics and radiolarites representing a considerably deeper part of the basin (Franke 1984a, b). The Bavarian facies is also spatially limited to the margins of Münchberg Gneiss Massif.

From the Middle Devonian onward, marine areas became gradually more restricted with the development of localized basin and swell facies. While fine-grained sediments, partly with high organic content, were deposited in the basin areas, **carbonate platforms** and reefs developed simultaneously on the swells (Fig. 10.7). Large **fringe** or **barrier reefs** were particularly well developed along the southern edge of the Rhenohercynian Basin and in the shallow marine basins of Armorica. In the area of Taunus and Hunsrück, a swell developed with deposition of thick clastic sequences during the Early Devonian. Located to the south was the suture zone between Laurussia (Avalonia) and the Armorican Terrane Complex, today represented by the Mid-German Crystalline Zone.

During the middle and early Late Devonian, extensional movements affected the passive margin of Laurussia and triggered intense volcanic activity in the area of the Rhenohercynian Basin. Abundant acidic and basic melts erupted from volcanoes at the surface, such as those of the **Lahn–Dill Depression** and in the Harz area. Mainly, keratophyrs, basalts ("diabase"), tuffs, and tuffites (basaltic tuff metamorphosed to greenstone: "Schalstein") were formed (Box 8: Diabase and Schalstein), which later underwent low-temperature metamorphism during the Variscan Orogeny. Very thick sequences of basalt, often preserved as submarine pillow lava, occur in the Lahn–Dill area where they are mined in large quarries. In some areas, volcanic edifices emerged above sea-level, and explosive eruptions resulted in the deposition of pyroclastic sediment (Fig. 10.8). During the late stages of volcanism, massive sulfide iron orebodies were generated with hematite, magnetite, and chlorite minerals. On the seafloor of the Rhenohercynian Basin, layer-parallel ores of hematite known as stratiform deposits formed containing metals precipitated from ore-bearing hydrothermal solutions.

Fig. 10.6 Early Devonian brittle stars (*Furcaster palaeozoicus*) in the Bundenbach Slate. Photograph with courtesy of the Bundenbach fossil museum [N49°50′54″/E 07°23′19″]

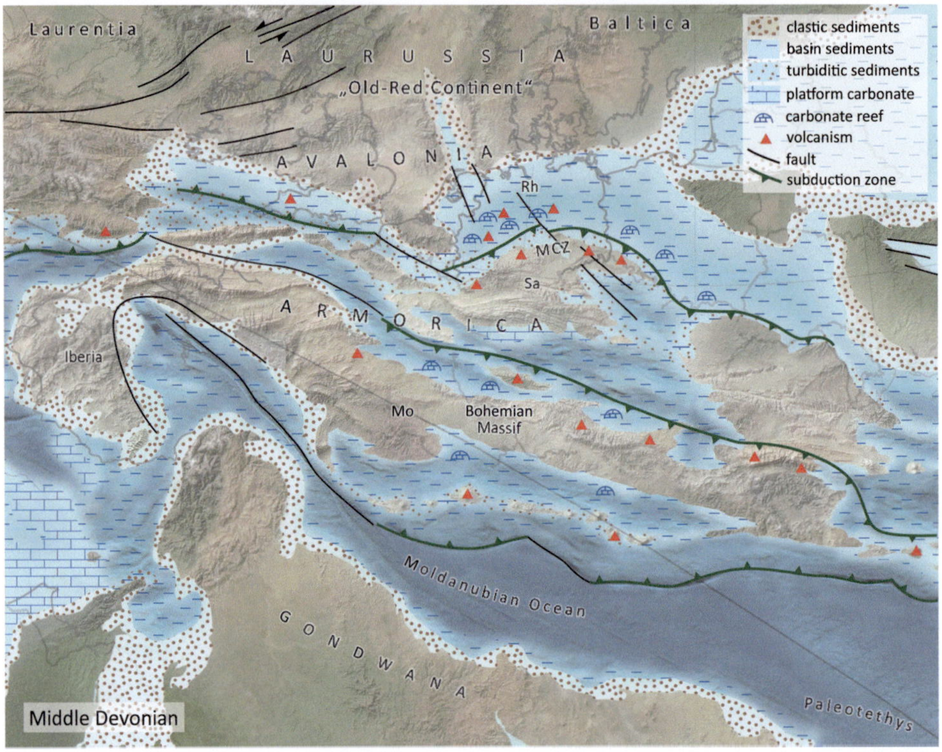

Fig. 10.7 Middle Devonian paleogeographic reconstruction of Central Europe. *MCZ* Mid-German Crystalline Zone, *Mo* Moldanubian, *Rh* Rhenohercynian, *Sa* Saxothuringian. Topographic base map from Blakey (2011), modified after Ziegler (1990)

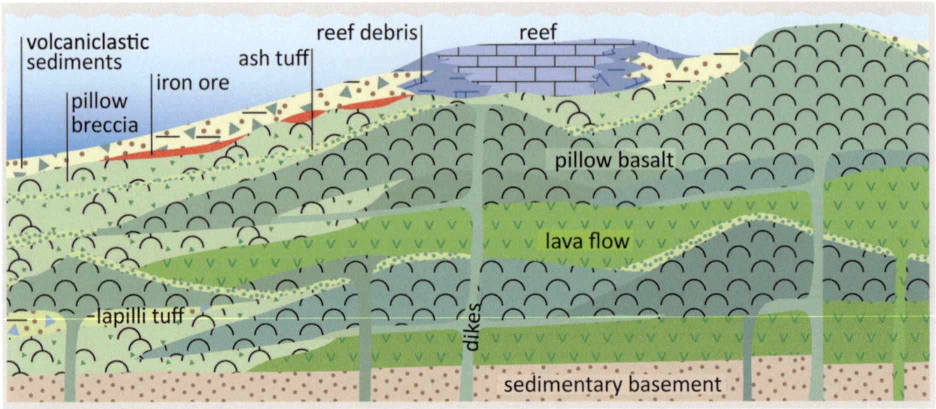

Fig. 10.8 Profile section through the Middle Devonian Philippstein volcano in the Lahn–Dill area. A reef composed of corals and stromatopores formed on top of the volcano and emerged at the sea surface. Modified after Königshof et al. (2010)

The ore formations of the **Lahn–Dill type** are generally considered to represent **sedimentary-exhalative** deposits (Box 11: Ore deposits). In addition to hematite ores, pyrite deposits of the **volcanic-hosted massive sulfide (VHMS)** type are also known in the Harz Mountains and include the particularly attractive *Harzer Blutstein* (Harz blood-stone), composed of pyrite, quartz, feldspar, and hematite (Fig. 10.9). These mineralizations resulted from solutions that were released during the diagenetic alteration of the pyroclastic host sediment (Flick et al. 1990). The circulation of seawater along fractures was also heated by the volcanic activity and thus able to dissolve the metals from the volcanic rocks. This occurred preferentially at the interface between volcanic rock and reefal limestone, with localized mineralization of the highly porous tuff and tuffite to produce layer-parallel (stratiform) orebodies. The occurrence of such nest-shaped hematite ores has been mined in the Lahn–Dill area and in the Harz Mountains for many centuries. In the Harz region, the mining of hematite ore ended in 1970 because of the declining price of raw materials. Mining activities, however, continued near Elbingerode until 1990 with the mining of pyrite for sulfur (S) that was used for the production of sulfuric acid. The last active mine of Oberbiel, located in the Lahn Valley of the Lahn–Dill area, closed in 1983.

Box 8: Diabase and Schalstein

Basaltic rocks from the Paleozoic areas of the Variscan Mountains are often referred to as **diabase** and volcaniclastic rocks together with vesicular lava as **schalstein**. Both names are now replaced by the more universally useful terms provided by international commissions. Nevertheless, the older terms continue to be used today in some regional scientific works.

The term diabase (Greek *diabainō*: transverse, oblique) described basalts originating in the Paleozoic that were altered by low-temperature metamorphism during the Variscan Orogeny. Today, they are mainly composed of minerals from the chlorite and hornblende group, formed by replacement of primary pyroxene (e.g., augite), as well as epidote originating from the partial replacement of feldspar. These minerals give the altered basalts their characteristic green coloration in a process referred to as "**greening**." Today, the name of diabase is replaced by the international term **metabasalt**.

The situation is similar with the term schalstein, which is used locally as a stratigraphic name. Schalstein represents volcaniclastic rocks (pyroclastic ash and tuff), commonly mixed with vesicular lavas that also originated in the Paleozoic. Schalstein is, in fact, an old miner's term for a rock that often splits in a shell-like fashion due to its well-developed cleavage. These rocks are typically weakly metamorphosed and formed near submarine volcanoes located in the Rhenohercynian Basin.

Contemporaneous with volcanism, reefs formed atoll-like structures on top of the volcanic edifices, such as those of the Elbingerode Complex in the Harz or at Gräveneck and Philippstein in the Lahn Syncline (Fig. 10.8). Other large reef structures formed as fringe or small barrier reefs, with examples in the Warstein area, in the Sauerland, or in the Eifel of the Rheinisches Schiefergebirge. These structures of organic origin are characteristic of the Middle Devonian and early Late Devonian and occurred at a time when reef-building organisms experienced their first major appearance to form some of Earth's largest reef complexes.

After a marine transgression during the Late Devonian, the reefs suddenly disappeared in the wake of the so-called **Kellwasser event**, characterized by rapid regression and a mass extinction event (see below). While basin sedimentation was dominated by pelagic shale, giving rise to **tentaculite slate** (tentaculita are extinct molluscs forming calcareous shells), and **cephalopode limestones** (from Greek *kephalē*: head; Greek *podós*: feet), the uplifted areas were characterized by **stromatoporoid and coral reefs**.

Fig. 10.9 *Harzer Blutstein* (Harz bloodstone) from the Schalstein Formation (late Middle Devonian, Givetian; Weller 2008) formed as part of the volcano and reef complex of Elbingerode. The vitreous keratophyre of this unique form of ore is intensely red colored due to the finely dispersed hematite and pervasive thin pyrite veins; locally referred to as *Pyrittrum* (*Trum* is an old German mining expression for a narrow vein)

Both the reef-building stromatoporoids, which are mostly classified as sponges, together with the Devonian corals of the orders rugosa and tabulate, were considerably affected by the Kellwasser mass extinction event and are now extinct. Several hundred-meter-thick reef structures occur as massive limestones in the Rheinisches Schiefergebirge, and the Harz, where they are well exposed in large quarries and mined for aggregate stone. The development and preservation of the reefs required a continuously subsiding substrate, provided by ongoing rift-related subsidence within the Rhenohercynian Basin. During more recent Tertiary uplift, these rocks were subject to karstification and the formation of large cave networks within the massive limestone formations. Occasionally, Eocene paleo-cone karst formations are preserved (Box 9: Karst areas in Germany).

The end of this distinct phase of reef-building carbonate sedimentation corresponded with the Late Devonian **mass extinction event**, which is included among the big five Phanerozoic extinction events of this eon (Fig. 10.12). The Late Devonian extinction actually consisted of two events of notable intensity, namely, the Kellwasser Event at the Frasnian–Famennian boundary (374 Ma; Buggisch 1991) and the Hangenberg Event at the Devonian/Carboniferous boundary (358 Ma). During these events, the reef-building organisms disappeared almost completely from the Earth and nearly 60% of the currently known genera of that time became extinct. The precise reason for the two crises is a topic of debate. Some researchers propose asteroid impacts and others global volcanic events as the cause, although there remains little evidence linking any singular event to this stage of mass extinction. Others point to the more gradual development of land plants and the decrease in

Box 9: Karstification in Germany

Karstification is widespread in Germany (Fig. 10.10; Box 23: Karstification) and can be observed everywhere where massive limestone, dolomite, anhydrite, or gypsum are exposed to extensive rainwater runoff. Limestone and dolomite areas are located in the Rheinisches Schiefergebirge, in the Sauerland or near Limburg (Fig. 10.24). They are also found in the Harz by Elbingerode (Fig. 10.37), and at Iberg near Bad Grund. Devonian limestone reefs occur in the Eifel close to Blankenheim. In almost all these areas, extensive cave systems reflect intensive karstification: the Atta, Klutert, and Dechen caves of the Sauerland, the Kubach Crystal Cave in the Taunus, and the Rübeland caves (Baumanns and Hermanns caves) together with the Iberg Stalactite Cave of the Harz Mountains. Karst formations also characterize Zechstein rock (Late Permian), with several caves developed in anhydrite, gypsum, and dolomite along the southern edge of the Harz Mountains. Similar formations are found in the Thuringian Forest and as isolated

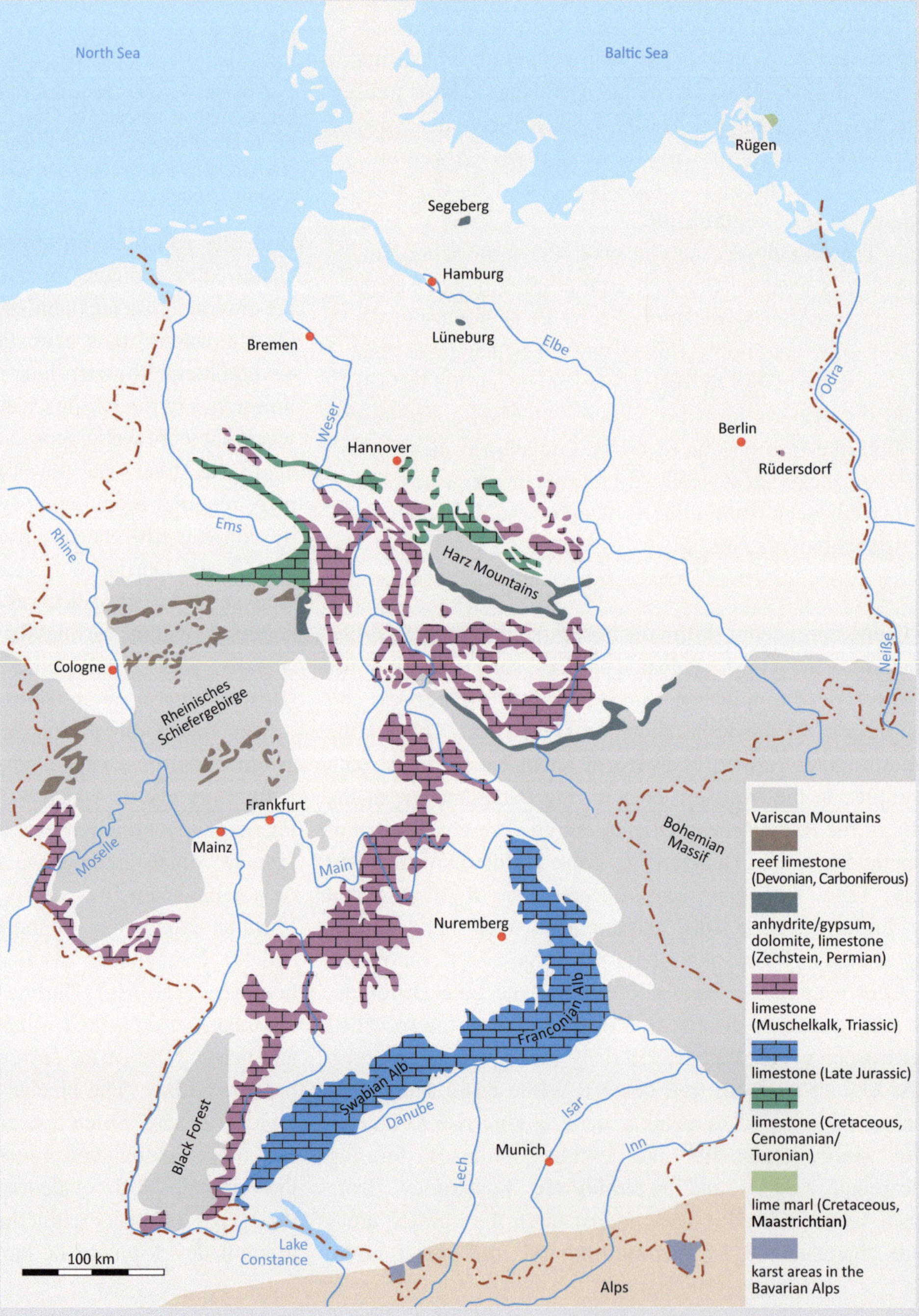

Fig. 10.10 Overview of the areas of limestone karstification in Germany

Fig. 10.11 Eocene paleo-cone karst developed in massive Devonian limestone (quarry Schneelsberg at Runkel-Hofen, near Limburg). The gaps between the karst cones are filled up with Oligocene sediment [N 50°25′51″/E 08°08′ 52″]

occurrences in Bad Segeberg. Other examples of caves in Zechstein rock are the Unicorn Cave at Herzberg (in the Harz) formed in dolomite, the Barbarossa Cave at the Kyffhäuser developed in anhydrite, and the Kalkberg Hill Cave in Bad Segeberg located in both gypsum and anhydrite. At the northern rim of the Harz, in the steeply dipping "*Aufrichtungszone,*" a narrow belt of almost vertical sedimentary beds is marked by numerous sinkholes along the Northern Harz Boundary Fault (Fig. 11.75). These are also caused by subsurface dissolution of Zechstein anhydrite and gypsum.

Karstification of Triassic rocks with cave formations also developed in Muschelkalk limestone, such as the Eberstadt Stalactite Cave, near Heilbronn. One of the most extensive cave areas in Europe is found in the White Jurassic rocks of the Swabian and Franconian Alb. Notable locations are the Fog Cave (*Nebelhöhle*) and Bears Cave (*Bärenhöhle*; Fig. 12.13; Box 23: Karstification), both near Reutlingen. Others are found in the area of Ulm, such as the Laichingen Deep Cave and the Blautopf Cave, which contains a large chamber known as the Mörike Cathedral (*Mörikedom*). Additional examples are Bing Cave, near Bamberg, or Sophie's Cave and the nearby Devil's Cave of Pottenstein (between Nuremberg and Bayreuth). Altogether, it has been estimated that there are as many as 3000 caves in the White Jurassic of southern Germany. Many of these caves are easily accessible and used for tourism, whereas many more caves have yet to be explored. Karst formations in Cretaceous limestone and marly limestone exist at the edge of the Münsterland Cretaceous Basin and on the Paderborn Plateau. Other examples are found in the Teutoburg Forest and the Wiehengebirge of the Lower Saxony Mountains. However, accessible caves are rare in this area due to their small size. On the Island of Rügen, recent karst formations with sinkholes and travertine formations can be observed on the Jasmund Peninsula located in an exposure of Late Cretaceous chalk (Fig. 16.14).

Most karst features in Germany have formed only within the last 5–10 Myr, but there are also older systems that were generated shortly after uplift of limestone areas in the Mesozoic. Early Cretaceous limestones and dolomites of the White Jurassic were karstified soon after their formation as tropical cone karst. The cones were subsequently covered by Cretaceous sediment and thus preserved as paleokarst structures. A well-known example is the Big Hole Rock (*Großer Lochstein*) in the Franconian dolomite of the Franconian Alb, near Pegnitz. However, not only the Jurassic limestones and dolomites were subject to early karstification. Massive Devonian limestones at Runkel-Hofen, near Limburg, underwent karstification during the Eocene (Fig. 10.11), with erosion structures formed as tropical cone karst that was subsequently filled with Oligocene sediment.

Fig. 10.12 Phanerozoic mass extinction events showing the five biggest episodes (the *big five phanerozoic extinction events*) corresponding to the Ordovician/Silurian boundary, the Late Devonian and the boundaries of the Permian/Triassic, Triassic/Jurassic, and Cretaceous/Tertiary periods. The percentages indicate the amount of extinction related to the number of genera living in the respective times. *CAMP* Central Atlantic Magmatic Province. Modified after Reitner (2011)

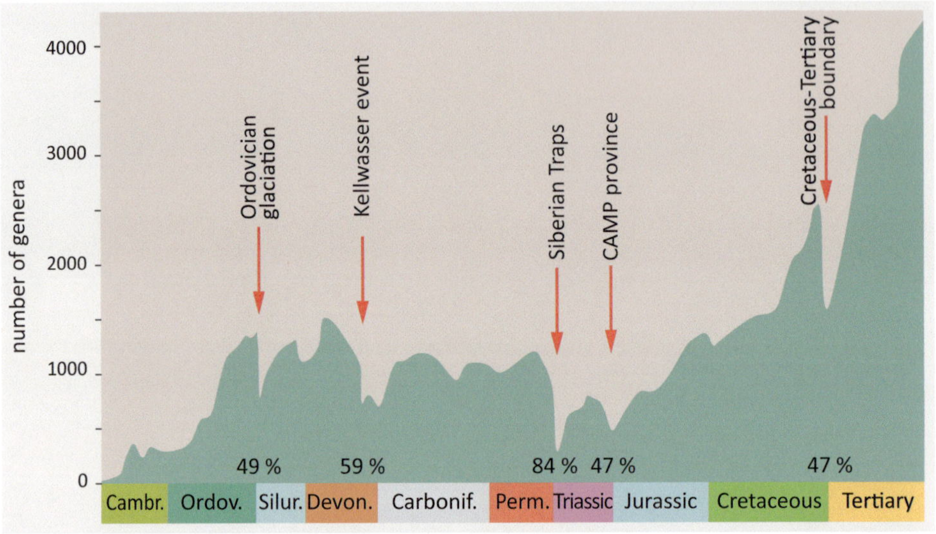

Fig. 10.13 The Kellwasser Horizon exposed in the Kellwasser Valley along the south side of the Oker dam lake in the Harz. This upper part of the horizon represents a mass extinction of about 60% of the known genera. Profile section modified after the wall chart of the Regionalverband Harz (2012) [N 51°49′12″/E 10°26′58″]

Box 10: Silicate weathering

Nearly all rocks at the Earth's surface are subject to weathering, particularly in wet environments where anhydrous minerals, typically formed at high temperatures and pressures, break down by chemical reactions with water in a process known as **hydrolysis** or hydrolytic decomposition. As siliceous minerals are dissolved, they release a range of metallic ions such as K, Na, Ca, Fe, or Mg. The hydrolytic decomposition of silicate minerals, such as feldspar or mica, usually ends with the formation of clay minerals, in particular, kaolinite or mixed-layered phases, and may result in the complete dissolution of the primary silicate rock. In the extreme case, only Al hydroxide ($Al[OH]_3$), Fe oxides, and dissolved Si acid ($3H_4SiO_4$) remain. Under favorable conditions, namely, wet tropical to subtropical climates, this leads to the formation of Al-enriched bauxite deposits.

Carbon dioxide (CO_2) plays an important role in silicate weathering. When CO_2 dissolves in water, it produces carbonic acid, which enhances the rate of mineral dissolution. For example, during the decomposition of Ca-feldspar (anorthite), Ca^{2+} and $2HCO^{3-}$ ions are released, and kaolinite ($Al_2[OH]_4Si_2O_5$) is formed as the alteration product.

$$2H_2CO_3 + CaAl_2Si_2O_8 + H_2O \\ \rightarrow Al_2(OH)_4Si_2O_5 + Ca^{2+} + 2HCO_3^-$$

Various aquatic organisms use the ions released by hydrolysis to build shells and skeletons, with $CaCO_3$ incorporated either as aragonite (only metastable at low pressures and temperatures) or calcite.

Silicate weathering is an extremely important process and has a strong influence on the global **carbon cycle**. The weathering of silicate-containing rocks over long periods removes large amounts of CO_2 from the atmosphere and binds the carbon in carbonate rocks. When global warming occurs, the enhanced temperatures and higher rainfall increase the rate of weathering and the degree to which this greenhouse gas is bound. This leads to global cooling and colder climatic conditions until the release of CO_2 will again exceed that bound by weathering and reaccumulate in the atmosphere. Thus, the weathering of silicate rocks acts rather like a "thermostat" for the global climate, with the concentration of CO_2 always moving toward an equilibrium state between CO_2 release and CO_2 sequestration related to the degree of crustal weathering. However, this regulating system works rather slowly and reacts over geological timescales characterized by global climate change. The rapid change that we are witnessing today by the extensive burning of fossil fuels is occurring at too fast a rate to be regulated by this process.

carbon dioxide levels of the atmosphere related to increased photosynthetic activity, which led to intense carbonate sedimentation and ultimately, extinction events. The Devonian saw the evolution of the first tree-like land plants with profound root systems that enhanced the rates of physical and chemical weathering and resulted in widespread soil formation. The intense **silicate weathering** also contributed to the reduction of carbon dioxide in the atmosphere. This type of weathering (Box 10: Silicate weathering) dissolves Ca^{2+} ions from silicate minerals (such as feldspar), which together with the dissolved carbon dioxide in the water (HCO_3^-) can be used by shell and skeleton-forming organisms to build up the required calcareous material ($CaCO_3$).

In the Late Devonian and Carboniferous, the increased rates of carbon dioxide fixation and carbonate rock production, enhanced by more vegetation and weathering, led to a phase of global cooling that terminated in a glaciation. The increase in soil formation during the Late Devonian furthermore increased the supply of nutrients into the oceans, causing over-fertilization of the waters and reduction in the number of reef-building organisms. The decomposition of organic matter derived from the dying reefs and in particular the increase in nutrients together with changes in the ocean circulation system, led repeatedly to oxygen-poor and life-hostile seafloor conditions (Box 17: Euxinic environment). This resulted in the deposition of sediments with high organic content in which the mass extinction events were recorded (Fig. 10.12). It was the sedimentation of large amounts of organic material into this carbon sink that reinforced global cooling by the removal of significant amounts of atmospheric carbon dioxide.

The two sedimentary layers that make up the **Kellwasser horizon** in the Devonian sequence are characterized by distinctive black shales and dark limestones that provide excellent marker horizons for stratigraphic correlation. The type locality for this horizon is situated in the Kellwasser Valley of the Harz (Fig. 10.13; Gereke et al. 2014), where the Frasnian–Famennian boundary lies directly above the upper Kellwasser horizon. The slightly younger Hangenberg Event is considered to have a similar origin as the Kellwasser horizon.

10.2.2 Carboniferous

The Carboniferous era of Central Europe, as well as North America, was marked by the collision of the Gondwana and Laurussian continents. In Europe, this closure resulted in the **Variscan Orogeny** and formation of the Variscan Mountains. The convergent movement between the two large continents continued into the Carboniferous and culminated at the end of the orogeny with the Late Carboniferous formation of the global supercontinent of Pangea (Fig. 7.1f). The Variscan Orogeny affected the entire area between the northern edge of Gondwana and the southern edge of Laurussia, including all intervening terranes and microcontinents. From the Early Carboniferous onward, extensive areas were deformed by folding and thrusting. The rocks underwent varying degrees of metamorphism at different temperatures and pressures, dependent on the amount of structural burial during the stacking of the fold-and-thrust nappes. By the Early Carboniferous, the once extensive Rheic Ocean was reduced to a narrow marine basin characterized by the deposition of turbiditic flysch sediments (Box 20: Turbidites). The orogeny culminated with a foreland basin (molasse) stage of sedimentation, with the accumulation of thick clastic sedimentary deposits in the downflexured foredeep of the Variscan Mountains, generally referred to as the Subvariscan Foredeep.

The term "Carboniferous" was introduced at the beginning of the nineteenth century in England as a geological time period named from the worldwide occurrence of **coal seams** in the clastic sedimentary rocks of this age (Latin *carbo*: charcoal). Coal actually comprises only about 2–3% (by volume) of all Late Carboniferous rocks, with the remaining part composed of sandstones, conglomerates, mudrocks (including shales), and some calcareous deposits.

The stratigraphic subdivision of the Carboniferous is not uniformly applied across Europe and America. The Early Carboniferous in Europe is referred to as the Dinantian (named after the town of Dinant in Belgium) and does not precisely coincide with the slightly longer Mississippian subperiod used in America (Fig. 10.14). The Late Carboniferous in Europe is known as the Silesian (from the Latin name, *Silesia*), which in America corresponds partly to the Pennsylvanian. The international Commission on Stratigraphy names the individual stages of the Carboniferous after Russian cities close to their type locations in the Urals. The exceptions are the Tournaisian and Viséan used for stages of the Early Carboniferous, which are named after stratigraphic occurrences near the Belgian cities of Visé and Tournai. The stages of the European Late Carboniferous are named after a number of diverse locations and regions, such as the Namurian from Namur in Belgium, the Westphalian from Westphalia in Germany and the Stephanian after the Latin name of *Stephanus* for St. Etienne in France.

Early Carboniferous

Sedimentation continued in the Early Carboniferous with deposition of predominantly marine sediment. Two facies areas developed in the Rhenohercynian Basin: a deep marine facies characterized by black shales and slates and a shallow marine facies with the formation of very fossil-rich limestone with a rich flora and fauna (including brachiopods, crinoids, corals, foraminifera, and calcareous algae). The latter accumulated on fault ridges and horst blocks of the rifted shelf area that lay on the southern flank of the Laurussia continental shelf ("Old-Red Continent"). This shelf was located to the north of the London–Brabant Massif and extended from Ireland to Poland (Fig. 10.15). It also crossed the border of the Rheinisches Schiefergebirge, with sedimentary rocks of the shelf been exposed today on the northwestern flank of the Velbert Anticline.

Although not containing coal seams, the Carboniferous limestone does in places have a high organic content due to the deposition of dark carbonaceous sediment under oxygen-poor conditions on the seafloor bottom (Box 17: Euxinic environment). Typical fossils found in the Carboniferous limestone are mainly shallow water species, such as crinoids (mainly crinoid columnal segments). These are visible as white fragments in the gray limestone, together with bryozoans, corals, brachiopods, and goniatites: the latter representing a Paleozoic form of ammonite (Box 3: Ammonites). This limestone is used as a natural stone for facade cladding, tiles, and sculptures and remains in great demand. The natural stone comprised of Carboniferous limestone from Belgium is known under the traditional trademark "Belgian Granite," even though it is not an igneous rock.

South of the limestone facies lay a deeper water basin where calcareous **turbidites** and **olisthostromes** formed from the sediment that slid off from the outer edge of the Carboniferous limestone platform (Fig. 10.15). Olisthostromes (Old Greek *olisthema*: gliding; Greek *stroma*: the outstretched) are submarine deposits produced from the chaotic, gravitational sliding or slumping of sedimentary material whereby huge blocks of rock are relocated downslope. Whereas the Carboniferous limestone has thicknesses between 200 and 700 m, the basinal facies of the Early Carboniferous is often dominated by turbiditic series and reaches up to 3000 m thick.

The basinal facies, sometimes locally referred to as the "Culm facies," can be divided into two parts. A northern and northeastern basin area is comprised mainly of fine-grained sedimentary rocks in the form of shales and radiolarian-bearing cherts, and a southern and western part is dominated by coarser flysch composed of bedded and massive graywackes. The latter was derived from the advancing Mid-German Crystalline Zone. The occurrence of chert, which in its pure form is referred to as **radiolarite**, and when distinctly black-colored is known as **lydite**, indicates a deepwater and calm depositional environment. The black lydite was deposited under conditions of reduced oxygen on the seafloor in a similar environment as during the Kellwasser event.

The Early Carboniferous **Posidonomya Shale** was deposited in these basinal areas. These strata are characterized by eponymous fossil shells of *Posidonomya becheri* (previously named *Posidonia becheri*) that are often found in large quantities. These shales are not to be confused with the Early Jurassic Posidonia Shale, which is very similar in appearance and contains large quantities of similar shells (Sect. 11.6.1). Previously, both formations were called Posidonia shale. However, today this term is only valid as a stratigraphic name for the Early Jurassic deposit.

The development of the Thuringian and Bavarian facies continued in the area of the Saxothuringian and Moldanubian. A change from calm basin sedimentation with

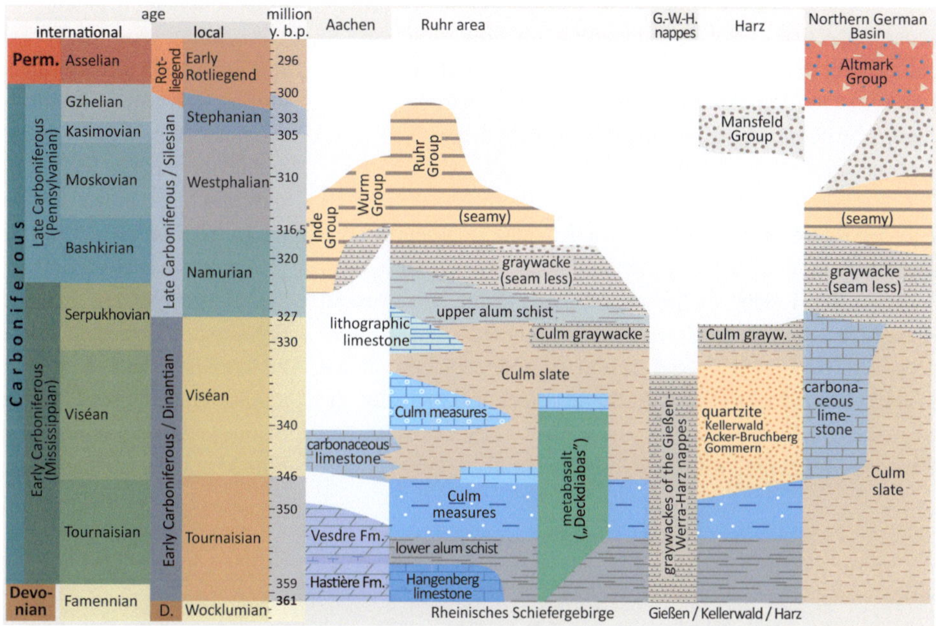

Fig. 10.14 Overview of the Carboniferous stratigraphic sequence of the Rhenohercynian, Germany. Modified after Kombrink et al. (2010) and STD 2016, Deutsche Stratigraphische Kommission (2016)

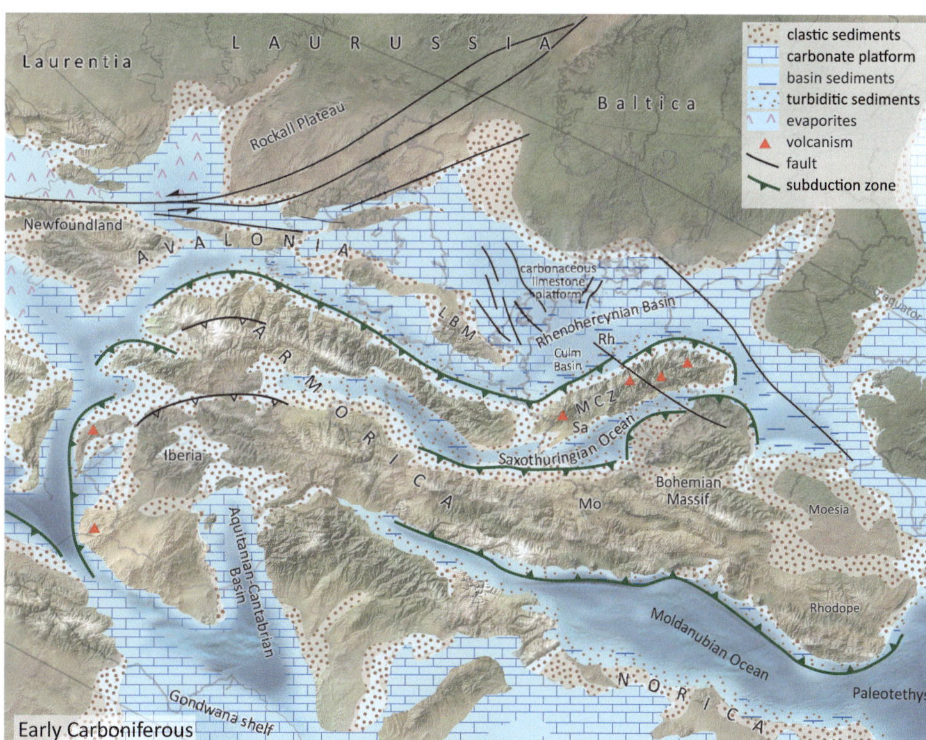

Fig. 10.15 Early Carboniferous paleogeographic reconstruction of Central Europe. *MCZ* Mid-German Crystalline Zone, *Mo* Moldanubian, *Rh* Rhenohercynian, *Sa* Saxothuringian. Topographic base map from Blakey (2011). Modified after Ziegler (1990) and Kombrink et al. (2010)

fine-grained dark shales to flysch deposits dominated by clastic components was similar to that recognized in the Rhenohercynian. The upper Early Carboniferous and Late Carboniferous sequence are almost exclusively composed of alternating graywacke and slate. In the Early Carboniferous, the contrast between the Thuringian and Bavarian facies is most distinctive. The flysch deposits of the Bavarian facies are significantly coarser and document the proximity to the area of origin. In the flysch lithology, brecciated layers of olisthostrome with olistholite of Carboniferous limestone occur, reflective of the development of shallow shelf areas in the vicinity during deposit. The flysch deposits commonly contain several meter thick layers of olisthostrome with a chaotic structure, typically referred to as wildflysch.

Late Carboniferous

The deposition of the basin and shelf marine sediments ended at the beginning of the Late Carboniferous. By this time, thick layers of flysch deposits had accumulated in the northwestern Rheinisches Schiefergebirge, while fine-grained shales were deposited over a prolonged period of sedimentation in the eastern Rheinisches Schiefergebirge. Sedimentation in the Rhenohercynian was increasingly influenced by the ascending Variscan Mountains, with prevailing terrestrial conditions in the form of fluvial and lacustrine deposits (Fig. 10.14). The **Subvariscan Foredeep** developed as a **molasse-type foreland basin** of the Variscan Mountains. Already by the end of the Namurian, the first Late Carboniferous terrestrial wetlands formed in the northwestern areas of Germany. These wetlands accumulated large quantities of peat, which with time and burial were converted into the coal seams characteristic of the Carboniferous.

The large wetland areas spread out under the prevailing tropical climate conditions and were not limited to the external Variscan molasse basins, but also developed further south in smaller, intramontane basins. Many of these were closed off from the sea and formed limnic basins (Greek *limnē*: lake). Notable examples of coal deposits of this type are found in the Saarland within the eastern Saar–Nahe Basin, in Zwickau, Lugau–Oelsnitz and Flöha in the western Erzgebirge Basin, as well as in the Upper Silesian area around Katowice/Poland (Fig. 10.16).

The extensive wetland environment led to the deposition of thick peat layers on the gradually subsiding flat surface of the Subvariscan Foredeep. The marshes were in direct contact with the molasse sediments of the Variscan Mountains. Regular marine transgressions occurred resulting in the deposition of marine sediments on top of the peat horizons. This type of coal formed in brackish marine areas is known as **paralic coal** (Greek

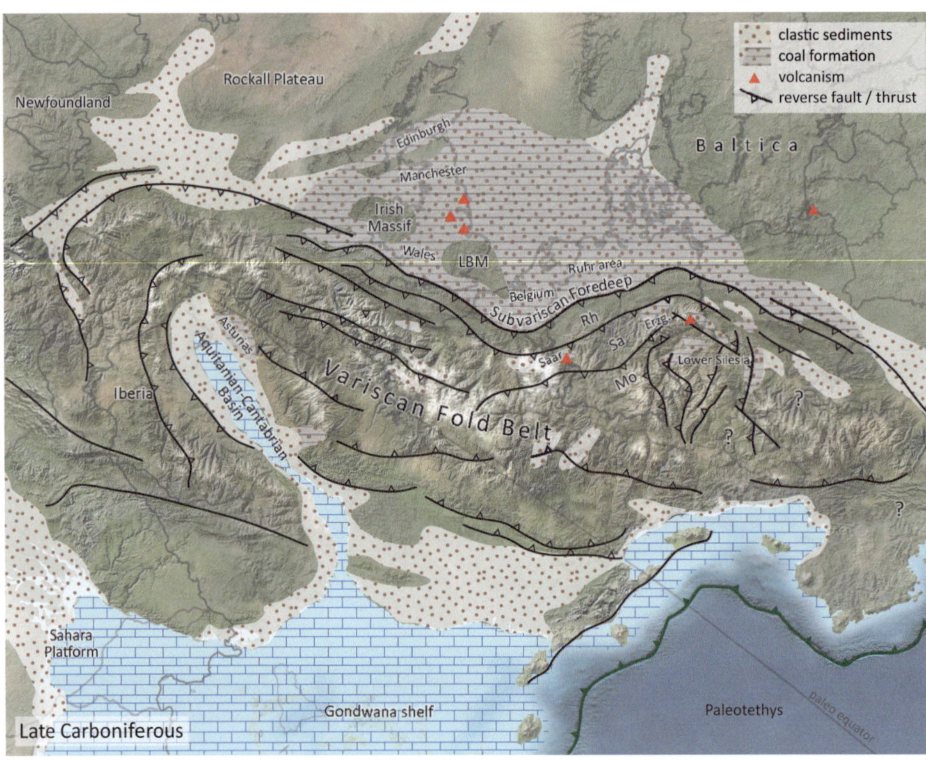

Fig. 10.16 Late Carboniferous paleogeographic reconstruction of Central Europe. *LBM* London–Brabant Massif, *Mo* Moldanubian, *Rh* Rhenohercynian, *Sa* Saxothuringian. Topographic base map from Blakey (2011). Modified after Ziegler (1990) and Kombrink et al. (2010)

paralios: near the sea) and occur in repeating sedimentary successions collectively referred to as **cyclothemes** (Fig. 10.17).

Today, cyclothems are more frequently described as **sedimentary sequences**. The systematically recurring Late Carboniferous strata contain lithologies ranging from fine mudstone to coarse sandstone with intercalations of coal seams. Marine and terrestrial parts of such sequences are also distinguishable. The cyclic sequence typically starts with coarse, partly conglomeratic, thick-bedded sandstones, which often erosively cut into the underlying substratum. Parallel laminated fine sandstones, siltstones, and mudstones follow on top, with remnants of roots and washed-in plant fragments that increase in abundance upward within the sequence. Sometimes, distinct **root horizons** are preserved, above which coal seams occur reaching thicknesses ranging from a few centimeters to several meters (Fig. 10.18). To form a 1-m-thick coal seam requires the diagenetic lithification and coalification of peat with an original thickness of about 7 m. Such a thick peat actually requires a period of about 10,000 years to accumulate. The final layer of the cyclic sequence is typically made up of shales that overlie the coal seam. In most cases, these mudrocks were deposited under marine conditions during **transgression** phases of the sea that introduced fine-grained sediment. The rapid influx of sediment was a prerequisite for the formation of coal, as the peat layers had to be quickly buried to exclude oxygen and thus avoid aerobic decomposition.

The transgressions were presumably related to sea-level fluctuations caused by the Late Carboniferous glaciations—a phenomenon well known from the more recent Quaternary period that resulted in rapid changes of coastline morphology in a geologically very short period of time (Chap. 16). Somewhat different was the formation of coal seams in the intramontane basins of the Saar–Nahe Depression, the Erzgebirge Basin, and younger sections of the Lower Silesia. Here, the peat layers were disconnected from any marine

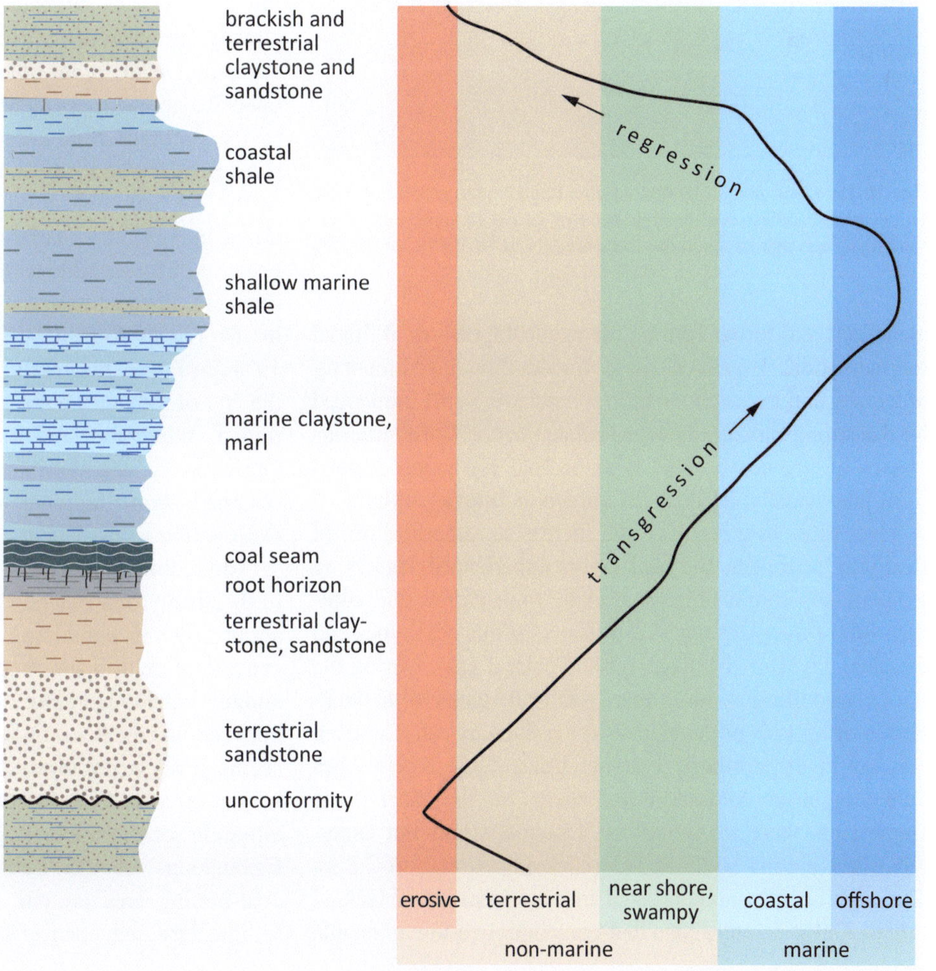

Fig. 10.17 Schematic representation of a typical cyclical sedimentary sequence of Late Carboniferous molasse deposited in the Subvariscan Foredeep. These sequences were previously described as cyclothemes (diagram compiled from various sources)

Fig. 10.18 Coal seam exposed in the Rauen quarry, near Witten. These are the oldest coal-bearing beds of Late Carboniferous (latest Namurian) age that occur in the Ruhr area [N 51°25′18″/E 07°21′23″]

incursions and formed during the cyclic deposition of fluvial sedimentation. However, these limnic–fluvial sedimentary processes and the accumulation of peat were still influenced by the strong climatic changes related to the Carboniferous phases of glaciation. This type of coal formation formed from lake and fluvial deposits known as **limnic coals**.

Over time, as a result of the increasing superposition of overlying sediment, the peat layers experienced increasing temperatures responsible for organic **maturation** and **coalification** of the plant material to form various coal types and methane gas (Box 14: Coal, oil and natural gas). During the Late Carboniferous, more than 100 coal seams with thicknesses of up to 3 m developed in the Subvariscan Foredeep that lay in the northerly external front of the Rheinisches Schiefergebirge fold-and-thrust belt. In the Saar–Nahe Depression, there are more than 150 coal seams but generally, these are not so thick, and reach a maximum of 2.2 m. The significantly smaller coal-mining areas of Zwickau, Lugau–Oelsnitz, and Flöha that contain only a few mineable coal seams were deposited in sub-basins of the Erzgebirge Basin. As these basins were not connected with each other, they cannot be correlated stratigraphically. The Upper Silesian coal-mining district contains some of the richest coal deposits in Europe. From more than 400 coal-bearing layers, nearly 100 have been mined, and due to their shallow depth compared to the Ruhr area and the Aachen–Erkelenz and Belgian coal-mining areas, the level of production has been high. Above a thickness of 60 cm and depending on the quality of the coal, the seams are generally regarded as mineable. In the Carboniferous of the Ruhr area, about 75 coal seams have been mined, from which only a few are operational today. Already at the beginning of the nineteenth century, more than 200 coal mines existed in the Ruhr area. During the most intense production phase in the middle of the last century, more than 600,000 miners were employed in the region.

Today, only one mine is still in operation in the **Ruhr area**. Another one exists in Ibbenbüren at the Schafberg, in a horst structure that was uplifted in the Cretaceous (Sect. 11.7; Fig. 11.71). The Auguste Victoria coal mine in Marl was already closed in 2015. However, by the end of 2018, both remaining coal mines, the Prosper-Haniel in Bottrop that opened a new sole just in May 2011 and the Ibbenbüren lying to the north of Münster will cease operation as the coal subsidy by the Federal Republic of Germany expires in 2019. In the Aachen coal-mining area, coal production was shut down in 1997. In Saarland, the last coal mine was closed because of foreseeable technical difficulties in June 2012. In Zwickau, the coal reserves were exhausted in 1978, whereas in the Lugau–Oelsnitz coal-mining area this happened in 1971. In Flöha, the mining activity had already finished by the end of the nineteenth century, where 12 coal seams were mined. While the coal industry is drawing to a close in Germany, in neighboring Poland 90% of its electricity is obtained from national coal resources, with increased exploitation of the Upper Silesian coal-mining area.

The sedimentary rock sequence of the Late Carboniferous in the **Ruhr area** reaches thicknesses of over 4000 m, with about 3000 m containing coal-bearing horizons. The percentage of coal in the sequence, however, only amounts to about 2–3%. Dividing the approximate time interval between the coal seam-bearing layers by the number of seams gives a duration of about 100,000 years per cycle which more or less corresponds to the fluctuation in the ellipticity of the Earth's orbit around the sun (see Box 25: Fluctuations of the Earth's orbital parameters). In the coal-mining area around Aachen, the Late Carboniferous sedimentary sequence is slightly thinner at about 1700 m.

The sedimentary rocks are mainly composed of mudrocks and siltstones, together with sandstones and conglomerates. In addition to past coal-mining activities, hard sandstones are still extracted from open-cast mines at various locations. The quarry at **Piesberg**, near Osnabrück, is today one of the largest hard rock quarries in Europe, which is located in a Cretaceous horst block similar to that at Ibbenbüren (Fig. 11.71). In earlier times, coal was also mined at this location but today only the thick uppermost Westphalian sandstones are excavated. The quarry is also known for its rich variety of Carboniferous fossils.

The **climatic conditions** during the Carboniferous differed significantly from the Earth's present climate. The oxygen content of the atmosphere was as high as 35%, which is significantly greater than the current 21% (Fig. 11.7). The strongly oxygenated atmosphere resulted in intense plant growth and notably **large insect species**. Under today's conditions, the insects of the Carboniferous period would not survive, since their trachea-based respiratory system would not be able to transport sufficient oxygen to the body. Some species of dragonfly had wingspans up to about 70 cm. There were also giant spiders with nearly one-meter-long legs and millipedes that reached lengths of two and a half meters. The vertebrates also became increasingly abundant across Pangea, leaving tracks such as those only recently discovered in sandstones that were deposited around the Namurian-Westphalian boundary in the Ruhr area, near Bochum. These are in fact some of the oldest trace fossils found in Germany and probably originate from the reptilian-like *Ichniotherium praesidentis*, which represents an ancestral link between the amphibians and reptiles.

In addition to the large amount of oxygen generated from the Carboniferous tropical rainforests of Pangea, a vast amount of CO_2 was fixed into plant matter to form a significant geological carbon sink. A large proportion of the carbon was not cycled back into the atmosphere after the death of the plants, but instead was stored as thick coal seams and disseminated organic matter within the Earth's upper crust. Important for the preservation of the organic material was the widespread coal swamps where the wood of dead trees accumulated under waterlogged anoxic conditions and was periodically covered in sediment (see above). These conditions were sealed off from oxygen so that subsequent oxidation and thus the release of CO_2 were prevented. Another contributing factor may have been the lower abundance of wood-degrading organisms, such as the lignin-degrading fungi, which did not evolve and diversify until the end of the Permian.

10.3 The Variscides in Germany

The deformed Variscan basement rocks of Germany are exposed in a number of low-lying mountainous areas, with the larger portions of basement hidden under thick sequences of Mesozoic and Cenozoic sedimentary and volcanic rock. A comprehensive knowledge of the underground tectonostratigraphy is available from the numerous deep boreholes and seismic surveys that have been undertaken across the country. Paleozoic and partly Proterozoic rocks of the Rhenohercynian are exposed at the surface in the Rheinisches Schiefergebirge, the Subvariscan Foredeep lying to the north, the Harz Mountains, and some small outcrops of the Flechtingen Heights. The Saxothuringian rocks are exposed further south at the northern edge of the Black Forest and in the Odenwald and Spessart area. Other areas include the Ruhla Crystalline of the Thuringian Forest, the hill range of Kyffhäuser, the Thuringian Slate Mountains (s.l.) of the Erzgebirge, the Saxon Granulite Massif, and the Lusatian Hills. These rocks also occur toward the east in the Sudetes and the Moravo-Silesian areas. To the south lie the crystalline basement rocks of the Moldanubian in the Black Forest, the Fichtel Mountains, and the Bohemian and Bavarian Forests (Fig. 10.19).

10.3.1 Regional Geology of the Rhenohercynian

The **Rheinisches Schiefergebirge** and the **Harz Mountains** represent the Rhenohercynian fold-and-thrust belt that formed the northern edge of the Variscan Mountains (Fig. 9.1). The Rheinisches Schiefergebirge is characterized primarily by northwest-verging fold-and-thrust structures, whereas the main part of the Harz belongs to the complex of the Gießen-Werra-Harz Nappes. These were thrust northward from the southeast onto the Devonian and Carboniferous strata of the Rhenohercynian during the Variscan Orogeny (Engel et al. 1983; Walliser and Alberti 1983; Huckriede et al. 2004; Fig. 10.20). At the base of the nappe stack, which consists mainly of deepwater sedimentary rocks of the Devonian and Early Carboniferous (including the Gießen graywacke), basalts are found in thin tectonic slivers, near Gießen. According to their composition, these correspond to mid-ocean ridge basalts (MORB; Box 2: Basalt). The origin of these MOR basalts is controversially discussed. Franke (2000) considers them as evidence for the existence of a separate "Rhenohercynian Ocean," whereas others suggest they represent a branch of the Rheic Ocean. The Gießen-Werra-Harz Nappes represent the uppermost

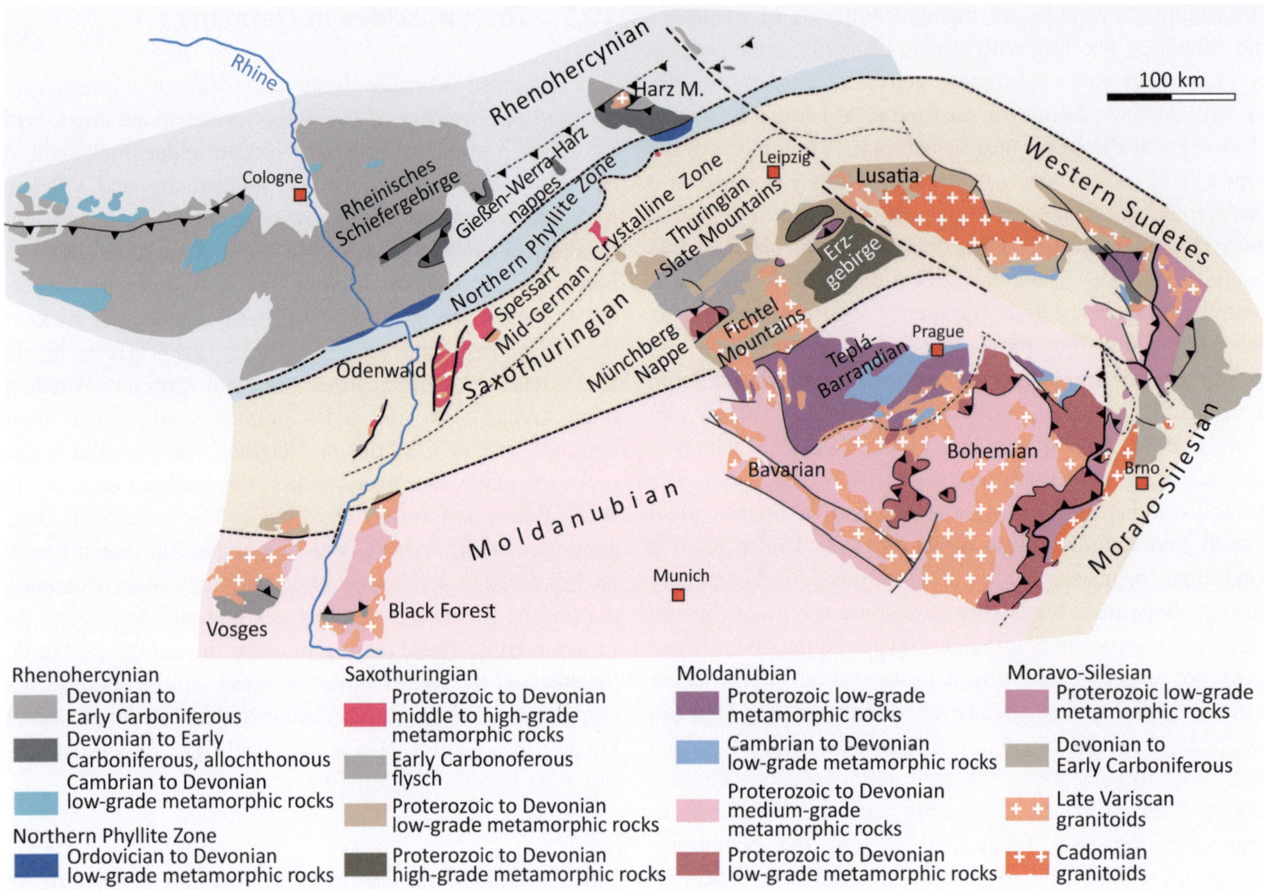

Fig. 10.19 Overview of the Central European Variscides. Modified after Dallmeyer et al. (1995)

tectonic level of the Rhenohercynian that originated at the southern edge of Laurussia (Fig. 10.21). They offer one of the very few opportunities to study basalts of the oceanic crust representing the remaining strands of the Rheic Ocean complex in the European segment.

Rheinisches Schiefergebirge

The Rheinisches Schiefergebirge and its westward continuation into the **Ardennes** stretch from the Kellerwald, near Bad Wildungen, through the Westerwald and Eifel, to Mons in Belgium (Fig. 10.22). It is structurally made up of a large number of syncline and anticline structures. To the southeast lie the Taunus and Hunsrück and to the north it extends into the Sauerland. Although not part of the Rheinisches Schiefergebirge, the Late Carboniferous rocks of the Ruhr area lie along the northern border of the Rhenohercynian. These rocks formed as Variscan molasse deposits in a foreland basin typical for the Subvariscan Foredeep. The Rheinisches Schiefergebirge is mainly composed of Devonian and Carboniferous rocks. At some places, however, older rocks occur uplifted to the surface in the center of anticlinal folds or within thrust nappes (Fig. 10.22). Despite its name, the Rheinisches Schiefergebirge is not only composed of slate (*Schiefer*). In terms of abundance, sandy mudstones (sandy slates), sandstones, graywackes, and quartzites dominate over pure slate lithologies. The term slate is, in fact, a metamorphic term used to describe a penetratively cleaved mudrock that were deposited as clayey sediment and converted to slate during **low-temperature metamorphism**, deformation, and recrystallization. Despite these transformations, sedimentary structures, and fossils are often preserved. A well-known example of excellent fossil conservation is found in the Devonian Bundenbach Slate of the Hunsrück area (Fig. 10.6). This formation was deposited under stillwater conditions in the central area of the Rhenohercynian Basin and belonged to the Hercynian facies (Fig. 10.5).

The oldest rock unit of the Rheinisches Schiefergebirge, known as the Wartenstein Gneiss, is exposed in the southern Hunsrück area and is located in the fault zone of the

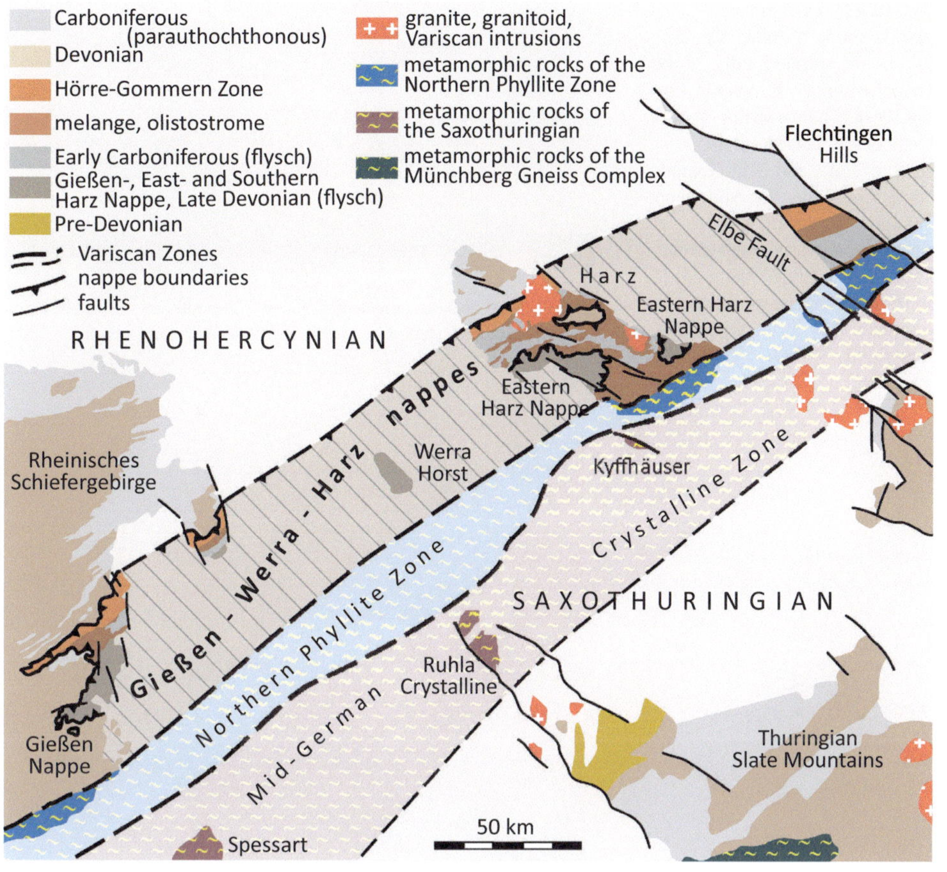

Fig. 10.20 Location of the Gießen-Werra-Harz Nappes at the boundary between the Rhenohercynian and Saxothuringian. Modified after Franke and Żelaźniewicz (2000) and Katzung (2001)

Northern Phyllite Zone (Sect. 10.3.2). This gneiss, with a Late Proterozoic metamorphic age (Molzahn et al. 1998), represents the Cadomian basement rock of Avalonia that was connected to Baltica during the Caledonian Orogeny (Sect. 9.2.2). In the eastern Rheinisches Schiefergebirge, the oldest rocks are of sedimentary origin and are of Ordovician and Silurian age. These occur in the **Ebbe Mountains**, near Plettenberg, where they are located in the core of an anticline containing strata that pass upward into the normal Devonian/Early Carboniferous sequence of the Rheinisches Schiefergebirge. In the **Stavelot–Venn anticline** of the western Eifel, the oldest rocks exposed are from the Cambrian (Late Proterozoic on the Belgian side) and were uplifted to the surface in thrust nappes (Fig. 10.22). This example is one of the few regions in the Rheinisches Schiefergebirge dominated by tectonic nappe tectonics, with fault displacements in the order of several tens of kilometers. The largest structure is the **Faille du Midi/Ardennes Thrust** that can be traced from Belgium to the Aachen area where it continues as the Aachen Thrust (Fig. 10.23). The French geologist Marcel Alexandre Bertrand (1847–1907) was the first to propose that the Stavelot–Venn anticline represents a nappe, i.e., a structure that has been laterally displaced over a large distance. He developed the basic principles of the nappe theory using the example of the **High Venn** in the late nineteenth century (Bertrand 1884). The nappe theory thus had its beginning not in the Alps, as one might suspect, but in the Rhenohercynian.

The Devonian to Carboniferous rock sequence of the Rheinisches Schiefergebirge records the paleogeographic evolution of the Rhenohercynian basin and shelf complex. The older series are located mainly in the southern areas of the Rheinisches Schiefergebirge in the Taunus and Hunsrück, where the sedimentary sequence starts with upper Late Silurian and Early Devonian rocks in the form of colorful shales and red-colored sandstones, as well as the thick deposit of the **Taunus Quartzite** (Fig. 10.5). In the area of the Hunsrück, the southern Eifel, and the northern Taunus region, fine-grained basin sedimentary rocks dominate with shales, siltstones and also sporadic sandstones, as well as occasional layers of rhyodacitic volcanic rock. On the northern edge of the Taunus swell, **coral and stromatopore reefs** developed in the Middle Devonian. The Lahn syncline in the north and its northeastern continuation as the Dill syncline are mainly characterized by rocks of Middle and Late Devonian age.

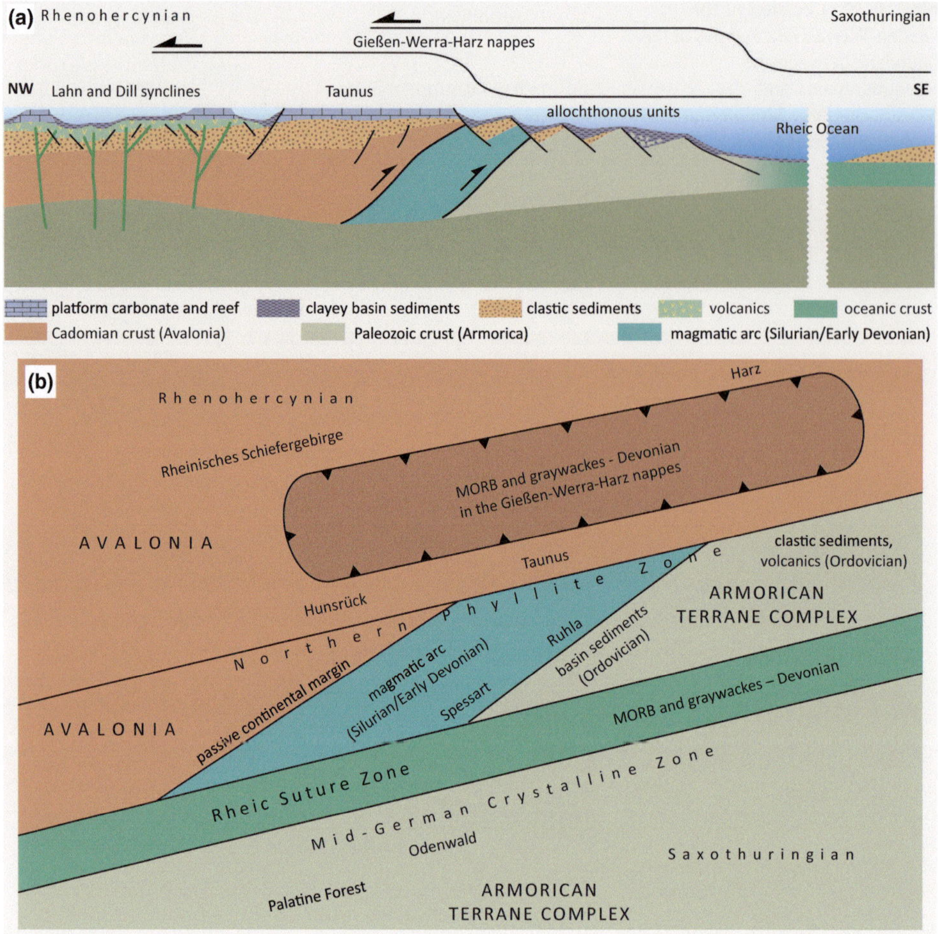

Fig. 10.21 **a** Schematic reconstruction of allochthonous units at the southern edge of the Rhenohercynian; **b** tectonic relationship between the Rhenohercynian, Northern Phyllite Zone, and Mid-German Crystalline Zone. The Siluran and Devonian magmatic arc represents an early stage of the Rheic Suture Zone with a northward-dipping suture zone (compare Fig. 10.1). Modified after Franke (2000)

Due to the extensional tectonics of the Rhenohercynian basin and shelf area, volcanic activity was frequent with volcanoes emerging from the water in association with the development of atoll-like structures around them. The volcanic rocks that formed are mainly composed of basalt ("diabase"), trachytes, and rhyolites (keratophyre) with geochemical compositions indicative of continental volcanism. The green coloration of the volcanics most commonly reflects the presence of minerals of the chlorite group that formed during low-temperature alteration of pyroxenes, feldspars, and olivine crystals in the basalt. Unlike the MOR basalts near Gießen (see above), these basalts did not erupt at a mid-ocean ridge but were extruded under submarine conditions with the development of **pillow lava**, which can be commonly observed in surface outcrops today. Volcaniclastic rocks occur frequently together with the basalt and originated as part of the volcanic cone structure ("Schalstein"; Box 8: Diabase and Schalstein). Soluble ferrous iron, which was transported to the surface in hydrothermal solutions, precipitated as ferric iron minerals in these porous rocks to form iron ore (see above). The mining of **red iron ore** (containing hematite) ended many years ago in the Lahn–Dill area, but examples of ore-bearing strata can be found in surface exposures.

Volcanic rocks of Devonian and Early Carboniferous age occur in the Rheinisches Schiefergebirge in the form of small, often isolated bodies. The largest deposit of mostly submarine basalt, volcanic breccia, and tuff is located in the area of the Lahn–Dill Syncline. The Middle and Late Devonian basalt of the Sauerland is called "*Hauptgrünsteinzug*" (Fig. 10.24) and in older publications, the obsolete term "*Deckdiabas*" was used for basalts of the Early Carboniferous (formed contemporaneous with the basin and swell facies, cf. Fig. 10.14).

Reef-building organisms settled around volcanic edifices in the Lahn–Dill area, leading to the formation of several hundred meters thick reef limestones. Some of the limestones are well known for their display of colors due to

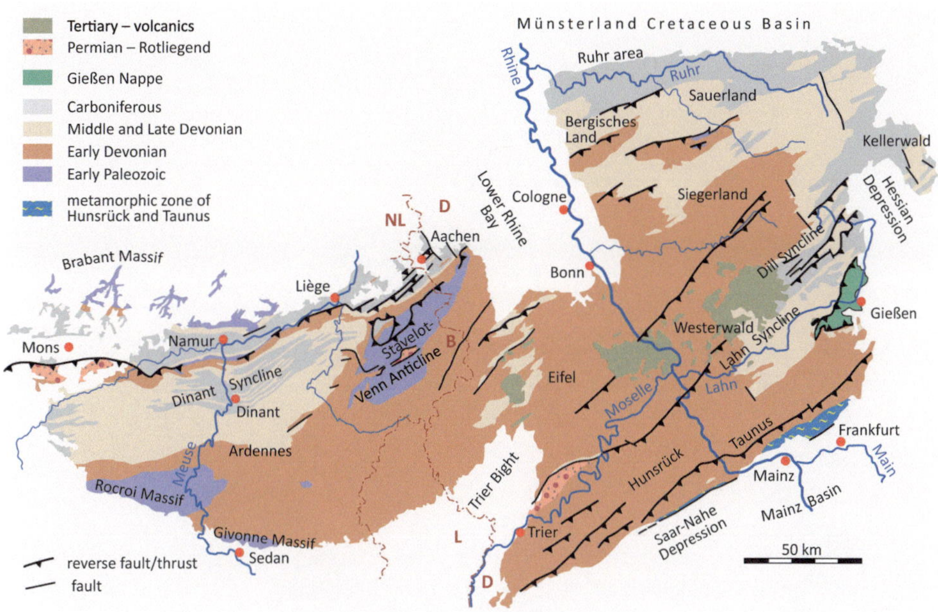

Fig. 10.22 Generalized geological map of the Rheinisches Schiefergebirge showing the distribution of Paleozoic sequences. Modified after Walter (2007) and the 1:200,000 generalized geological map of the BGR (1981)

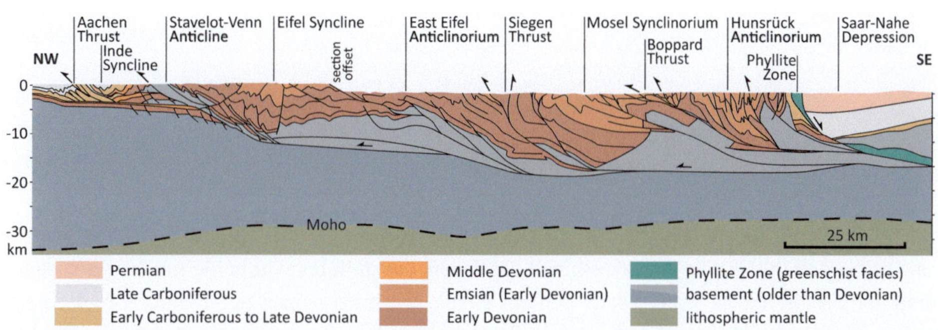

Fig. 10.23 Cross-section of the Rhenohercynian along a line following the left bank of the Rhine River from Bad Kreuznach to Düren (after Oncken et al. 2000)

admixtures of iron and carbon and are marketed under the trade name "Lahn marble" (although clearly not a marble). In the disused **Unica** quarry of Villmar, a unique exposure has been declared a natural monument and "national geotope" (protected site). Here, a Devonian reef can be observed virtually in three dimensions on the smooth, cut walls of the quarry. It displays layers of stromatoporoids mixed with the debris of coral, echinoderm remnants, and other limestone the debris (Fig. 10.25). This rock was mined until 1970 near Limburg and used abroad in many historic buildings such as the Empire State Building in New York and the Kremlin in Moscow. Back in Germany, it is also found in the Baroque church of Amorbach and in a number of cathedrals, notably in Würzburg, Mainz, and Berlin.

The diverse rock types found in the Middle and Late Devonian stratigraphy of the Lahn–Dill area are also seen in the northern parts of the Rheinisches Schiefergebirge. In addition to reef limestones, platy limestones occur that represent deposition in shallow lagoons. Clayey and calcareous sediments were deposited simultaneously in deeper water basinal environments, which were transformed after lithification and low-temperature regional metamorphism into **calcareous spotted slate** (*Kalkknotenschiefer*). The Kalkknoten–Schiefer is widespread across the Rheinisches Schiefergebirge, the Harz Mountains and the Thuringian Slate Mountains (s.l.). Frequently, this rock is referred to as *Kramenzelkalk* or ant limestone, the name been derived from the Low German language of the Sauerland area, where *Kramenzel* means ant. This rock type is a dark gray, often reddish-colored slate, with nodular limestone occurring in specific layers that are sometimes dissolved away from the more resistant surrounding rock by differential weathering. The remaining rock contains centimeter-sized holes resembling an ant's nest in wood.

Devonian reef limestones occur as atoll-like formations above the submerged volcanic edifices (seamounts) described earlier and as **barrier and fringing reefs** developed along the margins of the Rhenohercynian Basin. Large coral

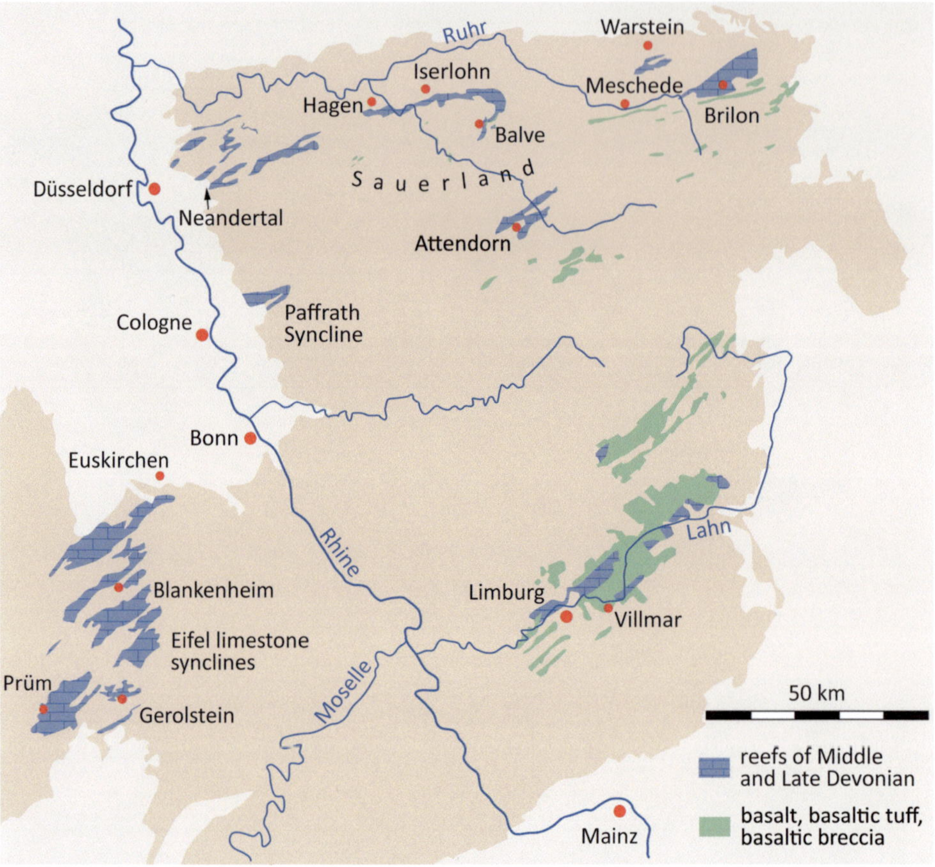

Fig. 10.24 Distribution of Middle and Late Devonian limestone, basalt, basaltic breccia, and tuffaceous rock (diabase and schalstein, Box 8: Diabase and Schalstein) in the Rheinisches Schiefergebirge

reefs also occur in the Eifel limestone preserved in troughs in the area south of Euskirchen, around Blankenheim, as well as at Gerolstein and Prüm. Massive limestone is found across the Sauerland, stretching from the Neander Valley (*Neandertal*), near Düsseldorf, through Wuppertal, Hagen, and Iserlohn to Balve. This belt includes limestones of the Paffrath trough and the prominent limestone deposits of Warstein, Brilon, and Attendorn (Fig. 10.24). The massive Devonian limestone, which exceeds thicknesses of 1000 m, often contain abundant fossils with numerous reef-building organisms responsible for the construction of the Devonian reefs, including brachiopods, clams, or snails. The fossil assemblages also contain numerous ammonites (goniatites) and fish bones.

The main reef builders were corals, stromatoporoids, and sponges. One of the most famous fossil sites worldwide is located in the Eifel limestone trough, known as the **trilobite fields** of Gees, near Gerolstein (Kowalski 1978). At this locality, trilobites of the widespread genus *Phacops* are particularly well preserved, although fossil collecting at this site was prohibited many years ago. At a few locations, such as in the area of Warstein at Suttrop and Hohenlimburg-Oege, weathered cap layers and fissures within the massive limestone beds contain small, centimeter-sized authigenic (Greek *authi*: thereat; Latin *generare*: generate, create) crystals of quartz with characteristic **double-ended crystal** shapes, known as the "Suttrop diamonds" (Fig. 10.26). These probably originated during Late Jurassic hydrothermal fluid circulation associated with metasomatic dolomitization (Goette 2004).

The main part of the sedimentary rock sequences of the Rheinisches Schiefergebirge is composed of **shales** and **graywackes**. The Early Devonian rocks consist mainly of shales, sandstones, and quartzites and altogether reach thicknesses up to 3000 m, whereas the Middle Devonian is dominated by dark gray shales. Homogeneous, fine-grained Devonian slate, which is still mined for roofing tiles, occurs in the area around Mayen (the *Moselschiefer* of Early and Middle Devonian), Disused quarries also occur in Bundenbach and Gmunden (Early Devonian, Emsian slate), as well as around Wissenbach in the Dill trough and in eastern Sauerland around the area of Meschede/Brilon (Early and Middle Devonian slate). In the Late Devonian and Early Carboniferous, the depositional environments were differentiated into distinct areas of varying sedimentary facies. In

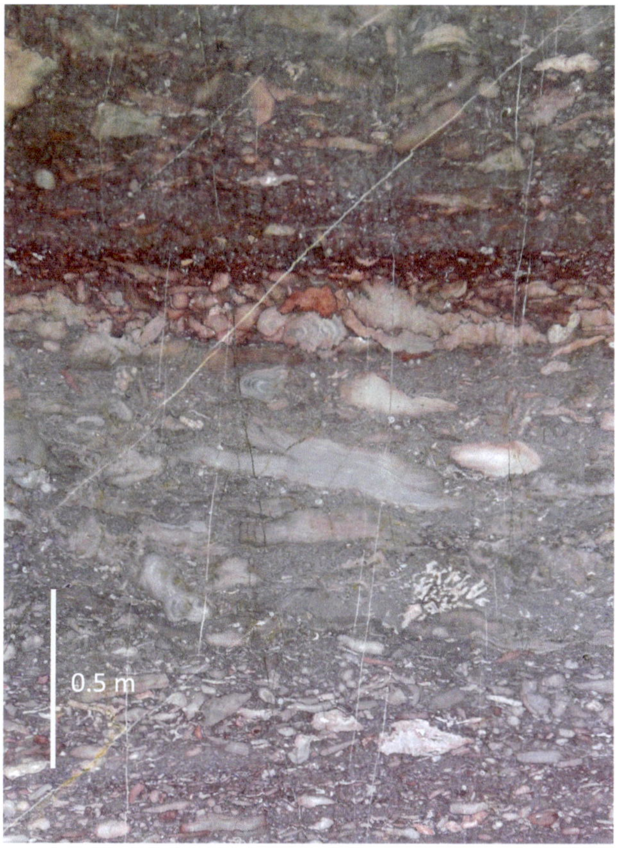

Fig. 10.25 Cut section of a rock wall from the former Unica quarry of Villmar. The wall of the quarry is a protected site and national geotope where the structure of a limestone reef of Givetian age (transition from Middle to Late Devonian) is well preserved. It shows layered stromatoporoids, corals, echinoderm remnants, carbonate debris, and sedimentary-filled fissures (not seen in the photo). The dark, very fine, and jagged lines on the block represent stylolite surfaces formed by pressure solution during compression of the rock, with the accumulation of insoluble components appearing as dark seams [N 50°23′39″/E 08°11′03″]

Fig. 10.26 The "Suttrop diamonds." These are authigenically formed double-ended crystals of quartz that crystallized in the cap layers of the massive Devonian limestone. Located at Suttrop near Warstein, in the Sauerland area [N 51°26′57″/E 08°22′50″]

addition to the reef limestones that developed on swells and in shallow water areas, deeper basinal facies accumulated clayey sediment with sands washed in at the peripheral areas of the basin. The clastic influx in the south of the Rheinisches Schiefergebirge, sometimes referred to as the Culm or flysch facies, is mainly represented by graywacke deposits, whereas in the northern regions it is dominated by shales and radiolarian-bearing cherts. Only in the northwestern part of the Rheinisches Schiefergebirge do Early Carboniferous limestones occur near Aachen, although these deposits are well developed across the westerly areas of the Rhenohercynian, as found in the Ardennes of Belgium, southern England, South Wales, and Ireland.

The rocks of the Rheinisches Schiefergebirge are only weakly metamorphosed (generally < 320 °C). The metamorphic grade generally decreases from south to north, with maximum temperatures reached at burial depths greater than 5 km. The very low-grade metamorphic rocks are now exposed at the surface due to the uplift of the Rheinisches Schiefergebirge during the Variscan Orogeny and erosion of several kilometers of overburden. The material eroded during the uplift of the mountains was shed off to the north and deposited as clastic sediment in the Variscan Foredeep basin. These deposits typically formed diagenetic graywacke sandstones or coarse molasse of Late Carboniferous to Rotliegend age (Sect. 11.2.1). In addition to the decrease in **metamorphic grade** toward the north, the radiogenic isotopic **age of regional metamorphism**, determined by the K–Ar method, also decreases northward. In the south, older ages of about 340–325 Ma occur, whereas in the north the ages are distinctly younger at 305–280 Ma (Ahrendt et al. 1983). This is considered to reflect the progressive migration of the active Variscan fold-and-thrust belt toward the external foreland. During the Late Carboniferous collision of the passive continental margin of Laurussia, the area of the Rheinisches Schiefergebirge was shortened by about 50% compared to its original length (Fig. 10.27; Oncken et al. 2000). This deformation occurred mostly by **northwest-vergent folding** (Fig. 10.28), whereby the vergence reflects the tilt direction of an inclined fold and indicates the main direction of structural shear and tectonic transport, which in this case was toward the northwest.

Contemporaneous to the shortening of the Rhenohercynian, a series of narrow turbidite basins developed from the early Late Carboniferous onward. They opened in front of the northward advancing deformation front of the Variscan fold-and-thrust belt and were shifted continuously

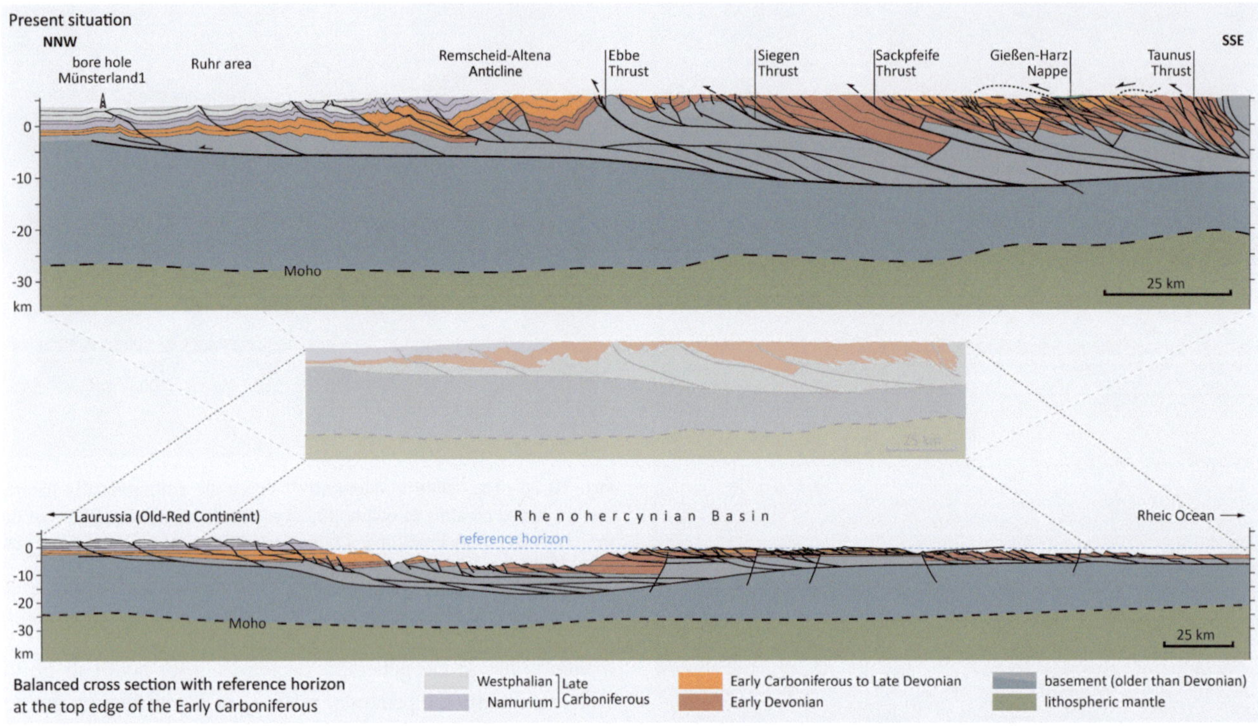

Fig. 10.27 Balanced (unfolded and unfaulted) cross-section across the Rhenohercynian following the line of the right bank of the Rhine River from Frankfurt (SSE) to Münster in Westphalia (NW). The schematic profile in the center corresponds to the upper profile. The diagram illustrates the 50% shortening of the Rhenohercynian that occurred during Variscan deformation (after Oncken et al. 2000)

Fig. 10.28 Steeply dipping, northwest-vergent fold developed in Early Devonian (Pragian) shale, siltstone, and sandstone lithologies at Teufelsley in the Kall Valley, lying south of Vossenack (northern Eifel). The northwest direction is toward the right side. The shear strain that caused the folding of the beds during the Variscan Orogeny was directed toward the north [N 50° 40′28″/E 06°22′32″]

northwestward, along with the migration of the basin depocenters within the Subvariscan Foredeep. The migrating basins were continuously filled by the Late Carboniferous synorogenic molasse sediment. Eventually, the pulse of deformation also migrated through the Subvariscan Foredeep, with folding and thrusting of the foreland basin

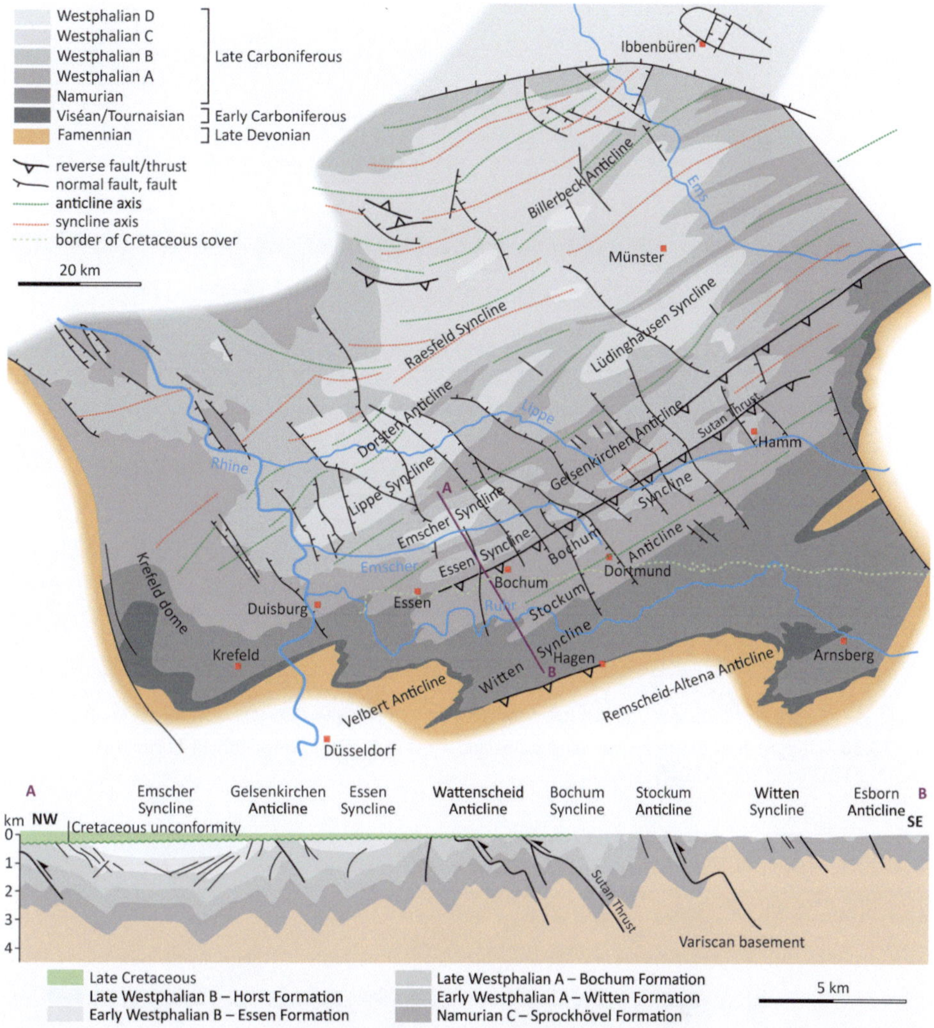

Fig. 10.29 Subsurface geological map of the Rhenish–Westphalian coal-mining area and adjacent regions, together with a geological cross-section through the southern part. Modified after Walter (2007), Drozdzewski and Wrede (1994)

sequences during the late phase of the Variscan Orogeny, although the intensity of rock deformation decreased northward. This deformation is evident in the many folds (anticlines and synclines) and thrusts developed in the Late Carboniferous coalfield of the Ruhr area (Figs. 10.29 and 10.30). One of the most prominent thrusts of the Ruhr area is the Sutan Thrust (also simply called the Sutan). This structure can be traced over 100 km and is characterized by thrust displacements of up to 1.3 km (Figs. 10.27 and 10.31; Geological Survey of North Rhine-Westfalia 2003).

In many places in the Rheinisches Schiefergebirge, aqueous hydrothermal fluids ascended into the deforming Variscan orogenic belt and generated numerous veins with enhanced concentrations of metals to form ore deposits. Characteristic are the hydrothermal **Pb–Zn mineralizations**, which may also contain economic concentrations of silver (Ag) and copper (Cu; Box 11: Ore deposits). Today, the ore deposits in the Rheinisches Schiefergebirge have been largely exhausted, although a number of the remaining mines are used for tourism. Some ore-bearing lodes postdated the Variscan Orogeny and formed as syntectonic extensional veins during lithospheric stretching associated with the opening of the North Atlantic. Such fissures may have also formed during the Tertiary rifting of the Upper Rhine Graben.

Fig. 10.30 Upright fold developed in sandstones and siltstones of the Sprockhövel Formation (Namurian age) at Pastoratsberg in Essen, Ruhr area [N 51°23′05″/E 07°00′24″]

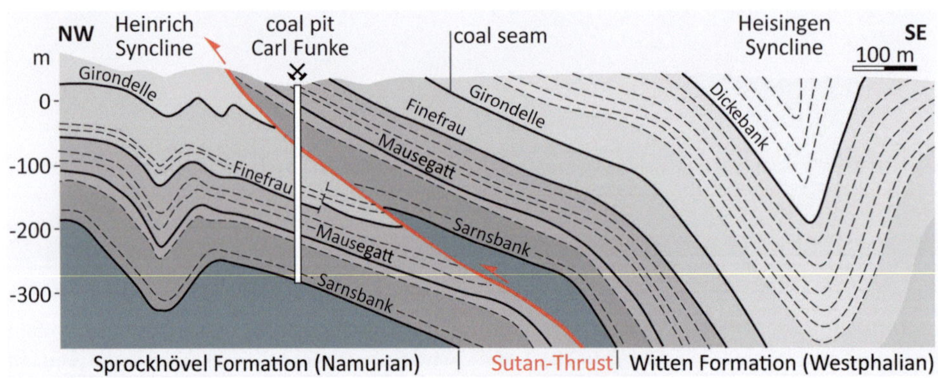

Fig. 10.31 Geological cross-section of the Sutan Thrust in the area of the old Carl Funke coal pit, shaft I (closed down in 1973). Modified from the information board of the Geopark Ruhrgebiet located at the exposure of the Sutan Thrust; Geological Survey of North Rhine-Westphalia (2003) [N 51°24′18″/E 07°02′49″]

Box 11: Ore deposits

Metallic ore deposits constituent an important georesource of considerable economic value and provide the diverse spectrum of metals that are used in almost all areas of our daily life. In Germany, only a few mines operate today, but since the Middle Ages and into modern times, ore mining has provided much wealth to monarchies and their dynasties. These materials formed the basis of the industrial revolution and today continue to support our modern technological development. Raw ore materials have been exploited mainly in the areas of the Variscan Mountains, in the mining areas of the Erzgebirge, Harz Mountains, Rheinisches Schiefergebirge, and the Black Forest. The exploration of these ore deposits began a long time ago and eventually led to the development of the mineralogical and eventually the geological sciences. One of the earliest and most comprehensive works on ore science published in the sixteenth century was written by the Saxon scholar Georgius Agricola (1494–1555), who is considered as the "father" of mineralogy: *De re metallica libri XII* (1556).

It is still difficult to classify the various types of ores in a simple and universal way. Ore deposits are most commonly grouped genetically by the process of formation, but this approach runs into problems when it is not possible to determine unequivocally the cause of mineralization or when more than one process was involved in its origin. In contrast, regionally based classifications have the disadvantage of not considering the formation process. There are also structural classifications that attempt to place the type of mineralization into the rocks deformation history and other subdivision schemes that group ores in relation to the type of metals present. A geologically useful approach is to classify ore deposits by plate tectonic setting (Fig. 10.32). However, the boundaries between the fields of formation are not always well defined and not all ore formations are included in such schemes.

Magmatic ore deposits form by the process of mineral segregation in the melt phase by which heavy minerals, such as chromite or phases bearing the platinum group elements, settle to form layered cumulates at the bottom of the magma chamber. Such deposits occur in large, mostly ultramafic or gabbro–noritic intrusive plutonic bodies, as well as in the deep parts of the oceanic crust preserved as ophiolites.

When a magmatic melt is subject to differentiation, there is a continuous change in composition (Box 1: Granites). Toward the end of the crystallization process, the elements that remain in the residual melt phase and enriched are the ones that are difficult to incorporate into the common rock-forming minerals. In rift settings, magmatic alkaline intrusive rocks are often found as differentiation products of upper mantle melt. Here (in order of increasing atomic number), trace elements are often enriched, such as zirconium (Zr), niobium (Nb), tin (Sn), tantalum (Ta), and U, as well as rare earth elements such as lanthanum (La), cerium (Ce), neodymium (Nd), and samarium (Sm). An example of this type of ore deposit is carbonatite, which is magmatically derived from carbonate rock with >50% carbonate minerals. These rocks appear at the surface usually in the area of rift graben systems, such as the German carbonatites of the Kaiserstuhl, which originated in the Upper Rhine Graben as a large Tertiary volcano typical of rift magmatism.

The passive enrichment of elements in the residual melt of a granitic pluton is also characteristic of pegmatitic and pneumatolytic processes. **Pegmatites** (Greek *pegma*: frame, binding) are intrusive igneous rocks that penetrate the country rock as dikes during the late stage of magmatic activity in a pluton. They sometimes contain several meters large, giant crystals bearing trace elements such as lithium (Li), beryllium (Be), boron (B), niobium (Nb), tantalum (Ta), and U, as well as a range of rare earth elements. **Pneumatolysis** (Greek *pneuma*: wind, air; Greek *lýsis*: dissolve) occurs in the immediate vicinity and at the top of the pluton, where supercritical and still very hot gaseous phases precipitate minerals with high contents of metals and rare earth elements; the latter of which have become extremely important for harnessing renewable energy sources. The host rock structure is also chemically altered by the highly reactive fluids and enriched with metallic substances. In this way, the roof of a granitic pluton may be altered to form **greisen** deposits that are often enriched in tin and tungsten. Greisen is a Saxon miners' expression for pneumatolytically altered granite. It consists mainly of quartz and mica, as well as less common minerals such as topaz or tourmaline partly formed by the alteration of feldspars. One of the well-known pneumatolytic deposits in Germany is the porphyry granite-hosted tin [Sn] ore of Altenberg in the Erzgebirge.

In the roof area of a pluton, **impregnation deposits** may also form when ore solutions percolate throughout the rock pores and grain boundaries. These precipitate fine-grained disseminated ore metals. One example is the disseminated Cu ore (Cu porphyry) that forms at active continental margins (Frisch and Meschede 2013). As anhydrous minerals generally crystallize first from the magma, the residual melt of a gradually cooling pluton becomes enriched in dissolved water and a range of soluble metals. If, however, these fluids suddenly escape, for instance, along faults into the country rock, the loss of water leads to rapid solidification of the residual magma and precipitation of the dispersed and enriched metals (especially Cu, molybdenum [Mo], Sn, and gold [Au]) within the solidified porphyritic rock.

Hydrothermal solutions play a particularly important role in forming nonferrous ore deposits. This type of aqueous fluid is typically enriched in anions and elevated in temperature compared to the background regional geothermal gradient. They generally carry a broad range of dissolved ionic species by which metals can be transported over long distances. Metals are often leached from the surrounding rock as meteoric water circulates down from the surface along fractures and as it heats up, starts to dissolve the dispersed metals from the rock and concentrates them in solution. It is the mixing of the meteoric groundwater with juvenile magmatic fluids that plays an important role in ore formation, where metals precipitate when the two fluids interact. As saturated hydrothermal fluids circulate into the surrounding rock along fractures, veins or open fissures, cooling leads to crystallization and formation of diverse types of ore and **gangue mineralizations**. If metal-bearing solutions penetrate into the country rock and modify or replace it, these are known as hydrothermal-metasomatic ores. **Metasomatism** is a type of chemical alteration that involves mass transfer and intense wall rock alteration. Although typical of hydrothermal systems of magmatic or metamorphic origin, this process may also occur during the circulation of diagenetic fluids, such as saline brine solutions (Pohl 2005).

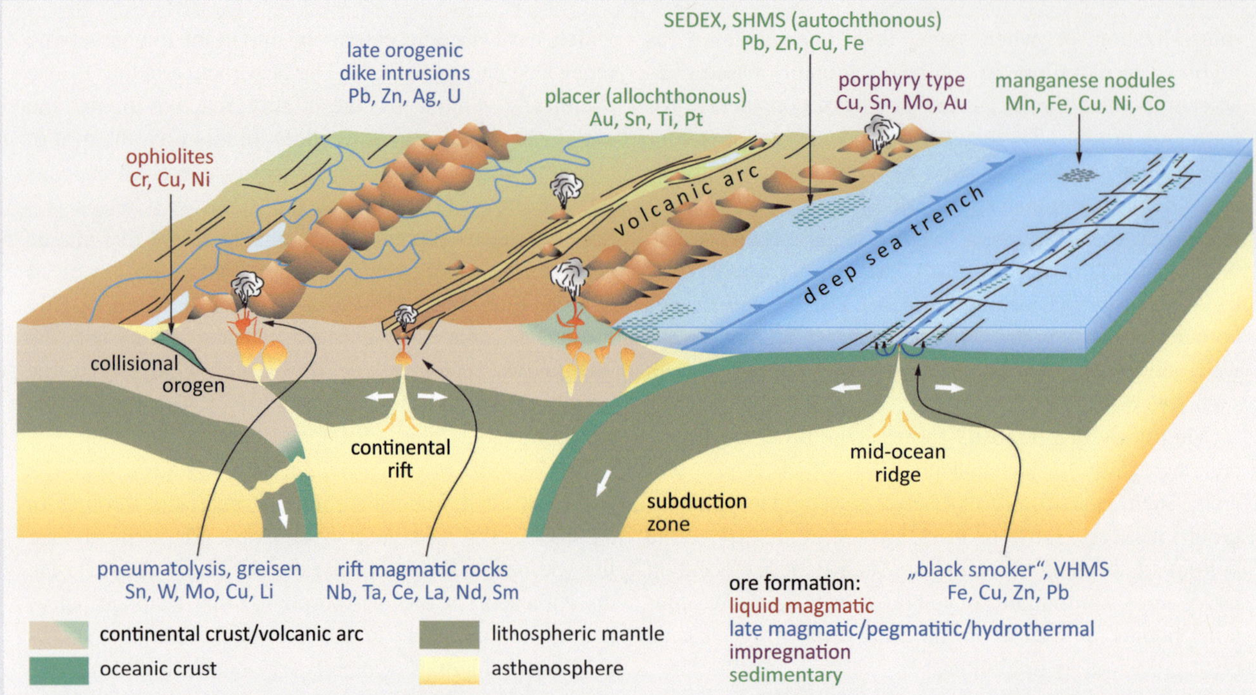

Fig. 10.32 Schematic overview of the occurrence and formation of ore deposits placed in their plate tectonic context (not to scale)

There are a number of hydrothermally derived ore deposits that form in sediments on the seafloor and are stratiform in nature. When the ore metals are associated with fluid flow and burial in continental sedimentary basins, they are known as **SEDEX** deposits (**sedimentary exhalative**, where exhale is derived from Latin: *exhalare*. This type of deposit forms when widely transported hydrothermal solutions pour out on the seafloor and mix with surface water. During mixing and cooling of the water, metal compounds are precipitated and concentrated in the sediment around the vent. Another major type of hydrothermally formed stratiform deposit are the **VHMS** deposits (**volcanic-hosted massive sulfides**; Fig. 10.32). These form thick sulphidic ore deposits integrated into the sedimentary layer sequence. Such deposits are typically formed from hydrothermal fluids that pass through the oceanic crust and emerge as **black** or **white smokers** at mid-ocean ridges or submarine volcanoes. Here, the hot (up to 350 °C) acidic hydrothermal fluid mixes with the saline water and as they cool, precipitate the metals (mainly Cu, Fe, and Zn [zinc]) picked up during the circulation of seawater through the oceanic crust.

Diverse examples of hydrothermal stratiform deposits occur in Germany. The iron ore deposits of the Lahn–Dill type were formed at submarine volcanoes by the extensive circulation of iron-charged hydrothermal fluids. These ore deposits have been mined for a long time in the Lahn–Dill area of the Rheinisches Schiefergebirge as well as in the Elbingerode Complex of the Harz Mountains. The **sediment-hosted massive sulfide (SHMS)** type ore field of the Rammelsberg, located in the Harz Mountains, is also considered to represent a SEDEX deposit (Fig. 10.32). The sulfidic, sedimentary basin derived, metallic ore was bound to surface sediments that originated in an oxygen-free zone at the bottom of a deepwater basin, where hydrothermal vents were located.

Sedimentary ore deposits occur in a variety of ways by weathering or near-surface metal precipitation. There is a diverse range of processes that occur in both continental and marine environments. Placers develop by the physical accumulation of specifically heavy metal-containing minerals by gravity separation in fluvial (river) or shoreline (beach) settings. Mineable ore-bearing soils, such as the Al-rich bauxite deposits, develop by intense chemical weathering of rocks under tropical conditions. In this environment, the metals are passively concentrated by leaching away the more soluble rock components. In the geological record, an abundance of sedimentary ore has formed by low-temperature precipitation of metal ions dissolved in water, such as in the marine precipitation of oolites, banded iron ores, manganese nodules, and many other examples of surface sediment metallic precipitations.

Harz Mountains

The Harz Mountains are located to the northeast of the Rheinisches Schiefergebirge and are included as part of the Rhenohercynian Zone. Despite a similar geology, there are some striking differences in tectonic structure. The main part of the Harz is today considered to belong to the complex of the **Gießen-Werra-Harz Nappes**. These rocks were structurally thrust onto lower tectonic units of the Rhenohercynian from the southeast (Fig. 10.20). According to the nappe concept (Engel et al. 1983), the **parautochthonous units**, similar to those of the Rheinisches Schiefergebirge, are found in Western Harz. This includes the Söse syncline, the Clausthal fold zone, the Upper Harz diabase zone (*Oberharzer Diabaszug*), and the Upper Harz Devonian Anticline (*Oberharzer Devonsattel*; Fig. 10.33). Geologically, this part of the Harz is subdivided into the Upper (*Oberharz*), Middle (*Mittelharz*), and Lower Harz (*Unterharz*). This parautochthonous tectonic unit can be correlated with similar lithological sequences in the Dill trough of the Rheinisches Schiefergebirge. The **Eastern Harz Nappe**, with the **Selke Syncline** and the **Southern Harz Syncline**, constitutes the uppermost tectonic unit that was thrust onto the Harz Nappe (Fig. 10.34). This uppermost unit is related to the **Gießen Nappe** on the eastern edge of the Rheinisches Schiefergebirge. The basaltic rocks found at the base of this nappe have been suggested to have formed at a mid-ocean ridge system (Franke 2000), whereas others consider these to represent intraplate basalts extruded from the strongly thinned continental crust (Gansslose 1996; Franzke 2012). In addition to the Middle Devonian shales and sandstones, siliceous shales and flysch lithologies occur representing deposition of clastic sediment derived from the Mid-German Crystalline Zone that lay to the south. The **Hörre zone** of the Rheinisches Schiefergebirge is located to the north of the Gießen Nappe and is geologically assigned to the complex of the Gießen-Werra-Harz Nappes. It continues in the northern Kellerwald and the **Acker–Bruchberg heights** as part of the Harz Nappe (Fig. 10.20; Walliser and Alberti 1983) and can be traced further to Gommern (as the Gommern quartzite), located on the Flechtingen heights.

In its present form, the Harz represents an uplifted segment of the Variscan orogenic fold-and-thrust belt that originated as a tilted half-horst block along the **Northern Harz Boundary Fault**: a fault with a total displacement of about 6000–7000 m (Franzke et al. 2004). The southern edge acted like a hinge and did not undergo any major uplift. Here, the Paleozoic sequence is unconformably overlain by Permian beds. The shape of the Harz Mountains was formed by erosion related to rivers that have cut into the uplifting block with the development of **antecedent river valleys**, i.e., deeply incised valleys where the rivers maintained their course despite variations in the underlying bedrock. The uplift and exhumation of the Variscan rocks took place in distinct stages, starting in the Late Cretaceous and again uplifted in the Tertiary. As a result, the uplift of the Harz Mountains has little to do with the Variscan Orogeny but is a consequence of Mesozoic and Cenozoic tectonics. The tectonic strike of the Northern Harz Boundary Fault, which runs in WNW-ESE direction (Fig. 10.33), therefore differs significantly from the NE-SW trending fold axial trend of the Variscan fold structures that formed perpendicularly in the compressional direction of Late Paleozoic deformation. The Northern Harz Boundary Fault, on the other hand, reflects the Cretaceous compressional direction that led to numerous parallel-striking reverse and thrust faults across Central Europe (Sect. 11.7.2).

The oldest rock formation of the Harz Mountains is the Ecker Gneiss exposed at the northwestern edge of the Bocken pluton (Fig. 10.33; Sect. 9.2.1). Detrital zircons within the gneiss yield Silurian and Devonian ages indicating that the protolith of this metamorphic rock must be younger than Early Devonian. Remnants of older rocks are also found in rock masses that are collectively interpreted as **olisthostrome**, such as those of the **Harzgerode** and **Blankenburg Fold Zone** (Hüneke and Ruchholz 2018). These sedimentary deposits are widespread in the Middle and Lower Harz and interpreted as syntectonic mass flows containing variably sized clasts of mixed origin (olistholites) and ages ranging from Silurian to the Early Carboniferous. The olisthostrome occurs together with graywacke in a belt (including the Blankenburg Fold Zone, Tann Zone, and Harzgerode Fold Zone; Fig. 10.33) that has a characteristic sigmoidal form (S-shaped) when viewed on the geological map. Ruchholz (1988, 1989) and Hüneke and Ruchholz (2018) suggested this structure resulted during displacement along a right-lateral strike-slip fault, which was active during the extension and subsidence of the Rhenohercynian Basin and reactivated later during the Variscan collision. The synsedimentary fault activity led to the formation of steep submarine escarpments along the **Central Harz Fault Zone**, with mass flow transport of older rock fragments and blocks moving downslope to areas of lower relief (Fig. 10.35).

The Upper Harz rock sequence begins with clastic beds of Early Devonian age (the Kahleberg sandstone), which are overlain at the top by the Wissenbach Slate that represent the classical type of Hercynian basinal facies. From the Middle Devonian onward, an increasing amount of volcanic material formed in areas such as the Upper Harz diabase zone and in the Blankenburg fold zone (Fig. 10.33). At the same time, thin **Hercynian limestones** (*Herzynkalke*) developed, which typically pass upward into **cephalopod limestones**. In the Lower Harz, the sequence begins with Silurian **graptolite schists** and passes stratigraphically

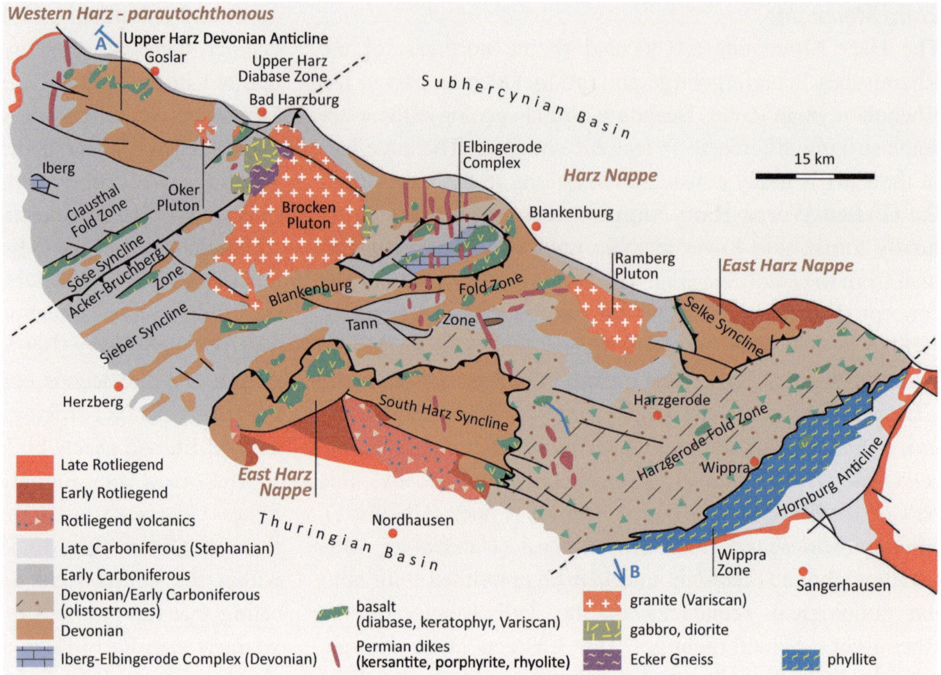

Fig. 10.33 Generalized geological map of the Harz area. Blue capitals (A, B) mark the location of the cross-section shown in Fig. 10.34. Simplified after the 1:200,000 generalized geological map of the BGR (1981). The nappe concept is after Engel et al. (1983) and Walliser and Alberti (1983)

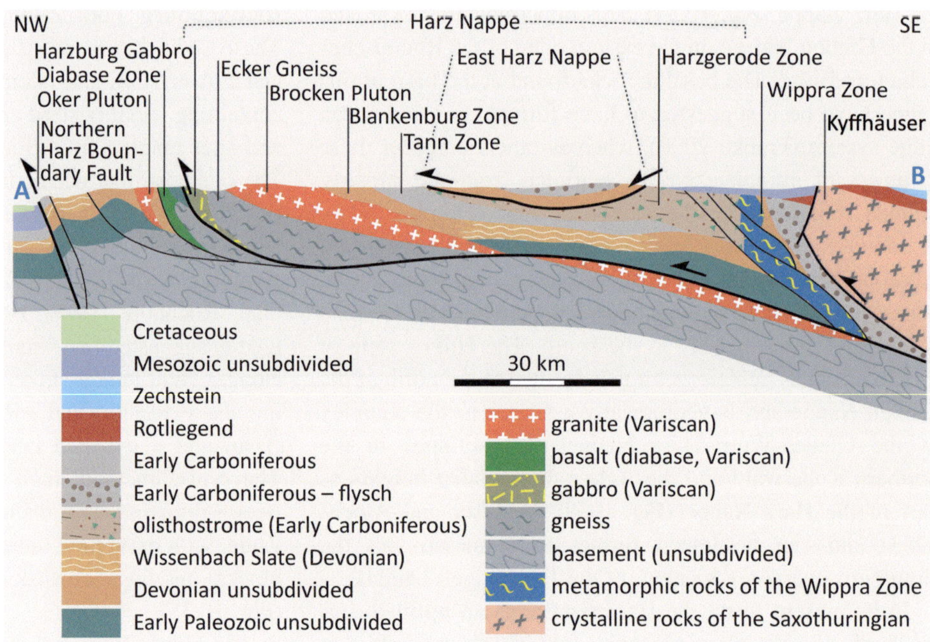

Fig. 10.34 NW-SE oriented geological cross-section of the northwestern Harz area. Modified after Franzke (2012)

upward into facially variegated Devonian rocks from the Lochkovian onward.

The Harz volcanism was often accompanied by the synsedimentary precipitation of red iron layers formed within or on top of the unconsolidated volcanic rock or porous sediment (Fig. 10.9). The iron that precipitated in the porous rock was derived from the volcanic rock and mobilized by hydrothermal fluids. A classic example of this type of iron ore deposit was developed at the northern edge of the Harz at **Rammelsberg**, near Goslar (Fig. 10.36). This ore deposit, where mining was already practiced in the third century, was listed as a UNESCO World Heritage in 1992. It is the world's largest known **synsedimentary-exhalative ore** (known as **SEDEX** deposit; Box 11: Ore deposits), although the sedimentary origin of the ore was controversially discussed up until the 1980s. The ore deposit originated in the Early Middle Devonian (Eifelian) along faults developed in the extensional area of the Rhenohercynian

Basin and was accompanied by volcanic activity. Ore-bearing hydrothermal fluid solutions emerged at the surface along the fault line at this location and upon cooling precipitated their metallic content into the surface sediment. The Rammelsberg ore is mainly composed of metallic sulfides (pyrite, galena, sphalerite, and chalcopyrite) and has an extremely high metal content reaching grades of 30%. Despite this, the mining operations in the Rammelsberg were abandoned in 1988.

In addition to thin limestone beds formed in swell areas, thick limestone reef facies also developed together with volcanic rocks in the Harz region and show similar reef-volcanic associations as in the Lahn–Dill area. The isolated reefs of Bad Grund, in the western part of the Harz, and the **Elbingerode reef complex** of the Middle Harz, which are separated by a longitudinal distance of ca. 30 km, presumably developed contemporaneously as atoll-like structures on the top of submarine volcanoes that protruded upward from the bottom of the Rhenohercynian Basin. However, such an origin has not yet been demonstrated for the **Iberg–Winterberg reef**, as this location has yet to be drilled through to the basement. In contrast, the underlying volcanics of the Elbingerode reef complex are well known and because of their ore content have been mined in a number of places (Fig. 10.37). The reefs that evolved in the Middle and Late Devonian are suggested to have been subsequently offset by large strike-slip faults (Ruchholz 1983, 1988, 1989). Hence, the two reef complexes of the Harz were probably displaced apart from each other during such fault movement. The Elbingerode complex belongs to the parautochthonous unit of the Harz, which is exposed today in a tectonic window within the Harz Nappe (Fig. 10.33). Its characteristic facies and stratigraphic development began in the Middle Devonian with volcanic activity, followed by the formation of the reef complex from early Late Devonian until Early Carboniferous (Weller 1991, 2008; Fig. 10.37).

Sedimentation continued in the Carboniferous with the formation of alum shales, siliceous shales, and slates of basinal facies intercalated with basaltic rock. However, sedimentation ended in the Harz during the Early Carboniferous at the start of the Variscan Orogeny, which led to widespread unconformities in the Carboniferous sequence. As a result, young Variscan sediments are not found in the area of the Harz, but Variscan molasse of Stephanian age and Rotliegend sedimentary rocks, such as conglomerates, sandstones, arkoses, and shales are located in the Eastern Harz and are known as the **Transitional Stockwerk** (*Übergangsstockwerk* or transitional crustal zone; Kästner et al. 1995).

In contrast to the Rheinisches Schiefergebirge, plutons intruded the Harz area, some of which are today exposed at the surface. In addition to the largest, the **Brocken Pluton**, which connects the Upper and Middle Harz, there are two other plutons known in the region: the **Ramberg Pluton** in the Lower Harz and the smaller **Oker Pluton** in the Upper Harz. The latter lies just a few kilometers to the northwest of the Brocken Pluton (Fig. 10.33). Moreover, in the vicinity of the Brocken Pluton, there is a **gabbro–norite intrusion** near Harzburg. This igneous rock type is unique to the Variscan Orogeny and is generally considered to have intruded at the same time as the other plutons, 293–295 Ma ago (Baumann

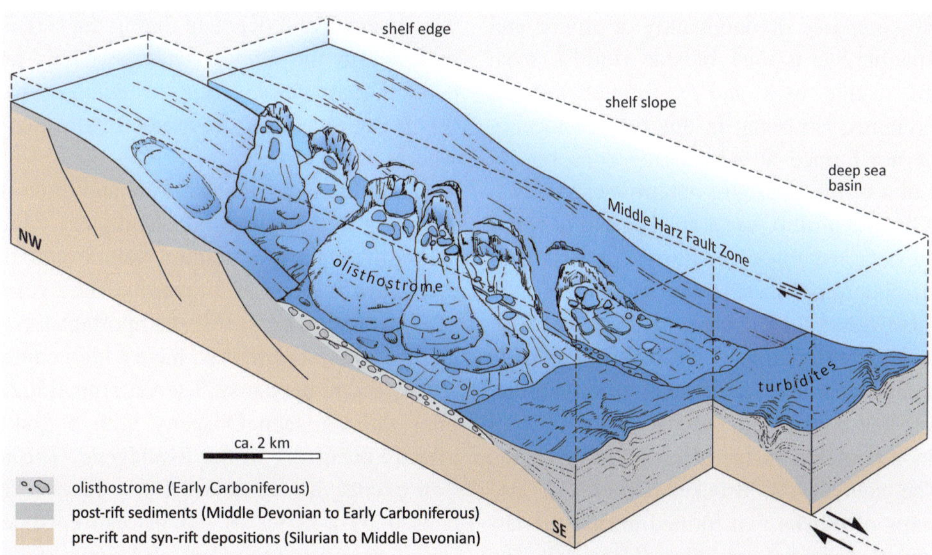

Fig. 10.35 Formation of olistostromes along the Central Harz Fault Zone: A transform fault that was initiated during Early Devonian rifting in the area of the extensional Rhenohercynian Basin and reactivated as a right-lateral strike-slip fault during the Early Carboniferous. Horizontal displacement resulted in a steep relief in the area of the continental margin, leading to the formation of debris and mudflows. The drawing was provided by Heiko Hüneke (Greifswald) and is based on the work of Hüneke and Ruchholz (2004)

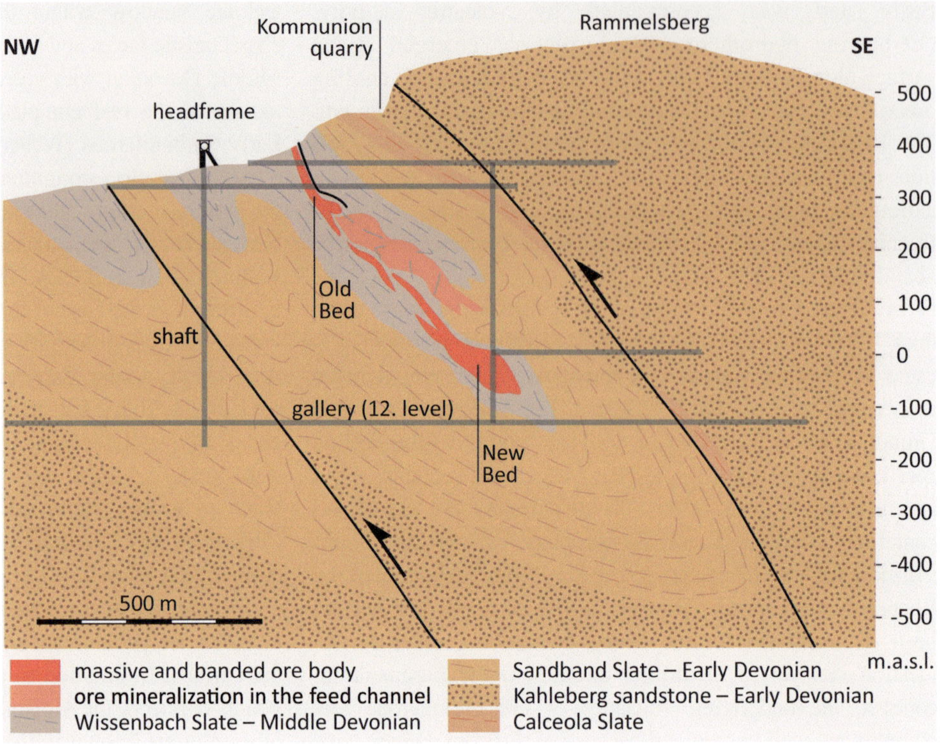

Fig. 10.36 Cross-section of the ore deposit at the Rammelsberg, near Goslar. Modified after Walther (1986)

et al. 1991). More recent studies (U/Pb age datings of zircons from the Brocken and Ramberg plutons (Ilgner et al. 2009) indicate a slightly younger age of 283 Ma, which raises the question as to whether the plutons were intruded during a late phase of the Variscan Orogeny or during extensional movements that began in the Early Rotliegend. The gabbro–norite intrusion contains the type locality for the rock known as **harzburgite**. This ultramafic igneous rock is a type of peridotite composed predominantly of olivine and orthopyroxene. Harzburgite is one of the world's most common types of mantle rock and is formed beneath mid-ocean ridge systems. However, in this type of locality occurrence, it was not formed at a spreading zone but by magmatic segregation of the Harzburg gabbro–norite.

The **Brocken Granite pluton** was emplaced at a depth of about 3–5 km and occurred together with the intrusion of dikes into the neighboring rocks. Prior to cooling, the intrusive bodies also resulted in contact metamorphism of the surrounding rock and formation of an ultra-hard **hornfels** metamorphic rock that is particularly resistant to weathering. This resistant rock that formed around the pluton is well exposed today on top of the Achtermannshöhe or at the Hohneklippen. The area of the Brocken Granite is also known for the **heaps of blocks** that formed during the last glaciation phase under periglacial conditions (Chap. 16). The surface granite was split largely by frost weathering along preexisting fissures to produce distinct block piles, instead of forming slope scree composed of sand or gravel. The size of the blocks and the high rate of water runoff also prevented the accumulation of soil matter.

The Harz region has been mined for metallic ore since the Bronze Age, about 3000 years ago. Following an initial period of mining activity in the Middle Ages (twelfth/thirteenth century), the Water Regale (*Wasserregal*) was invented and operated during the sixteenth to nineteenth centuries in the Upper Harz area. This preindustrial engineering masterstroke allowed mining down to depths of 600 m by regulating the water supply that drove the water wheels in the mines.

In addition to the Middle and Late Devonian Rammelsberg and Elbingerode Complex, which originated as synsedimentary stratabound sulfide and iron ore deposits (see above), there are numerous Harz vein mineralizations of economic interest. Most important are the **Pb–Zn mineralizations**, which often have a high content of Ag and Cu (Box 11: Ore deposits). The vein mineralizations developed after the Variscan Orogeny with a first fissure opening already occurring in the Rotliegend. The main mineralization events, however, are Early Jurassic in age and were caused by a series of extensional episodes related to the initial opening of the Central Atlantic (Sect. 11.6; Stedingk 2012).

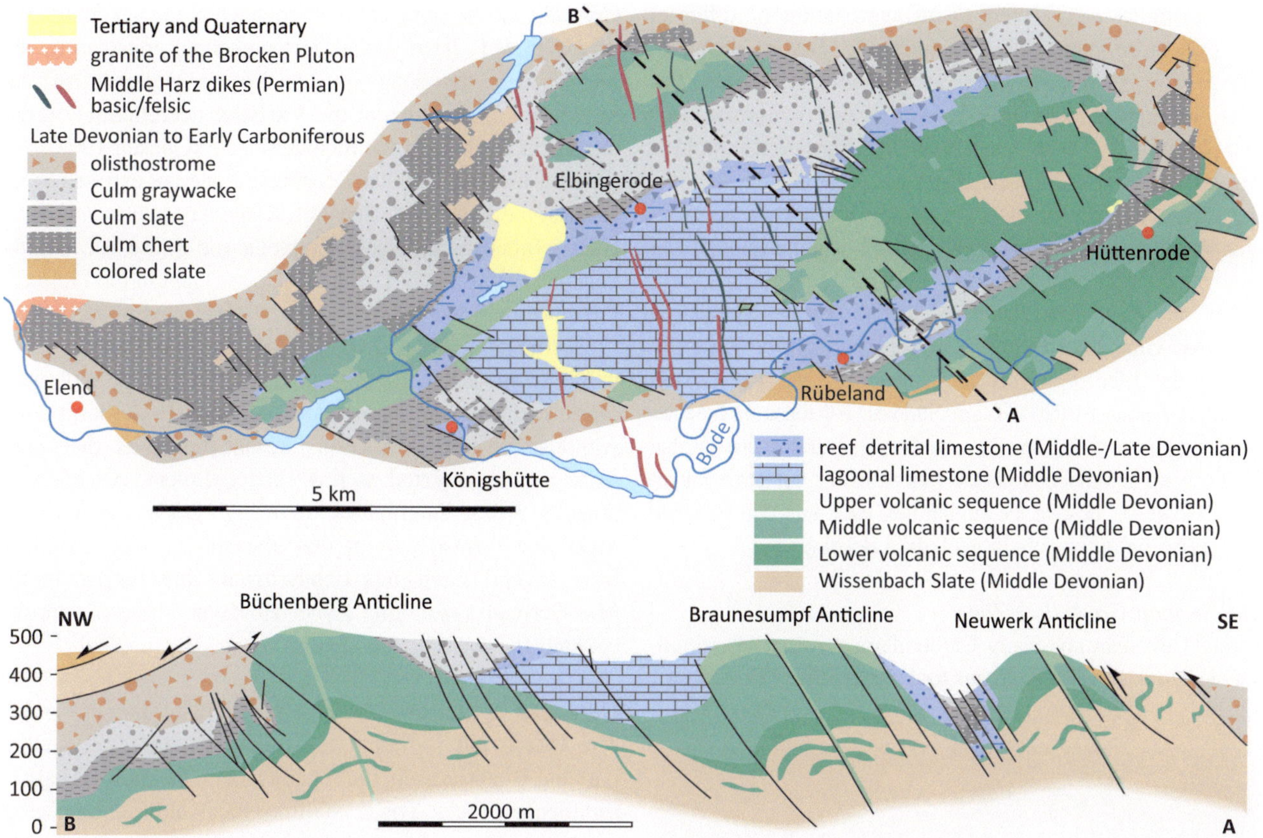

Fig. 10.37 Geological map and cross-section of the Elbingerode Complex (simplified after Weller 2012)

10.3.2 Northern Phyllite Zone

The Northern Phyllite Zone is located along the **suture line** between the Rhenohercynian and the Saxothuringian (Figs. 9.1 and 10.20). This zone extends from the southern edge of the Hunsrück to the Taunus and continues into the Eastern Harz as the Wippra Zone (Fig. 10.33). It has also been detected at depth further east based on borehole information. The zone is characterized by low-grade metamorphic sedimentary and volcanic rocks (greenschist facies; Oncken et al. 1995) that originated mainly on the **Rhenohercynian shelf**, located on the northern edge of the Rheic Ocean (Klügel et al. 1994). However, some rocks in this zone have affinities to the Saxothuringian, making a classification sometimes difficult. The contact to the Rhenohercynian is characterized by thrust faults, whereas the southeastern boundary to the Saxothuringian is not exposed (Anderle et al. 1995). In the Hunsrück area, metamorphic equivalents of the Rhenohercynian shelf are present, whereas the Taunus is dominated by Silurian/Devonian volcanics formed in a magmatic arc setting (Fig. 10.21). The Wippra Zone has an Ordovician core and, in the northwest part of the Phyllite Zone, it contains a small slice of relatively low-temperature metamorphic rock, equivalent to the olisthostromes of the Harzgerode Fold Zone.

During the early stage of Variscan collision, sediments were subducted along the Northern Phyllite Zone and thus mark the suture zone between the Rhenohercynian and the Mid-German Crystalline Zone as part of the Saxothuringian. During subduction, the rocks were subjected to a low-grade high-pressure metamorphism (Klügel 1997). Although greenschist facies metamorphism (Fig. 10.38) is still classified as a state of low-grade metamorphism, it is higher than the very low-grade (diagenetic to anchizonal) metamorphism of the Rheinisches Schiefergebirge (Sect. 3; Fig. 3.14). Overall, the Northern Phyllite Zone is considered to represent a deeper tectonic crustal level of the Variscan collision zone, where parts of the Rhenohercynian shelf, the Silurian/Devonian magmatic arc and sedimentary sequences of the Armorican Terrane Complex are represented (Anderle et al. 1995).

10.3.3 Regional Geology of the Saxothuringian

The entire Saxothuringian Zone is comprised of two major units: the Mid-German Crystalline Zone representative of an active continental margin during the Variscan Orogeny and the Saxothuringian sensu stricto, which is composed of a heterogeneous mixture of strongly schistose and tectonically displaced rocks ranging in age from Proterozoic to Early Carboniferous. The latter consists predominantly of shallow to deep marine sedimentary rocks, largely slate, sandstone, and limestone that were deposited on elevated swells, as well as igneous and metamorphic rocks of different ages such as gneiss, amphibolite, and granitic rocks. Overall, the

Saxothuringian is regarded as an aggregation of different zones of an orogenic belt. It includes the remains of Paleozoic marine basins, magmatic arcs, subduction zones, and older continental material derived from the Armorican Terrane Complex. These various components were welded together during the Variscan Orogeny.

A large number of units are recognized within the Saxothuringian, which in addition to isolated occurrences of older basement complexes of the Mid-German Crystalline Zone, occur in the Thuringian Slate Mountains (s.l.), the Fichtel Mountains, the Saxon Granulite Massif, the Erzgebirge, the Elbe Zone (including the Elbe Valley Slate Mountains), and the Lusatia Mountain Range of the Western Sudetes. All these units share some common characteristics, such as similar metamorphic grades, contemporaneous granite intrusions, or the same parent rocks. However, despite the similarities, each unit can be separated based on differences in geological development.

Mid-German Crystalline Zone

During Devonian and Early Carboniferous volcanism, simultaneous emplacement of plutons took place in the Mid-German Crystalline Zone along the northern edge of the Saxothuringian zone (Fig. 9.1). Two distinct tectonothermal units can be distinguished in this zone: (i) Relics of the **volcanic arc** and (ii) Metamorphic rocks of the **Variscan accretionary wedge** that lay at the base of the Saxothuringian (Oncken 1997). The rocks of the Mid-German Crystalline Zone are exposed in the areas of the Odenwald, Spessart, Ruhla Crystalline, and Kyffhäuser. Small occurrences of plutonic rocks also occur in the Palatinate Forest, near Delitzsch (Dessau) and in the Pretzsch–Prettin Massif; the latter of which was detected by drilling.

Sediment material derived from the Mid-German Crystalline Zone indicates this elevated swell area underwent erosion during the Late Devonian and Carboniferous. Although volcanic formations are not preserved, their existence can be detected as reworked sediment. A variety of igneous rocks, migmatites, and high-grade metamorphic rocks that formed part of the volcanic arc were exhumed from several kilometers depth toward the surface in the Mid-German Crystalline Zone. However, these are mostly covered by younger sedimentary strata.

Fig. 10.38 Folded Ordovician phyllites of the Northern Phyllite Zone exposed in a road-cut section through rocks of the Wippra Zone, southeastern Harz, east of Wippra [N 51°34′07″/E 11°17′46″]

Odenwald

The **Odenwald** is traditionally divided into its crystalline part and Buntsandstein part characterized by Early Mesozoic sedimentary rocks. The crystalline Odenwald is composed of several rock complexes of different ages, each with its own geological history. The oldest rocks of the crystalline complex are exposed in the **Böllstein Odenwald** (Fig. 10.39). Here, orthogneiss occurs derived from granitoids intruded during the Late Silurian to Early Devonian and metamorphosed in the Late Devonian. In the northern Odenwald, gabbros of the Frankenstein complex were intruded in the Late Devonian. These are interpreted as intrusives formed above a subduction zone (Kirsch et al. 1988). In addition to the gabbro, amphibolites (metamorphosed basalts) are also found together with schists, gneisses, and calc–silicate rocks.

At the time the Rheic Ocean was subducting beneath the Armorican Terrane Complex, a volcanic arc formed with the intrusion of plutonic rocks at deeper crustal levels along the northwestern edge of the Mid-German Crystalline Zone. In the central and southern portion of the Odenwald known as the **Bergsträßer Odenwald**, predominantly granitic rocks (flaser granitoid, granite, tonalite, and granodiorite) were intruded during the Early Carboniferous collisional phase of the Variscan Orogeny. Examples are the Weschnitz Pluton, the Tromm Granite, and the Heidelberg Granite, which are found in association with schist and isolated occurrences of marble. The magmatic rocks of the Mid-German Crystalline Zone are today only represented by rocks that intruded at a deep crustal level. However, in the northern Odenwald (Sprendlinger Horst), which is covered by Permian sedimentary layers, basaltic and rhyolitic rocks occur that originated at the beginning of the Rotliegend in connection with widespread continental volcanic activity (see Sect. 11.2.1).

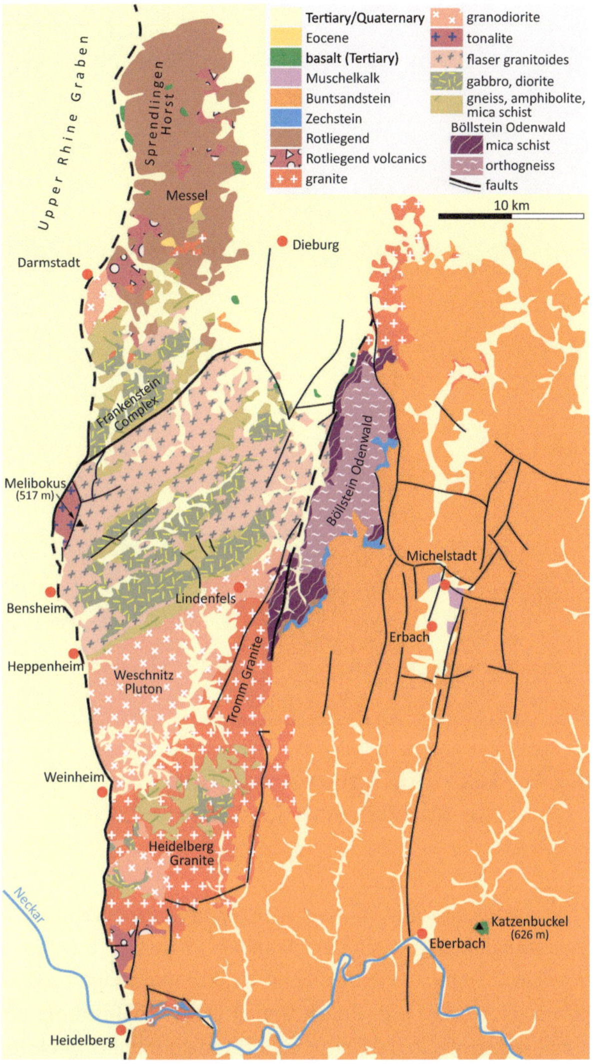

Fig. 10.39 Geological map of the Odenwald area. Modified after the 1:200,000 generalized geological map of the BGR (1981)

At the southern edge of the Odenwald, only rhyolites are found forming up to 150-m-thick layers.

Spessart

Crystalline rocks of the **Spessart** hills are located only a few tens of kilometers northeast of the Odenwald (Fig. 9.1). They are in part similar to the gneissose rocks of the Böllstein Odenwald and represent granitic to granodioritic rocks intruded in the Early Silurian and metamorphosed in the Late Devonian. In the Spessart area, the crystalline rocks are exposed in a **gneiss dome** structure (Fig. 10.40). The paragneiss in the northwest at Alzenau corresponds to the same rock types in the southeast of Aschaffenburg. There are also similarities in mica-schists, amphibolites, and quartzites that occur on both sides of the staurolite **paragneiss** and **orthogneiss** series. The likely parent rocks of the paragneisses are Early Paleozoic sandstones and slates with some intercalated basalts (Cambrian to Silurian). These rocks underwent Late Devonian Variscan metamorphism: an event that occurred at the same time as formation of the crystalline rocks in Böllstein Odenwald. It is assumed that the high-grade metamorphic rocks of the Spessart hills belong to the underthrust units of the Mid-German Crystalline Zone. These rocks were probably formed as sediment on the oceanic crust of the Rheic Ocean.

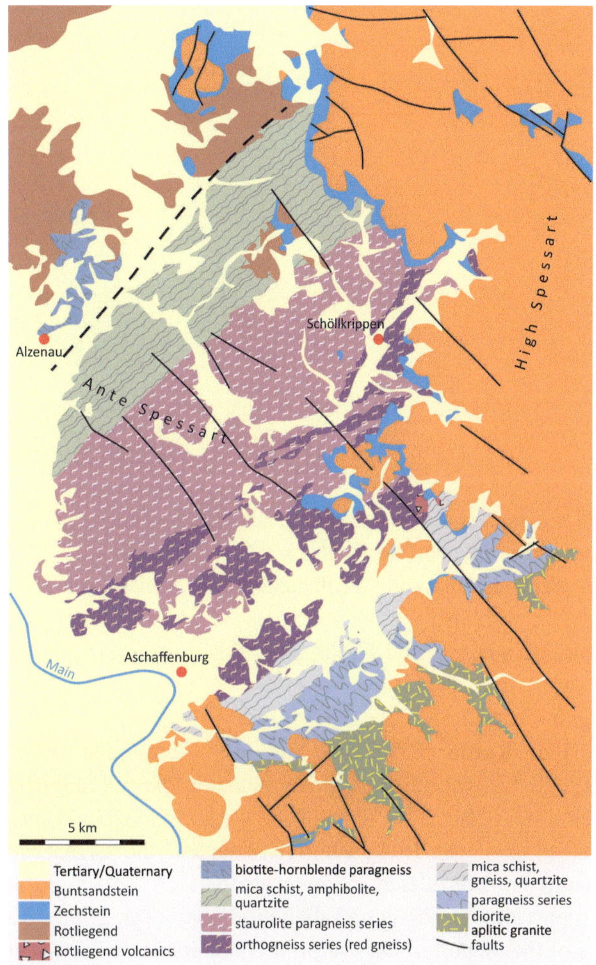

Fig. 10.40 Geological map of the Spessart area. Modified after the 1:200,000 generalized geological map of the BGR (1981)

Kyffhäuser

The **Kyffhäuser** Crystalline is a small (about 1.5 km^2) area composed of igneous and metamorphic rocks of the Mid-German Crystalline Zone. It is located a few kilometers from the Northern Phyllite Zone but the contact is not exposed. The crystalline rocks of the Kyffhäuser are mainly composed of gneiss (Fig. 10.41) whose parent rocks originated in the Ordovician and subduction-related magmatic rocks (granites and gabbros) of Early Carboniferous age. The latter were intruded into the older, highly metamorphosed host rock. The crystalline rocks are unconformably overlain by non-metamorphic sedimentary rock representing the Variscan

Fig. 10.41 Dioritic gneiss containing leucogranite dikes in the Kyffhäuser Crystalline complex, exposed in an abandoned quarry in the Oberes Steintal [N 51°25′22″/E 11°04′ 08″]

molasse that contains Stephanian and Rotliegend strata. Late Cretaceous uplift of the Kyffhäuser Crystalline occurred along the Hercynian trending Kyffhäuser Fault, which was active together with many other structures in the Harz Mountains, Thuringian Forest, and Flechtingen Hills.

Ruhla Crystalline

In the **Ruhla** Crystalline of the Thuringian Forest (Fig. 10.42), various high-grade metamorphic and magmatic rock units occur. Two units of the Central Crystalline rocks consisting of the **Liebenstein Group** and the **Brotterode Group** are considered to represent part of a volcanic arc complex that lay above a subduction zone (Zeh 2005). The age of the metamorphic overprint is not precisely known but probably occurred during the Early Carboniferous stage. Two other units (the Ruhla and Trusetal groups) originating as graywacke and other types of clastic sediment also occur as high-grade metamorphosed gneiss and quartzite, together with layers of metavolcanic rocks in the form of amphibolite. These rocks have been radiometrically dated and are considered to have cooling ages corresponding to the Cambrian, Silurian and Early Devonian. They are thought to have formed as relics of a former ocean floor area that was overthrust by the volcanic arc complex of the Mid-German Crystalline Zone and during the process were welded to the base of the arc (Oncken 1997). In this crystalline unit, metamorphism took place during the Late Devonian and Early Carboniferous.

10.3.4 Saxothuringian (excluding the Mid-German Crystalline Zone)

Thuringian Slate Mountains (s.l.)

Variscan deformation of the sedimentary sequences of the Saxothuringian had a similar character to that of the Rhenohercynian. The rocks are in general weakly metamorphosed and strained. However, with the exception of the coal-bearing continental sediments in the Erzgebirge Basin and in the Stockheim Basin of northern Bavaria, no large molasse basins developed following the infilling of the flysch basins. The direction of fold vergence is, however, opposite to that in the Rhenohercynian fold-and-thrust belt. Whereas the Rhenohercynian is dominated by northwest-vergent folds, in the Saxothuringian they are generally southeast vergent. A Late Carboniferous unconformity overlies older Paleozoic strata of the Devonian and Early Carboniferous and thus constrains the age of folding as occurring around the transition from Early to Late Carboniferous (i.e., the boundary between the Viséan and Namurian). This deformation phase is known as the **Sudetic Phase**.

The Thuringian Slate Mountains (sensu lato, including the Vogtland area) are characterized by an NE-SW trending sequence of anticlinal and synclinal fold structures (Fig. 10.43). The **Schwarzburg Anticline** is located in the northwest with Late Proterozoic slates occurring in the fold core that represents the oldest rocks of the region. In the southeast, follows the **Teuschnitz–Ziegenrück Syncline** where, similar to the **Mehltheuer Syncline**, folded and foliated rocks of Early Carboniferous age are exposed.

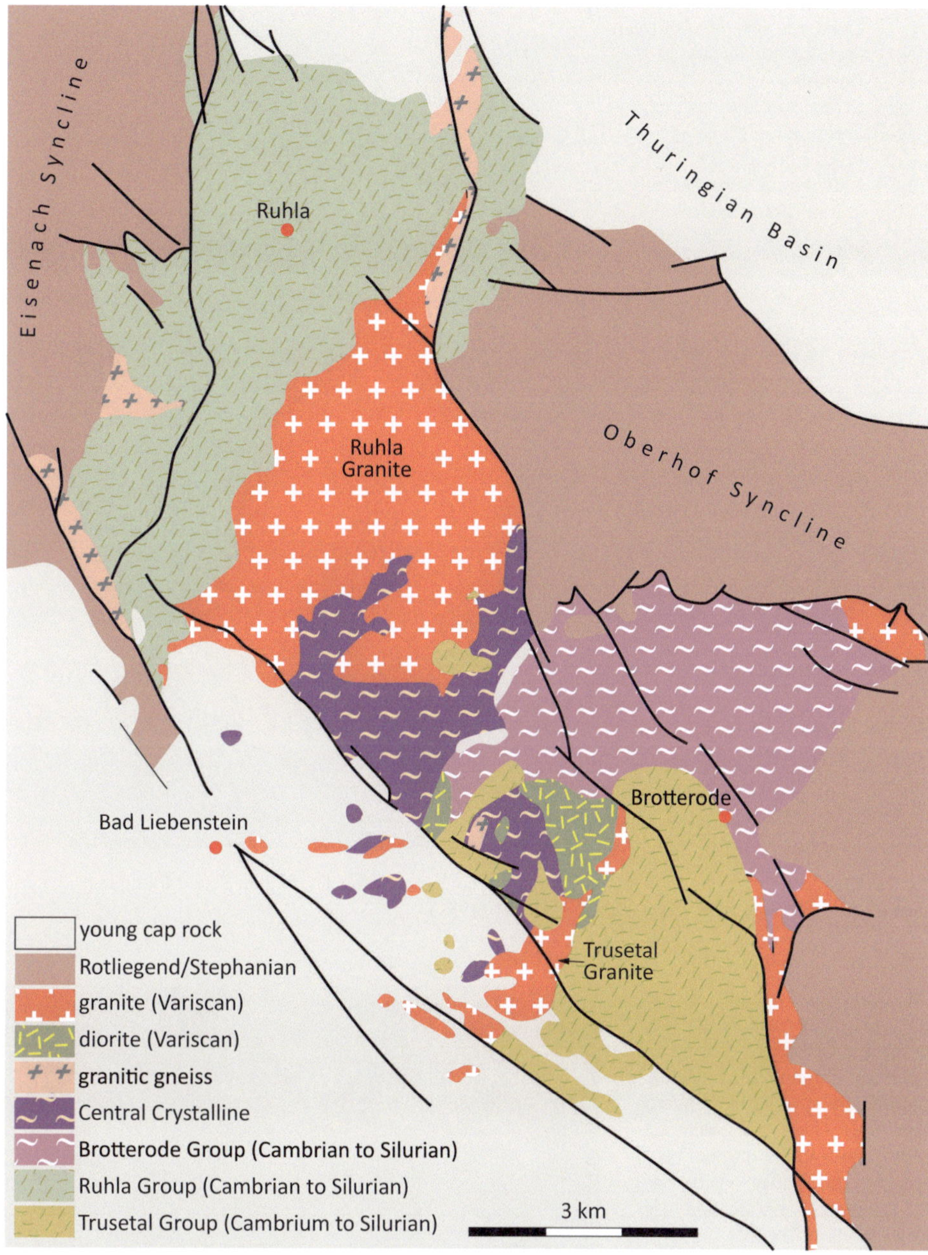

Fig. 10.42 Geological map of the Ruhla Crystalline. Simplified after Zeh (2005) and Zeh and Wunderlich (2003)

Numerous roof slate pits still provide roofing tiles and house facade cladding (Fig. 10.44). The roofing slate is in places overlain by several kilometer thick sequences of clastic flysch (graywacke). The **Berga Anticline** and the **Vogtland Syncline** also expose progressively deeper crustal units of Devonian, Ordovician, and Silurian slates.

Saxon Granulite Massif

The **Saxon Granulite Massif** is a special feature of particular geological interest. It represents a so-called **metamorphic core complex** where high-grade metamorphic rocks that were once buried deep in the Earth's crust are now located at the surface and surrounded on all sides by less metamorphosed units (Fig. 10.45). **Granulites** represent the highest possible degree of metamorphism in the field of regional metamorphism (Fig. 3.13). At higher pressures and temperatures, these rocks melt to produce granites. A granulite consists mainly of K-feldspar and quartz, but also commonly contains garnet, which is conspicuous by its red coloration in the otherwise white to grayish rock. This rock type typically forms under anhydrous conditions because the melting temperature of rock is significantly reduced in the presence of water: a condition that favors granite formation.

The Saxon granulites were formed at temperatures above 1000 °C, indicating the complete absence of water and high pressures corresponding to a burial depth between 50 and 70 km (Rötzler and Romer 2010). Despite the great depth, these granulites were uplifted to the surface within a short time

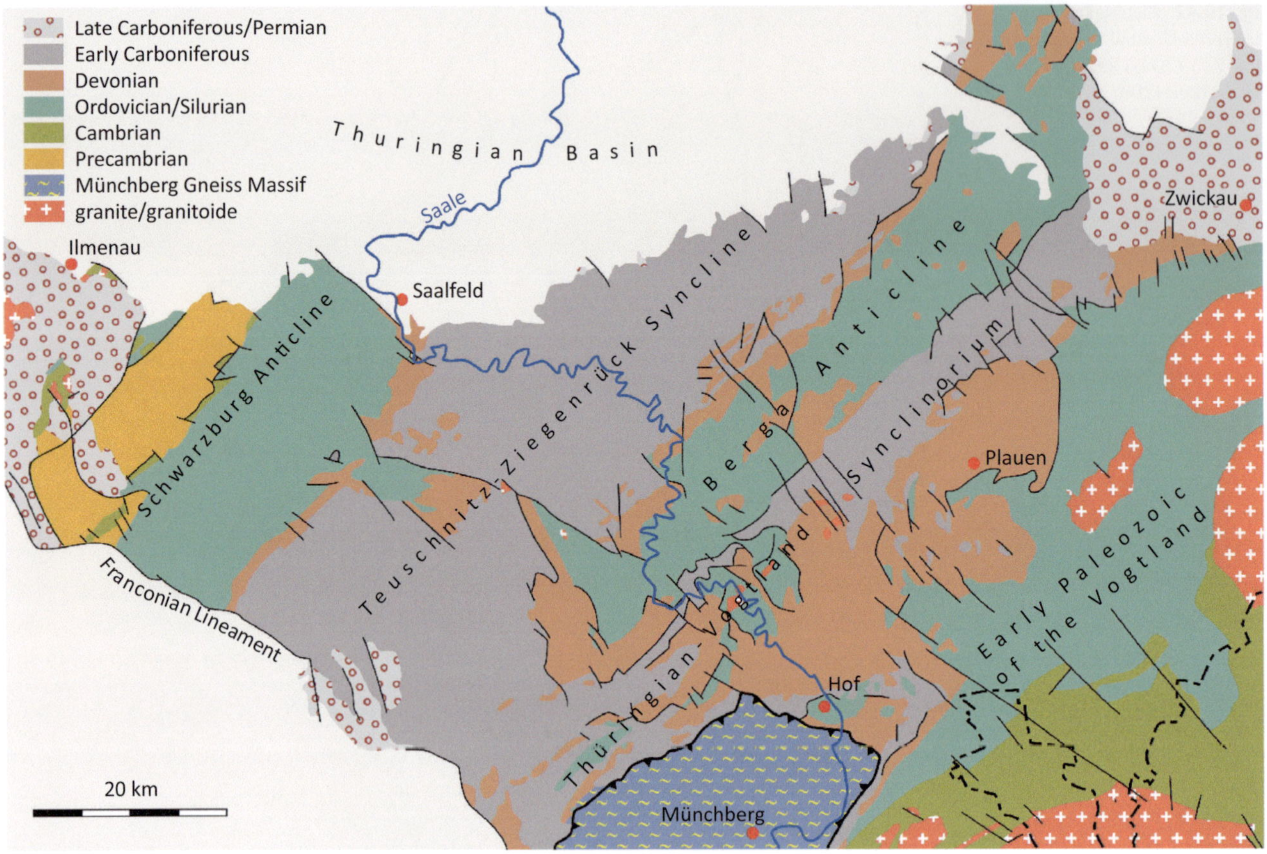

Fig. 10.43 Geological map showing the Paleozoic sequences of the Thuringian Slate Mountains (s.l.). Modified after Seidel et al. (2002) and the generalized geological map 1:200,000 of the BGR (1981)

Fig. 10.44 Cut section in the roof slate quarry at Schmiedebach, near Lehesten (Thuringia). The Early Carboniferous (Middle Tournaisian to Middle Visean) slates split along their cleavage fabric, although the original bedding is still partly visible as fine-laminated stratification [N 50°29′19″/E 11°28′05″]

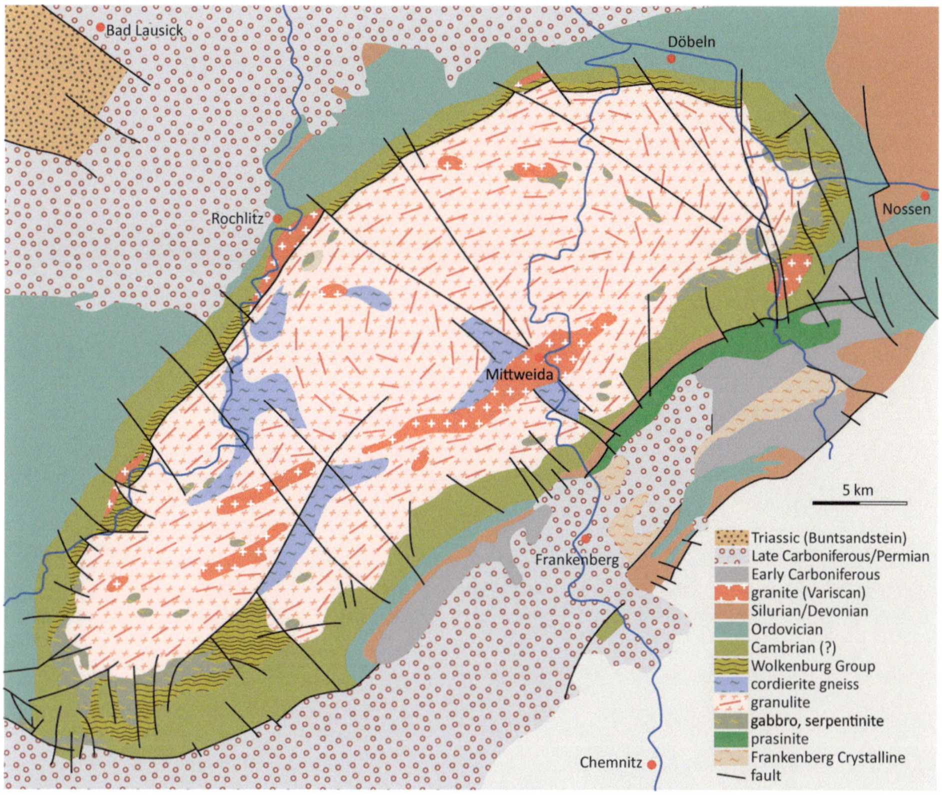

Fig. 10.45 Geological map of the Saxon Granulite Massif. Modified after Berger et al. (Berger et al. 2008)

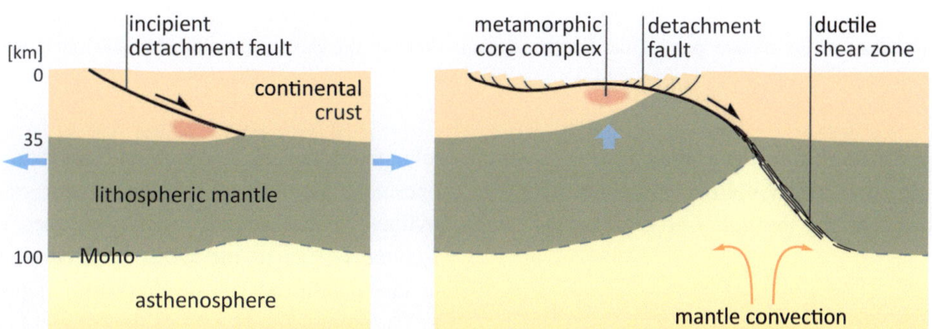

Fig. 10.46 Uplift of a metamorphic core complex, based on the simple shear model of Wernicke (1985). In this model, the rapid uplift of a high-grade metamorphic complex from the deep lower crust results from the extension of the lithospheric plate (*blue arrows*)

of about 5 Myr during the late extensional phase of the Variscan Orogeny. Such high rates of exhumation are only possible under specific conditions and cannot be explained by isostatic uplift and erosion alone. One possible model for the rapid exhumation is that the Granulite Massif complex lay in the footwall of a large normal fault that reached down into the lithospheric mantle (Fig. 10.3). Along such a structure, known as a detachment fault, very deep-seated crustal levels can be uplifted to the surface within a short time, in response to the thinning and upwelling of the mantle (Wernicke 1985; Fig. 10.46).

The metamorphic granulite core of the Saxon Granulite Massif represents continental rocks that were subducted beneath another continental plate within the Armorican Terrane Complex to great depths during the Variscan Orogeny (Fig. 10.3). Based on Proterozoic radiometric ages and the occurrence of biogenically formed carbon, it is assumed that the parent rocks of the granulite were Proterozoic to Early Paleozoic sedimentary and magmatic protoliths. As the metamorphic core complex was uplifted and reached the surface, the dome-like ascent of the Granulite Massif resulted in an oval shape (Fig. 10.45). The ring-like belt is surrounded by a **slate envelope** (*Schieferhülle*) composed of the **Wolkenburg Group** and Cambrian metamorphic rocks that surround the granulite complex. The rocks are less metamorphosed than the granulite and mark the location of the detachment horizon (Wolkenburg Group) or the lowest

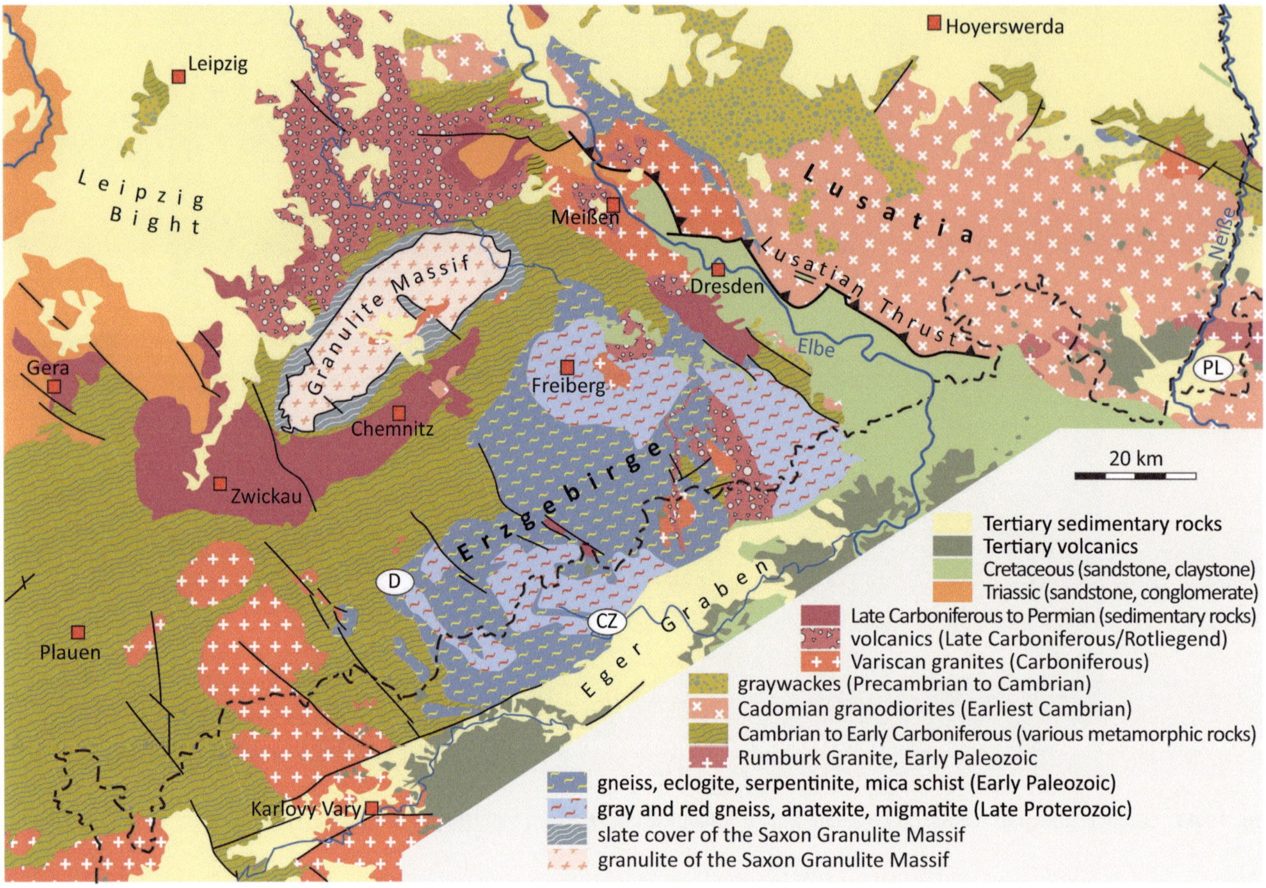

Fig. 10.47 Geological map of the eastern part of the Saxothuringian, showing the Saxon Granulite Massif, the Erzgebirge, and Lusatia. Simplified after Goth et al. (2001)

layers of the upper plate during the collision. The granitic intrusions penetrated the already uplifted granulite complex and its slate envelope during the Late Carboniferous.

Erzgebirge

The Erzgebirge contains a range of metamorphic rocks of varying type, grade, and pressure–temperature–time history that occur in close proximity to each other (Fig. 10.47). The crystalline rocks represent a highly complex stack of nappes containing in part high-grade metamorphosed rocks formed during the Variscan collision of the Saxothuringian and the Teplá-Barrandian of the Moldanubian Zone (Figs. 10.1 and 10.2). Cadomian crustal units of the lowermost structural level (Sect. 9.2.3) are represented by more moderately metamorphosed gneiss (**gray and red** Latest Proterozoic to Earliest Cambrian **gneisses** of the East Erzgebirge Group). Separated by shear zones, they are overlain by high-grade metamorphic units composed of orthogneiss, paragneiss, eclogite, serpentinite, and quartzite. In the central and western Erzgebirge, the degree of metamorphism decreases upward into higher tectonic units. Here, mainly younger rocks from the Ordovician and Silurian are found. The uppermost tectonic unit contains low-grade metamorphic phyllites that resemble rocks of the Thuringian Slate Mountains (s.l.) to the west. In general, the typical sedimentary parent rock of the Erzgebirge nappes was related to the Thuringian facies type of depositional environment. The final stages of the Variscan Orogeny in the Late Carboniferous led to the intrusion of a number of granites in the Erzgebirge (Zinnwald, Eibenstock, and Kirchberg), which host economically important tin (Sn) ore deposits.

The *Erzgebirge* (Ore Mountains) is, as its name suggests, a historical region for the source of metallic ore deposits. Ore veins and **granite mineralizations** have been mined for more than 800 years. The most common types of ore are secondary, hydrothermal Sn, and W deposits located within the Variscan granite plutons that formed as residual phases from magma-related fluids (Box 11: Ore deposits). There are also indications the Sn originated from nearby Ordovician sedimentary rock as it was enriched and mobilized during granite intrusion (Mingram 1998).

The age of **hydrothermal vein mineralizations** in the Erzgebirge have not yet been precisely dated, but as they mostly truncate the metamorphic foliation, they are either of late- or post-Variscan (Mesozoic or Cenozoic) origin. A large number of mineralization events, however, have been dated further afield and can be linked to known plate tectonic events in Central Europe (Sebastian 2013). Uranium (U) and polymetallic ore deposits occur in the Rotliegend during formation of the Central European Basin (Sect. 11.2.1) due to younger remobilization events. Widespread mineralization also formed during Early Jurassic crustal extension as the Atlantic Ocean started to open (Sect. 11.6). Hydrothermal fluids circulated along fault and fracture zones, resulting in the formation of Ag ores, Pb–Zn mineralizations, fluorite–barite, and U ores as well as polymetallic deposits. Notable hydrothermal mineralizations occurred around the Jurassic/Cretaceous boundary and during the Early Cretaceous. However, Late Cretaceous compressional tectonics in the Erzgebirge prevented further circulation of these types of hydrothermal fluids (Sect. 11.7.2).

Lusatian Mountains and Western Sudetes

The basement rocks of **Lusatia** and the **Western Sudetes** are altogether referred to as the Lugikum or Lusatia–Izera Block. These rocks form the northeastern foothills of the Saxothuringian and represent the Cadomian basement lying beneath the Paleozoic Thuringian facies. These rocks consist primarily of Early Proterozoic turbiditic graywackes, which were intruded by numerous granodioritic plutonic rocks during a late phase of the Cadomian Orogeny in the Earliest Cambrian (Sect. 9.2.3; Fig. 10.47). Until the Devonian, the Lugikum was dominated by crustal extension reflecting the northbound movement of the Armorican Terrane Complex. At the beginning of the rift phase, Early Ordovician sediments were deposited unconformably on Late Proterozoic graywackes and granodiorites. After a hiatus, the rock sequence continues with Ordovician quartzites and sandstones as well as Silurian to Early Carboniferous shales and cherts. In the Late Cambrian/Early Ordovician, incipient rifting gave rise to bimodal magmatism (e.g., the **Rumburk Granite**; Fig. 9.4). An Early Devonian metamorphic event occurred in the eastern and southeastern part of the Lugikum (border of the Riesengebirge in the Czech Republic and in the Eulengebirge in southwestern Poland) and probably resulted from the collision between the Saxothuringian and the Moldanubian. As in other areas of the Saxothuringian, the Late Carboniferous emplacement of late orogenic granite plutons also occurred in the Lugikum, the largest been the **Riesengebirge Granite**. In the Cadomian basement rocks of Lusatia, Variscan granites are, in contrast, found only occasionally within the granodiorite units (Fig. 10.47).

Fichtel Mountains

The **Fichtel Mountains** are located in the southwestern prolongation of the Erzgebirge (Fig. 9.1). They are mainly composed of Paleozoic rocks but include Early Proterozoic units of varying metamorphic grade. The metasedimentary sequences contain thick layers of metabasic volcanic rocks frequently characterized by the occurrence of pillow basalts. Examples are found in Bad Berneck, where Late Devonian

Fig. 10.48 Late Devonian pillow basalts from the Saxothuringian exposed at the entrance to the spa gardens of Bad Berneck [N 50°03′02″/E 11°40′37″]

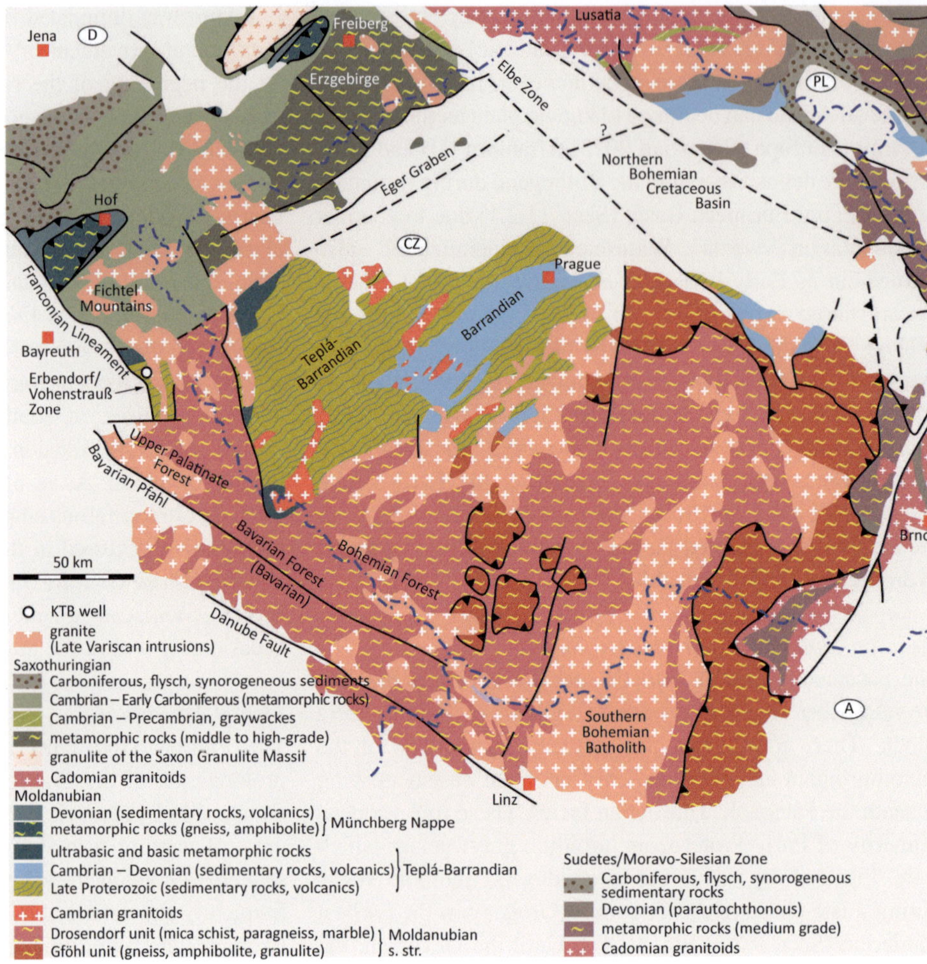

Fig. 10.49 Geological map of the Variscan rocks of the Bohemian Massif. KTB Continental Deep Drilling Program of the Federal Republic of Germany. Simplified after Franke (2000)

low-grade metamorphic pillow basalts are exposed (Fig. 10.48). By regional comparison, the sedimentary rocks correlate with the Thuringian facies. In the upper part of the sequence, calcareous rocks with limestone and dolomite marbles become increasingly abundant. During the Variscan Orogeny, the rocks of the Fichtel Mountains were overthrust by the Münchberg Gneiss Massif, which today forms a nappe in the northwestern area of the Fichtel Mountains and lies on top of the Saxothuringian (Figs. 9.1 and 10.3; see below Sect. 10.3.4). In the Late Carboniferous, numerous granite plutons intruded into the previously deformed and metamorphosed rocks. Today, these intrusions comprise more than half of the area of the Fichtel Mountains.

10.3.5 Regional Geology of the Moldanubian

The Moldanubian of Germany is exposed in the Black Forest and in the Bavarian and Upper Palatinate Forest (Fig. 9.1). The **Black Forest**, together with the **Vosges** of France, is a unit that was uplifted by several kilometers during the evolution of the Upper Rhine Graben (Sect. 13.2). The Moldanubian basement, which was covered by thick Mesozoic strata, has been re-exposed at the surface following extensive erosion. In the **Bavarian Forest**, the Moldanubian occurs in the western foothills of the Bohemian Massif with crystalline rocks of the old European basement (paragneisses, amphibolites, graphites schists, gneisses, and granulites), and granites and granitoids that intruded during the late phase of the Variscan Orogeny. This western boundary area of the Bohemian Massif is also referred to as the **Bavaricum**.

Bohemian Massif

The Bohemian Massif in the broader sense includes the basement units of southeastern Europe, with the Thuringian Slate Mountains, the Fichtel Mountains, the Erzgebirge, and the Sudetes as part of the Saxothuringian. The massif also contains the Teplá-Barrandian, the Bavaricum, and the Bohemicum of the Moldanubian (Fig. 10.19). The core areas of the Moldanubian were in part uplifted during the Paleozoic and served as a sediment source area. The Bohemian Massif formed an island or was part of the mainland during the Mesozoic and Cenozoic and thus was extensively eroded to expose the deep crustal levels seen at the surface today. The basement rocks of the Bohemian Massif belonged to the **root zone** of an Early Paleozoic mountain range and are comprised of metamorphosed sedimentary, volcanic, and plutonic rocks of Late Proterozoic to Silurian age (Chap. 9).

A variety of igneous rocks formed at the Late Proterozoic to Cambrian transition and during the Ordovician, the younger of which were probably subduction-related (Teipel 2003). In contrast to the rest of the Bavarian Forest, the **Künic Mountains** (*Künisches Gebirge*) north of Lam (northern Bavarian Forest) are composed mostly of mica-schists derived from Silurian sedimentary rocks, including some quartzites of possible Ordovician age. At the edge of the Saxothuringian, the **Erbendorf–Vohenstrauß Zone** is located in the northwest of the Bohemian Massif. Here, one of the deepest wells in the world was drilled in the 1990s with a total depth of 9101 m (Fig. 10.49). This was known as the Continental Deep Drilling Program of the Federal Republic of Germany (KTB). The deepest drill hole worldwide was made in the Kola Peninsula in 1989, which reached a final depth of 12,262 m. The KTB hole passed through an alternating sequence of paragneiss and metabasite (amphibolite), whose affiliation to the Saxothuringian or Moldanubian has still not been satisfactorily resolved.

Rocks of the Bohemian Massif suffered in part high-grade Variscan metamorphism of Late Devonian and Early Carboniferous age to form gneiss, granulite, and migmatite. Two structural levels can be distinguished in the Bohemian Massif and are separated from each other by thrust faults: (1) the **Drosendorf unit** composed of parautochthonous metasedimentary rocks (mica-schist, gneiss, graphite-schist, and marble), (2) the **Gföhl unit** with high-pressure metamorphic orthogneiss, paragneiss, granulite, and amphibolite (Fig. 10.48), which structurally overlie the Drosendorf unit. The two levels are interpreted as a continental microplate (Drosendorf unit) and a terrane (Gföhl unit) that were originally divided by an ocean, the **Moldanubian Ocean**, which closed during the collision of the two units. Remnants of the ocean can be found as ultramafic and mafic rocks at the base of Gföhl unit. A similar tectonic situation can also be observed in the Black Forest, where the **Badenweiler–Lenzkirch Zone** is considered to represent the suture zone of the Moldanubian Ocean (see below).

Granites intruded during the Variscan Orogeny are widespread in the Bohemian Massif and have been estimated to make up about a quarter of the total volume of the rocks present (Behrmann and Tanner 1997). Most of the granites were intruded at the end of the Early Carboniferous and in the Late Carboniferous (340–310 Ma) and thus represent late orogenic intrusions (Klein et al. 2008). The largest pluton is the South Bohemian Batholith, which occurs in the eastern part of the Bohemian Massif in Austria and the Czech Republic. It has a lateral extension of more than 100 km in the northeasterly direction (Fig. 10.49). In the Bavarian and Upper Palatinate Forest, several smaller plutons are also exposed. In **Flossenbürg**, the granite is characterized by surface-parallel pressure release joints with shell-like separations producing granite blocks of several meters thickness. This process of surface-parallel separation is known as **exfoliation**. The granite summit of Flossenbürg, where this type of crevasse formation is well exposed, is also called a *Schalendom*. Today, it is a protected nature reserve and represents one of the finest geotopes of Bavaria. The natural horizontal fracturing was used in the eighteenth century for the mining of the large granite blocks. In other places in the area around Flossenbürg, granite mining continues today, and it is still used as a natural building stone.

Following the Variscan Orogeny, Permian, and Triassic extensional movements occurred along several fault systems, such as the Franconian Lineament, the Bavarian Pfahl, and the Danube Fault at the southwestern edge of the Bohemian Massif (Fig. 10.49). During dilation of the Pfahl Fault, silicate-bearing solutions circulated in this deep-seated zone of structural weakness. Minerals (mainly quartz) were deposited in open fractures that reached widths of 120 m and extended to depths of several kilometers. The hard quartz crystals within the mineralized veins that are exposed at the surface, following uplift of the basement, have been dissected by erosion to produce a wall of rock up to 30 m high and 50 m wide. With some minor interruptions, this zone of mineralization, known as the **Bavarian Pfahl**, can be followed at the surface for more than 140 km. It is known to be composed of 95–98% quartz and contains some ore minerals, such as barite, fluorite, and a uranium mica, as accessory phases. At Viechtach in the Bavarian Forest, a part of this impressive quartz vein was mined in earlier times for the local glass industry. This site was made a national geotope in 2006. Two other quartz vein structures run parallel to the Bavarian Pfahl. These are the Rundingen Pfahl Zone, near Bodenmais, and the Aicha-Halser-Nebenpfahl, both of which are marked morphologically by bends in the river Ilz and the Danube, near Passau.

Münchberg Gneiss Massif

An unusual structure is the **Münchberg Gneiss Massif**, which is surrounded by metamorphic rocks extending from Hof, in Upper Franconia, to about 30 km to the southwest (Fig. 10.50). These units are further away surrounded by rocks of the Saxothuringian. The massif is considered to represent a Moldanubian metamorphic unit that was thrust on top of the significantly lower metamorphic grade rocks of the Saxothuringian as part of a tectonic nappe pile (Figs. 10.3, 10.49 and 10.47). Since it is no longer in direct contact with the area of origin, it represents a tectonic klippe. The metamorphic grade of the **Münchberg nappe** pile increases toward the top as

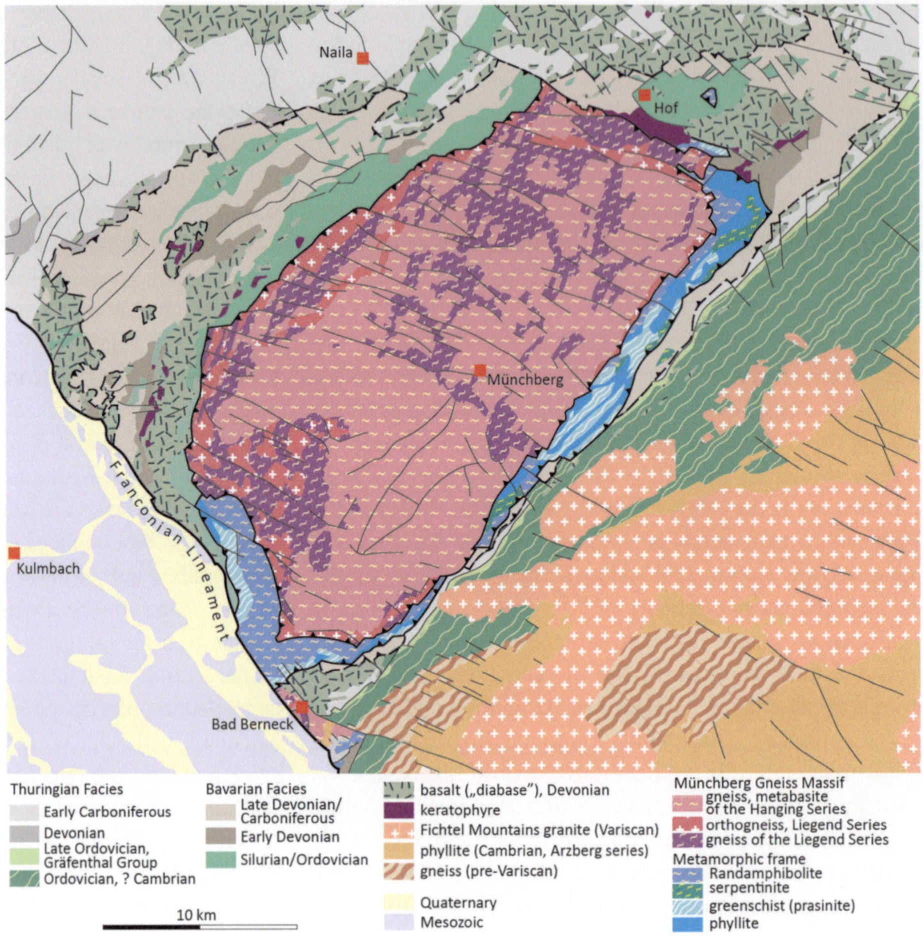

Fig. 10.50 Geological map of the Münchberg Gneiss Massif and adjacent regions. Simplified after the 1:200,000 of the generalized geological map the BGR (1981)

presented in textbook examples. The nappe pile is all around bounded by faults dipping toward the center of the gneiss massif. However, the southwestern boundary is truncated by the **Franconian Lineament**, which juxtaposes the rocks of the massif against the Franconian Mesozoic. The two smaller units of the so-called **Saxon Zwischengebirge**, the crystalline areas of **Frankenberg** and **Wildenfels**, form a northeastern prolongation of the Münchberg Nappe (Fig. 9.1) and have strong similarities with the gneiss massif. They contain high-grade metamorphic gneisses and mica-schists of the Moldanubian that structurally overlie slates of the Saxothuringian.

The autochthonous rocks below the Münchberg nappes belong to the Saxothuringian and are composed of weakly metamorphic Ordovician to Early Carboniferous graywackes and shales of the Thuringian facies and crystalline rocks of the Fichtel Mountains (Fig. 10.3). At the base of the nappe occur strongly sheared and intensely folded, low-grade metamorphic rocks of the Bavarian facies with Silurian to the Early Carboniferous flysch deposits and conglomerates. The rim of metamorphic rocks consists mainly of **phyllites** and **prasinites**. A prasinite is an original high-pressure metamorphic basalt that has been overprinted by greenschist facies metamorphism. The nappe pile directly above is composed of slightly higher metamorphic Ordovician metasediments and amphibolites of the so-called *Randamphibolit*. It is overlain by the upper nappe pile containing the highest metamorphic units; first, the *Liegendserie* (footwall series) with Precambrian and Cambrian orthogneisses, followed upward by the *Hangendserie* (hanging-wall series) with high-pressure metamorphic mica-schists, eclogites, and garnet amphibolites (Fig. 10.50). The eclogites of the uppermost structural level of the Münchberg nappes make up one of the largest occurrences of this rock type in Central Europe. The age of metamorphism is similar to that of the Bohemian Massif and occurred during the Devonian and Early Carboniferous stages of the Variscan Orogeny. The root zone of the Münchberg nappes is located in the northwestern border area of the Moldanuban between the anticlinorium of the Fichtelgebirge and the Teplá-Barrandian (Fig. 10.3).

Black Forest

In the **Black Forest**, medium- and high-grade metamorphic rock units and plutonic intrusions of the Moldanubian are exposed in the uplifted shoulders of the Upper Rhine Graben as they were stripped of their sedimentary cover (Fig. 10.51;

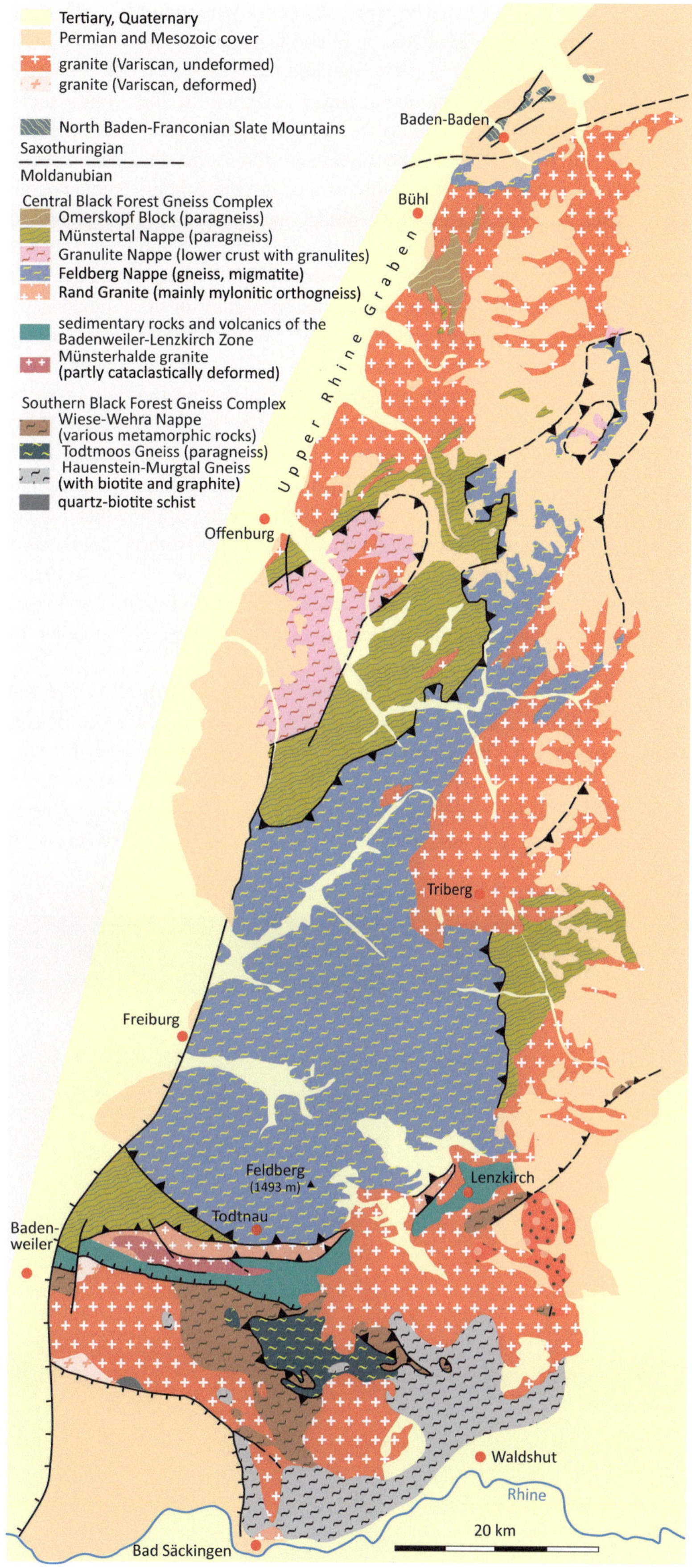

Fig. 10.51 Geological map of the Black Forest crystalline rocks. Simplified after Hann and Zedler (2008)

Geyer and Gwinner 2011). The **North Baden–Franconian Slate Mountains**, located in the northernmost part of the Black Forest, consist of weakly metamorphosed schists (phyllites, mica-schists, and metabasites) assigned to the Saxothuringian based on similarities in rock types. In this segment, which is known as the **Baden–Baden Zone**, the metamorphic grade increases southward up to the amphibolite facies. The boundary to the Moldanubian, however, is not exposed, since it is hidden under younger Permian and Mesozoic sedimentary rocks. The north-dipping boundary fault continues on the other side of the Rhine Valley in the Vosges. Also here, a shear zone separates similar, slightly weaker metamorphic Saxothurigian rock units of the Northern Vosges from gneiss and granite complexes of the Moldanubian (Fig. 13.9).

The Moldanubian Black Forest is subdivided into four geological distinct areas: (1) the Northern Black Forest granite intrusions, (2) the Central Black Forest gneiss complex, (3) the Badenweiler–Lenzkirch Zone, and (4) the Southern Black Forest granite–gneiss complex.

The area of the **Northern Black Forest granites** (1) is dominated by a large contiguous igneous body composed of several intrusions. Radiometric age data and structural criteria indicate that the granites evolved at the turn of the Early to Late Carboniferous following completion of Variscan nappe thrusting. A few gneiss blocks within the granites probably originated from Proterozoic or Early Paleozoic parent rocks before been affected by Variscan metamorphism in the Early Carboniferous.

The **Central Black Forest gneiss complex** (2) is dominated by highly metamorphosed gneiss and is subdivided into the tectonically highest unit of the Granulite Nappe (Nordrach unit), including several other smaller gneiss complexes, the underlying Münstertal Nappe with the Randgneiss Group, and the structurally lowest unit of Feldberg Nappe. The latter contains the Kerngneiss Group and the Randgranite unit with mylonitized orthogneisses.

The **Kerngneiss Group** is comprised of several thousand meters thick paragneiss complexes, the main rock type being a biotite plagioclase gneiss. A special occurrence of cordierite-bearing garnet gneiss known as **kinzigite**, named after a location in the Kinzig valley of the Black Forest, serves as a type locality for this rock type worldwide. Proterozoic and Early Paleozoic shale and graywacke were intruded by felsic melts during the Cambrian and Ordovician. They probably formed the parent rocks to the paragneiss and orthogneiss units that also contain remnants of ophiolites and eclogites. Based on radiometric age dating, it is assumed that the ocean floor that gave rise to the ophiolitic remnants formed in the Ordovician. The Variscan metamorphism that partly reached eclogite and granulite facies conditions took place in the Early Carboniferous (Chen et al. 2003).

The **Randgneiss Group**, which envelopes the Kerngneiss Group, contains different metamorphic units of paragneiss and leptinite (a very fine-grained gneiss with felsic volcanics, tuffs, or arkose as the parent rock). The highest-grade metamorphic rocks are located in the Granulite Nappe of the Nordrach unit, whereas the Münstertal Nappe does not contain any

Fig. 10.52 Randgranite exposure, near Gschwendt, ca. 4 km south of Todtnau (Black Forest). The mylonitic overprint is clearly marked by a distinct schistose foliation and the alignment of large feldspar crystals [N 47°48′28″/E 07°56′ 09″]

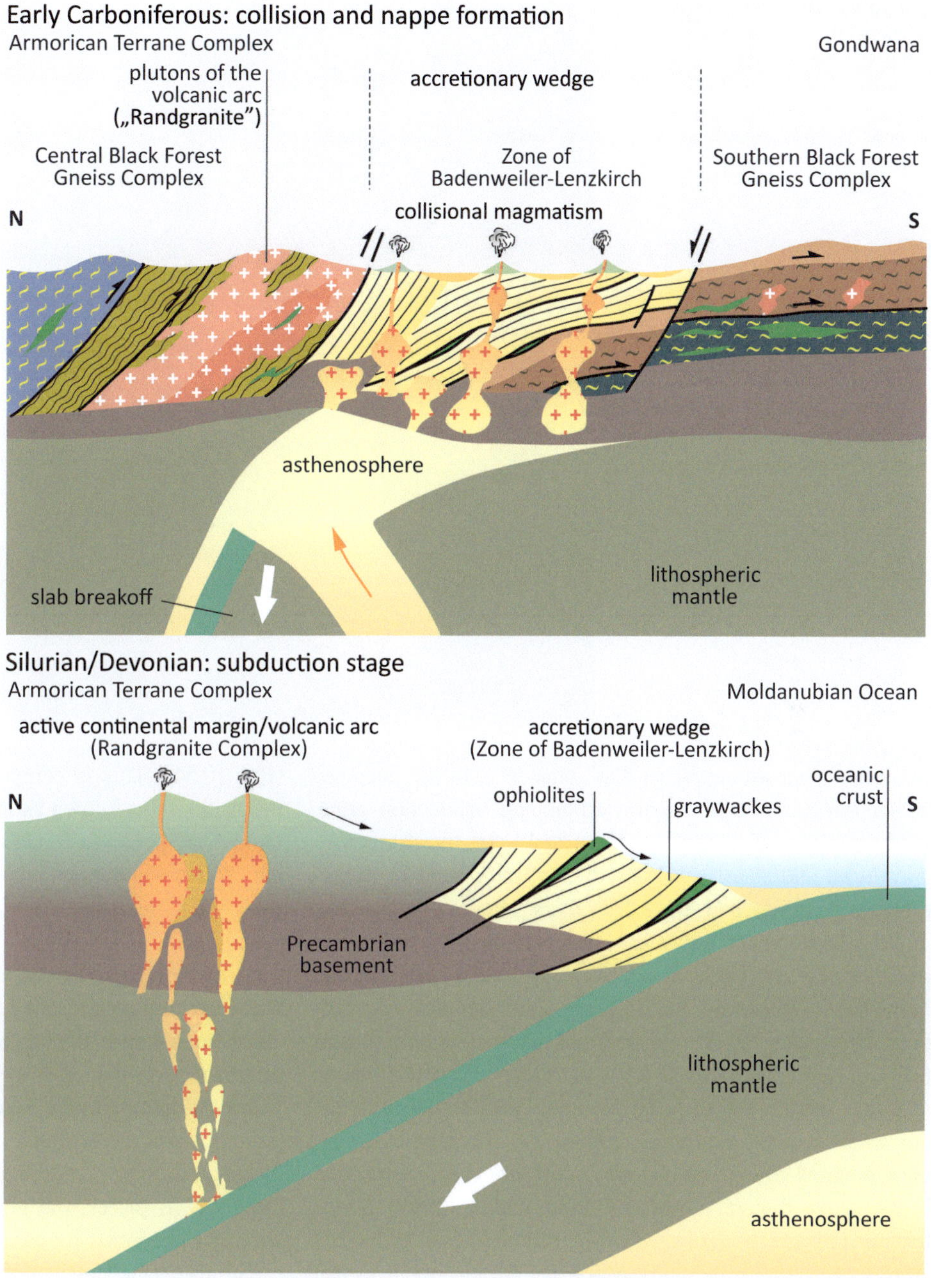

Fig. 10.53 Evolution of the Moldanubian suture in the Badenweiler–Lenzkirch Zone of the southern Black Forest. Modified after Hann et al. (2003a) and Frisch and Meschede (2013)

high-pressure metamorphic rocks. Mainly, Proterozoic and Early Paleozoic parent rocks are assumed for these gneisses, which were also metamorphosed in the Early Carboniferous.

The **Randgranite unit** is located at the boundary of the Badenweiler–Lenzkirch Zone (Fig. 10.51). It consists primarily of granites, granodiorites, and trondhjemites that were metamorphosed to orthogneisses and flasergranites during the Variscan Orogeny (Fig. 10.52). Hann et al. (2003a) consider the granites, which have Silurian and Devonian intrusion ages and geochemical compositions typical for subduction-related magmatic rocks, to represent relics of the plutonic crustal level of a magmatic belt (Fig. 10.53). The trondhjemites evolved later as melting products of the asthenosphere, which probably ascended to the surface because of a **slab breakoff** (Box 22: Slab breakoff). This detachment caused strong heating of the deeper crust, resulting in the formation of large amounts of granitic melt.

The **Badenweiler–Lenzkirch Zone** (3) lies to the south of the Randgranite zone and separates the highly metamorphosed gneiss complexes of the Central Black Forest from those of the Southern Black Forest. The zone has been partly overthrust by the Randgranite of the Central Black Forest gneiss complex (Fig. 10.51). The rocks of this zone consist of low-grade metamorphic clastic sedimentary lithologies that were deposited from the Ordovician to the Early Carboniferous in the

Fig. 10.54 Partial melt textures (anataxis) in a migmatite rock within the Wiese–Wehra Nappe of the Southern Black Forest, exposed in a road-cut section north of Todtmoos [N 47°44′58″/ E 07°59′56″]

environment of a convergent plate boundary. In the deep sea trench lying above the northward-dipping subduction zone, sandy to clayey sediment was deposited (later transformed to graywacke and slate), and subsequently imbricated in an **accretionary prism** and stacked as thrust nappes (Fig. 10.53). These rocks, which today are deformed and folded in the Badenweiler–Lenzkirch Zone, were intruded by granites immediately following the collision (Münsterhalde Granite; Figs. 10.51 and 10.53). The sedimentary rocks contain remnants of volcanics derived from the magmatic belt (such as andesitic detritus of meta-rhyolitic ash and boulders in the graywacke units; Hann et al. 2003b) and incorporated ophiolitic material containing high proportions of chromium spinel in the heavy mineral fraction (Güldenpfenning 1998). These features indicate the Badenweiler–Lenzkirch Zone was located close to a suture zone along which the Moldanubian Ocean closed during the Variscan Orogeny (Fig. 6.1). The suture marks the boundary between the Armorican Terrane Complex in the north and the northern edge of Gondwana to the south, including the Variscan units of the Alps. Although no ophiolitic bodies are known in the Badenweiler–Lenzkirch Zone, their presence during the time of sedimentation in a deep sea trench setting is inferred from the preserved remnants of volcanic rocks and detrital minerals.

The **Southern Black Forest granite–gneiss complex** (4) is also characterized by high-grade metamorphic rock units. Three units are recognized, the uppermost of which is interpreted as a tectonic slice known as **Wiese–Wehra Nappe** (Fig. 10.54). This is composed of almost completely melted migmatite (also known as anatexite or diatexite) derived from Ordovician parent rock. Underlying the Wiese–Wehra Nappe is the **Todtmoos Gneiss** unit that experienced regional metamorphic temperatures of 1000–1100 °C and thus indicates burial depths of 40–50 km. These represent the highest grade metamorphic rocks of the Black Forest in terms of the maximum temperatures reached without melting. The third unit is the **Hauenstein-Murgtal Gneiss**, which also contains migmigmatites developed under significantly lower, amphibolite facies conditions. In the Southern Black Forest, several granite plutons intruded in quick succession into the highly metamorphosed gneisses during the late phase of the Variscan Orogeny. In places, the granitic melts ascended close to the surface and formed rhyodacitic and lamprophyric dikes.

Similar to the Harz and the Erzgebirge, the Black Forest was also affected by hydrothermally formed **Pb–Zn mineralizations** with high Ag content, which have been mined since the Middle Ages (Box 11: Ore deposits). The last Black Forest mine where nonferrous metals were excavated closed in 1954. The mining of **fluorite** and **barite**, however, still continues today in the Clara Mine of the Rankach Valley, and represents the last active mine of the Black Forest. A number of former mines are now open to the public as interesting sites to visit. The vein mineralizations of the Black Forest originated during post-Variscan extensional tectonics related to the Jurassic and Early Cretaceous opening of the Central Atlantic, as well as to Tertiary rifting of the Upper Rhine Graben (Pfaff et al. 2009).

11 Permian and Mesozoic Geology of Germany

Following the Variscan Orogeny, the supercontinent of Pangea began to break apart in the Permian: an event that initiated extensive continental volcanism. In the northern part of Central Europe, the Central European Basin started to subside in the Late Permian. The southern part of the basin, and later the northern part were dominated by a multifaceted history that began with small depressions where continental sediments were deposited. These areas later evolved into epicontinental shallow seas and eventually developed open marine conditions. Large parts of Germany were covered with sediments of Mesozoic age. Some areas of the Variscan basement, however, were reexposed during subsequent tectonic movements and erosion.

11.1 Post-Variscan History

The end of the Variscan Orogeny marked a turning point of geological events worldwide. During the Permian, almost all continental plates merged to form the large supercontinent of **Pangea** (named by Alfred Wegener; Chap. 7). In the east, Pangea was surrounded by the **Paleotethys Ocean** (named after the Greek sea goddess *Tethys*, the sister of *Okeanos*), which was already subject to subduction. This ocean was joined by island arc systems with the globally extensive **Panthalassa Ocean** and to the south with the still narrow **Neotethys Ocean**, which reached its greatest extent during the Triassic (Fig. 11.1). Due to its large size, the supercontinent had already developed instabilities that formed directly after the Variscan Orogeny. The continental mass was affected by initial crustal strike-slip movements along fault and fracture zones, and stretching of the lithosphere, which led to extensional basin development and finally to the breakup of Pangea. This final event marked the beginning of a new Wilson Cycle (Chap. 5) that continues to this day and will be completed in the distant future only by the formation of a new supercontinent.

At the end of the Variscan Orogeny during Late Carboniferous to Early Permian times, subducted oceanic lithosphere of the Rheic and Moldanubian oceans is considered to have broken off from the continental lithosphere (Fig. 11.2.). This detachment is known as **slab breakoff** whereby the heavy oceanic lithosphere sinks into the asthenosphere and the continental lithosphere remains buoyant due to its lower density (Box 22: Slab breakoff). The sinking oceanic slab is replaced by molten asthenospheric material that moves upward to increase the heat flow. This leads to thermally driven thinning of the lithospheric mantle beneath the continental crust and eventually to intense volcanic activity. However, since the intruding hot asthenospheric material is lighter than the lithospheric mantle, **isostatic compensation** initially results in the uplift of the continental plate (Box 21: Isostatic compensation and mountain uplift).

Such uplift enhanced the rate of erosion of the newly formed Variscan Mountains prior to local extension. This occurred in conjunction with strike-slip faulting and local subsidence in graben structures where eroded detritus accumulated. Following the extensional collapse, the thickened Variscan continental crust was reduced back to its normal thickness of 30–35 km by about the Middle Permian, following a period of thinning that lasted for 30–40 Myr. The addition of new lithospheric mantle to the base of the lithosphere, however, continued for a considerably longer period of time, with thermal subsidence leading to almost continuous sedimentation during the Mesozoic (see explanation in Sect. 11.2.1). From about the Middle Permian onward, large-scale basin development dominated the geological history of Central Europe throughout the remaining Mesozoic.

M. Meschede and L. N. Warr, *The Geology of Germany*, Regional Geology Reviews,
https://doi.org/10.1007/978-3-319-76102-2_11

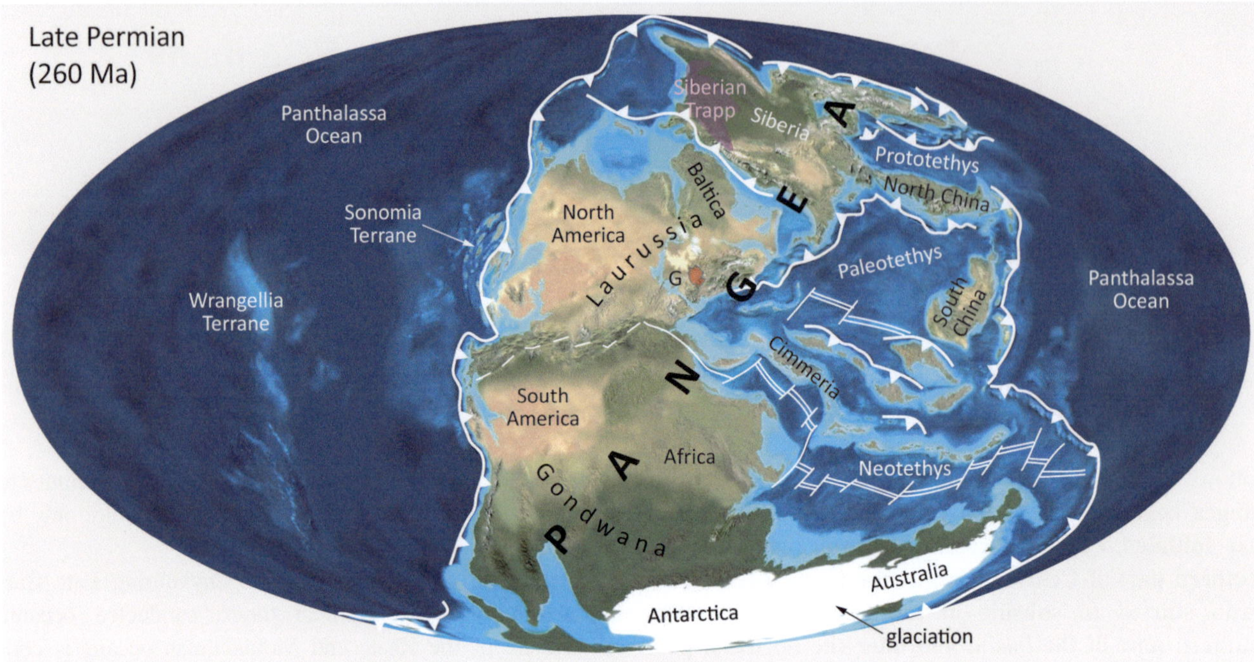

Fig. 11.1 Distribution of continents and oceans during the Late Permian (260 Ma). G—approximate location of Germany. The topographic base for the paleogeographic map was made available by Ron Blakey, Flagstaff, Arizona (Blakey 2003)

11.2 Permian

11.2.1 Rotliegend

In Central Europe, the transition from a compressional to extensional tectonic regime occurred in the Permian. Until the end of the Carboniferous, folding and compression dominated with formation of the Variscan Mountains during the Devonian and Carboniferous. The first signs of crustal stretching were initiated at the end of the orogenic phase in the Early Rotliegend (earliest Permian). At this time, a NNE–SSW oriented graben system was formed in the area north of the Variscan Thrust Front (Fig. 11.3; Gast 1988) and was also active later during the Mesozoic when additional subsidence occurred. In the Hessian Depression, a more than 1000 m thick sequence of Cornberg sandstone (top Rotliegend) was deposited. These rocks contain a variety of reptile fossils together with their traces, which provide an insight into the living environment at that time during the Rotliegend. Numerous smaller transtensional **intramontane basins** formed further south in the central elevated parts of the Variscan Mountains. In these depositional sites several hundred meters of thick terrestrial sediments and volcanics accumulated. These spatially narrow basins are distributed in large numbers all over Central Europe and are often located at the intersections between extensional and lateral displacement faults. These structures are considered to originate from the relative westward movement of the African Plate during the final stages of the Variscan Orogeny. Examples of some of the larger intramontane basins can be found in the Saar-Nahe Basin, the Saale Basin, the Kraichgau Depression, the Schramberg Trough, and the Burgundy Basin. Other examples found in the northeastern region of the now eroded Variscan Mountains are the Sudetic Basin and the Barnim Basin (Fig. 11.4).

At the beginning of the Rotliegend (Early Permian), intensive and widespread continental volcanic activity occurred across Central Europe as a result of slab breakoff and the ensuing continental fragmentation that resulted in the rise of magmatic melts (Fig. 11.2). Volcanism was mainly located along extensional fault systems and graben structures. This phase of extensional tectonics occurred in direct connection with changes in the plate tectonic regime following the Variscan Orogeny. Convergent movements between Gondwana in the south and Laurussia in the north ceased when the continental plates collided. The main stress directions then fundamentally changed as deep-seated faults, which formerly acted as strike-slip faults were reactivated as extensional faults. The lithosphere below the Variscan fold belt that was thickened during collision was then thinned with brittle upper crustal faults and fractures providing pathways for the rising magmatic melts. In the large northern basinal areas, the Permian volcanics are today buried under a thick sedimentary cover, but they are encountered in boreholes and detectable in geophysical surveys (Timmermann 2008). In the area of the uplifted Variscan basement to the south, they are accessible in many locations at the surface,

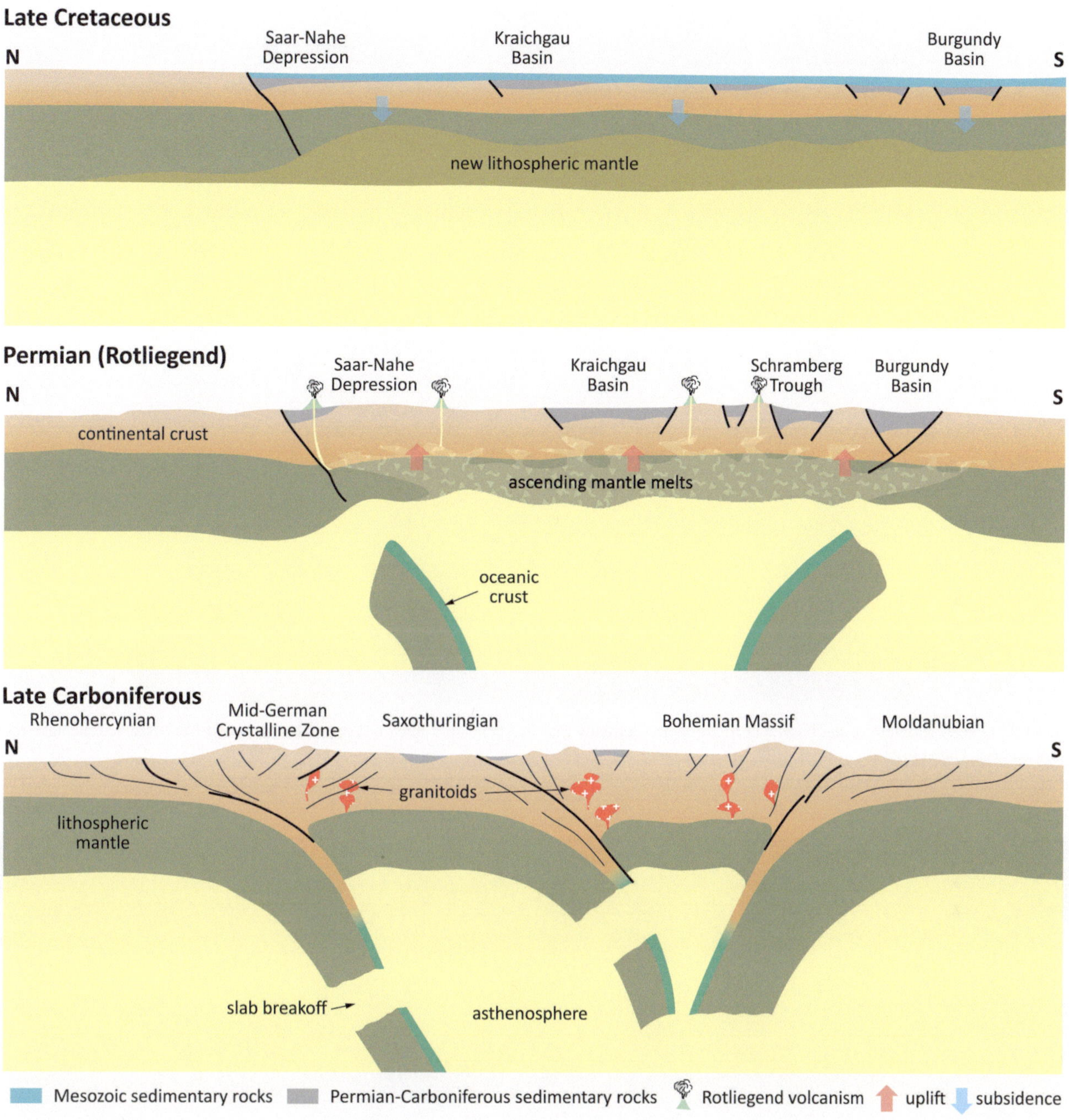

Fig. 11.2 Conceptual model for the development of the Variscan lithosphere from the Late Carboniferous to the Late Cretaceous along a cross-section from northern Germany to the French and Swiss Jura (not to scale). Adapted from Ziegler and Dèzes (2006)

as, for instance, in the Thuringian Forest, around Halle and Chemnitz, or in the large quarries near Flechtingen (Fig. 11.5). Another major occurrence of Rotliegend volcanic rocks is located in the Saar-Nahe Depression, near Bad Kreuznach.

Rotliegend **volcanic activity** was dominated by the surface extrusion of acidic igneous rocks. In addition to the more common rhyolites and ignimbrites, quartz-rich porphyries were also extruded or intruded at shallow levels as lava domes or laccoliths. These rocks were commonly formed together with andesites and basaltic andesites and sometimes with basalts (Breitkreuz et al. 2009). Lava domes are typical features of volcanic formations characterized by acidic rocks because the lava becomes more viscous with increasing SiO_2 content and during flow tends to solidify just below the surface. A **laccolith** is a plutonic body that intrudes and solidifies at shallow depth concordant (layer parallel) with the country rock and is generally characterized

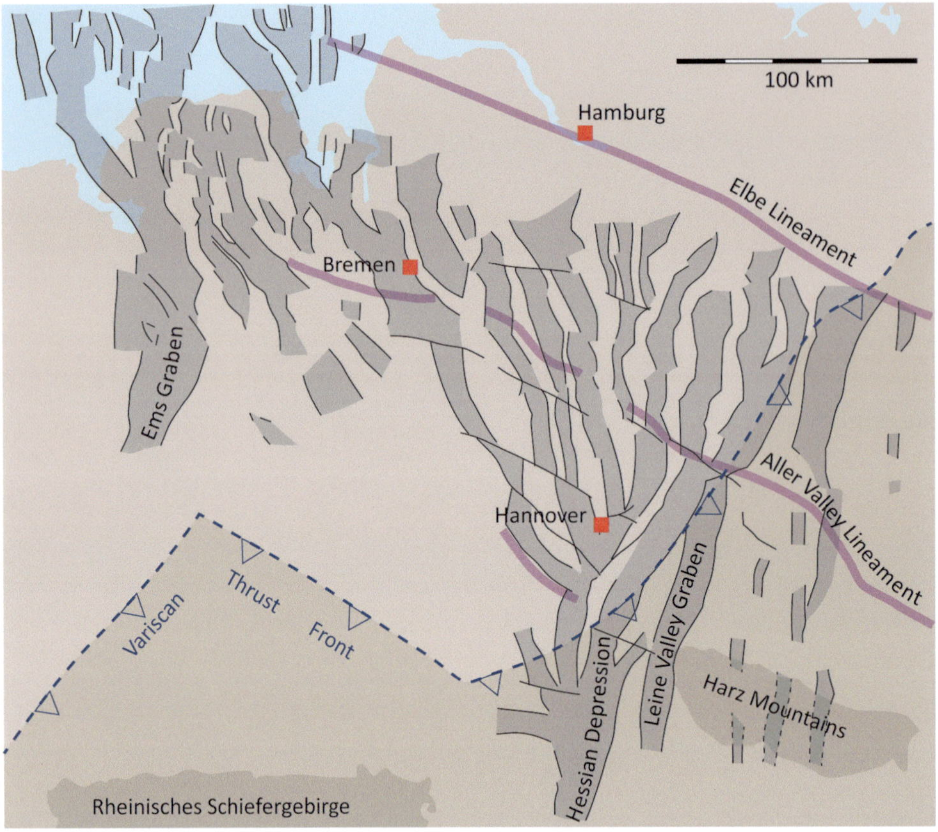

Fig. 11.3 Graben structure of the basement in Northern Germany that lies beneath a thick cover of Mesozoic and Cenozoic sedimentary rock. Modified after Gast (1988)

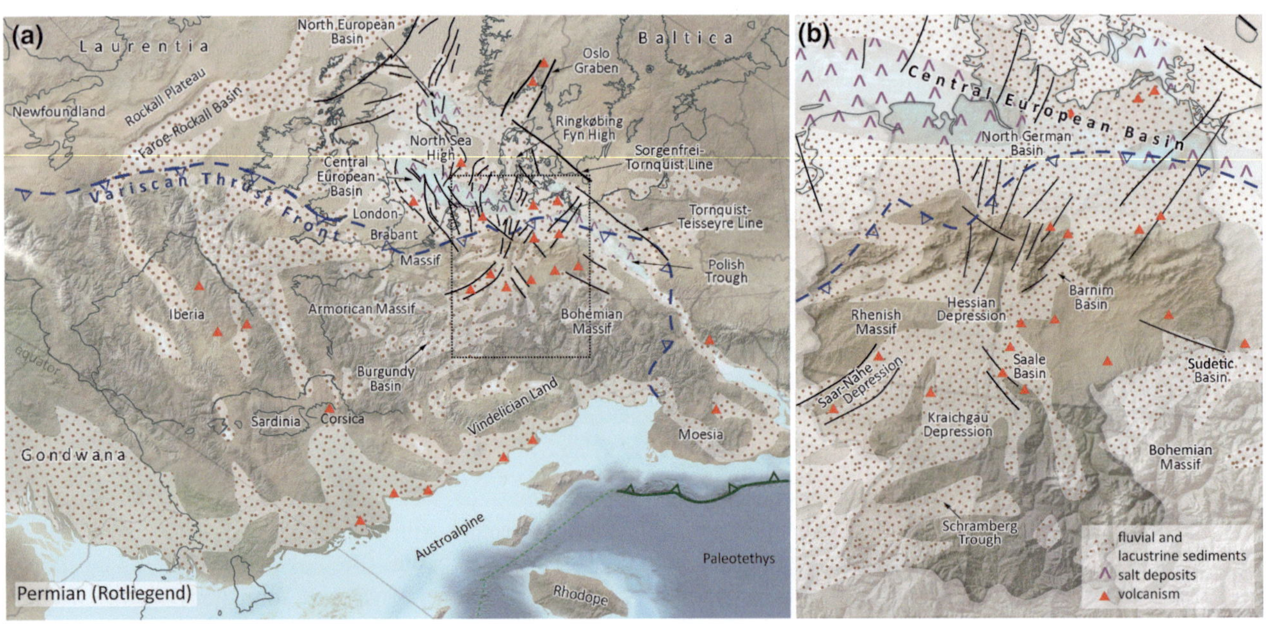

Fig. 11.4 Paleogeographic reconstruction of Central Europe (**a**) and Germany (**b**) at the time of the Rotliegend. Topographic base map from Blakey (2011), modified after Ziegler (1990)

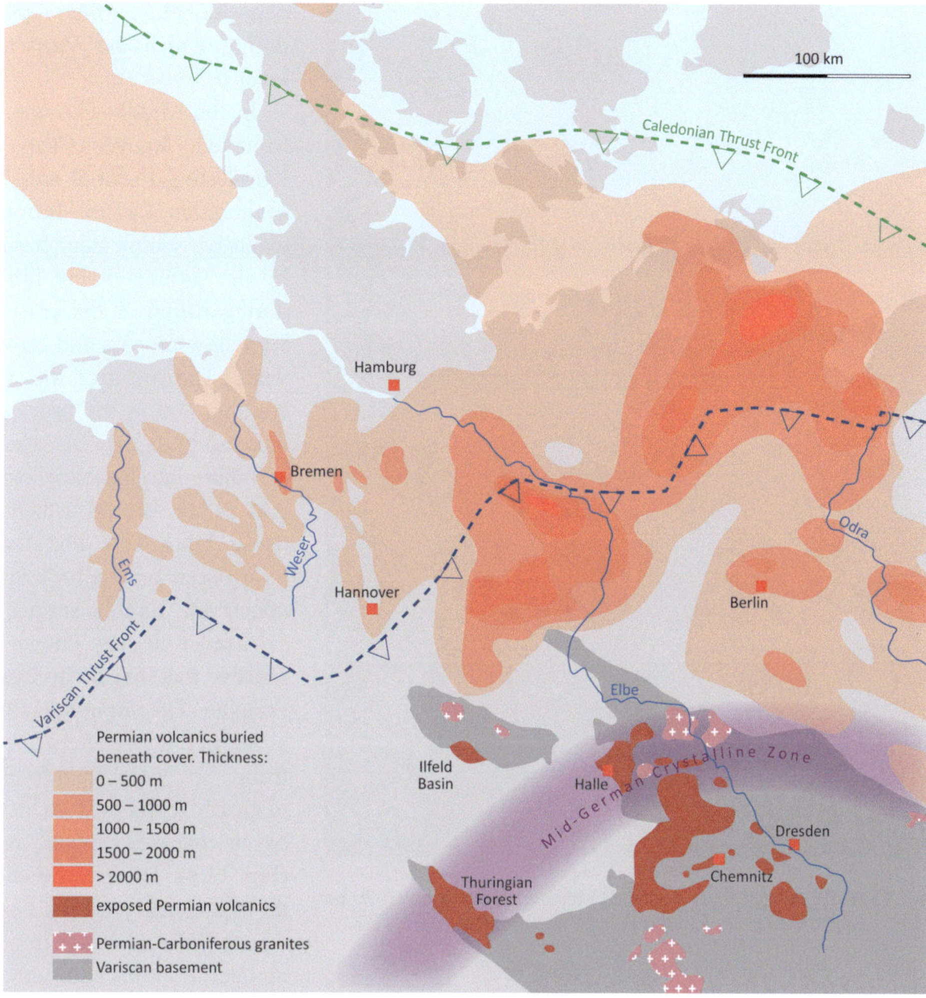

Fig. 11.5 Distribution of Rotliegend volcanic rocks in the North German Basin and adjacent areas to the south. Modified after Breitkreuz and Kennedy (1999); Timmermann (2008)

by a flat base and an arched roof. Locally, some pumice stone and welded glass-bearing tuff are generated by the highly explosive volcanic activity, which is extruded as pyroclastic flows. Sometimes granitic magmas crystallize at depth and in some rare cases gabbros form by fractionation.

Coinciding with the rhyolitic dominated volcanism of the North German Basin and in Central and West Germany, the **Oslo Graben** began to form (Fig. 11.4). The Oslo Graben is dominated by widespread intrusive rocks such as monzonite (including a variety known as larvikite, named after its location at Larvik) or syenite. Extrusive alkaline magmatic rocks are also abundant whereas rhyolites are less common. Typical volcanic rocks of the Oslo Graben are **rhombe-porphyry** (Fig. 11.6), which are frequently encountered as glacial boulders in northern Germany. They are distinguished by their characteristic diamond-shaped feldspar crystals and petrographically are classified as trachyte (the volcanic counterpart of larvikite or syenite; Fig. 3.5). The Oslo Graben was active mainly in the Permian, although subsidence continued into the Triassic. Sediments in the graben reached a total thickness of several thousand meters.

To the south, the rift system of Neotethys extended toward the northwest and remained active into the Jurassic, before leading to the separation of the Adriatic–Apulian plate from the large, at that time still intact, southern continent of Gondwana. This rifting was associated with strong volcanic events in the Early Permian, such as the **Etschtal Volcanic Complex** of the Southern Alps, which is made up of a thick volcanic and intermediate sedimentary succession dominated by rhyolites at the base of the Neotethys sequence. This volcanic complex was previously named the Bozen Quartz Porphyry, but this term is today only used locally. The term quartz porphyry is now obsolete as a rock term and is referred to as a rhyolite.

Another striking example of continental volcanism in Central Europe can be found in the northwestern part of Corsica in the caldera of Monte Cinto. These rocks originated at the base of the Permian when rhyolitic magmas

Fig. 11.6 Rhomb porphyry from the Oslo Graben. Location ca. 20 km northwest of Oslo at Krogskogen Hill [N 60°03′17″/E 10°19′35″]

were extruded at the margin of the Neotethys Ocean during spreading (Fig. 11.4). The cliffs along the Calanche at Porto, with its scenic bays and picturesque weathering features, belong to this volcanic complex. They were declared a World Heritage Site by the UNESCO in 1983.

Chronologically, the magmatic activity was concentrated along the boundary between the Carboniferous and Permian. It began in the Stephanian (ca. 302 Ma) and terminated in the Early Rotliegend (between around 280–290 Ma). Across Central Europe, this phase of strong magmatic activity was followed by sporadic but significant volcanic events during the rest of the Permian period. The magmatic rocks of Permian age are dominated by volcanic formations. Intrusive rocks are significantly less common, although they are likely to be present in larger volumes concealed in the upper crust. Sedimentary rocks formed at that time were preserved only sporadically because the Early to Middle Permian was marked by erosion and associated volcanic eruptions. The base of the Late Rotliegend sedimentary sequence is often marked by the so-called Saalian Unconformity, where rocks of the Late Rotliegend overly rocks deformed during the Variscan Orogeny resulting in an angular unconformity. The depositional gap often represents more than 40 Myr in time, during which the Variscan Mountains were eroded and leveled.

In the Middle Permian, after the uplift of the Variscan Mountain Range (Chap. 10.2.2; Fig. 11.2), a stage of long-lasting **thermal subsidence** began. This was centered in northern Central Europe and led to the formation of extensive sedimentary basins. Thermal subsidence is driven by the cooling of hot lithosphere that has moved upward during rifting of the crust. Once initiated it is generally a very slow process and can last for many 10's of millions of years. The rocks of the upper lithosphere contract during cooling, and in the process become more compressed and increase in density. As a result, the heavier rock sinks more and more into the asthenosphere. In addition, the uppermost part of the plastic asthenosphere is converted into solid, dense lithosphere that helps bring the thickness of the lithospheric mantle back to its normal level (Fig. 11.2). This affects the Earth's surface by the gradual but continuous subsidence at sites known as sedimentary basins. It is this process that began in Central Europe during the Middle Permian. The intramontane basins, which primarily occurred as isolated depocenters, were filled with the detritus derived from the Variscan Mountains. These depositional sites gradually combined to a few larger and more expansive basins that were spread over large parts of Central Europe (Fig. 11.4). However, due to their initial isolation, each of the sedimentary basins has its own history, making correlations difficult, if not impossible. As a result, quite a large number of basins and subbasins exist in which very different formation names are used. According to the recent geological nomenclature, the sedimentary fill of the Permian basins is to be regarded as **molasse of the Variscan Mountains**. It consists of erosional detritus primarily composed of terrestrial or shallow marine deposits that accumulated either at the foothills of a mountain range or in intramontane basins.

The assemblage of separate landmasses to form a single large continental area and the simultaneous northern drift of the Pangea supercontinent led to far-reaching **changes in the global climate**. Following the widespread development of the Carboniferous **tropical rainforest**, the climate became drier with **arid** (from Latin *aridus*: dry) **conditions** prevailing. Water no longer reached the now far-distant inland, as rainfall became more confined to regions closer to the coast. Extreme seasonal differences resulted in the onset of the arid climate along the inner part of the large continental masses. Wet, monsoon-dominated regions developed only along the coastal areas of the Paleotethys and Neotethys oceans. As Pangaea moved 10–20° toward the north after the Carboniferous, the tropical rainforests changed to the drier climates characteristic of the Permian.

Despite the Late Carboniferous/Early Permian climatic changes that arose around equatorial latitudes, a period of

glaciation lasting over several tens of millions of years occurred at the South Pole and probably also at the North Pole. The maximum extent of the glaciation was reached during the Westphalian stage of the Late Carboniferous (Chap. 10.3), with the ice cover continuing into the Permian and locally also the Triassic. Since the North Pole was located as it is today in an oceanic area, any evidence of glaciation at the northern polar cap was not preserved. The main glacial phase ended in the Early Permian before about 290 Ma when the global climate became considerably warmer. The onset of the Late Carboniferous glaciation is considered to have resulted from the large-scale removal of atmospheric CO_2. This reduced the CO_2 **greenhouse effect**, causing the climate to cool and to eventually enter a glaciation phase. The reduction in greenhouse gas had already started well before the emergence of the wetlands that led to the formation of Carboniferous coal and removal of significant volumes of carbon from the atmosphere by plant growth (Fig. 11.7). Therefore, reasons other than the binding of carbon in dead plant matter have been proposed to explain the early removal of atmospheric CO_2. According to Berner (1999), the main cause was related to the extensive weathering of silicate rock. With the progressive rise of vascular plants and the increase in deep-rooted vegetation, there were higher concentrations of humic and carbonic acids that progressively enhanced the alteration of Mg-Ca-silicate minerals. The subsequent release of Ca and Mg ions during weathering then reacted with CO_2 to form carbonate minerals. As a result, the CO_2 was permanently bound into calcareous rock and removed from the atmosphere (Box 10: Silicate weathering).

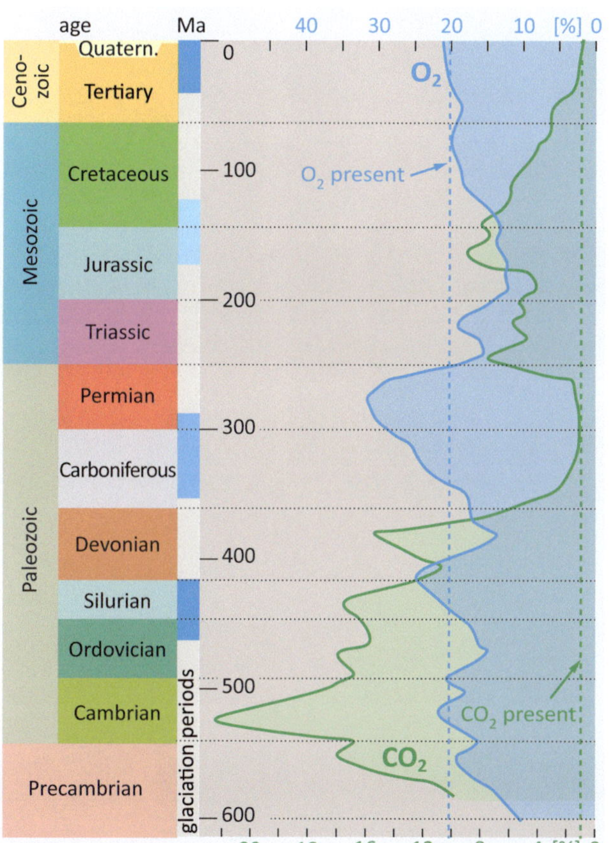

Fig. 11.7 Development of O_2 and CO_2 concentrations in the atmosphere during the last 600 Myr compared to today's actual values (*dashed lines*). Redrawn and modified after Berner (1990), Berner et al. (2000), Berner and Kothavala (2001), Brenchley and Harper (1998)

In addition to the gradual northward drift of Pangea, the massive volcanic events at the beginning of the Permian were responsible for the gradual re-accumulation of CO_2 and other important greenhouse gases, namely, methane (CH_4) and water vapor (H_2O) in the atmosphere. Thus, the global temperature slowly increased and the heating subsequently led to the melting and retreat of the polar ice. While the CO_2 content gradually increased, the concentration of O_2 progressively decreased during the Permian (Fig. 11.7). The largest O_2 drop occurred at the Permian/Triassic boundary in connection with the greatest mass extinction in the Earth's history (Sect. 11.3).

As the German name of the formation indicates, the continental sediments of the Rotliegend are characterized by a distinctive red coloration. The term *Rotliegend* originates from a mining area in the Copper Slate region of Mansfeld. At this location, the geological formation was found as waste rock referred to by the miners as the *Rothes Todtes Liegendes* (old German, red dead footwall), because the Copper Slate always occurred in the footwall of a mineralized layer. Already at the beginning of the nineteenth century, it was introduced as a stratigraphic term to the geological literature. The red colored clastic sediments were deposited as conglomerates, sands, and muds under arid climatic conditions. The characteristic red color is due to the presence of finely distributed iron oxides (trivalent iron, Fe^{3+}) in the form of the mineral hematite (Fe_2O_3). Red beds are important climate indicators in the geological record and reflect sedimentation in tropical or subtropical zones. In arid continental climates, strong oxidizing conditions prevail such that most of the iron released by weathering is incorporated as hematite, a mineral that is particularly resistant to alteration under these conditions. Strong oxidation in the arid environment is also the reason that, unlike other depositional environments, few fossils are preserved, although the conditions are not always hostile to life.

The lithostratigraphic unit of the Rotliegend in Central Europe is in rare cases referred to as the **Early Dyas** (Dyas means "split into two" and Trias "split into three"). Internationally, completely different terms are used that rely on biostratigraphic classifications from the Urals (Cisuralian, Early Permian), Texas (Guadalupian, Middle Permian), and China (Lopingian, Late Permian). These names, however, are not widely used in the description of the geology of Germany. The Rotliegend represents a lithostratigraphic classification that encompasses the Lower and Middle Permian when Central Europe was almost exclusively dominated by the deposition of continental red bed sediments. In these sedimentary facies, a biostratigraphic classification is often not possible due to the lack of index fossils and the difficulty in recognizing the divisions used. Locally, the sedimentary rocks of the Rotliegend are known to start in the Late Carboniferous (Stephanian) and extend into the Late Permian. The unit, therefore, encompasses 45 Myr ranging from 302 to 257 Ma. This very long period was marked by more local sedimentation with detritus derived from a variety of source areas that resulted in varying sediment compositions deposited in singular basins, usually separated from each other. In Central Europe, this stage was characterized by severe erosion and molasse formation during the **leveling of the Variscan Mountains**. Only during the subsequent Zechstein period did the rate of subsidence increase over wider areas, such as that recorded in the North German and Polish basins. An additional change at this time resulted from the melting of the ice sheet at the South Pole. This caused a global rise in sea-level and contributed to the flooding of some continental areas.

At the beginning of the Rotliegend, the climate was still strongly influenced by the tropical humid rainforest that developed at the end of the Carboniferous. Argillaceous (clay-rich) sediments were deposited in large lakes in which numerous fish, freshwater sharks, and large-size amphibians were preserved as fossils. As the supercontinent of Pangea continued to drift northward and the more continental land was located at an increasing distance from the shoreline, the climate changed to dry and hot desert conditions with the drying up of water environments and reduction in fish and amphibian habitats. The land changed from damp, tropical rainforests to steppes and deserts occupied by plants that could withstand the arid climatic conditions. In place of the fish now came the reptiles. These spread throughout the mainland and established themselves as a highly successful group of animals that dominated the next 200 Myr. Among them were the dinosaurs.

During the Rotliegend, there were poor conditions for the preservation of fossils. In water, dead bodies and plants are often quickly incorporated in sedimentary material and under anoxic conditions become preserved. However, in the terrestrial environment organisms decompose within a short time under oxygenated conditions and are thus rarely preserved. Therefore, the continental sediments of the Rotliegend generally contain few fossils. The most common are the imprints of individual plants and occasional trace fossils left by reptiles. The precise species of animals responsible for such tracks are not easy to determine. A special rarity preserved during this time is the **"Petrified Forest"** of Chemnitz (Fig. 11.8), with some fossil trunks on exhibition in the Chemnitz Museum of Natural History. The fossil material is composed of silicified wood preserved by a volcanic event about 291 million years ago. The conserved logs were distorted by the eruption of the **Zeisigwald volcano** (Rössler et al. 2009) when they were stripped of their branches and partially uprooted in a similar way to what occurred during the North American eruption of Mount St. Helens in 1980 (Fig. 11.9; photograph from 2012). During subsequent silicification, the Chemnitz forest was extraordinarily well preserved and in some cases, even the cell structure remained intact. Some specimens were even found with their roots still attached.

Rotliegend sediment is most frequently characterized by **fanglomerate**. An easily accessible outcrop of this rock type exists at the Wartburg Castle in Eisenach (Fig. 11.10). Such rocks are formed as sheet flood deposits in arid desert regions where it rains very rarely and the lack of vegetation allows debris flows. Sheet flows form during heavy rainfall and occur only once or twice a year or sometimes only once each decade. The flows lead to the accumulation of large

Fig. 11.8 The Rotliegend "Petrified Forest" on display inside the Chemnitz Museum of Natural History. The logs originally exhibited in the open air were found in the urban area surroundings of Chemnitz (photograph by courtesy of the Chemnitz Museum of Natural History) [N 50°49′51″/E 12°55′26″]

Fig. 11.9 Flattened trees located above Spirit Lake on the Mount St. Helens volcano in Washington, USA. Traces of the volcanic eruption can be observed even 30 years later in the form of broken and frayed tree branches that lay in the line of the blast [N 46°16′18″/W 122°06′37″]

Fig. 11.10 Exposure of a Rotliegend fanglomerate at the entrance of the Wartburg Castle, near Eisenach [N 50°57′58″/E 10°18′22″]

amounts of unsorted clastic sediments. Fanglomerates are conglomeratic rocks composed of very different grain sizes and with fragments that are typically poorly rounded and unsorted in composition. If the source area is comprised of feldspar-rich rocks, **arkoses** form as the erosional product under arid climate conditions with often more than 30% feldspar content. If water is constantly available as in humid climates, arkoses do not form because the feldspars are highly susceptible to weathering by dissolution. Arkoses are, therefore, a good indicator of arid climates and are common in the sedimentary succession of the Rotliegend.

North German Basin

The North German Basin forms part of a large system of interconnected, intracontinental basins grouped together under the encompassing term: the **Central European Basin**. This sedimentary basin complex developed during the Permian across much of northern Central Europe and during the entire Mesozoic played an important role in the geological development in this area. Several large deposition cycles, in which significant amounts of sedimentary and volcanic lithologies accumulated, are recognized spanning from the Permian to today (Bachmann 2010). These deposits reached an impressive total thickness of up to 10,000 m in the deepest part of the basin lying north of Hamburg (Fig. 11.11). The Central European Basin is the largest and most important basin of Central Europe and extends from the North Sea to Poland. The middle area of the basin includes the **North German Basin** and the **Polish trough** forming its eastern continuation (Fig. 11.4). The Central European basin developed in an area between the Variscan Mountains to the south and the **North Sea High**, the **Ringkøbing-Fyn High** and the **Tornquist–Teisseyre zone** in the north. Its northern boundary is hidden beneath thick sedimentary cover and can be traced across northern Germany along a line extending from Usedom at the Baltic coast, south of the Rügen Island, across to the Danish

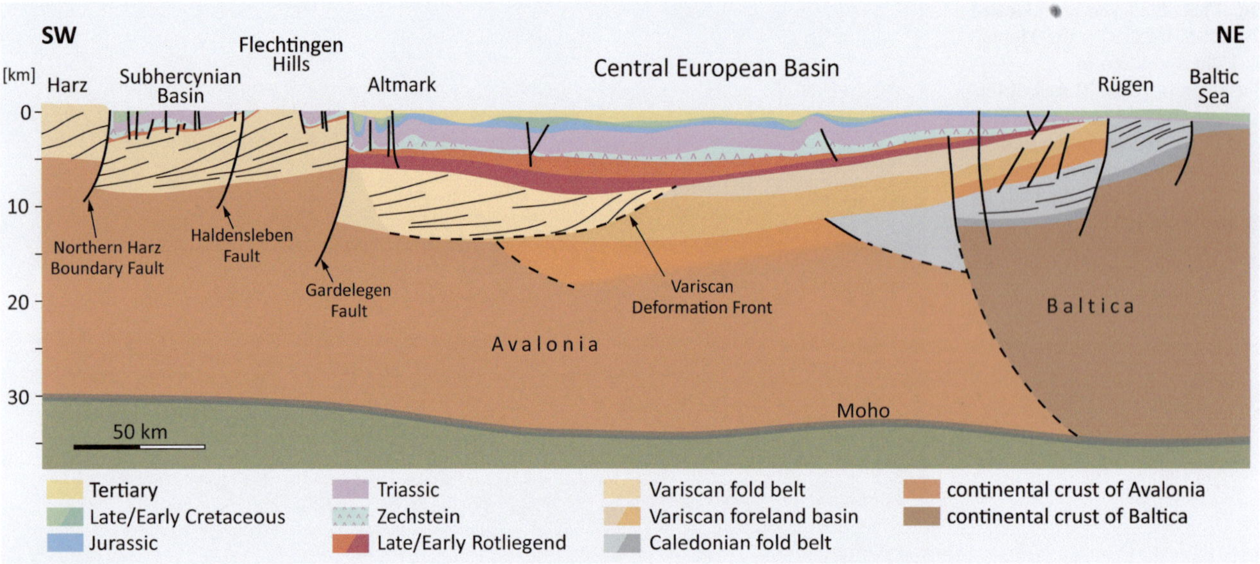

Fig. 11.11 Cross-section through the Central European Basin from the Harz Mountains to the Baltic Sea (strong vertical exaggeration). Redrawn and modified after Krawczyk et al. (2008)

island of Møn. In this region, the sedimentary rocks of the Rotliegend are today partly buried at depths more than 3000 m beneath younger sedimentary sequences. However, the present basin sediments of the Baltic Sea have nothing to do with the subsidence that occurred during the Permian and Mesozoic as the southern edge of the Baltic Sea is located almost directly above the northern edge of the Permian North German Basin. Surface outcrops of the oldest deposits of the North German Basin are in fact rare. Only on the southern edge of the basin, following a WNW–ESE line extending from Lusatia to the cities of Halle and Magdeburg, do some outcrops of the Rotliegend occur. These lie approximately along the old Variscan Thrust Front. This southern edge is referred to as the area of the Central German major escarpments.

After an Early Permian rift stage that affected the northwestern edge of the Central European Basin, with the opening of the **Faroe Rockall Basin** (Fig. 11.4) and related extension occurring between Greenland and Norway, more widespread basinal subsidence began in the Middle Permian (Glennie et al. 2003). As previously highlighted, the earliest episode of rifting resulted in N–S oriented graben systems in northern Lower Saxony and further north in the Oslo Graben. At these sites, localized subsidence and deposition of thick deposits of clastic sediments occurred at the end of the Carboniferous and beginning of the Permian (Fig. 11.3). In contrast, the **Glückstadt Graben** in Schleswig-Holstein and the **Horn Graben** in the eastern North Sea did not start subsiding until after the Early Rotliegend. These episodes of graben formation and basin subsidence were linked to the movements of the African Plate, which during the late stage of the Carboniferous/Permian collision drifted westward resulting in deep-seated strike-slip faulting across Central Europe (Ziegler 1994). Small, en echelon arranged basins developed that are considered characteristic of the later Central European Basin complex.

Thick layers of volcanic rocks occur at the base of the Central European Basin succession. These are widely distributed and typically of andesitic composition. At a later stage, the composition changed to mainly rhyolitic magmas. The volcanic rocks of the Altmark Group at the southern edge of the basin, radiometrically dated to an age between 295 and 299 Ma (Breitkreuz et al. 2007), reach an impressive thickness of more than 2500 m. Following this period of volcanism, and from about the Middle Permian onward, a long-lasting stage of thermal subsidence occurred, related to isostatic movements of the lithosphere (see above, Fig. 11.2). During this subsidence, an up to 1500 m thick succession of continental sediments was deposited in the depocenter of the basin during the Late Rotliegend (Fig. 11.11). In the North European Basin, which extends from the northern North Sea to Scandinavia (Fig. 11.4), the Permian sedimentary rocks reach thicknesses of up to 600 m (Gast et al. 2010). In contrast, all other Central European Rotliegend basins are smaller than the Central European basin that contains the thickest sedimentary and volcanic sequences of this period—a trend that continued into the Zechstein.

Initially, continental sedimentation dominated all of the Permian basins and was typical for the arid climate conditions. Tectonically controlled small-scale subbasins formed as half-graben structures bounded on one side by active extensional faults that controlled the rapid deposition of alluvial fan deposits and river sediments. Later during the onset of thermal subsidence, shallow sedimentary environments characterized by **playa sediments** (mudstones and evaporites; Box 12: Playa sediments) developed with alternating deposits originating from braided rivers and wind-blown dune systems. This environment was periodically flooded by short-lived marine ingressions. In some locations, a marine influence is documented by the occurrence of calcareous rocks containing numerous tetrapods (Walter and Schneider 2008; Eichler and Werneburg 2010), for instance, in the Döhlen Depression near Dresden and in the northern foothills of the Sudetic Basin (Fig. 11.4). At the margins of the flat basins, **dune sediments**, fluvial sand, and fanglomerates predominated, the latter containing boulders composed mostly of rhyolitic rocks of Early Rotliegend origin. Within these sedimentary sequences, channel fills known as **wadis** (Arabic *al-wādī*: dry valley) have been well documented. The wadis dry river valleys of arid areas only contain water derived from heavy rainfall, resulting in flash floods capable of moving large amounts of coarse detritus.

Rocks of the Rotliegend sedimentary facies crop out only in a few places in the Central European Basin. In most cases, the rocks are buried beneath younger thick layers of sedimentary rocks and are encountered in some boreholes. In the **Flechtingen Hills**, which were uplifted in the Cretaceous by compressive tectonics (Chap. 11.7.2), volcanic rocks from the Early Rotliegend are actively mined in several quarries. The rocks occur as porphyry deposits formed as ignimbrites and lavas. Sedimentary layers of the Rotliegend with interbedded tuff are found as local intercalations at the base of the sequence. Andesitic sills (layer parallel intrusive bodies) also intruded into the surrounding weakly solidified sediments of Stephanian age (Latest Carboniferous). These were related to the larger scale intrusion of Permian granites that were generated during the late stages of the Variscan Orogeny (Flechtingen and Roxförde granites). The younger rocks of

Box 12: Playa sediments

The term **playa sediment** is frequently used to describe sedimentary deposits formed in the alkali flat (sabkha) environment, characteristic of the arid climatic conditions of desert regions. The term has nothing to do with a beach, as the name suggests (*playa* is Spanish for beach). Rather, it represents a very shallow depression, desert or near coastal environment that is intermittently flooded during heavy rainfall, whereby only a few centimeters of water depth is sufficient to form a large but temporary lake (Fig. 11.12). In this environment, primarily argillaceous sediments are deposited that are often mixed with salt and some sand. Due to the high salt content, the depressions are referred to as salt flats. After rainfall, a shallow silt flat or playa lake forms in the middle of the basin. This occurs only sporadically and evaporates quickly due to the arid climate. Playa sediments are typically characterized by alternating layers of muds, silts, dolomite, gypsum, and sand, with occasional pure layers of salt. The salt deposits frequently display typical depositional sequences ranging from Ca and Mg carbonate ($CaCO_3$, $MgCO_3$), gypsum ($CaSO_4 \cdot 2\ H_2O$), Na-sulfate or Na-carbonate (Na_2SO_4, Na_2CO_3) to halite (NaCl, normal table salt). In rare cases, it may lead to the precipitation of exotic minerals such as borax ($Na_2B_4O_7 \cdot 10\ H_2O$). Another common characteristic of playa sediments is the formation of desiccation cracks that form as the evaporites dehydrate.

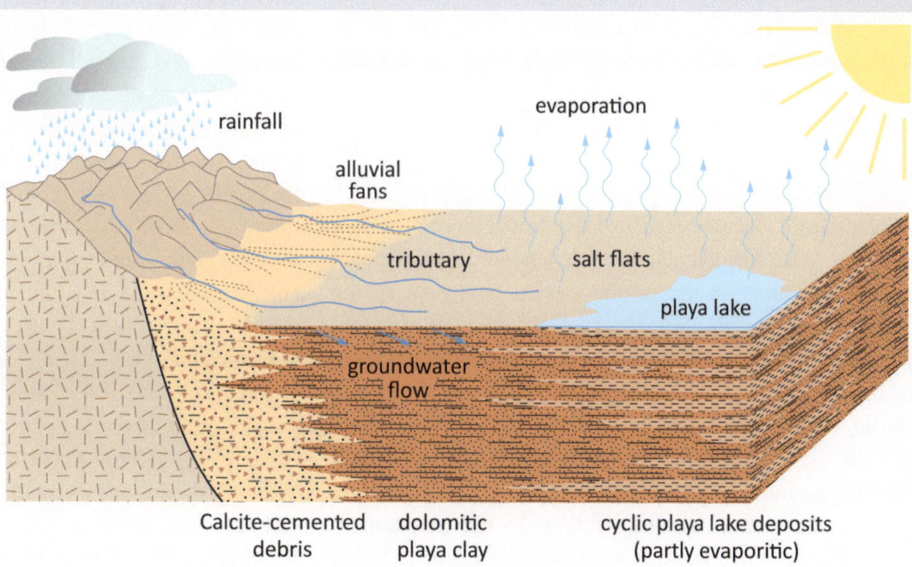

Fig. 11.12 Evolution of playa sediments under arid climatic conditions characterized by high evaporation rates, sporadic water supply and the formation of ephemeral, cyclically occurring playa lakes. Modified after Eugster and Hardy (1975)

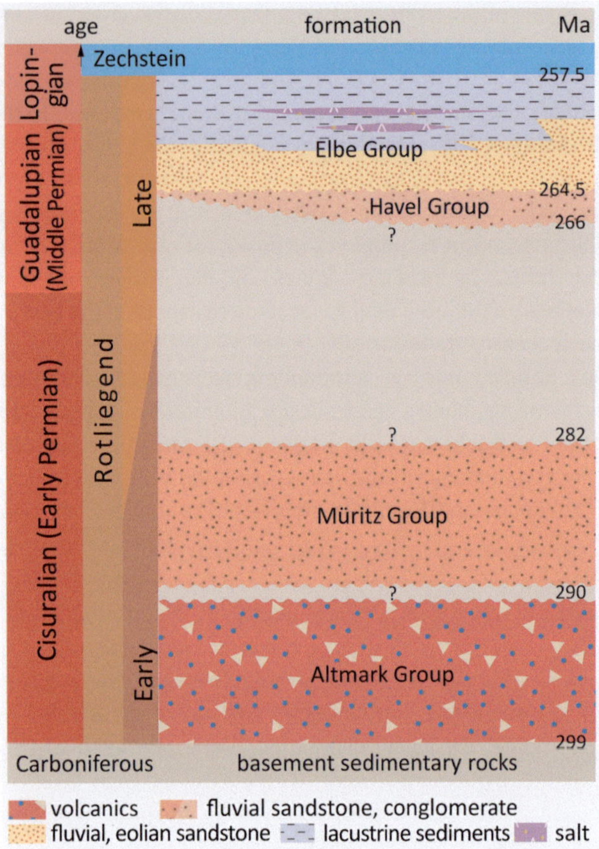

Fig. 11.13 Rotliegend stratigraphy of the North German Basin and adjacent regions. Modified after Menning et al. (2002), and the Deutsche Stratigraphische Kommission (2016)

the Middle to Late Rotliegend, comprised of sandstones and conglomerates, were deposited above an unconformity (Fig. 11.13). In the southeastern prolongation of the Flechtingen Hills lies the Magdeburger Domfelsen. At this location, flat-lying sandstones of the Early Rotliegend can be observed below the Magdeburg Cathedral at low water in the riverbed of the Elbe. These rocks play an important role in stabilizing the banks of the river. Volcanic rocks are notably absent in this region.

Saale Basin

The Saale Basin, located south of the North German Basin (Fig. 11.4), was formed as an independent intramontane sequence deposited on top of the eroded basement of the Variscan Mountain Range. Most of the detrital material entering the basin was therefore derived from the denudation of the orogenic mountain belt. Relatively rapid subsidence in the basin evolved along structures defined by lineaments within the basement. The basin-fill comprised of clastic sediments, including conglomerates of Westphalian age, have been encountered during drilling. In the Stephanian, rapid basin subsidence was accompanied by the formation of coal seams. All these sediments were deposited as molasse sediments of the Variscan Mountains during a period spanning from the Late Carboniferous to Permian and therefore collectively known as Permo-Carboniferous strata. As the formation of these basinal sequences was strongly controlled by the underlying basement structure with sedimentary facies very similar to the overlying rocks, these deposits are described as **transitional level strata** (*Übergangsstockwerk*; Kastner et al. 1995).

The Early Rotliegend sedimentary strata are typically found interbedded with volcanic rocks. In the central Saale Basin, sedimentation continued throughout the entire time of the Rotliegend. A connection to the Central European basin is only known during the deposition of the Late Rotliegend Eisleben Formation. In many places, nondeposition hiatuses or angular unconformities are observed. Most notable is the **Saalian unconformity**, such as that exposed in the *Tal der Heiligen Reiser*, near Hettstedt, in the area of Mansfeld (Mansfelder Land; Fig. 11.14). Here, shales and sandstones of the Latest Carboniferous are overlain by conglomerates and sandstones of the Late Rotliegend (Eisleben Formation) and are separated by a distinct angular unconformity. The time gap between the two layers is about 40 Myr. During this time, various extensional and compressional movements occurred in Central Europe due to the effects of isostatic rebound. These tectonic movements induced tilting of previously deposited layers and subsequent erosion of the uplifted blocks. The subsequent discordant deposition of clastic sediments in the Late Rotliegend reflected the previously mentioned long-lasting phase of thermal subsidence.

The Saale Basin contains an extensive suite of igneous rocks. In the northeastern part of the basin, large quantities of porphyry overlie a sequence of several volcanic and subvolcanic formations comprised of andesitic rocks. Due to the relatively coarse grain size of the porphyry rocks, they may be partly considered as microgranites. However, these rocks are more accurately described as porphyritic rhyolite laccoliths and known under the general name: the Halle Laccolith (or Volcanic Complex of Halle). There is evidence of some explosive volcanic events with the formation of pyroclastic rock deposits. The laccoliths, with ages varying from 301 to 294 Ma (Breitkreuz and Kennedy 1999; Breitkreuz et al. 2009), were intruded at very shallow depths into the sediments of the Latest Carboniferous and Early Permian (Fig. 11.15). In the south of the Saale Basin, pyroclastic rhyolites, lavas, and subordinate basic volcanics mostly occur.

According to Breitkreuz et al. (2009), five stages of development can be recognized in the **Volcanic Complex of Halle** (Fig. 11.16). These are considered to be typical of most of the basins formed in the Early Permian and are summarized as follows: (1) Formation of Westphalian and Stephanian clastic sedimentary sequences with

Fig. 11.14 Rotliegend sedimentary rocks near Hettstedt (Mansfeld area) in the *Tal der Heiligen Reiser*. Late Carboniferous red shales are discordantly overlain by Late Rotliegend sandstones and conglomerates forming the Saalian unconformity. The hiatus represents about 40 Myr of erosion [N 51°38′59″/E 11°31′24″]

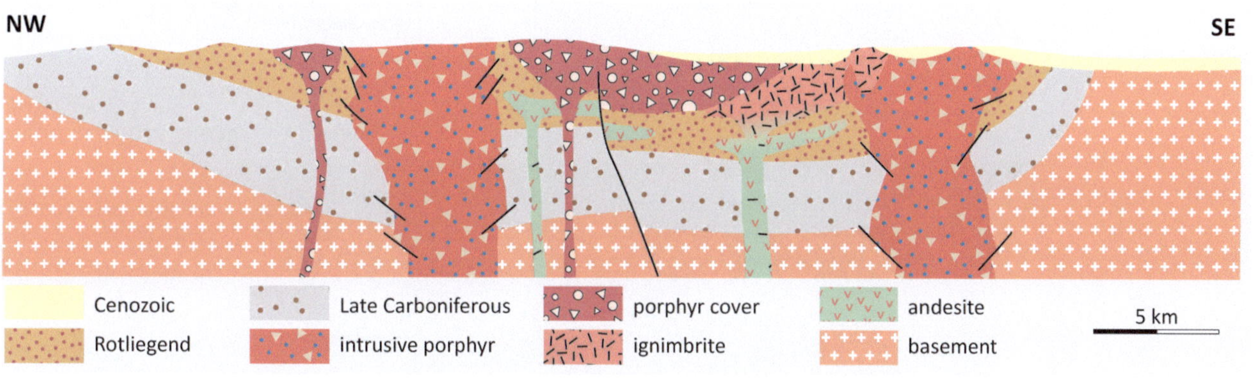

Fig. 11.15 Cross-section through the Halle Volcanic Complex After Breitkreuz and Kennedy (1999)

conglomerates, sandstones, and mudstones, as well as isolated carbonate layers and intermediate coal beds. (2) Development of subvolcanic andesitic intrusions of Carboniferous/Permian age that were emplaced into shallow buried sedimentary sequences. (3) The occurrence of the main phase of volcanic activity consisting largely of rhyolitic magmas intruded as laccoliths and erupted superficially forming porphyritic lava flows and ignimbrites. These volcanics are well exposed in the more northerly located outcrops of the Flechtingen Block, west of Magdeburg. (4) Emplacement of laccoliths and volcanic domes resulting in a highly differentiated surface relief with rhyolitic breccias and conglomerates accumulating in depressions. (5) Final late stage volcanic activity of Permian age (partly explosive) with aphanitic (=fine-grained, glassy) lava rich in silica.

Ilfeld Basin

Another large complex of Rotliegend sedimentary and volcanic units occurs in the Ilfeld Basin at the southern edge of the Harz Mountains. It comprises an 800 m thick sequence of basalts, andesites, and rhyolites that are for the most part discordantly overlain by Rotliegend sedimentary rocks (Fig. 11.17). The basaltic volcanics are often referred to as "melaphyre": a term that is used locally for basalts of

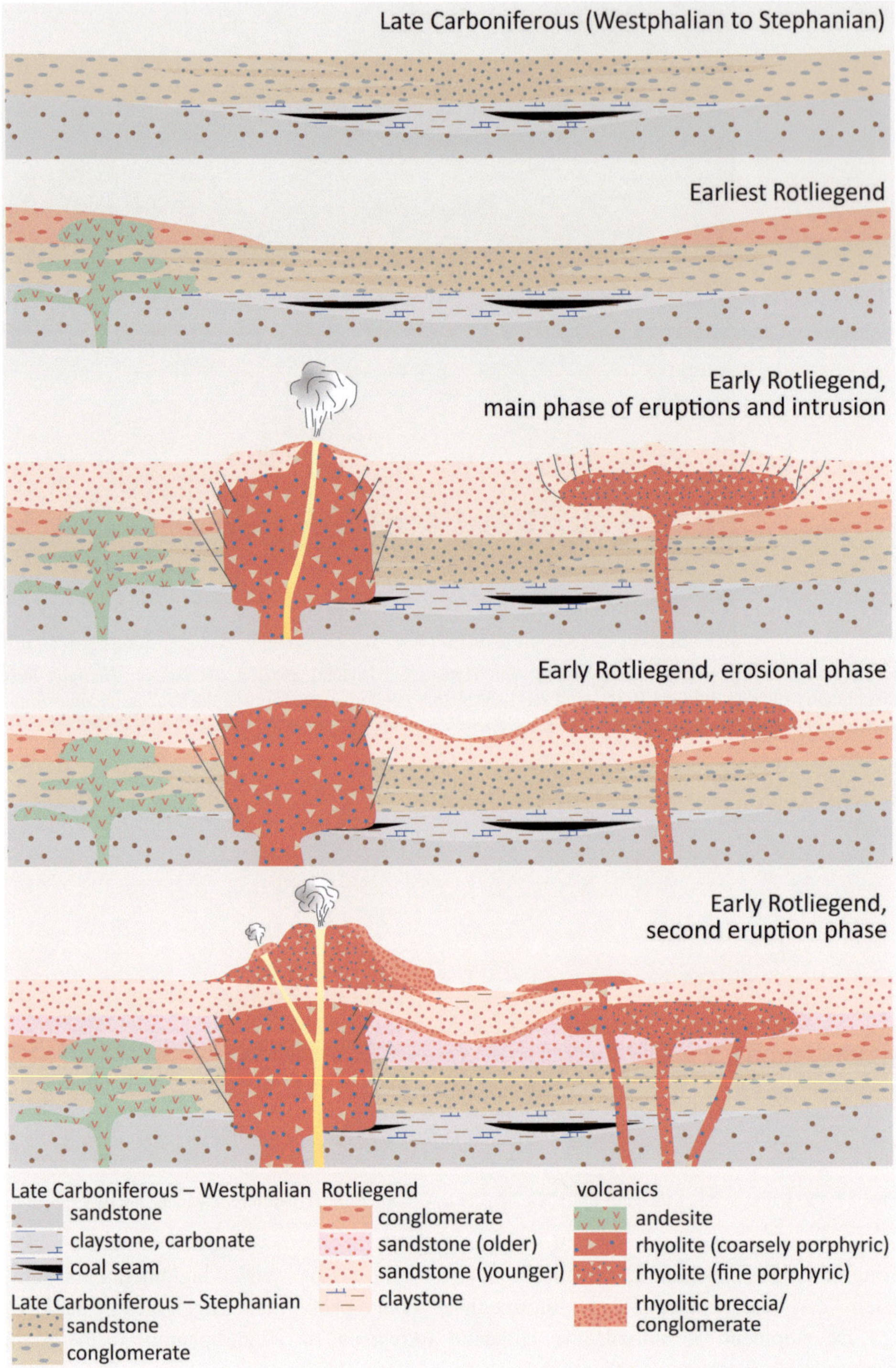

Fig. 11.16 Evolution of the Halle Volcanic Complex marked by several phases of eruption and erosion. After Breitkreuz et al. (2009)

Permian age. Radiometric age dating of the rhyolite yields ages of around 298 Ma, and therefore, they are grouped together with the general age range of the Permian volcanic rocks occurring mainly at the base of the Rotliegend sequence. It is assumed that the Ilfeld Basin evolved as a **pull-apart basin** along older transtensional faults within the Variscan basement. The thick volcanic layers produced at this time were also connected to underlying faults (Breitkreuz and Mock 2004). Directly above the fanglomerate located at the base of the sedimentary sequence, lie partly mineable coal seams with thicknesses of 30–40 cm. Coal mining ceased in this area toward the end of the nineteenth century but was revived for a short time after the Second World War due to the extreme fuel shortage.

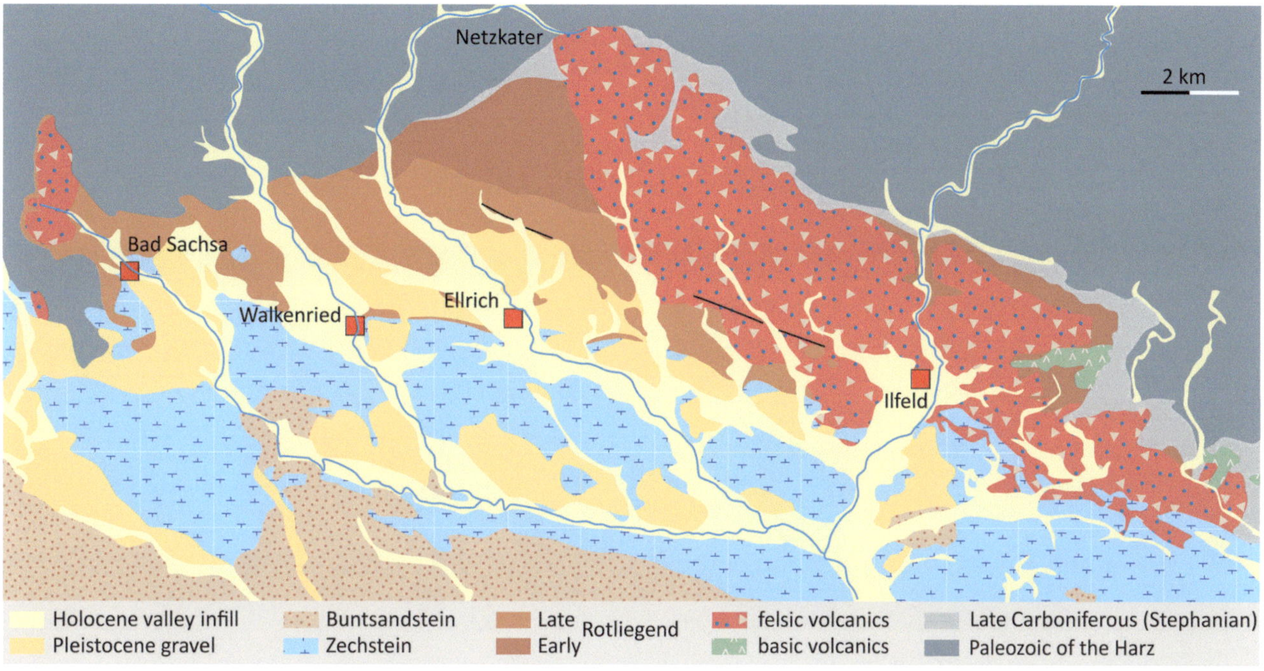

Fig. 11.17 Geological map of the Ilfeld Basin at the southern edge of the Harz Mountains. Simplified after the 1:1.000.000 geological map of the Federal Republic of Germany, BGR (1993)

Saar-Nahe Basin

The Saar-Nahe Basin is the largest of the Central European basin complex and has the best-exposed late- to post-orogenic sedimentary sequence that developed within the actively eroding Variscan Mountain Belt (Franke 2000). It formed as an **asymmetrical graben** structure located along the tectonic boundary between the Rhenohercynian and Saxothuringian, which also marks the location of the **Rheic Suture** (Chap. 10.2). The basement underlying the Saar-Nahe Basin dips toward the northwest and is composed of crystalline rocks of the Saxothuringian that represents part of the Mid-German Crystalline Zone. Directly to the north of the Southern Hunsrück Boundary Fault lies the Rhenohercynian Zone, regionally known as the Rheinisches Schiefergebirge (Fig. 11.18). South of the boundary fault, rocks of the Northern Phyllite Zone (Chap. 10.4.2) occur which form a steep slope to the Saar-Nahe Basin. The basin history indicates that subsidence commenced during the compressive phase of the Variscan Orogeny, as marked by non-metamorphosed conglomerates of Late Carboniferous (Namurian) age found at the base of the sequence. The basinal depression was formed by a right-lateral (dextral) transtensional strike-slip fault motion, with its depocenter lying parallel to and in the immediate vicinity of the fault line (Schäfer 2011). Extensive subsidence occurred with an accumulation of a ca. 10 km thick sedimentary sequence extending into the Late Rotliegend (Permian). After partial erosion, almost 8 km of this sequence is preserved today. To the east, the Saar-Nahe Basin is overlain by Tertiary deposits of the Mainz Basin and the Upper Rhine Graben (Fig. 13.9). However, several deep boreholes have revealed that the basin sequence extends beneath the Upper Rhine Graben, with a thickness of about 1.5 km, and continues on the opposite side of the rift structure in the Sprendlingen Horst (Becker et al. 2012).

The oldest sedimentary rocks of the Saar-Nahe Basin are continental conglomerates of Namurian age and are only encountered in boreholes. These are overlain by a thick clastic series that originated as alluvial plain deposits composed of river and lake sediments, together with widespread lacustrine coal (Chap. 10.3) and delta deposits. These sedimentary sequences represent molasse sediments deposited to the north of the uplifted Rhenish Massif.

As in many other places in Central Europe, widespread and partly explosive volcanism occurred during the Early Rotliegend alongside with shallow intrusive activity. Subsequently, in the Late Rotliegend, continental sedimentation resulted in the deposits of fanglomerates, arkoses, sands, and muds. Occasional aeolian strata occur, such as the dune sediments of the Rote Lay in Bad Kreuznach (Schäfer and Korsch 1998).

Although the volcanic rocks of the Saar-Nahe Basin resemble those of other areas in terms of composition (such as in the North German Basin or Saale Basin), they do have some notable characteristics. Well-exposed rhyolite can be found at the Rotenfels, in Bad Münster am Stein, and at the Donnersberg lying along the southeastern edge of the basin. Some of the fine-grained acidic rocks formed as shallow intrusions that thermally overprinted the host rock during contact

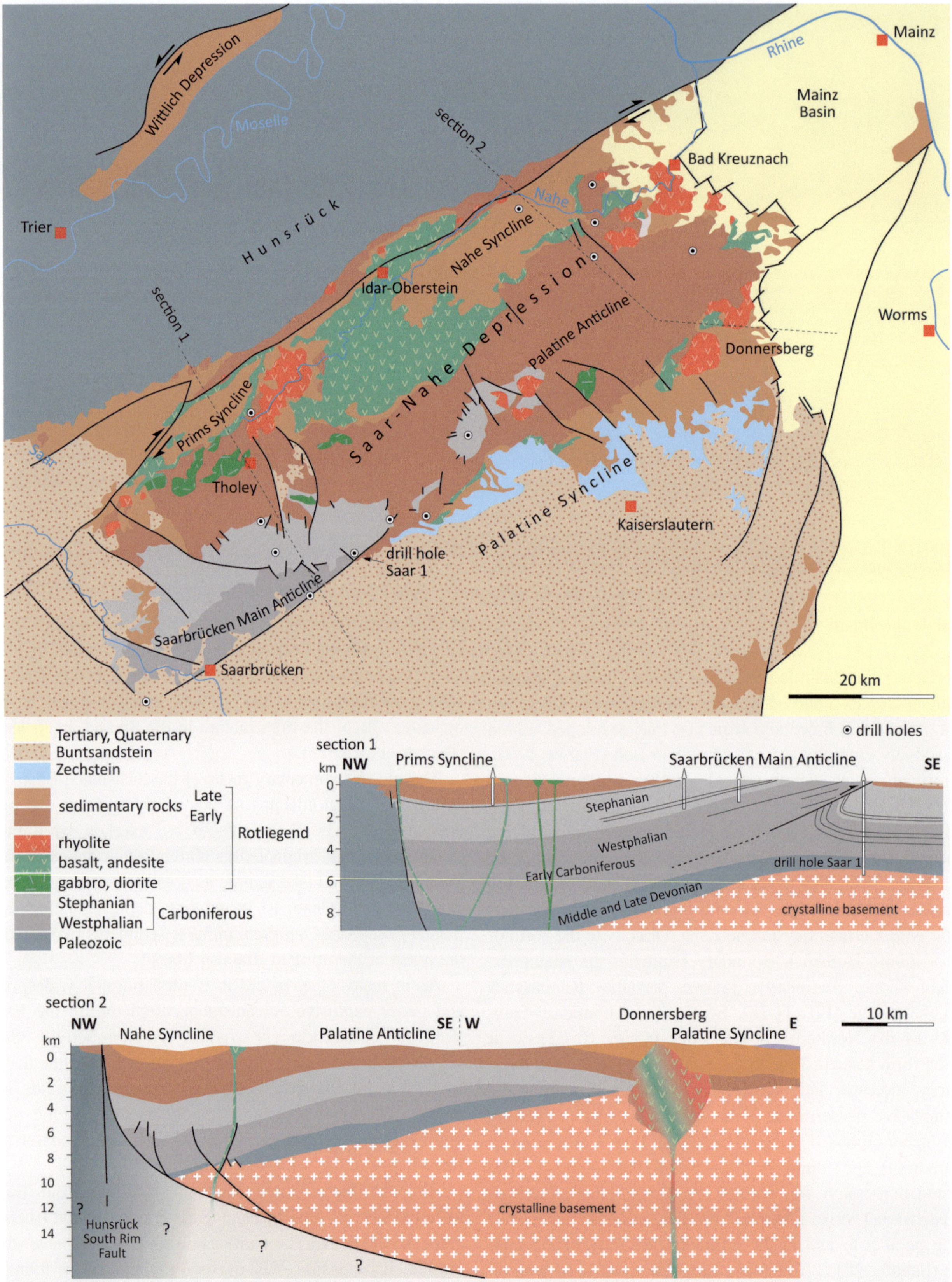

Fig. 11.18 Geological map of the Saar-Nahe Basin between Mainz and Saarbrücken Simplified after the 1:1.000.000 geological map of the Federal Republic of Germany, BGR (1993); cross-section modified after Walter (2007)

metamorphism. In the area of Kaiserslautern, liquid magma heated the fossil-rich sediments of the Rotliegend to produce a particular kind of fossiliferous rock. Normally, the fossil remains of fish are black colored because of the carbon present. However, heating of the calcified cartilaginous skeleton of sharks resulted in crystallization of apatite. This reaction induced a color change whereby the fossils were preserved as light-colored inclusions to produce the world-renowned "White Sharks" of the North Palatine Uplands.

In addition to acidic rhyolitic volcanics, a number of basic to intermediate rocks occur in the form of basalts and andesites. A location of special importance is found near **Tholey** (Fig. 11.18), which gave its name to the most abundant rock type of the Earth's crust: the **tholeiites**. However, it has since been found that the composition of tholeiites from the Schaumberg locality, close to Tholey, does not quite correspond with the typical composition of a basalt formed at a mid-ocean ridge, despite still bearing its name (Box 2: Basalt).

A special rock type of the Saar-Nahe Basin is the andesitic lava. Formerly often referred to as **melaphyre or mandelstein**, these basic to andesitic volcanics contain a large number of gas vesicles that were later filled by the mineral **chalcedony** (Fig. 11.19). These almond-shaped fillings gave the rock its old name (*Mandel*, almond; *Stein*, rock). In particular, in the area between Idar-Oberstein, Freisen, and Kusel, there are numerous occurrences of **agate** (Box 13: Agate) upon which the development of the gemstone industry in this region was based during the seventeenth century. Today, Idar-Oberstein depends on mineral imports from abroad (especially Brazil), but the jewelry industry remains the town's main commercial activity.

Other ore deposits in the Saar-Nahe Basin were at times important. By the end of the nineteenth century, **mercury** (Hg) **ore** was mined in Kusel from the rhyolitic and Rotliegend sedimentary rocks. This ore occurs in the form of cinnabar and amalgams with Ag, but also in pure form as native Hg-inclusions in gas vesicles. For a limited time, U-ore was mined until 1965, which was located in rhyolitic rocks at Nohfelden. At the Donnersberg, Cu-Co-Ag-Mn mines operated over several centuries until the 1920s. Today, this mining activity can be visited in the tourist mine of the Weiße Grube in Imsbach. All these deposits were deposited from hydrothermal fluids associated with Permian volcanism, where the metal ore veins precipitated from solutions that flowed through the open faults and fractures.

Fig. 11.19 Rotliegend basalt from the Saar-Nahe Basin with chalcedony-filled vesicles, formerly known as mandelstein (location in the Fischbach Valley near Idar-Oberstein, Juchem quarry) [N 49°45′26″/E 07°20′23″]

Thuringian Forest

The Thuringian Forest is a mountainous area made up of a basement block that was uplifted in the Late Cretaceous and Tertiary in a similar way as the Harz Mountains (Chap. 11.7.2). Today, rocks of the Rotliegend are mostly exposed at the surface. The underlying Variscan basement can be observed, as the **Ruhla Crystalline** rocks, occurring in the northwestern part of the mountain range (Fig. 11.21) and in some smaller outliers further to the southeast. The Ruhla Crystalline is composed of a large variety of high-grade metamorphic rocks of the Early Paleozoic, together with Variscan granites that intruded during the Early Carboniferous. This is one of the few areas where the **Mid-German Crystalline Zone** is well exposed at the surface. The Mid-German Crystalline Zone belongs to the Saxothuringian tectonostratigraphic unit and marks the site of the Variscan suture line along which the Rheic Ocean closed during the last phase of the Variscan Orogeny (Chap. 10.4.3). Besides the Ruhla Granite, the Variscan basement includes the granites of Suhl and Ilmtal (known as the Thuringian Main Granite, or in German, the Thüringer Hauptgranit) located at the southeastern edge of the Oberhof Trough (Fig. 11.21). To the south, the Thuringian Forest is bordered by the Schwarzburg Anticlinorium, which forms part of the Thuringian Slate Mountains. Here, Precambrian and Early Paleozoic sequences of the Saxothuringian are exposed.

The Rotliegend rock sequence of the Thuringian Forest represents the connecting area between the Saale Basin in the northeast and the Saar-Nahe Basin in the southwest. This region also extends into the territory of the Marne in Lorraine. Both sedimentary and volcanic rocks of the Rotliegend occur across the connecting part and are very similar to the two adjacent basins. However, the area of the Thuringian Forest, including the Variscan granite intrusions, formed a zone of deep weathering and erosion at the beginning of basin formation that had already begun in the Westphalian (Late Carboniferous). Only in the late Stephanian did sedimentation commence in this area and, as in the two adjacent basins, a thick sequence of continental clastic sediments accumulated with

Box 13: Agate

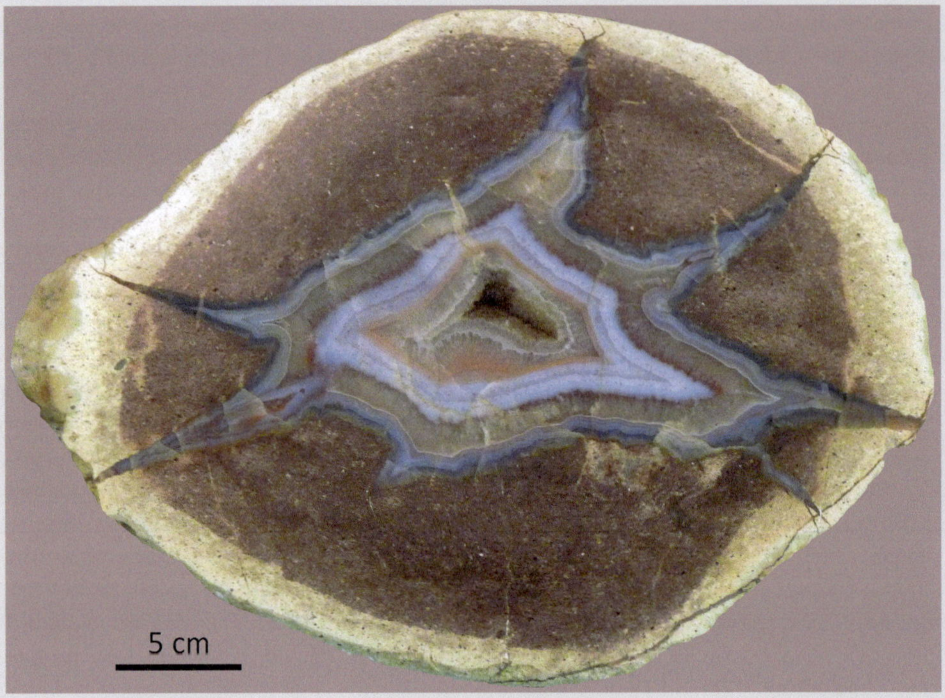

Fig. 11.20 Agate geode from a welded ignimbrite (rhyolite) of the Vorerzgebirge Basin. Photograph with the kind permission of the Chemnitz Museum of Natural History [N 50° 49′51″/E 12°55′26″]

Mineralogically, agate is a **cryptocrystalline** (non-amorphous) variety of SiO_2 with the same chemical composition as quartz. Its part crystalline nature was first discovered using the scanning electron microscope at a very high (about 40,000 times) magnification.

Agate is a variety of **chalcedony** that includes **flintstone**. It is distinguished from other chalcedonies largely by its colorful banding (Fig. 11.20) with shades of gray, yellow, or brown and sometimes greenish to pale blue. Intense colorations, such as pink or deep blue, do not occur naturally but are for commercial purposes artificially introduced. The individual layers of agate are often very thin and composed, like all chalcedonies, of microcrystalline SiO_2. The coloration is caused by varying amounts of additional ions or compounds, such as iron oxides and hydroxides of different chemical valences.

Agates precipitate mainly in irregular forms (also referred to as lithophyse) during the cooling of mobile fluids released from rhyolitic melts. They are commonly found in spherical cavities that develop as the volcanic lava depressures when the magma approaches the surface. In the uppermost part of the still red-hot molten, lava bubbles accumulate below the already solidified crust and unite to form larger cavities. After solidification of the lava, these cavities often reach several decimeters in diameter and contain agates in the form of geodes with a particular shape, known as **agate almonds** (*Achatmandeln*). As agate generally forms after cooling of the lava, they are not strictly volcanic products but result from the subsequent precipitation reactions involving aqueous solutions. They occur in a temperature range below 180 °C where the Si dissolved in water is transported and deposited in open rock cavities. The dissolved silicic acid (H_4SiO_4) originates in most cases from alteration of the rhyolitic ash that covers the lava. The complete filling of a cavity with agate can take several million years. However, it is common that the agate nodules are still hollow inside and contain druses with quartz crystals. In the Saar-Nahe region, these are frequently formed as amethysts: a violet variety of quartz.

thick volcanic sequences intercalated at the base of the Rotliegend. In the Oberhof trough, a ca. 2000 m thick sequence of continental sediments was deposited, while in the Eisenach trough, which forms a prolongation of the Hessian Depression, about 500 m accumulated.

In the Thuringian Forest, the sedimentary rocks of the Rotliegend show local variations in composition reflecting their origin as molasse deposits derived from various source areas of the Variscan Mountains. These rock units are given a variety of formation names, such as the Eisenach, Elgersburg,

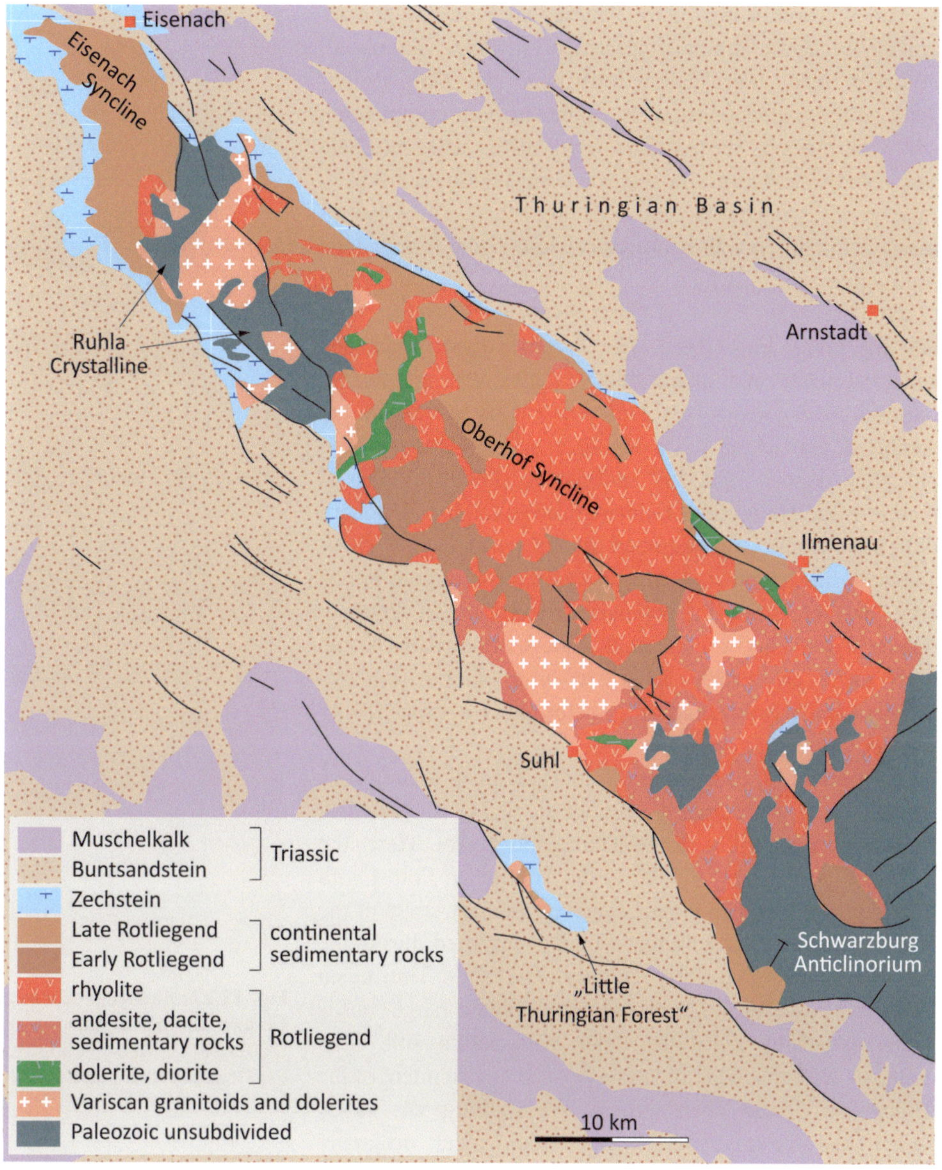

Fig. 11.21 Geological map of the Thuringian Forest Simplified after the 1:1.000.000 geological map of the Federal Republic of Germany, BGR (1993)

Georgenthal, Goldlauter, Ilmenau, Manebach, Möhrenbach, Oberhof, Rotterode, and Tambach formations. The sedimentary sequences often contain periods of nondeposition caused by phases of erosion that in some cases probably lasted for several tens of millions of years. In the Early Rotliegend, coal seams formed that were mined in the Thuringian Forest near Manebach and Gehlberg. In the lake deposits near Friedrichsroda (Gottlob quarry, Goldlauter Formation), which are today protected as a geological monument, a large number of fossils from the Early Rotliegend were preserved bearing plants, fish, amphibians, and dinosaurs.

The youngest sequence is represented by the **Tambach Formation** of Late Rotliegend age, which unconformably overlies the older sedimentary series. The unconformity can be stratigraphically correlated with the Saalian unconformity of the Saale Basin (see above). In the fluvial sandstones of the Tambach Formation occur extremely well-preserved dinosaur tracks and bones. Several complete dinosaur skeletons have been recovered from a fossil site in Tambach-Dietharz, which represents one of the most important sites of its kind in Europe. Among other saurian and amphibian species, two adjacent, fully preserved skeletons of the dinosaur *Seymouria sanjuanensis*, known as the Tambach lovers (*Tambacher Liebespaar*) were found at this location at the end of the 1990s. Prior to this find, this dinosaur species was known only from Early Permian deposits of Utah, USA. These discoveries are a clear indication that the two locations were much closer to each other at the time of deposition where it was possible for the dinosaurs to migrate across the large supercontinent of Pangea. The separating Atlantic Ocean between Europe and North America did not exist during the Permian.

The volcanic series of the Thuringian Forest is characterized by acidic rocks comprised of rhyolite, ignimbrite, volcanic breccia, and shallow intrusions. This Early Rotliegend suite of igneous rocks is similar in both composition and occurrence as the volcanic series in the adjacent Saale and Saar-Nahe basins. In the southern Thuringian Forest, two eruptive sequences can be distinguished. The Oberhof series contains rhyolitic porphyry, tuffs, and interbedded sedimentary rocks, which are exposed along the main ridge of the Thuringian Forest, at Great Beerberg, the Schneekopf and the Great Finsterberg. The rhyolites contain isolated spherical agate and quartz-filled cavities that are locally referred to as *Schneekopf balls* (Box 13: Agate). In the southeast, the Gehren series dominates with an abundance in porphyry deposits and few sedimentary layers. A special feature is the **Höhenberg dolerite** (also called the Hühnberg dolerite), which was intruded along an old NNE-SSW Variscan lineament. It extends across the Thuringian Forest area and represents the youngest occurrence of Permian igneous rock. This several hundred meters thick subvolcanic body forms an intrusive sill of gabbroic composition that intruded Early Rotliegend rocks before been overprinted by contact metamorphism.

Surrounded by Buntsandstein rocks, a small area known as the "Little Thuringian Forest" (Fig. 11.21) is located in the southern foothills of the Thuringian Forest. Here, the rocks are geologically similar to the Thuringian Forest with exposures of Variscan granites and sedimentary strata of the Rotliegend and Zechstein that were also uplifted during the Late Cretaceous and Early Tertiary. The Little Thuringian Forest is, however, not a prominent mountain range, but is characterized by cultivated fields with fertile soil. This contrasts with the predominantly forested environment of the Thuringian Forest characterized by nutrient-poor, sandy soils. This structurally displaced tectonic block originally belonged to a system of fault ridges and basins in the area of the German low mountain ranges.

11.2.2 Zechstein

The Rotliegend was followed by a relatively short evaporitic phase that lasted just over 7 Myr long. During this time, an extraordinary thick rock series was deposited that due to its particular rheologic property significantly influenced the future structural development of the basinal area. Multiple sedimentary sequences composed of clay, limestone, and evaporitic rocks such as gypsum, anhydrite, and various other types of salt were deposited (Fig. 11.22). The name *Zechstein* is linguistically an old mining term for a tough rock (*zäher Stein*), but also refers to the pits (*Zeche*) as a synonym for mine shafts used to excavate the *Kupferschiefer* (Copper Slate). The term today is only used in a lithostratigraphic sense since the transition from the Rotliegend to the Zechstein is diachronous, i.e., occurred at different times in different places. In addition, the upper limit does not match the international Permian/Triassic boundary, but instead, it lies just beneath. The term Zechstein is still widely used in Northern Central Europe to represent rocks deposited in a restricted marine epicontinental sea during the Permian.

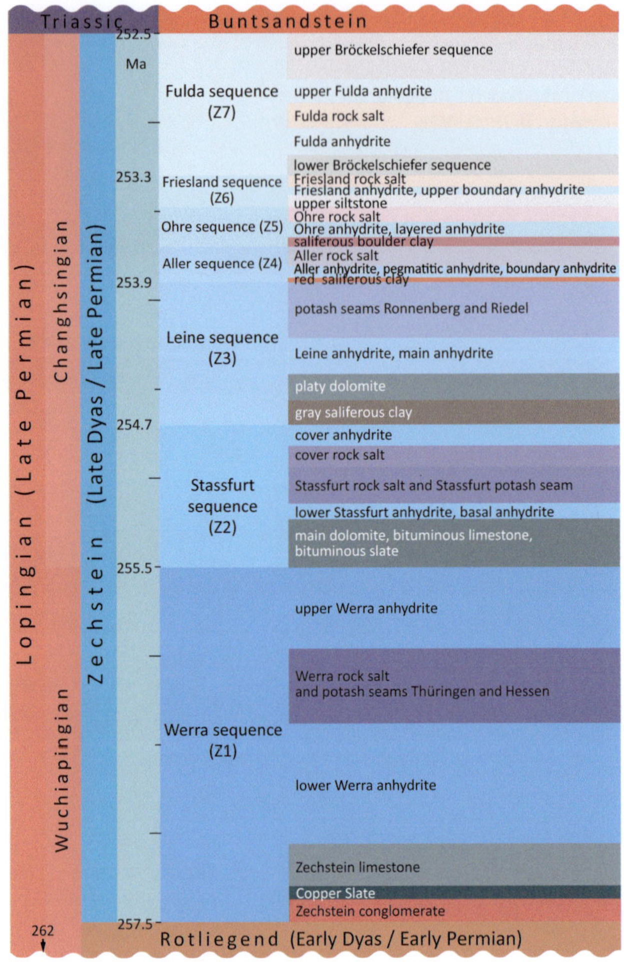

Fig. 11.22 Stratigraphic column of the Zechstein series in the North German Basin and adjacent regions. Modified after Menning et al. (2002, 2011), Menning and Hendrich (2005), Peryt et al. (2010), and the Deutsche Stratigraphische Kommission (2016)

Following the Variscan Orogeny, the beginning of the Zechstein was characterized by fully marine conditions which during evaporation led to deposition of thick rock salt deposits across northern Central Europe (Fig. 11.23). The Zechstein transgression marked a new phase of long-term deposition with subsidence and filling of the **Central European Basin** to form regionally thick sedimentary sequences. The transgression occurred in several cycles, which varied quite considerably in the different basin areas. The main depositional area was located in the center of the **North German Basin**, where the Zechstein series reached

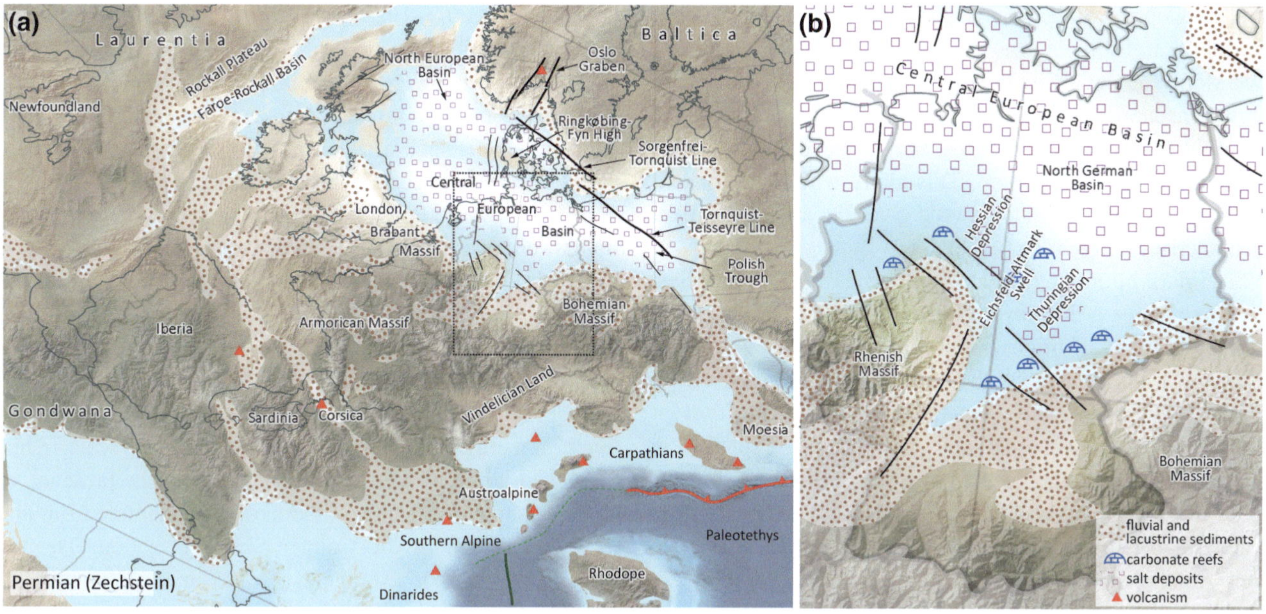

Fig. 11.23 Paleogeographic reconstruction of Central Europe (**a**) and Germany (**b**) at the time of the Zechstein. Late Permian; topographic base map from Blakey (2011), modified after Ziegler (1990), and Peryt et al. (2010)

over 2000 m thickness (Olsen 1987). Due to subsequent salt movement and partial erosion, the current thickness of Zechstein deposits has been strongly modified, thus preventing determinations of its original stratigraphic thickness. In addition to the deposition of evaporitic salt deposits, the Zechstein contains cyclic sequences of mudrock, limestone, and gypsum, the latter of which was later transformed to anhydrite during burial diagenesis.

The base of the **Zechstein** sequence is represented by a **conglomerate** layer that was deposited during a sea-level rise and subsequent transgression of the sea across the North German Basin. The conglomerate is widespread and represents a distinct chronostratigraphic marker horizon that in places discordantly overlies the Rotliegend to form an unconformity. Directly above the conglomerate lies the **Kupferschiefer**. The uppermost 10–50 m of the Rotliegend is at some locations characterized by sandy sedimentary rocks recognized by their white coloration and known as *Weißliegend*. There are different interpretations as to the sedimentary environment of deposition with the suggestion of them representing aeolian sands, marine clastics, or fluvial and estuarine deposits. Another sedimentary horizon separate from the **Weißliegend** is that of the **Grauliegend** (explanation of German terms: *rot*: red, *weiß*: white, *grau*: gray). This represents the uppermost Rotliegend sediments that were later discolored by diagenetic alteration. During this process the red colored iron oxide mineral known as hematite was dissolved and removed by water, resulting in the gray color, which sometimes changed to bright green if any remaining iron was reduced to its ferrous state.

The incursion of the Zechstein Sea spread rapidly due to **catastrophic flooding** at about 257.5 Ma as it entered the Central European Basin from the northwest via the Barents Sea. This event was considered rapid due to the low level of the basinal area, which was already several hundred meters below sea-level due to thermal subsidence that began in the Late Rotliegend. The flooding was triggered by rifting between Norway and Greenland in combination with eustatic sea-level fluctuations caused by the melting of ice on the southern Gondwana continent. The barrier in the northwest of the basin was subsequently breached and the Central European Basin flooded until normal marine conditions were attained. The water depth of the newly formed shallow sea basin is assumed to have lain between 200 and 300 m (Ziegler 1990). The rapid flooding loaded the lithosphere causing isostatic adjustment with further subsidence and expansion of the Zechstein Basin (McCann et al. 2008).

Copper Slate (Kupferschiefer)

A thin but very characteristic layer known as the **Copper Slate Formation** (Kupferschiefer) directly overlies the Zechstein conglomerate as well as the sedimentary rocks of the Rot-, Weiß-, or Grauliegend and older deformed basement rocks. It represents a widespread and important marker horizon within the Permian sequence. The term "slate" given to the rock because of its ability to split into platy fragments is actually a misnomer, as the rock is not a metamorphic slate but is comprised of a shale or more strictly speaking a diagenetically modified mudstone. The term "Copper Slate" is still maintained, however, because of its widespread usage in the literature.

The Copper Slate was deposited in a shallow closed marine basin following rapid flooding and development of a rich flora and fauna in the upper oxygenated waters. The high concentration of organic carbon that settled onto the basin floor was preserved in the oxygen-free conditions that prevailed on the sea bottom. This along with high concentrations of S and associated metals gives the mudstones their characteristic gray color.

The dark bituminous muds that formed the Copper Slate accumulated as a thin layer on the bottom of the Zechstein Sea and ranged in thickness from a few decimeters up to 1 m. Only some areas of the basin were characterized by shallow well-oxygenated waters with local deposition of calcareous facies. The bituminous sediment was deposited in a closed stillwater part of the basin under **euxinian** (anaerobic) **conditions** (Box 17: Euxinic milieu). As the bottom water stagnated, the concentration and exchange of oxygen were dramatically reduced. Any organic matter that settled on the seafloor was not decomposed under the oxygen-poor conditions but instead accumulated in the sediment causing it to blacken. In the Zechstein Sea, the euxinian conditions lasted only for a relatively short period of about 50,000 years. Due to the high rate of evaporation in the extremely arid environment of the North German Basin, strong salinization of the waters followed by the intense deterioration of the living conditions in the upper water layers.

The Copper Slate Formation is an important and distinctive stratigraphic marker horizon due to its unique fossil assemblages that are exceptionally preserved. Well known are the many types of fossil fish, including the commonly occurring *Palaeoniscum freieslebeni*, known as the "Permian herring" (Fig. 11.24). Other well-conserved fossils found in the black shales are coelacanths, sharks, reptiles, and plant remains. At the beginning of the last century, the world's earliest reptile capable of gliding flight (*Coelurosauravus jaekeli*; named after the Greifswald paleontologist Otto Jaekel, 1863–1929) was found in the Copper Slate Formation at Eisleben.

The Copper Slate Formation has a special economic importance because of its content of Cu and some other valuable metals such as Ag, Zn, and Pb. These metals were partly deposited together with the sediment during the Late Permian to Early Triassic as hot mineral-rich hydrothermal fluids were discharged at submarine springs and reacted with hydrogen sulfide (H_2S) on the seafloor. The majority of metal, however, migrated later into the rock as chloride solutions and precipitated from the circulating water (Jowett et al. 1987; Bechtel et al. 1996). During the Triassic and younger times, the metal-charged fluids reached the relatively impermeable mudstones that acted as seals. This, combined with the reducing conditions and the presence of sulfur, resulted in the precipitation of Cu in sulfide form and its accumulation in the Copper Slate Formation. The Cu content, which is on average 2–3% and in exceptional cases exceeds 5%, is quite high and profitable despite the small thickness of the Copper Slate layer.

Mining of Cu, Ag, Pb, and Zn was carried out for over 1000 years in the Mansfeld region, which was still in operation until 1990. In former times, underground quarrying was often carried out under rather inhumane conditions with mine shafts not even one meter high. During the so-called *Duckelbergbau* (duck down mining), shafts of several meters depth were drilled and the Copper Slate seam extracted in all directions so far as was allowed by the supply of fresh air (max. 7–8 m). Numerous mine dumps reaching up to 50 m high appeared over the mining epochs, the periods of which can be recognized by their characteristic shapes that mark the activities of former Cu production (Fig. 11.25). With recent shortages of metal resources and the associated sharp rise in prices, there have been thoughts of reviving Cu mining in the Mansfeld region. In Lusatia, the reopening of mining activity has moved beyond the planning stage and applications have been made for official planning

Fig. 11.24 *Palaeoniscum freieslebeni*, the "Permian herring" recovered from the Copper Slate Formation (Late Permian) at the southern edge of the Harz Mountains next to the "Lange Wand" (German: long wall) near Ilfeld [N 51°34′06″/E 10°47′04″]

Fig. 11.25 Waste dump from Copper Slate mining, east of Mansfeld [N 51°34′58″/E 11°33′ 40″]

permission (*Raumordnungsverfahren*). In this area, significantly thicker mineralized layers occur compared to that of the Mansfeld deposit (Kopp et al. 2006). In addition to Cu mining, the Mansfeld region also saw some interesting by-products in the form of building materials that were cast in cube or cuboid blocks produced from the slag (sintered, partially melted rock) formed during the Cu recovery process. During the period of the German Democratic Republic, such building blocks became a successful export product and were used for road and wall construction. These blocks can still be found today in many places throughout Europe. However, they were not considered suitable for the construction of houses due to radiation risk from the elevated concentrations of U.

Carbonate sediments of the Zechstein

Above the Copper Slate Formation occurs a cyclic repetition of thick dolomites and evaporitic rocks containing gypsum or anhydrite deposits, as well as various salt layers periodically interrupted by clay layers. Altogether, the seven cycles that can be distinguished are shown in order of their stratigraphic sequence in Fig. 11.22. The original division into four classical cycles (Werra, Stassfurt, Leine, Aller sequence; Z1–Z4) is meanwhile complemented by an additional three younger cycles (Ohre, Friesland, Fulda sequence; Z5–Z7). Although the extent and completeness of these units are still under discussion, local deviations from an ideal sequence, which is a synthetic compilation of all possible cycles, are widespread. For instance, in northern Mecklenburg-West Pomerania only the lower five cycles are present, whereas, towards the deeper part of the basin area, all cycles are typically present (Fig. 11.26). In contrast, south of the Harz Mountains only the first four cycles are represented.

The Copper Slate Formation belongs to the lowermost cycle of the Zechstein known as **Werra sequence** (Z1) (Fig. 11.22). Directly above the mudstone formation follows the first **Zechstein limestone**. This was deposited as a marine platform limestone that in a number of places in the Central European Basin contains some dolomitic limestone. The carbonate rocks of the later cycles are mostly dominated by dolomite but depending on the facies of deposition also contain oolites, algal laminites, and bioclastic limestones. At the lower edge of the platform and in the deeper part of the basin, the Zechstein limestones and their equivalents pass laterally into dark bituminous limestones. This lateral change in facies also occurs in subsequent cycles. Carbonate reefs developed within the Zechstein limestone (Z1, Fig. 11.22) are found on the southern edge of the basin and in emergent areas such as the Eichsfeld–Altmark swell. This swell had a width of several tens of kilometers and subdivided the Central European basin into a western (the Hessian Depression) and an eastern part (the Thuringian Basin; Fig. 11.23b). Besides the deposition of shallow water carbonates in lagoon areas, more than 50 m thick carbonate reefs mostly composed of bryozoans and rudist bivalves formed. Well-preserved Zechstein reefs are exposed in the Orla depression around Pößneck (Fig. 11.27) and at the edge of the Thuringian Forest in Bad Liebenstein and Thal. They were deposited partly on the folded Paleozoic basement (Fig. 11.28), but locally they grew directly upon

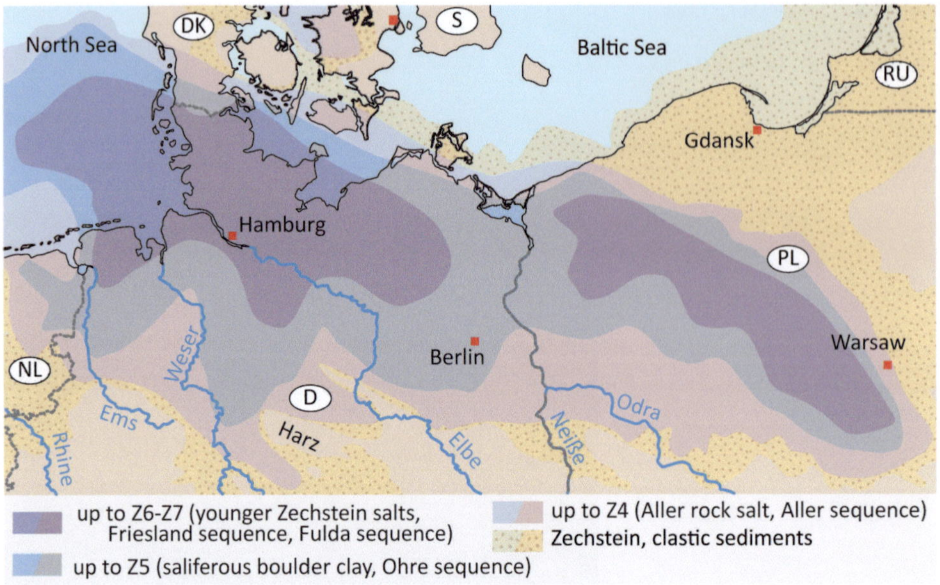

Fig. 11.26 Occurrence of Zechstein salt and sedimentary rocks in the area of the North German and Polish basins Modified after Peryt et al. (2010)

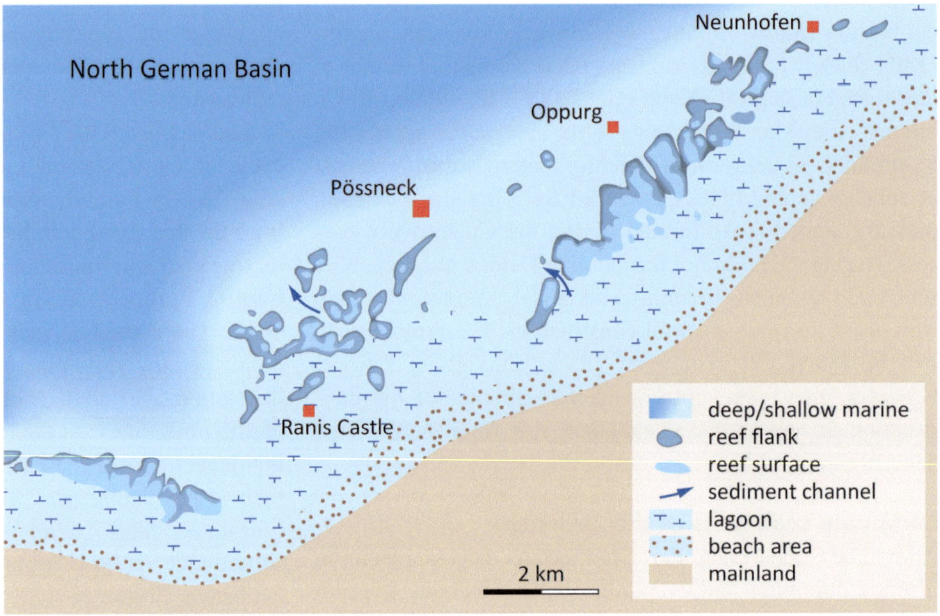

Fig. 11.27 Reef structures and the distribution of sedimentary facies at the southern edge of the North German Basins during the Late Permian (Zechstein, Werra cycle). The fossil reefs are located north of the Thuringian Slate Mountains in the Orla depression, near Pössneck. After Paul and Huckriede (2004)

sedimentary strata of the Rotliegend. At the southern edge of the Harz Mountains, lagoonal Zechstein limestone partly overlies the Copper Slate Formation and the Permian volcanics, such as exposed at Ilfeld near the ***Lange Wand*** (long wall) (Fig. 11.29).

A location of paleontological importance is the **Korbach crevice**, near Kassel, northern Hesse. Exposed in a 1 km long crevice, numerous fossils of terrestrial vertebrates are well-preserved that formed coevally with deposition of the Copper Slate Formation (Sues and Munk, 1996). The fossil assemblages were in fact transported from their primary sites of deposition and rearranged during redeposition into the several meter wide crevice. Highly fragmented bones of pristine reptiles (*pareiasaurs*), ancestors of the dinosaurs (*archosaurs*) and mammal-like reptiles (*therapsida*), were preserved within this crevice that developed in the Zechstein limestone of the Werra sequence (Z1, Fig. 11.22) shortly after its formation. Since the fossils are slightly older than the rocks in which the crevice resides, the bone fragments are considered to have been redeposited several times (Bökenschmidt 2006).

The second Zechstein cycle (Z2, **Stassfurt sequence**, Fig. 11.22) begins with the Main Dolomite (*Hauptdolomit*) horizon. This layer passes laterally into thinner bituminous limestones and shales toward the interior basin. The

Fig. 11.28 Zechstein limestone of lagoonal facies discordantly overlying folded Carboniferous graywackes of the Variscan basement. Locality is directly below the Totensteinriff (Zechstein limestone) at the Harrasmühle near Lausnitz, Orla depression [N 50°43′08″/E 11°41′ 56″]

Fig. 11.29 The "Lange Wand" (German: long wall) near Ilfeld. Bedded Zechstein limestone discordantly overlies rhyolites of the Rotliegend that are partly bleached in the upper part. At this location, the Copper Slate is tectonically thinned out but occurs in the neighborhood of the Lange Wand, only a few meters from this exposure [N 51°34′06″/ E 10°46′59″]

bituminous limestones are dark colored due to a high proportion of organic matter and the rock samples give off an unpleasant smell due to the release of H_2S. In platform areas, the Main Dolomite of Northern Germany is an important oil and gas reservoir (the so-called Ca2-Stassfurt carbonate). The third cycle (Z3, **Leine sequence**) is characterized by a platy dolomite lithology (*Plattendolomit*) that reaches up to 90 meters thickness toward the edges of the basin. In the interior of the basin, this sequence changes into thinner argillaceous sedimentary layers. No more carbonate rocks occur in the remaining cycles (Z4–Z7).

The cyclic changes in the Zechstein sequence are attributed to global sea-level fluctuations related to **glacial activity in the Permian**. Widespread indications of Carboniferous and Permian glaciations occur in the form of tillites (consolidated glacial deposits) or glacial striations that have been detected in South America, South Africa, India, and Australia. The traces of a Permian glaciation are today distributed over several continents, but at the time of formation, all areas lay close to the ice-capped Permian South Pole (Fig. 11.1). It is not known if the Permian North Pole also developed an ice cap at this time due to the lack of underlying continental crust where tracks could be preserved. Evidence of a Permo-Carboniferous glaciation, which today is partly found in equatorial positions, was one of the key arguments proposed by Alfred Wegener in support of his 1912 theory of continental drift (*Kontinentalverschiebungstheorie*: Wegener 1912, 1915, 1929). We know from the Pleistocene that several consecutive glaciation phases (Chap. 16) induced strong sea-level fluctuations of more than 100 m. It, therefore, appears reasonable to expect similar cyclic fluctuations in sea-level to have occurred during the Late Carboniferous and Permian resulting in multiple repetitions of the sedimentary sequence during the Zechstein. No such cyclicity is recognized in the Rotliegend sedimentary record that was formed largely under continental conditions in an arid environment and thus was unaffected by sea-level changes.

Evaporitic sediments of the Zechstein

The average seawater contains a broad range of dissolved ions that make up about 3.5% of its mass, with its largest portion of nearly 80% attributable to NaCl (table salt; Fig. 11.30). During the evaporation of seawater, **evaporitic minerals** crystallize sequentially according to their solubility, and consequently, a typical sequence of precipitates is deposited on top of each other. The least soluble carbonate mineral that precipitates first is **calcite ($CaCO_3$)**, followed by **dolomite** ($CaMg[CO_3]_2$), then the sulfates of **gypsum** ($CaSO_4 \cdot 2H_2O$). Finally, the salts solidify, starting with large amounts of **halite** (NaCl) and by the end of a complete evaporation cycle, potash salts such as **sylvite** (KCl) and **carnallite** ($MgCl_2 \cdot KCl \cdot 2H_2O$; Fig. 11.31) form. This so-called **salinar cycle** is not always complete in Zechstein sequences since evaporation was frequently interrupted by marine ingressions (Fig. 11.22).

The lowermost three Zechstein cycles begin with fine-grained clayey sedimentary rocks followed by a carbonate layer. These evaporitic series always start with an anhydrite layer (Greek *anhydros*: waterless); a Ca sulfate mineral ($CaSO_4$) formed from the diagenetic alteration of gypsum that first precipitated during the evaporation of seawater and was later converted to **anhydrite ($CaSO_4$)**. Unlike anhydrite, gypsum contains water bound in its crystal lattice ($CaSO_4 \cdot 2\ H_2O$). Only under elevated temperatures (>66 °C) does anhydrite develop directly.

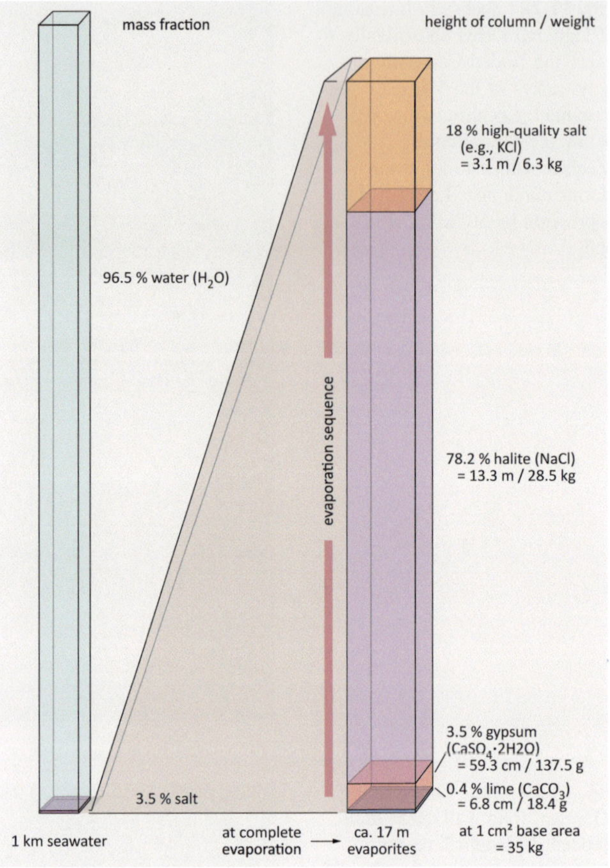

Fig. 11.30 The salt content of seawater calculated for a 1 km long column with a 1 cm^2 base area. If such a water column is completely evaporated, a 17 m high pillar of salt remains with minerals precipitated in their ideal evaporative sequence. In natural evaporates, carbonates and gypsum are the most abundant mineral phases formed

The first **Werra anhydrite** (Fig. 11.22.), with thicknesses of more than 600 m in deep basinal areas, formed when the Zechstein Sea was cut off from the inflow of fresh seawater during lowering of the sea-level. During the first cycle (Werra sequence, Z1), however, deposition of rock salt occurred only sporadically in the southern part of the North German Basin in the area of the Emsland and in the Hessian Depression. The sequence ends with an upper, thinner layer of anhydrite. A temporary rise in sea-level at the beginning of the Stassfurt sequence (Z2) increased the inflow of fresh seawater and renewed deposition of carbonate rocks occurred (mainly in form of dolomite and **bituminous limestone)**. These carbonates formed mainly along the edge of the basin, whereas anhydrite was more common toward the interior part. During this cycle, the marine seawater evaporated

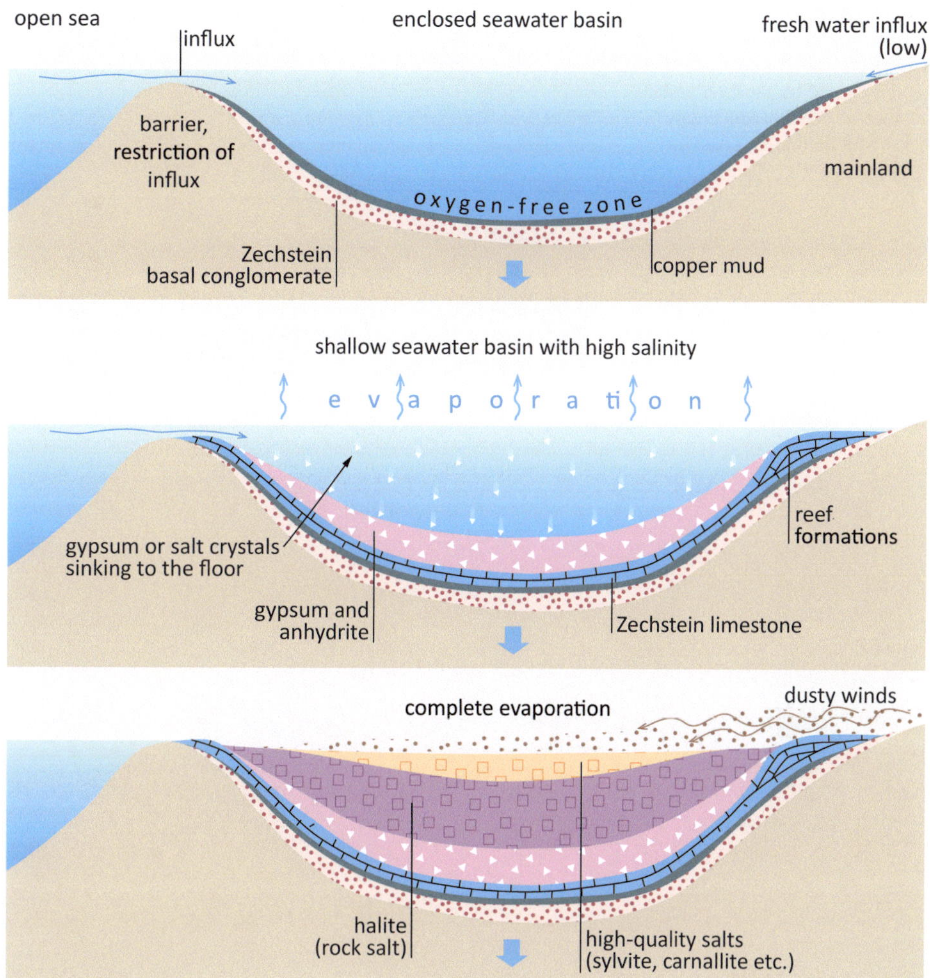

Fig. 11.31 Simplified presentation of a typical evaporation sequence that was formed at the beginning of the Zechstein series (Werra and Stassfurt sequence)

completely to form a thick sequence of halite and potash salts. In the deepest areas of the basin in the area of the present North Sea, the salts reached a thickness of more than 1500 m. Toward the edge of the basin, these layers are generally thinner, but in the Polish basin they do still reach thicknesses of over 400 m. The Leine sequence (Z3) is characterized by another ingression of marine waters. This resulted in clay deposits similar to those found in the Stassfurt sequence (Z2). Altogether, this sequence is thinner, although complete evaporation with the formation of potash salts occurred. The following cycles (Aller, Ohre, Friesland, Fulda sequences; Z4–Z7; Fig. 11.22) also contain deposits of argillaceous sedimentary rock at the base, followed by anhydrite and halite but no potash salts. Beginning with the deposition of the Ohre sequence (Z5), the basin became more restricted, with incomplete sequences developed in the peripheral region that ended with the Aller cycle (Z4; Fig. 11.26). Although the cyclic sequence of the Zechstein salts was primarily controlled by sea-level changes, other geomorphological and climatic influences also had an effect, as did the flow of paleo-currents in the basin. At the basin edges, the youngest Zechstein deposits contain clastic lithologies that mark the beginning of terrestrial-dominated sedimentation characteristic of the Triassic period.

Salt domes and salt pillows

The present-day salt deposits of the Zechstein are rarely found in their original thickness due to modification by ductile flow. Salt has, independent of depth, a specific weight of $\rho \leq 2.2$ g/cm^3 and thus is significantly less dense than the surrounding sedimentary rock, which after significant burial has densities typically above 2.5 g/cm^3. With an overburden of about 1000 to 1500 m, a distinct **density inversion** develops whereby salt becomes unstable due to gravitational forces and the buoyancy effect. In addition, the salt has a rheological property that causes it to flow under increased pressure, reaching flow rates as high as 0.3 mm per year. At a first glance, this looks like a low rate but adds up to 300 m every one million years. Due to its lower specific weight, salt typically migrates from its original

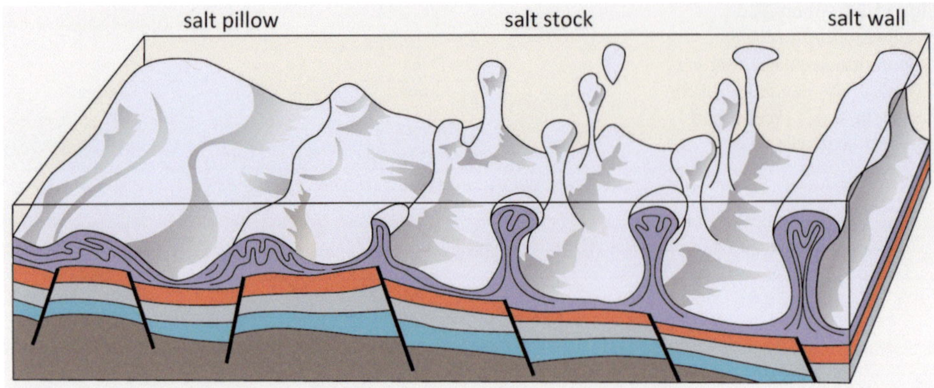

Fig. 11.32 Evolution of salt pillows, domes, and walls that result from lateral migration of the salt caused by the overburden pressure of denser sedimentary rocks. Modified after Trusheim (1960)

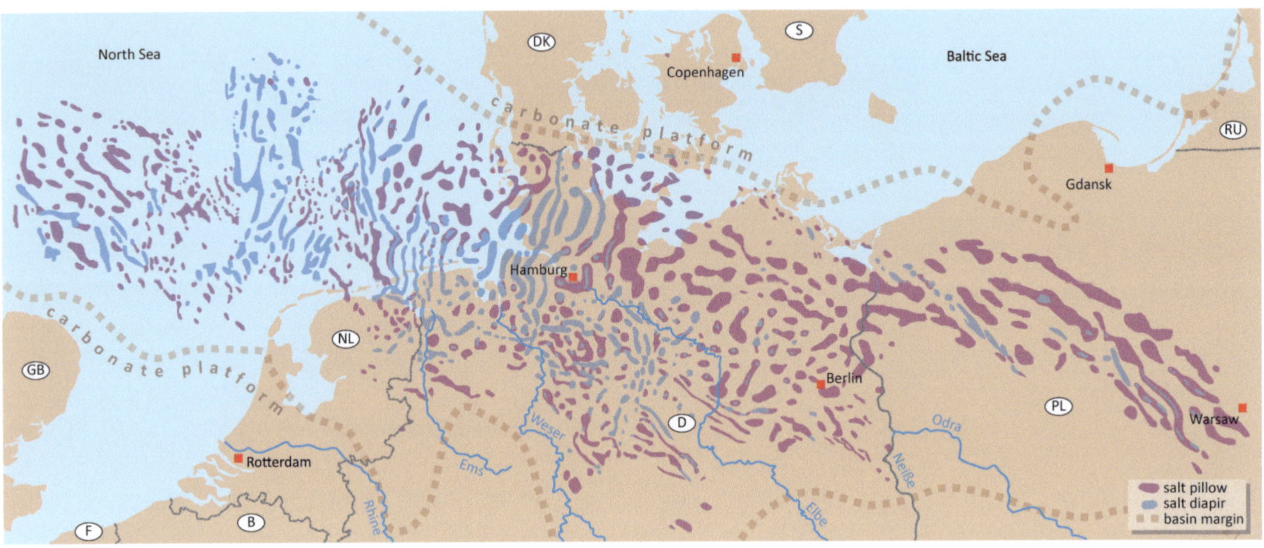

Fig. 11.33 Salt pillow and diapir structures in the upper crustal levels of the North Sea, Northern Germany, and Poland. The greatest thickness of salt occurs beneath the North Sea and Schleswig-Holstein where salt diapirs are connected to prominent salt walls. After Lokhorst (1998) and Pharaohet et al. (2010)

stratigraphic position and accumulates as **salt pillows** in the uplifted segments of fault zones. Continued salt supply results in the formation of **diapirs** (Greek *diapeirein*: permeate), which may flow up to the surface (Fig. 11.32). Depending upon the original thickness of the salt deposits, entire **salt walls** can form. Like **salt domes**, they may become completely disconnected from their original stratigraphic formation.

Following the deposition of the Zechstein salt, further subsidence and sedimentation in the North German and Polish basins continued, with the salt reaching its critical burial depth for migration by the Early Triassic (Middle Buntsandstein). At this time, the overburden pressure was sufficient enough that the salt became mobile and flowed upward along fault zones that typically formed by reactivation of older, inherited structures in the upper crust. Such structures also continued to influence the patterns of sedimentation above the sites of ascending salt. Today, almost all salt deposits of the Zechstein represent **migrated salt**, which depending on their thickness can be detected at depth beneath the entire North German and Polish basins (Fig. 11.33). The largest thicknesses are encountered in the area northwest of Hamburg in the Lower Elbe trough and in the Glückstadt Graben, where several salt walls are developed lying side by side. To the south, especially in northern Lower Saxony, numerous salt domes also occur, whereas the abundant salt pillows are located in the upper crust of Mecklenburg-West Pomerania and the area of the Polish Basin.

As salt migrates upward to form salt pillows, the overlying rock sequence often leads to doming of the overlying rock series and, when approaching the surface, to the development of a basin floor topography. The latter forms basin and swell areas that are often characterized by changes in both sediment thickness and sedimentary facies (Fig. 11.34). Not only can the uplift above salt diapirs result

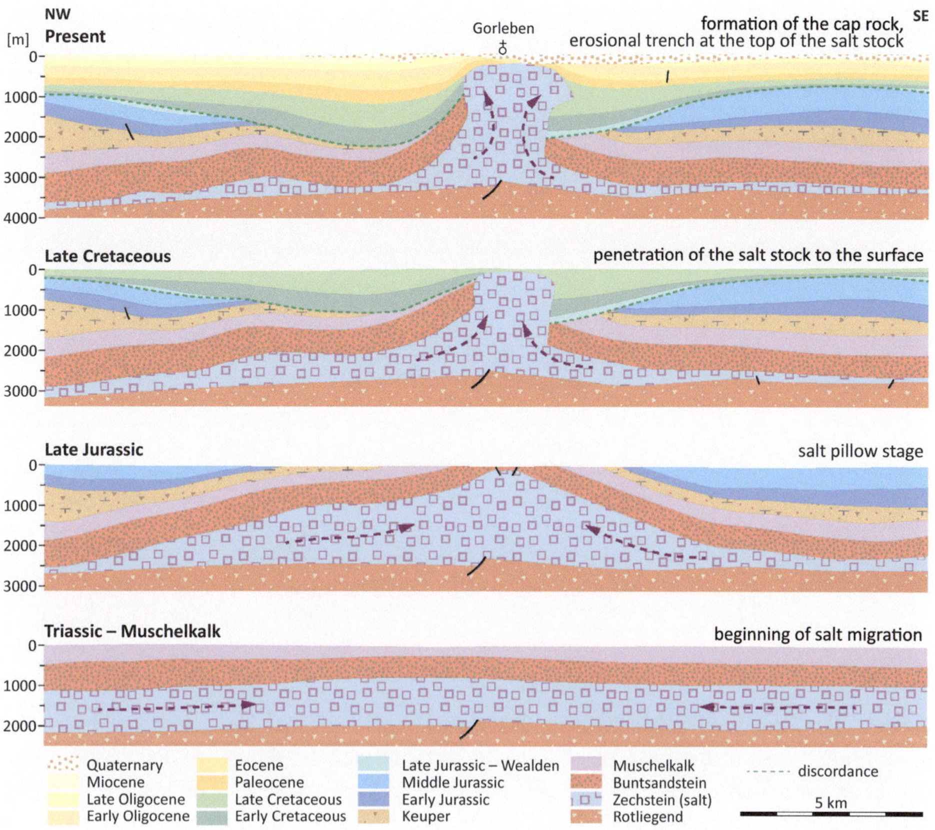

Fig. 11.34 Evolution from a salt pillow to a dome, such as the well-studied Gorleben salt dome. Modified after Zirngast (1991), Bornemann et al. (2008)

in local elevation and erosion of the overlying strata, but also affects the salt itself. The term salt tectonics is used to describe all of the associated structures, facies differences, thickness changes, hiatuses and unconformities that affected the upper crustal stratigraphic sequence of the Mesozoic and Cenozoic. Based on detailed analysis of the basin geometry and the structural/stratigraphic changes over time, the history of a salt diapir can be usually reconstructed.

As salt structures enter the groundwater zone, the soluble components may dissolve in the fresh water causing leaching of the uppermost part of the salt dome. What remains are the insoluble components like anhydrite and gypsum mixed with clayey sediment. The leached upper leading edge of a salt dome is known as the **caprock zone** (in German called the *Gipshut*: **gypsum hat**). In Northern Germany, these are exposed at the surface of several salt domes where these structures are still characterized by persistent tectonic movements. A good example of a caprock zone is exposed in Bad Segeberg, at the so-called *Kalkberg* (=limestone hill). This location is well known for its Karl-May festival performed in a disused gypsum quarry. The Kalkberg is in fact not composed of limestone but of gypsum and anhydrite, which was mined here until well into the 1930s. The Sperenberg diapir (exposed in a quarry south of Berlin) is also composed of gypsum, where in 1867 the world's first deep borehole of over 1000 m (total depth 1272 m) was drilled. This borehole was of particular importance for obtaining one of the first direct measurements of the Earth's geothermal gradient (determined with 3 °C/100 m).

In the Kalkgrube Lieth (Kalkgrube: limestone pit) in Schleswig-Holstein, near the town of Elmshorn, rocks of the Rotliegend and Zechstein were uplifted to the surface by the updoming of salt to form the northernmost surface exposure of Copper Slate (s. above this chapter) in Europe. The effects of **underground leaching** of the salt beneath the city of Lüneburg have caused numerous cases of urban subsidence with tilted buildings and cracks in walls, such as the heavily damaged "gateway to the underworld", where the two gates of a fence are today overlapping by several decimeters. Another example is the leaning columns of the gothic Michaelis Church. Both salt leaching and the mining activities over the last centuries still result in continuing subsidence within the city limits, causing massive fissures and significant structural damage that in some cases has led to the total destruction of buildings.

More commonly large depressions form above the uprising salt dome because of prolonged leaching. These structures are referred to as **subrosion depressions** that can reach several kilometers in diameter. In some places, they are filled by a lake, such as the Zwischenahner Meer in northern Lower Saxony or the Arendsee in the Altmark

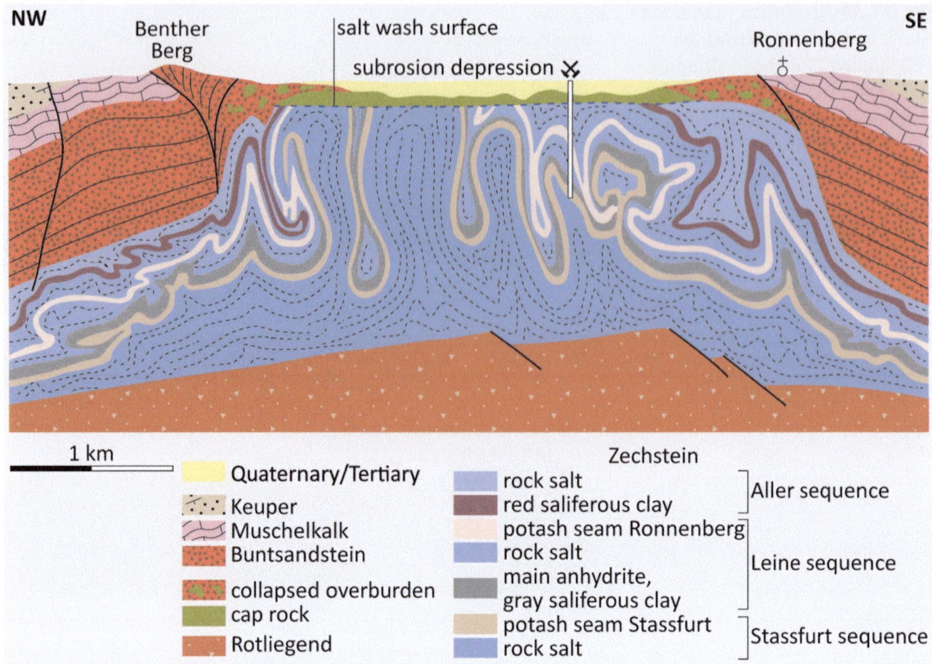

Fig. 11.35 Cross-section through the salt dome of Ronnenberg near Hannover. In the flank of the dome, on top of the Benther Berg, the Buntsandstein strata are tilted vertically. A subrosion depression with a diameter of several kilometers developed above the salt dome. Modified after Ahlborn and Richter-Bernburg (1953)

region. The Steinhuder Meer, near Hannover, although resembling a lake of the same origin, was in fact caused by glacial erosion during the last ice age. At the edges of a subrosion depression, the country rock is typically dragged along with the rising salt dome and rotated into the vertical to subvertical position. These structures emerge at the surfaces as elevated hills, such as the Benther and Gehrdener Berg near Hannover. Similarly uplifted, steeply dipping rocks of the Buntsandstein are found at the crest of the Benther Berg (Fig. 11.35), at the Asse near Braunschweig, and in the hills of Huy and Hakel located in the Subhercynian basin on the northern edge of the Harz Mountains. Some smaller ridges also developed above salt pillow structures, such as the Hildesheim Forest and the Elm near Helmstedt, where layers of the Mesozoic were uplifted in gently inclined anticlinal structures. Also of the same origin are the limestone hills near Rüdersdorf, where a large limestone quarry is located in obliquely tilted limestones of the Lower Muschelkalk (Middle Triassic). This location lies in the middle of a flat glacial valley filled with Quaternary sediments.

Zechstein salt deposits have been economically important for many centuries. Many names of places in the North German Basin, such as Halle an der Saale, Salzgitter, Salzhausen near Lüneburg, or Salzwedel, originate from former salt mining activities that brought importance to these areas. The deformed country rock surrounding the salt domes also provided geological reservoirs responsible for trapping petroleum in the form of crude oil and natural gas. Today, artificial caverns in the salt domes also provide a useful option for storage of these hydrocarbons.

Zechstein salt has been extensively mined from the salt domes of the North German Basin, with over a hundred former mines, ten of which are still active today mainly for the extraction of potash salt used primarily for fertilizers and the chemical industry. Rock salt is mined only in a few places and is used mainly for producing Na and Cl, in addition to table and road salt. An unpleasant side effect of salt mining is the sizable gray and whitish spoil heaps that mostly lack vegetation. These consist almost exclusively of impure rock salt produced as waste during the extraction of potash salt. The spoil heaps are gradually leached by rainwater (annually about 10 cm), but—despite some protective measures—can lead to salinization of the groundwater in the area. This problem may persist for several hundreds of years until complete dissolution of the waste.

At 40 locations in Germany, large **caverns** in salt domes are used for storage of natural gas and crude oil. Cylindrical cavities generally up to 100 m in diameter and 500 m in height are leached out artificially with water at depths of more than 2.5 km. These reservoirs with volumes of more than 100,000 m^3 are excellent for the long-term storage of liquids and gases due to the plastic properties and tightness of the salt. The statutory stockpiling of energy resources for the time of crisis in Germany is relatively easy and safe. It has the great advantage of requiring no large protective buildings or tanks on the surface. Such large underground reservoirs are mainly used to compensate for the high seasonal variations in oil and gas consumption.

For decades, the suitability of salt formations have been investigated for the **disposal of high-grade radioactive waste**, but so far without satisfactory result. Whereas the properties of salt, such as its low thermal conductivity and low permeability, are favorable for the safe storage of the waste, its usage as a host rock material has run into problems and is been critically reconsidered.

Salt domes such as the Asse salt mine are now considered unsuitable for such storage due to the infiltration of groundwater into this low to medium-grade waste repository site. Although the groundwater is not radioactively contaminated and the leakage could be potentially solved by refilling the old mine, the German government has decided to remove all of the radioactive waste from the site and to reevaluate all options for the underground storage of waste. Effective sealing of excavated repositories continues to remain a topic of extensive research and debate, particularly as any fresh excavations into an undisturbed salt dome could lead to mechanical instabilities and potential groundwater infiltration. Although it is internationally agreed that the best option for the safe disposal of extremely hazardous nuclear waste is underground in mined repositories, it remains a challenge to contain safely this material and to prevent radionuclides entering the biosphere over long periods of time. Looking at the poor condition of the remains of buildings built by our ancestors at the beginning of the cultural development 5000–10,000 years ago gives us some idea as to the challenge of engineering repositories that should remain intact from at least 1000–10,000 years. This is the length of time where all nuclear waste decays to the level of the ore material from which it was derived.

The structural emplacement of salt domes has led to the formation of numerous **oil and gas traps** that are typically located in the flanks or crests of the domes surrounding the salt (Fig. 11.36). The oil and gas source rocks are typically Jurassic and Cretaceous sedimentary oil shales, but formations with high proportions of organic matter also occur in the Permian, Carboniferous, and Devonian. As these sedimentary rocks became more deeply buried, the organic matter known as kerogen underwent **thermal maturation** to produce hydrocarbons. These organic compounds were mobile and transported to higher structural reservoir sites where they were trapped (Box 14: Coal, oil, natural gas). Trapping occurred when an effective seal was encountered in the form of an impermeable layer, such as a shale or evaporate horizon. Otherwise, the mobile hydrocarbons of low density seep out at the surface. Natural seeps of petroleum were already known in the preindustrial era. These oils were used as lubricants or ointments and were obtained from so-called *Teerkuhlen* (**tar pit**) at various locations in northern Germany. The first oil well in the world was drilled in 1858 in Wietze, near Celle; one year before the drilling and discovery of underground oil at Titusville, Pennsylvania (USA) that led to the first oil boom in America. However, the borehole in Wietze did not have an immediate lasting effect on the local oil production in Germany because the discovery was followed up by mining of the oily sand. It was not until 40 years later that oil production began in this country, which by the beginning of the twentieth century covered about 80% of Germany's oil needs. Following the drilling of 2000 holes, 1600 of which successfully encountered oil, production in Wietze was eventually shut down in 1963 for economic reasons. Despite this, Lower Saxony remains a center of German oil and gas production.

Similar hydrocarbon discoveries were also made in other regions, such as in the North Sea or in Mecklenburg-West Pomerania. In the 1960s and 1970s, petroleum was encountered in the area of the Grimmener Wall, south of the islands of Rügen and Usedom. Today, these wells still produce

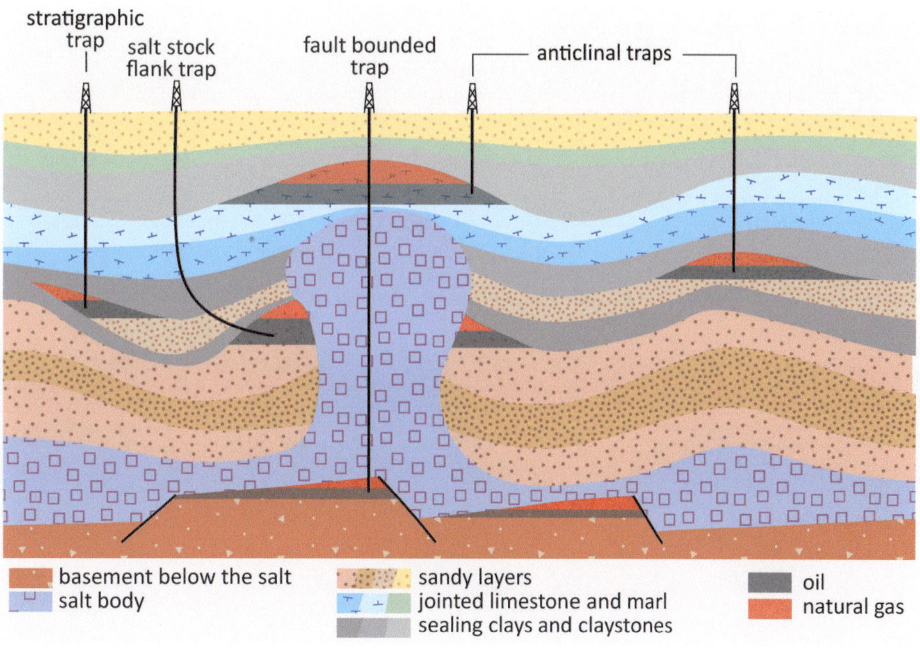

Fig. 11.36 Schematic reconstruction of the possible sites of oil and gas traps near a salt dome

Box 14: Coal, oil, natural gas—the cornerstones of our modern industrial society

In today's modern society, coal, oil, and natural gas are indispensable resources in our lives. They are used in almost all of our activities, forming the key ingredients in many of the materials we use including fuels and medical supplies. Currently, coal, oil, and natural gas are the world's main source of energy (altogether >80%) and they are likely to remain so for some decades. Oil also represents the basis for most synthetic materials used today with an increasingly wide range of products. These include plastic bottles, waterproof fabrics, heart valves, dressing material, pigment for colors, creams, and candles, to name just a few. Due to the importance and large spectrum of hydrocarbon products, it is of importance to understand the geological origin and compositional differences between coal, oil, and natural gas.

All these raw materials belong to the group of hydrocarbons known as fossil fuels. These were formed in the geological past and are today mined in open pits or extracted underground. As the amount of available resources constantly decrease, and they are not renewable, there will come a time when they are exhausted. Estimations from the energy industry calculate that there is enough oil left for at least another 50 years. However, it has to be considered that these calculations are related to the current state of technology and the future rate of usage. Based on today's eligible coal reserves, there is enough for at least 100–200 years assuming similar consumption rates.

Coal, oil, and natural gas originate from diagenetically transformed organic-rich sediment. Coals are formed from plant remains that accumulated in swamps, lakes or coastal environments. In contrast, oil and natural gas are formed mainly from marine sediment containing a high percentage of organic remains that were deposited under very special conditions (plants and animals, mostly microorganisms). As organisms die, their remains are usually degraded within a short time. During this process, the abundance of O_2 in the environment plays an important role since it oxidizes the carbon bound in the organisms so that it is returned to the atmosphere in the form of CO_2. Only the remains of the organisms that accumulated in oxygen-free settings immediately after their death are therefore preserved and converted by thermal maturation to hydrocarbons.

Coal formation: Coal represents a carbon-rich sedimentary rock derived from the remains of terrestrial land plants. In wetlands, plant residues from trees, ferns, grasses, and mosses accumulate where they are gradually bogged down in the swamp to form increasingly thick packages of dead plant material. This leads to the formation of peat. An important prerequisite for this development is that the water in the swamp contains minimal O_2 that may contribute to decomposition of the organic matter. Under these anaerobic conditions, chemical processes convert the plant materials of cellulose and lignin into humic substances.

Following deposition, organic sediments are gradually covered by more sedimentary layers during which pressure and temperature increase with burial depth, and the **coalification process** begins (Table 11.1). First, the water is expelled from the peat to form brown coal with lignite which develops at the transition from peat to **brown coal** and contains about 70–75% water. This contrasts with a sub-bituminous coal that represents the transition to **hard coal** with only 10% water. The formation of brown coal is

Table 11.1 German coal classification

Material	Water content (%)	Carbon content (%)	Caloric value (ca.; kJ/kg)
Wood	35–45	45–50	20,000
Peat	50–80	50–60	23,000
Lignite	50–75	60–65	25,000
Hard brown coal	30–50	65–70	27,000
Subbituminous coals	10–30	70–75	28,000
Flame coal	4–7	75–82	32,500
Gas flame coal	3–6	82–85	33,500
Gas coal	3–5	85–87	34,500
Fat coal	2–4	87–89	35,000
Forge coal	2–4	89–90	35,300
Non-baking coal	1–3	90–91.5	35,600
Anthracite	<2	>91.5	36,000
Graphite	<1	100	–

mainly characterized by the decrease of water and the conversion of lignin, whereas during hard coal formation significant amounts of volatile gases are lost. Depending on the residual gas content, different types of coal develop, such as lean coal or high-volatile coal. With increasing pressure and increasing temperature, the gas content decreases and the calorific value (and carbon content) increases. The most valuable coal is, therefore, the highest grade category of **anthracite** with the lowest gas content (<10%) and the highest carbon content (>90%). If the coal is subjected to higher metamorphic pressures and temperatures, the coal minerals (known as macerals) are converted into a pure carbon mineral, namely, **graphite**.

Recoverable coal occurs in the form of layers known as **seams**. A seam is a raw material deposit of sedimentary origin with variable thickness. Brown coal seams may reach thicknesses of several tens of meters, whereas hard coal seams are usually no thicker than 3 m. In the Ruhr region, about 100 coal seams are known in the coal-bearing successions, but not all are exploitable. Every single seam represents the development of a wetland and its subsequent flooding during the Late Carboniferous. This reflects the cyclic repetition of regressions and transgressions in a slowly subsiding molasse basin, which was filled with sediment derived from the advancing Variscan Mountains. During regression, the shoreline receded causing the swamps to silt up and spread landward. In contrast, transgressions flooded the peat areas previously formed in the swamp environment and buried them during the rapid deposition of new sediment. In this way, a large number of coal seams were preserved during the Carboniferous era.

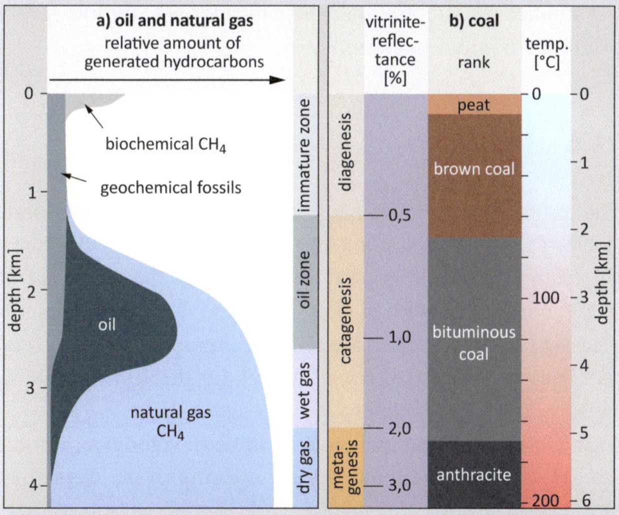

Fig. 11.37 Formation conditions of a) oil and natural gas, and b) coal, showing their dependence on the depth of burial and temperature. The range where oil and gas are formed (**a**) is known as the oil window and natural gas window, respectively. The vitrinite reflectance parameter (**b**) provides a measure of the degree of coalification attained by a rock. It is based on photometric measurement of the degree of green light reflected by the vitrinite maceral, which is the organic matter in coal derived from woody tissue

Oil and natural gas formation: oil and natural gas represent liquid and volatile hydrocarbon compounds originating as maturation products from the organic remains of both animal and plant origin. These mobile hydrocarbons contrast to the solid coal derived exclusively from terrestrial plant remains. In oil and natural gas generation, the zooplankton and algae provide the main source of organic matter. As dead marine organisms sink down in the water column, they are usually rapidly decomposed in well-aerated waters. However, when O_2-poor conditions prevail on the seafloor, **organic-rich muds** with high carbon content accumulate in a similar way as peat formation in terrestrial conditions.

During compaction under the weight of the overlying sediment, large volumes of free water are first squeezed out of the mud. This occurs prior to the organic material been converted into an insoluble residue material known as kerogen. As the sediments are slowly buried over periods of many 10's to 100's of million years, the kerogen slowly matures and releases hydrocarbons in response to the gradual increase in pressure and temperature. This process occurs under reducing conditions, and therefore, these reactions do not involve oxidation. Sedimentary rocks having high content kerogens are referred to as oil source rocks.

As the converted hydrocarbons are lighter than water they readily migrate within the pores of the rock and move upward until they reach an impermeable layer (known as a cap rock) thus preventing any further rise or escape to the sides. If the porosity of the host rock is high (porous sandstones, fractured limestones, etc.), large amounts of liquid and/or gaseous hydrocarbons may accumulate to form a reservoir of petroleum or natural gas in what is known as an oil or gas trap. Such deposits form, for instance, in anticlinal fold structures, in truncated strata at the edge of salt domes or within porous beds isolated by faults (Fig. 11.36).

For the conversion of the organic precursor material (kerogen) into hydrocarbons, certain conditions need to be fulfilled. For this process, the temperature is the most important factor (Fig. 11.37). For a normal geothermal gradient (about 30 °C) oil formation occurs at a depth between about 1500 m (from about 60 °C) and 3500 m (around 110 °C): a temperature range known as the **oil window**. Natural gas generally forms at higher temperatures (110–200 °C), which is why the **gas window** occurs at greater burial depths.

small amounts of oil, marked by the characteristic horse-heads (or nodding donkeys): a term that describes the shape of the oil pump. Examples of these pumps are encountered in several places on the island of Usedom, where oil production is ongoing. The oil reservoirs of the German sector of the North Sea and in Mecklenburg-West Pomerania are typically located at more than 2.5 km depth in the Stassfurt carbonate of the Zechstein. These reservoir rocks formed on the northern slope of the North German Basin as reef and lagoonal carbonates during the second evaporitic cycle (Stassfurt sequence; Fig. 11.22). The oil originated mainly from the basinal facies of the Stassfurt sequence, which produced a petroleum source rock rich in organic matter that lay paleogeographically not far to the south. Recently, new exploration holes have been drilled in these areas, using modern methods such as horizontal drilling. These activities have led to the discovery of new, small hydrocarbon-bearing reservoirs in the Stassfurt carbonate. However, today the focus of interest is drilling near salt domes for the production of natural gas (Fig. 11.36), with the largest reserves remaining in the central area of the North German Basin, in the southern part of the North Sea and in northwest Germany, as well as areas of Holland to the west. Here, the source rocks were buried to depths below that of the oil window but remained within the gas zone to release significant volumes of methane that were trapped in these regions (Box 14: Coal, oil, natural gas).

11.3 Permian/Triassic Boundary

Globally, the Permian/Triassic boundary is marked by one of the largest volcanic events in geological history: the eruption of the **Siberian Traps**. Originally, up to 3500 m of thick lava and tuffaceous sequences covered an area of more than 7 million square kilometers in the area that is now Siberia (Fig. 11.1; for comparison, the area of the Federal Republic of Germany is just 357,000 km^2). However, an accurate estimation of the number of volcanic deposits that erupted is difficult due to extensive erosion and part coverage by younger sedimentary strata (Czamanske et al. 1998). The extrusion of the enormous quantities of volcanic material took place in an extremely short period of time. It is assumed that this **flood basalt event** lasted only about 200,000 years and occurred immediately prior to the Permian/Triassic boundary at 251 $\pm$ 0.3 Ma (Bowring et al. 1998).

In addition to this globally important volcanic event, the Permian/Triassic boundary is characterized by the largest documented **mass extinction event** in Earth's history and is commonly known as the Great Dying. It has been suggested that the flood basalt event was responsible for, or at least contributed to, the mass extinction. Almost 96% of marine invertebrates and 70% of terrestrial vertebrates died out at the Permian/Triassic boundary, and it was the only event that resulted in the mass extinction of insects (Sole and Newman 2002; Benton 2005). It has been proposed that the intense volcanism resulted in a significant increase in atmospheric CO_2 (Fig. 11.7). There was also a complex relationship between the formation of a mantle diapir, ascending fluids, and a change in seawater chemistry that led to the increased release of greenhouse gases, in particular, methane. The changes in atmospheric gases then resulted in deteriorating climatic conditions and ultimately the mass extinction of a broad range of animal and plant groups. The relationship between ascending mantle melts, volcanism, tectonic activity and the evolution of the living world is particularly evident in this example, but the causes and interrelations are still the subjects of fierce debates. The role of a possible meteorite impact on the Permian/Triassic mass extinction has also been proposed despite the lack of globally elevated levels of iridium left after a meteorite impact that characterizes the well documented Cretaceous/Tertiary boundary extinction (Sect. 11.8). Possible locations for a Permian/Triassic impact crater have been investigated in the marine area off the northwest coast of Australia and in Wilkes Land of Antarctica (Australia-facing side).

By the end of the Permian, the global climate had significantly changed from the coal-forming tropical conditions at the beginning of the period to the much drier conditions of an arid climate by the Permian/Triassic boundary. This change was reflected by a major decrease in the concentration of oxygen from more than 30% in the well-oxygenated conditions of the Early Permian to about 15% by the end of the Permian period (Fig. 11.7). At the same time, the CO_2 level increased slightly after the characteristically low levels of the Carboniferous period.

11.4 Triassic

The term "Triassic" originates back to Friedrich von Alberti (1795–1878), who studied the geology around Heilbronn. He recognized a three-part rock series in the area that is today considered the model landscape for the Germanic Triassic (s. below). The three lithostratigraphic subgroups of the Triassic (old greek *trias*: trinity) are the Buntsandstein, Muschelkalk, and Keuper.

In Central Europe, the plate tectonic development during the Triassic was dominated by the opening of the Neotethys Ocean toward the northwest and the beginning of the breakup of Pangea. The Triassic **Neotethys Ocean** had two precursors: the **Prototethys Ocean** that was mostly subducted by the end of the Paleozoic, and the **Paleotethys Ocean**, which although also mostly consumed by subduction, existed as an oceanic rift system into the Triassic (Fig. 11.1). Sedimentary deposits and igneous rocks originating in the areas of Prototethys and Paleotethys are located today along the northeastern edge of the Neotethys in the mountain ranges of the Himalayas, Pamirs, and Karakorum, as well as in the Zagros Mountain Belt of Iran. The branch of the Mesozoic Neotethys began to open like a telescope toward the northwest around the time of the Carboniferous/Permian boundary. This marine extension was located to the west and south of the southern part of the Pangea supercontinent, which originally belonged to Gondwana. In the northeast and east, the Cimmerian Block separated from Pangea but during the Mesozoic was welded to the Eurasian continent, along with other continental splinters and blocks. The collision of the Cimmerian continent with Eurasia and the associated closure of the Paleotethys is known as the **Cimmerian Orogeny**. It began in the Late Triassic and lasted until the Early Cretaceous.

Seafloor spreading ended in the Paleotethys at the beginning of the Triassic, while the youngest branch of the Neotethys system continued to spread further. In the area of the Alps, lithospheric extension, graben formation, and intense continental volcanism had already reached its peak during the Permian. During the Triassic, an approximately NW-SE striking partially oceanic rift system developed, which in combination with the subsequently evolving Penninic Ocean (Chap. 11.6), separated the Adriatic–Apulian block from southern Pangea (respectively, Africa) in the Jurassic (Fig. 11.38).

The northwestern edge of the Neotethys was of particular importance for the geological development of Central and Southern Europe. It was repeatedly flooded by marine waters and extensive **shelf areas** developed with thick carbonate sequences deposited in the areas of the Alps, the Carpathians, and the Dinarides. These shelf areas resulted from extensional tectonics in the northern part of the Neotethys Ocean and were associated with the formation and gradual subsidence of passive continental margins on both the African and Eurasian plates. From the Jurassic onward, these margins were fragmented to form a new Adriatic–Apulian plate.

The marine transgression that occurred during the Triassic was also influenced by the melting of the polar ice caps following the end of the Permo-Carboniferous glaciation, with the gradual rise in sea-level extending across the area of the submerged shelf. Large parts of southern Central Europe were included in the subsiding area influenced by the Neotethys Ocean as it expanded toward the northwest. This key area of the Neotethys spreading zone was subject to diverse and complex tectonic changes that occurred in response to the incipient breakup of Pangea.

In Central Europe, thermal subsidence continued in the Central European Basin with Triassic continental sedimentation and eventually marine deposition related to the formation of an **epicontinental sea** that was only occasionally connected via narrow inlets. However, the rate of subsidence in this basinal area was not as high as in the open shelf areas. Despite being unaware of the plate tectonic context, geologists of the nineteenth century did realize the major difference between the relatively thin deposits of the **Germanic**

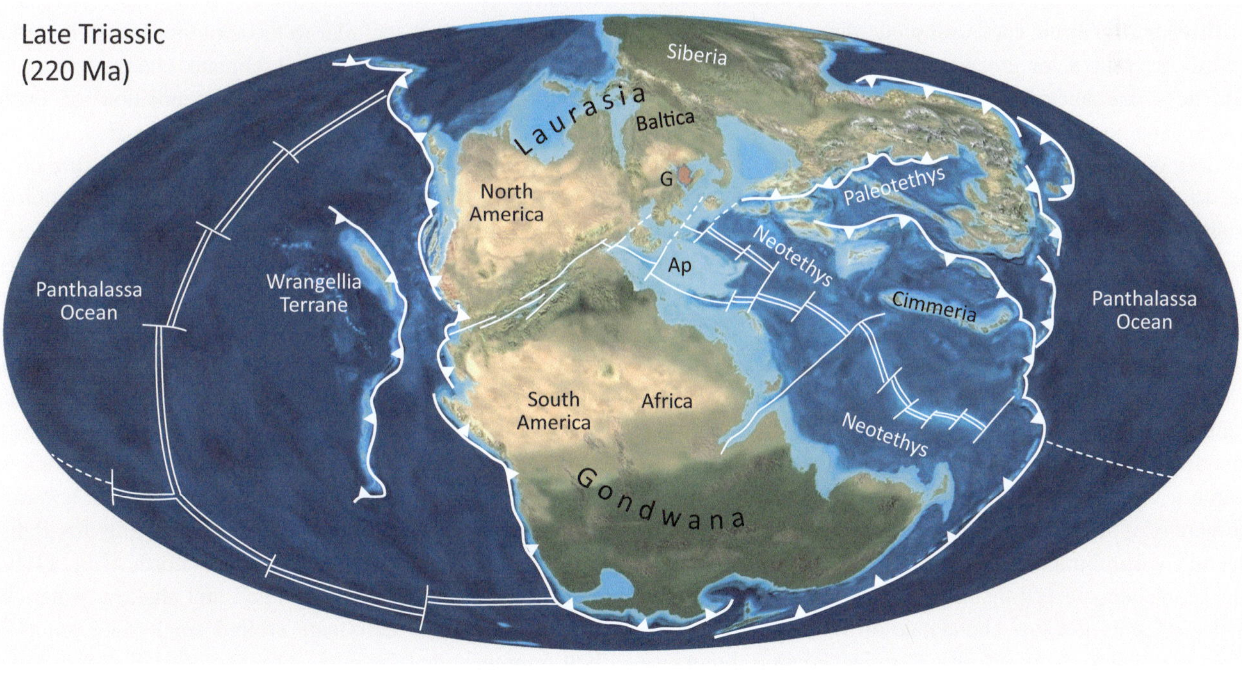

Fig. 11.38 Late Triassic (220 Ma) distribution of continents and oceans. Ap—Apulian Plate, G—approximate location of Germany. The topographic base for the paleogeographic map was provided by Ron Blakey, Flagstaff, Arizona (Blakey 2003)

Triassic compared to the **Alpine–Mediterranean Triassic** sequences. The Germanic Triassic was characterized by the deposition of the predominantly continental clastic sediments of the Buntsandstein, the shallow marine carbonates and partly hypersaline sediments of the Muschelkalk, and the marine-brackish and fluvial sediments of the Keuper. This contrasts to the often several kilometer thick marine successions of the Alpine–Mediterranean Triassic.

During the Early Triassic, a new structural pattern formed in the area of the Central European Basin with the development of largely N-S trending graben systems. This extensional phase marked the precursor stages to the opening of the North Atlantic and resulted in the prominent structure of the gradually subsiding **North Sea Graben**. Crustal extension and rifting became increasingly important during the Triassic period with the accumulation of large amounts of sediment that reached thicknesses of several kilometers. In the south of the basin, the tectonic evolution was associated with the development of a transtensional passive continental margin in the course of the opening of the Neotethys. This is documented by transtensional and lateral movements in the area of the Central European Basin. During the Late Triassic, the closure of the Paleotethys caused by the incipient collision of the Cimmerian continent with Eurasia resulted in a widespread unconformity (Sengör, 1984).

11.4.1 Buntsandstein (Bunter Sandstone)

The Buntsandstein (in English also Bunter or red sandstone) is a **lithostratigraphic rock unit** predominantly composed of red-brown, yellow, or greenish-gray sandstones in colorful sequences that alternate with shales and sometimes limestones. The term Buntsandstein, introduced by Abraham Gottlob Werner (1786), is therefore sometimes misleading due to the presence of other lithologies. In the Germanic Triassic, the term is used as a geological chronological unit for the time span ranging from 251 to 243 Ma (Fig. 11.39). Internationally, this period is subdivided into two stages: the Indusian (251–249 Ma) and the Olenekian (249–244 Ma). Although the exact geochronology is still under discussion, these stages are not precisely contemporaneous with the Buntsandstein (Menning and Hendrich 2005). The lower boundary with the underlying Zechstein varies in age. In southern Germany, the sequence begins with the Eck conglomerate of a slightly older age (Menning et al. 2002). Overall, a stratigraphic classification of the predominantly nonmarine deposits is difficult using fossils. At the beginning of the Triassic, the low abundance of fossils following the mass extinction event at the Permian/Triassic boundary resulted in fossil-poor lithologies that are of little use for stratigraphic correlations. Palynology has proven the most suitable approach by using pollen and spores as stratigraphic markers. In the marine environment, **conodonts** with their tooth-like hard parts can also be used as index fossils for Triassic stratigraphy. However, the eel-like conodont-bearing animal group, which belonged to the phylum Chordata that includes the vertebrates, became extinct at the end of the Triassic.

In northern Central Europe, the deposition of the Buntsandstein continued in a very warm and dry desert climate characteristic of the Permian. The arid conditions were caused by the paleogeographic position of the continental landmass that was located a few tens of degrees north of the Equator. Such a latitude today is characterized by the Sahara and the Gobi deserts. The extreme aridity was also related to the large distance of this continental facies from the shoreline of Pangea. This distance was sufficient for any moisture from the oceans to precipitate at the margins of the large continent so it did not reach the areas farther away from the coastline. This resulted in minimal rainfall in these regions as experienced today in the central continental area of Australia. The sedimentary environment of this period was characterized by **terrestrial deposits** that were transported far distances in river systems that underwent seasonally controlled flash flooding. When the rivers dried out, depressions without outlets formed, where wind-blown aeolian sediment accumulated. Moreover, following the extensive mass extinction at the Permian/Triassic boundary, there was a general lack of deep-rooted plants and, as a result, the soil was much more vulnerable to widespread erosion.

Sea-level changes similar to those of the Zechstein period and related to ongoing global climate change continued during the Buntsandstein with the deposition of cyclic sequences of sands, silts, and muds. Salty clay deposits occur at the base of the sequence. These were formed in saline repeatedly desiccating lakes as a transitional lithology following the Zechstein. These deposits preserved primeval crabs, known as the clam shrimps, whose descendants are found in salt lakes today or in other environments characterized by evaporating water. Above the salty clay follow sedimentary deposits formed in a shallow marine basin belonging to the Bernburg Formation (Fig. 11.39). This basin failed to dry out from time to time and as a result, covered large parts of northern Germany. It was responsible for deposition of dolomitic sands in the south and a variety of coarse-grained oolitic carbonates resembling fossil fish eggs (known as **Rogenstein)** in the north (Fig. 11.40). **Ooids** form in warm supersaturated and shallow waters by the precipitation of carbonate around small sand grains or shell fragments that aid nucleation and crystallization. Small oolites with diameters up to several millimeters are kept in

Fig. 11.39 Stratigraphic sequence of Triassic sedimentary rocks in the North German Basin and adjacent regions. Modified after Menning et al. (2002, 2011, 2012), Menning and Hendrich (2005); Peryt et al. (2010), and the Deutsche Stratigraphische Kommission (2016)

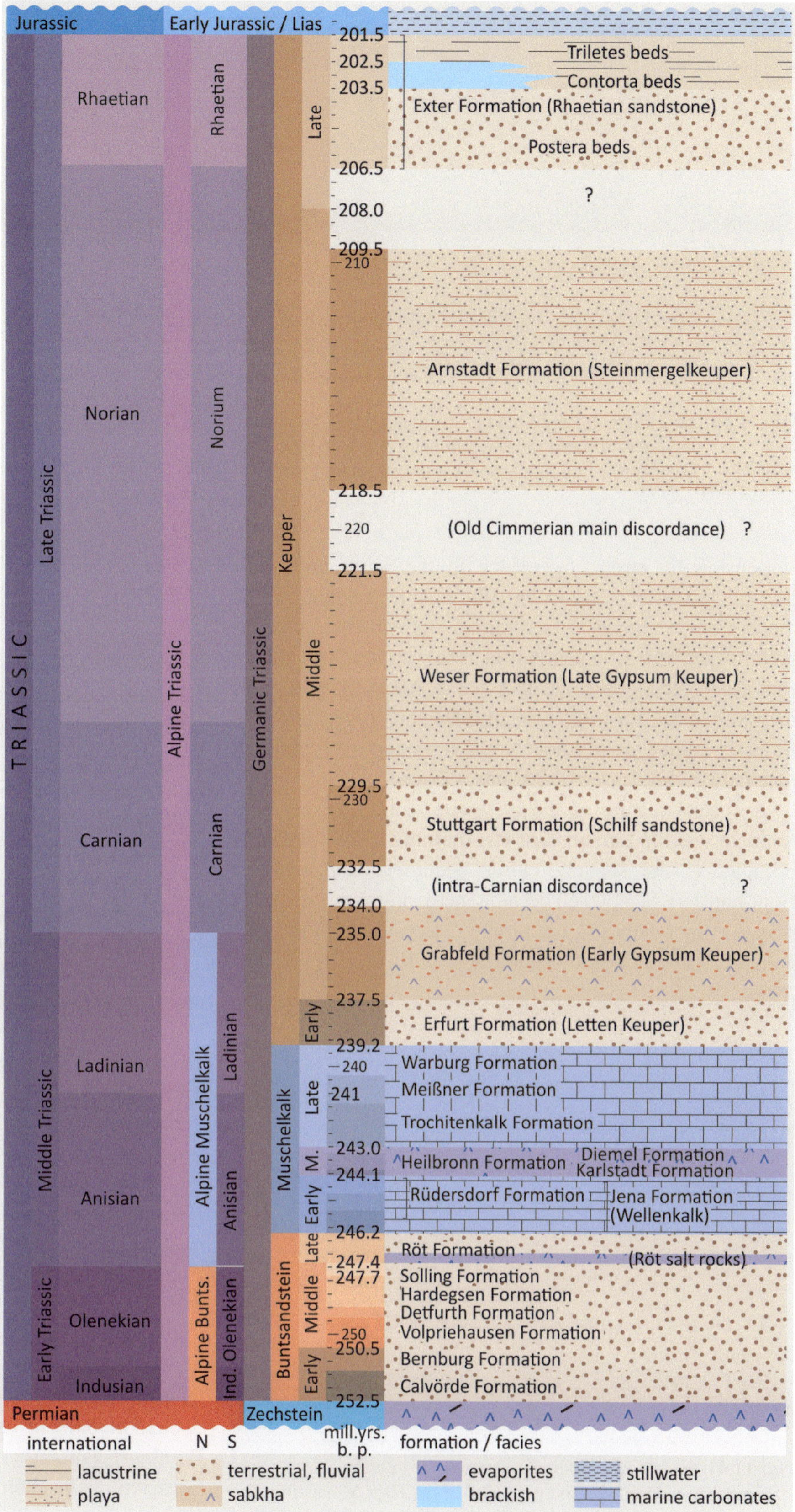

Fig. 11.40 Rogenstein from the Early Buntsandstein of the Hesseberg, near Jerxheim, located at the northern edge of the Harz Mountains. The diameter of single ooids is about 5 mm [N 52°05′ 01″/E 10°51′22″]

Fig. 11.42 Early Buntsandstein stromatolite colony in the wall of a disused quarry at the Hesseberg near Jerxheim, located at the northern edge of the Harz Mountains. This is the type locality for stromatolites where they were first described (Kalkowsky 1908) [N 52°05′03″/E 10° 51′55″]

suspension by strong turbulence in the water until they became too heavy and eventually sink to the bottom. In modern environments, it is common that carbonate precipitation is supported by algal or bacterial activity. The size of each Rogenstein oolite averages 5 mm, but individual ooids may also reach up to 15 mm in diameter. Rogenstein was formerly mined as a source of diverse types of building materials, but today no active pits remain.

A special feature of the Zechstein saline marine environment was the formation of **stromatolites**. They probably occupied a very shallow, closed endorheic branch of the Buntsandstein Sea, which became increasingly saline during strong evaporation and occasional desiccation. Under such extreme environmental conditions, where only a few creatures were able to survive, the stromatolites could develop, protected in their surroundings from grazing and physical destruction. Impressive examples of such stromatolites appear as alternating ooliths and rogenstein in the Subhercynian Basin along the northern edge of the Harz Mountains (Fig. 11.41). Since the stromatolite beds are not suitable for construction material, they are generally left in abandoned mining pits and can still be observed today. The term "stromatolite" is globally used for this type of carbonate precipitating blue-green algae that evolved in the Precambrian and is still present in a few localities around the world. It was introduced by Kalkowsky (1908) to describe columns and dome structures in Early Triassic carbonate beds of the Harz foreland area. The type locality contains a well-preserved colony of stromatolites originally described by Kalkowsky (1908). It is located in the disused quarry at the Heeseberg, near Jerxheim, and is a designated geotope (Fig. 11.42).

Fig. 11.41 Early Buntsandstein stromatolite colony from Benzingerode, located on the northern edge of the Harz Mountains. The stromatolite colony was supposedly found in a nearby disused rogenstein quarry between Heimburg and Benzingerode [N 51°49′ 31″/E 10°53′14″]. The specimen is today located in a front garden of a house in the village of Benzingerode [N 51°50′01″/E 10°51′35″]

Unique natural rock monuments composed of Early Buntsandstein occur in the Palatinate Forest a few kilometers west of Pirmasens. The sandstone deposits of the ***Dahner Felsenland*** (Dahn Rockland), which are widely known for their bizarre rock formations consist mainly of rocks of the Calvörde Formation. In this region, these rocks are locally known as the Trifels Formation. At the *Teufelstisch* (devil's table) in Hinterweidenthal (Fig. 11.43), there is a geological natural monument composed of sedimentary layers of the Bernburg Formation locally referred to as the Rehburg Formation. The monument is composed of silicified sandstone beds alternating with thinner laminated sandstones, siltstones, and conglomerates. Due to the lack of siliceous cement, these rocks are generally more susceptible to

Fig. 11.43 The "*Teufelstisch*" (devil's table) composed of silicified sandstone beds of the Early Buntsandstein near Hinterweidenthal in the Palatinate Forest [N 49°11′41″/E 07°44′37″]

weathering and erosion. This results in harder, tabletop-like blocks bounded by preexisting fractures that rest on top of thin pillars composed of less resistant material. On the other side of the Rhine graben in the Odenwald Forest, rocks of the Early Buntsandstein are exposed in the narrow valley of the Neckar River. The numerous cliff escarpments between Heilbronn and Heidelberg are composed of the Neckar Valley Sandstone. This stone found widespread application as a building material in the area and was used to build the Heidelberg Castle, as well a number of churches between Heidelberg and Eberbach. In addition to the Neckar Valley outcrops, rocks of the Early Buntsandstein are exposed in many valley sections and quarries in Thuringia and Hesse.

Toward the end of Early Buntsandstein deposition, the influence of extensional tectonics related to the beginning of the breakup of Pangea became evident. This resulted in the structural subdivision of the North German Basin into a series of basinal depressions and swells. Thick >3000 m sedimentary units of the Middle Buntsandstein accumulated in the Horn Graben, located in the North Sea area, and >2000 m of mainly fluvial and lacustrine sediment accumulated in the Glückstadt Graben of Schleswig-Holstein (Fig. 11.44). The best developed sedimentary sequences formed in graben structures, whereas swell areas, such as the Eichsfeld–Altmark Swell, resulted in multiple unconformities or hiatuses that reflect repeating tectonic movements. Overall, the thickness of the Buntsandstein deposits in the North German Basin reaches well over 1000 m. This is consistently thicker than in other regions where sequences thicker than 500 m are encountered only locally.

The Volpriehausen Formation at the base of the Middle Buntsandstein contains characteristic fossil shells (*Avicula*) and snails. These provide evidence of a temporary connection between the northern German inland lake and the Neotethys Ocean. Despite this, the Middle Buntsandstein is characterized by almost exclusively continental sediments from the Detfurth Formation upward. Typical river sediments deposited during this period formed several meter thick cross-bedded sandstones and thin clay and siltstones containing detrital mica. Desiccation cracks and sporadic raindrop traces occur in the clay and siltstone lithologies that reflect the periodic drying of the shallow lakes or tributary rivers.

Sediments deposited along the southern edge of the North German Basin transported weathered debris derived mainly from the erosion of crystalline basement in the southern neighboring region of **Vindelician Land**. Feldspathic arkose was deposited whose feldspars were readily transformed into **kaolin** by weathering under humid climatic conditions (Box 15: Kaolin). The porcelain industry in Kahla (near Jena, Thuringia), utilized sedimentary kaolin deposits originating from arkose lithologies of the Volpriehausen sequence of the Middle Buntsandstein, as well as similar deposits in the Upper Palatinate. In the early days of the European porcelain industry, arkose of the Solling Formation exposed on the plateau of the Solling Mountain Range (Fig. 11.39) was used in the 18th century to provide the coveted raw kaolin, although only in small quantities. These deposits provide the clay for the now world-famous Fürstenberg porcelain manufacturer. During the period of Buntsandstein deposition, the Solling area lay in a similar geographical setting to the Hessian Depression, with sedimentation of thick sand units along the margins of the North German Basin comprised of erosional detritus derived from the south. Early and Middle Buntsandstein strata are also widespread a little further south on the eastern slopes of the Odenwald and at the edge of the Thuringian Basin. Excellent outcrops can be examined in the narrow Unstrut Valley, near Nebra.

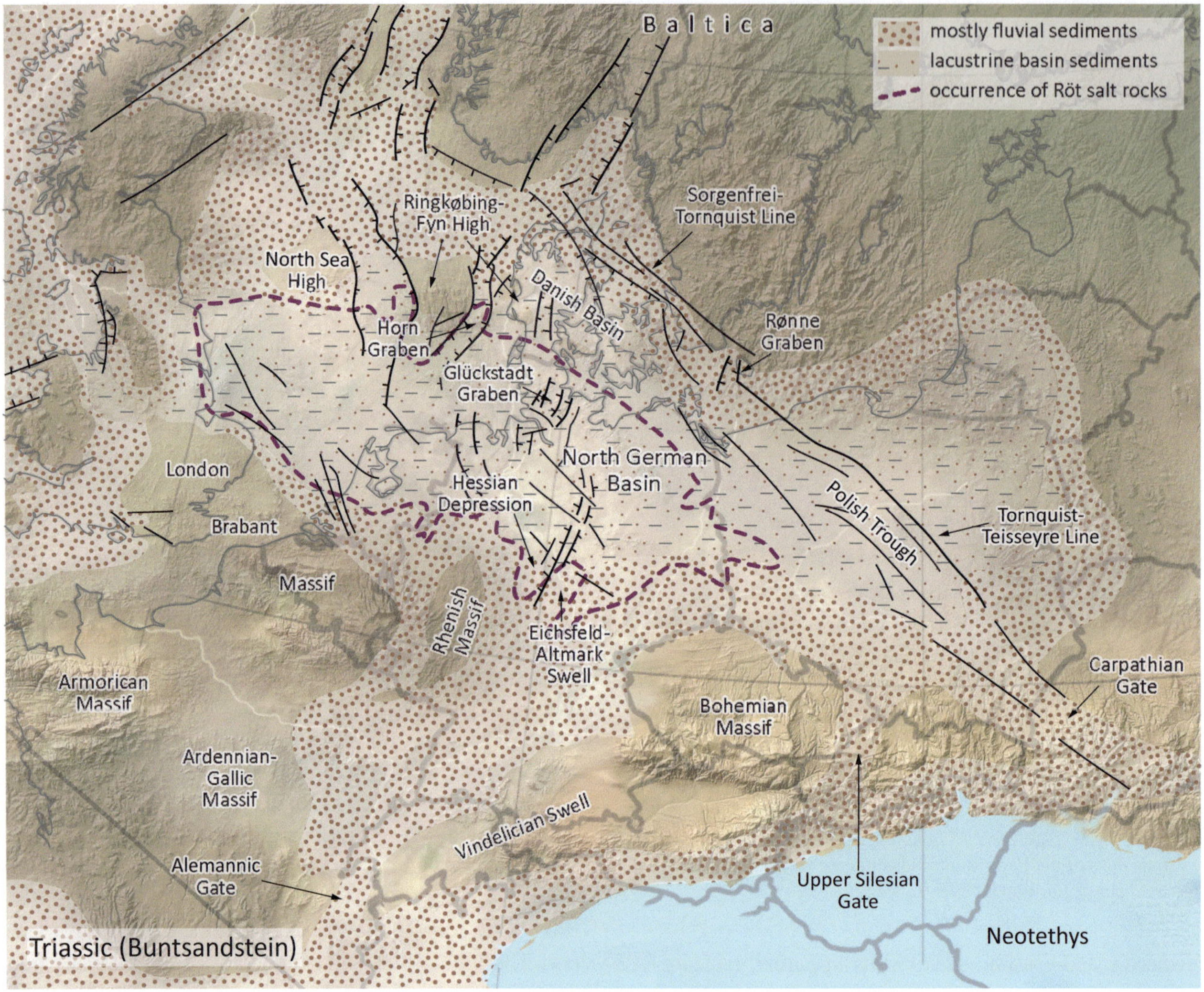

Fig. 11.44 Paleogeographic reconstruction of Central Europe at the time of Buntsandstein sediment deposition. Topographic base map from Blakey (2011) modified after Ziegler (1990), Hiete et al. (2006) and Bachmann et al. (2010)

Box 15: Kaolin

Kaolin (Chinese *gāo lǐng*: white earth), also known as china clay, is widely known as one of the main raw materials for the manufacture of porcelain. However, it is most commonly used for the production of paper as a filler and pigment. Kaolin is encountered in everyday life, for instance, as a carrier in tablets, in creams, and in cosmetic powders.

The main component of kaolin rock is the clay **mineral kaolinite**: a hydrated Al-silicate ($Al_2Si_2O_5[OH]_4$). This mineral commonly forms by the weathering of feldspar at the Earth's surface. Kaolinite also forms during supergene (near surface) alteration in humid climatic regions characterized by the circulation of meteoric groundwater. This mineral may also crystallize during the intense hydrothermal fluid alteration of rocks rich in alkali feldspar, such as granite, rhyolite, and other plagioclase-rich or glass-bearing volcanics.

Deposits of kaolin may form within the source rock itself, such as a hydrothermally altered granite, or in other cases, the material is eroded, transported, and deposited as sedimentary accumulations. The largest kaolin deposits in Germany, located around Hirschau in the Upper Palatinate, formed by the in situ alteration of Middle Buntsandstein feldspar-rich arkose rock. Other deposits were formed as weathering products of Permian rhyolites (Schwerdtner et al. 2007), such as the kaolin of Kemmlitz and Seilitz in Saxony, which provide the raw materials for the famous Meissen porcelain.

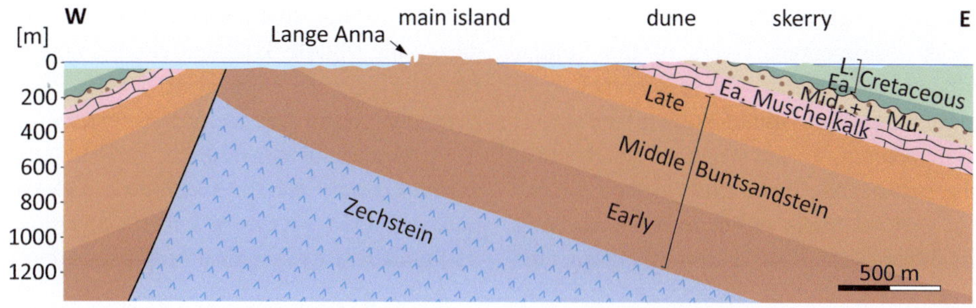

Fig. 11.45 Cross-section through the island of Helgoland showing the distribution of Middle Buntsandstein strata. The Helgoland dune is underlain by layers of Muschelkalk, which 250 years ago was exposed in a white cliff section. Modified after Grafenstein et al. (1991)

One of the best-known exposures of the Middle Buntsandstein is the towering rock of the island of **Helgoland** that forms prominent North Sea cliffs (Fig. 11.45). The *Lange Anna*, the famous landmark of the island, is in danger of collapsing due to intense weathering. Several dinosaur fossils have also been encountered in the Buntsandstein cliffs of the island that expose sandstones, siltstones, and mudstones spanning from the upper Volpriehausen Formation to the Solling Formation. These were deposited under the prevailing arid climate conditions of that time.

At the beginning of the Triassic, and corresponding to the deposition of the Middle Buntsandstein, the evolution of the **dinosaurs** began, leading to one of the most successful groups of animals that have ever lived on the planet (Fig. 11.46). A large number of **tracks** were left behind in the terrestrial mud deposits of the Middle Buntsandstein of Central Europe. Among the best-known trace fossils are those of the Chirotherium (ancient Greek *cheiros*: hand; Greek *therion*: wild animal) that resemble human hands. Notable finds of such traces occur in Thuringia, the

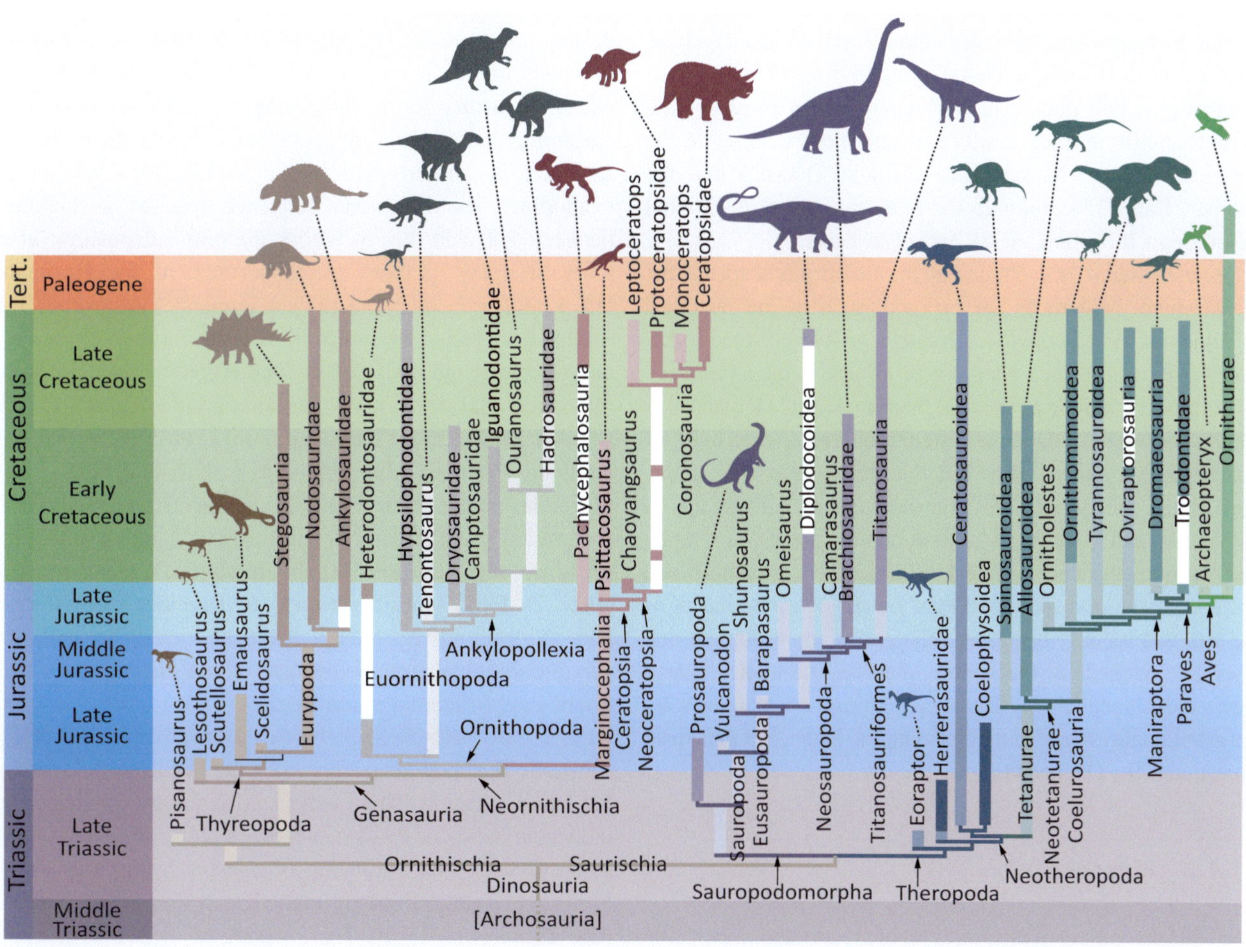

Fig. 11.46 Mesozoic evolution of the dinosaurs. After Sereno (1999)

Odenwald, and along the eastern edge of the Rheinisches Schiefergebirge in the sedimentary layers of the Solling Formation, where they occur together with many other small tracks of reptiles and some of the oldest known turtles.

The Chirotherium belongs to the group of **archosaurs** that includes the **dinosaurs** and other extinct groups such as the **pterosaurs**, but also the **birds** and **crocodiles** as living representatives of this group. The archosaurs were the dominating group of vertebrates both on land and in the air during the Triassic and in the following periods of the Mesozoic. The dinosaurs appeared for the first time during the Middle Triassic as a subgroup of the archosaurs and then quickly spread over the entire Earth.

The sedimentary sequence of the Late Buntsandstein (Röt Formation; also simply called **Röt)** begins with red clay and marlstones with only a few sandy interlayers. They originated in a large playa lake environment (Box 12: Playa sediments) that extended over the entire North German Basin. The marine influence increased with the inflow of water from the Neotethys Ocean that lay to the south, with water flowing through the Carpathian and Upper Silesian seaways into the Central European Basin before evaporating (Fig. 11.44). As a result, the sediments that were deposited gave rise to claystone, limestones, dolomites, gypsum, and rock salt layers following compaction and lithification. This marine influence can be traced into Central Germany in the form of the **Röt-Salinar** (salty layers in the Röt Formation) occurring as two thick rock salt horizons that in places are more than 100 m thick. Further marine incursions led to the deposition of the Muschelkalk (Middle Triassic) and the Keuper (Late Triassic), with the latter representing the final marine epoch of the Central European Basin.

Buntsandstein rock dominates the urban landscape in the areas where these rocks are exposed, such as the historical buildings of the old towns of Heidelberg (see above; Heidelberg Castle and others), Marburg or Erbach in the Odenwald. Also, modern buildings such as the Himbächel viaduct, near Marbach, which is a class listed monument, are built of Buntsandstein. Today, however, the active quarrying of this rock is rare. Generally, the Buntsandstein rock landscape is covered by extensive forest areas due to the poor soils that are low in nutrients and not very suitable for agricultural use because of the sandy subsoil. Often the geological boundary of the Buntsandstein, when lying adjacent to more nutrient-rich layers, is marked by the course of forest boundaries. Large forest areas underlain by Buntsandstein and characterized by poor soil quality occur in the Spessart, the Rhön, the Palatinate Forest, and the Odenwald.

11.4.2 Muschelkalk

The Muschelkalk is a lithostratigraphic term used to describe a sequence of Triassic carbonate rock strata in central and western Europe that formed between 246.5 and 239.9 Ma during the Middle Triassic period (Fig. 11.47; Hagdorn and Simon 2005; Menning et al. 2005). With a duration of 8 Myr, this period was as long as that required for deposition of the Buntsandstein (Fig. 11.39). The term Muschelkalk is now used only for the **Germanic Triassic** and was introduced in 1761 by Georg Christian Füchsel (1722–1773), a co-founder of the practice of stratigraphy.

Although the Triassic was initially divided into three parts based on the Germanic facies, this epoch is now subdivided on the basis of calcareous marls and clayey marls into the Lower, Middle, and Upper Triassic. These units do not precisely correspond to the classic facies division. Internationally, the Middle Triassic, for example, is now separated into two stages recognized in units of the **Alpine Triassic**: the Anisian and Ladinian. Adopting these stages means the lithostratigraphic unit of the Muschelkalk is of late Anisian and early Ladinian age. It should also be noted that the sedimentary rock sequence of the Muschelkalk is not exclusively composed of limestone, but also contains other lithologies, in particular, calcareous marls and clayey marls.

The Muschelkalk was deposited in a shallow marine basin connected to the Neotethys Ocean by various marine seaways (Fig. 11.48). In contrast to the "North German Basin", which only describes the northern portion of the basinal complex, the Muschelkalk was deposited across the complete marine basin known as the **Germanic Basin** that extended further south. Following a general sea-level rise, calcareous sediments were deposited, although the carbonate sequence was interrupted in the Middle Muschelkalk by evaporites. The limestone sequence reached a thickness between 200 and 400 m across the entire deposition area. Only in a few graben structures, such as the Horn and Glückstadt grabens, were thicker sequences deposited. The sedimentary facies was uniform over the entire marine basin, and therefore, the units can be well correlated, despite some minor regional differences. The stratigraphic subdivision of the Muschelkalk is based primarily on conodonts and ammonites (Kozur 1999, and others). Due to the rich fossil assemblages, the subdivision of this lithostratigraphic unit is so detailed that cyclic series are recognizable that reflect the influence of the Earth's 100,000 yearly cycle of eccentricity (Box 25: Earth's orbit parameters). Together with isotopic and biostratigraphic data, a detailed correlation with the geochronological time sequences of the Neotethys is well established (Bachmann and Kozur 2004).

The Early Muschelkalk was deposited over most of the Germanic Basin, as characterized by the **Wellenkalk layers** of the Jena Formation. These rocks can be further subdivided by individual oolitic or fossil-rich beds. In the northeastern part of the basin, the Jena Formation is replaced by the Rüdersdorf Formation characterized by numerous oolitic limestone beds (Bachmann et al. 1999). The lowering of the sea-level in the Middle Muschelkalk resulted in restricted water exchange with

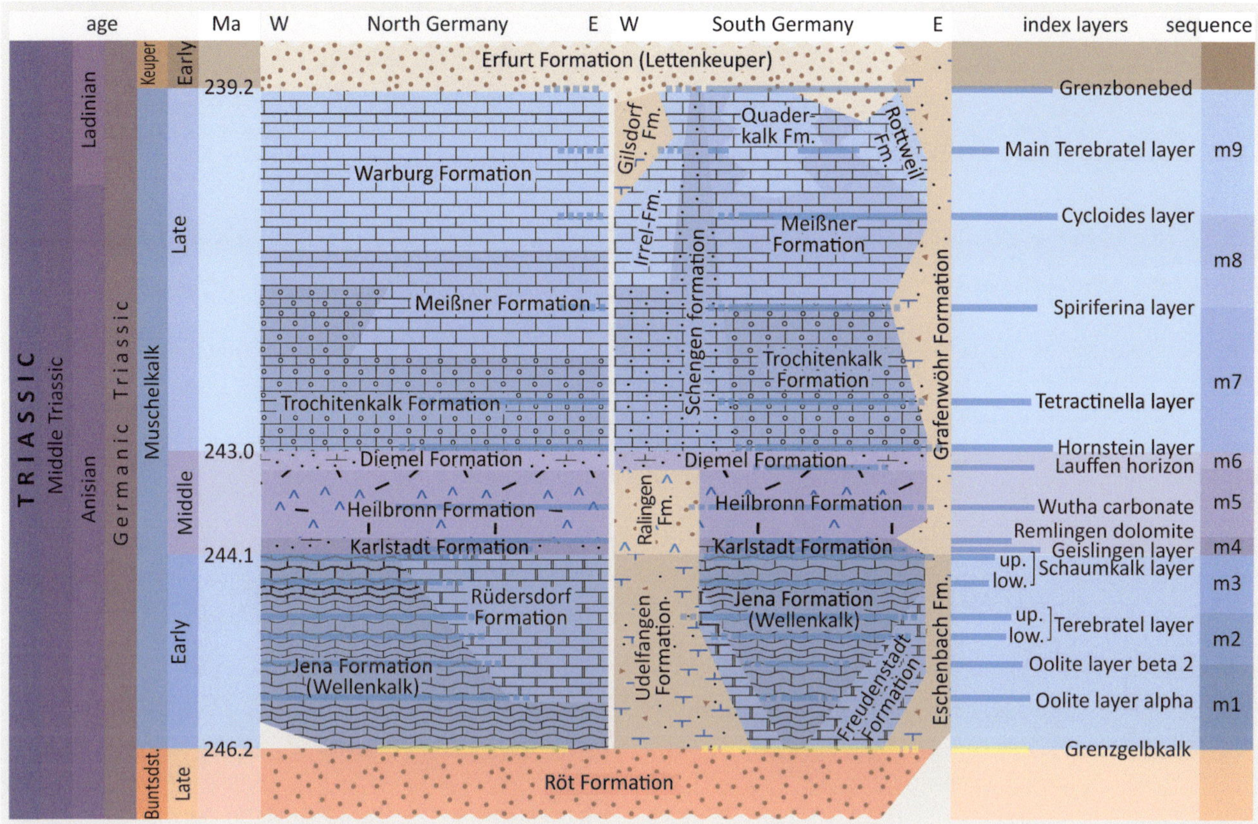

Fig. 11.47 Stratigraphic and facies subdivision of the Muschelkalk in northern and southern Germany modified after Hagdorn and Simon (2005), Menning and Hendrich (2005), and the Deutsche Stratigraphische Kommission (2016)

the Neotethys and triggered another sequence of evaporitic deposits in the form of the Heilbronn Formation. These are bounded by lower and upper dolomite horizons known as the Karlstadt and Diemel formations, respectively (Fig. 11.47). With the exception of a few areas representing topographic swells, the special hypersaline depositional conditions gave rise to the almost barren layers of the Middle Muschelkalk that contain virtually no fossil remains. The Late Muschelkalk, part of which is also referred to as the *Hauptmuschelkalk* (main Muschelkalk), is again distinguished by prominent limestone beds. The Trochitenkalk Formation is typically characterized by crinoid-bearing limestones that often contain the characteristic stems of crinoids: the so-called ***Trochiten*** (popularly also known as Boniface pennies or witches money). The crinoids found in the Late Muschelkalk are used as index fossils due to the diverse types that evolved at that time. Fossils from the Muschelkalk were in fact among the first documented in detail in the early essays of sixteenth- and seventeenth-century paleontology. One of the best known is the crinoid *Encrinus liliiformis* described as early as in 1546 by Georgius Agricola (1494–1555; Hagdorn 1999). The Meissner Formation which, in the central Germanic Basin, comprises the main part of the Late Muschelkalk, is more thin-bedded and marly compared to the Trochitenkalk Formation. It is characterized by the frequent occurrence of ammonites of the species *Ceratites* known only from the Late Muschelkalk (Fig. 11.49). At the edge of the basin, the sedimentary facies shows an increase in clastic input marking the beginning of the terrestrial environment of the Keuper with dolomitic–argillaceous sediments occurring both in the north in the Warburg Formation and in the south as the Rottweil and Grafenwöhr formations (Fig. 11.47).

In contrast to the Late Muschelkalk, where some layers are exceedingly rich in fossils due to suitable depositional conditions, the Early Muschelkalk is more characterized by its paucity of fossils. Seilacher (1999) attributed the lack of fossils to the closed nature of the Germanic Basin and the consistently shallow nature of the epicontinental sea. The Early Muschelkalk contains some fossils in the Wellenkalk facies located predominantly in southern Germany. However, these are not diverse assemblages due to the hypersaline and low oxygenated conditions of the bottom sediments where only a few organisms could exist. During deposition of the Late Muschelkalk, however, conditions became more oxygenated so that a rich fauna and flora could develop. At the same time, the low supply of terrigenous material favored the accumulation of shells and bones in the sediment.

Under normal conditions, the wave base, i.e., the depth where sediment is influenced by wave movement, is only a few

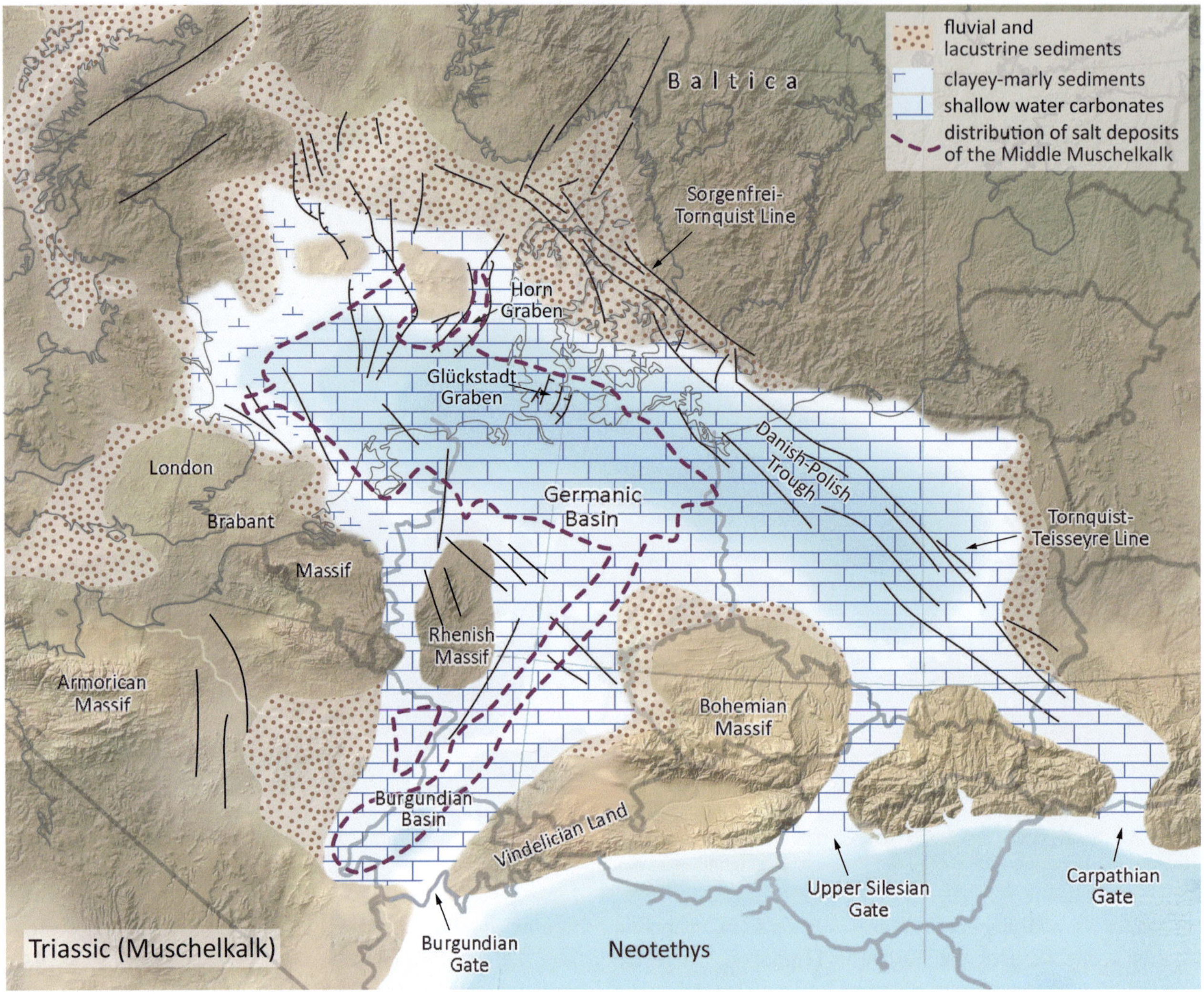

Fig. 11.48 Paleogeographic reconstruction of the Muschelkalk in Central Europe. Topographic base map from Blakey 2011; facies relations modified after Ziegler (1990) and Bachmann et al. (2010)

meters deep. However, during **storm events**, the seabed is disturbed down to depths of more than 100 m. Once a storm calms, the churned sediment is again redeposited and forms a sedimentary layer known as a **tempestite** (Latin *tempestas*: severe weather, storm). These beds can reach thicknesses of up to 30 cm with graded bedding marked by decreasing grain sizes toward the top and a frequently sharp erosive base at the bottom that is often characterized by broken fossil shells (Aigner 1985, 1999). Tempestites are notably abundant in the Muschelkalk and formed over the entire Germanic Basin due to widespread shallow water depths.

The Muschelkalk plays a significant role in shaping the landscape across a broad area known as the southern German cuesta landscape (*Schichtstufenlandschaft*; Fig. 13.9). In the far south, a complete section of the sequence of the Triassic sequence is exposed in the Wutach Gorge (*Wutachschlucht*). The Muschelkalk is also responsible for the fertile soils between the Kraichgau and the Rhön, and the Triassic Gäuland landscape with the river valleys of Main, Jagst, and Tauber. Numerous exposures of the Muschelkalk offer an excellent insight into the nature of the rock sequence. In particular, the area around Crailsheim is known for its richness in Late Muschelkalk echinoderms. An impressive compilation of the living world of the Muschelkalk collected by Hans Hagdorn is exhibited in the Muschelkalk museum of Ingelfingen, (Hagdorn 2004). Located in the southwestern edge of the area, the **Freudenstadt Graben** is a young Tertiary structure that structurally offsets the Late Muschelkalk against the Late Buntsandstein (Fig. 8.5). In Dornstetten, a few tens of meters beside the main fault of the Freudenstadt Graben, a **rollover anticline** (Fig. 11.50) is exposed in the layers of the Late Muschelkalk along with a normal fault that runs parallel to the main graben fault.

Fig. 11.49 *Ceratites nodosus* from the Late Muschelkalk Meißner Formation, found at Erkerode in the area of the Elm, near Braunschweig [N 52°12′17″/E 10°42′26″]

The Muschelkalk is also exposed in numerous locations in the Thuringian Basin (Fig. 11.51). The Naumburg Cathedral and other architectural monuments in this region have been constructed with limestone of the Early Muschelkalk Jena Formation (Koch et al. 1999). Limestone deposits of the Muschelkalk are also commonly exposed in areas where rising salt diapirs have uplifted the overlying Mesozoic strata to the surface. One example is located in Rüdersdorf, near Berlin, where fossiliferous limestone beds of the Rüdersdorf Formation are exposed and still quarried for the cement industry. In this deeply excavated quarry, which is unique for northern Germany, the complete Triassic sequence from Late Buntsandstein to Late Muschelkalk is exposed. In 1875, the Swedish geologist Otto Torell (1828–1900) recognized glacial striations developed on the upper surface of the Muschelkalk limestone and used this feature to help gain acceptance for the then-controversial ice sheet theory (Chap. 16.2). Late Muschelkalk exposures, located above a salt pillow, are also found in the hill range of the Elm, southeast of Braunschweig. These are still extensively quarried for building blocks. Notable monuments of distinction composed of this limestone are the Roland statue in Bremen and the tomb of Henry-the-Lion in Braunschweig (Katzschmann and Lepper 1999). At the northern border of the Harz Mountains, the Mesozoic rocks have been steeply tilted in a number of locations along the Northern Harz Boundary Fault. In the fault zone known as the **zone of structural steeping** (*Aufrichtungszone*; Chap. 11.7.2), a number of excellent outcrops of Muschelkalk occur along the prominent small ridges formed due to the higher resistance of this rock to erosion.

The **saline rock horizons** of the Middle Muschelkalk were used in several places, in particular in Baden-Württemberg for either brine extraction or for salt mining. From all the mines, however, only the Stetten salt mine in Haigerloch remains in operation. Until 1994, this salt was also mined in Bad Friedrichshall near Heilbronn. The salt deposit of Stetten is located in the northwestern continuation of the Hohenzollern Graben at the junction with the Freudenstadt Graben (Fig. 8.5), where the border fault is exposed in the mine. The salt that appears here has a thickness of up to 10 m and is particularly suitable for road or industrial salt because of its high purity.

Fig. 11.50 Rollover anticline developed in Late Muschelkalk limestones at the northeastern edge of the Freudenstadt Graben, near Dornstetten [N 48°28′16″/ E 08°30′53″]

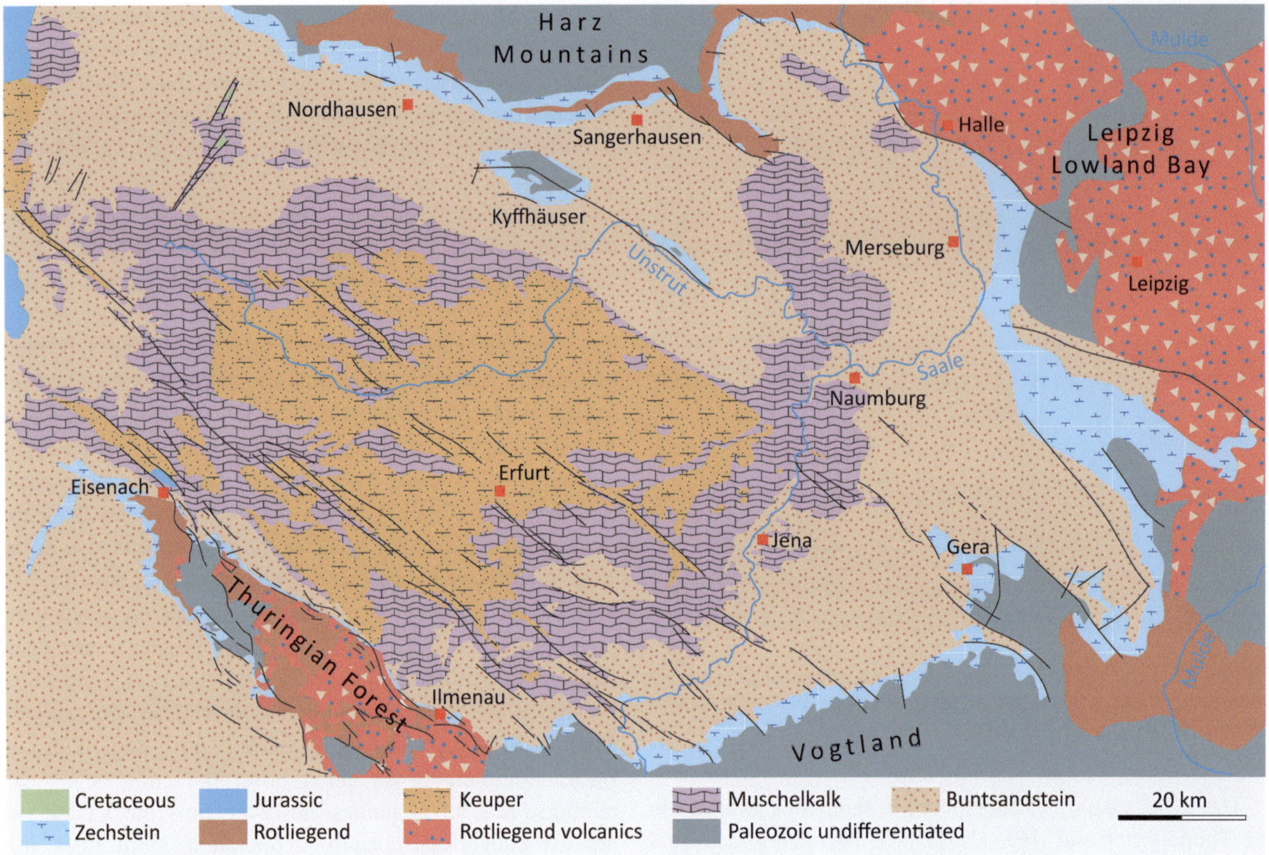

Fig. 11.51 Geological map of the Thuringian Basin. Simplified after Seidel et al. (2002), without Cenozoic formations

11.4.3 Keuper

The lithostratigraphic sequence of the Keuper encompasses almost 40 Myr of sedimentary record and is by far the longest period of deposition during the Triassic (Fig. 11.39). The sequence is, however, rather incomplete due to tectonically induced regression and erosional events that occurred in the area of the Germanic Basin. These events are frequently marked by a variety of paleosols occuring in almost all Keuper formations along several distinct unconformity horizons. In the strict sense, the Keuper represents a lithological rock facies and is not a recognized geochronological unit. The name is derived from the Franconian term *Keiper* or *Kieber* for a colorful, crumbly, and argillaceous rock. Leopold von Buch (1774–1853) introduced the term Keuper into the geoscience literature in 1822 (v. Buch, 1822) but today it is only used in describing the regional geology of the Germanic Triassic. The Keuper period lasted from the Early/Middle Ladinian (240 Ma; Menning and Hendrich 2005) until the Rhaetian stage at the end of the Triassic (200 Ma). Overall, the Keuper deposits typically range between 400 and 600 m in thickness but exceeds more than 5000 m in rift sequences such as the Glückstadt Graben.

During the Early Keuper, representing a short time period of only 1.2 Myr, sedimentation of the Erfurt Formation (traditionally called Lettenkeuper or Lettenkohlenkeuper) occurred in the same marine basin as the Muschelkalk. The style of sedimentation changed rapidly with the introduction of deltaic and terrestrial depositional conditions (Fig. 11.52). This led to the Grafenwöhr Formation that began as a marginal facies to the Muschelkalk (Fig. 11.47). In the southern part of the Germanic Basin, deltaic, lagoonal and estuarine deposits, together with some **playa sediments**, were influenced by the **tides** (Box 16: Tides), while in the northern regions fluvial and lacustrine sediments were deposited on a large alluvial plain (Bachmann et al. 2010). Locally, the rivers cut into the underlying layers of Muschelkalk as they flowed from the northeasterly adjacent Caledonian Mountains in Scandinavia and drained out toward the south before emerging at the Neotethys coast. The flow direction was thus opposite to that of the Buntsandstein. In the south, the Silesian Block was also uplifted to form an erosional source area.

During the deposition of the Keuper, local wetlands formed with a rich flora of horsetails, ferns, and cycads. Economically insignificant coal seams accumulated in Germany, although with some mineable coal seams occurring in neighboring Poland. In addition to bivalves, brachiopods, and ostracods, rich vertebrate assemblages are known from the lacustrine and fluvial deposits that include abundant fossiliferous, bone-bearing layers, known as **bonebeds**. Bonebeds are distributed throughout the whole of the

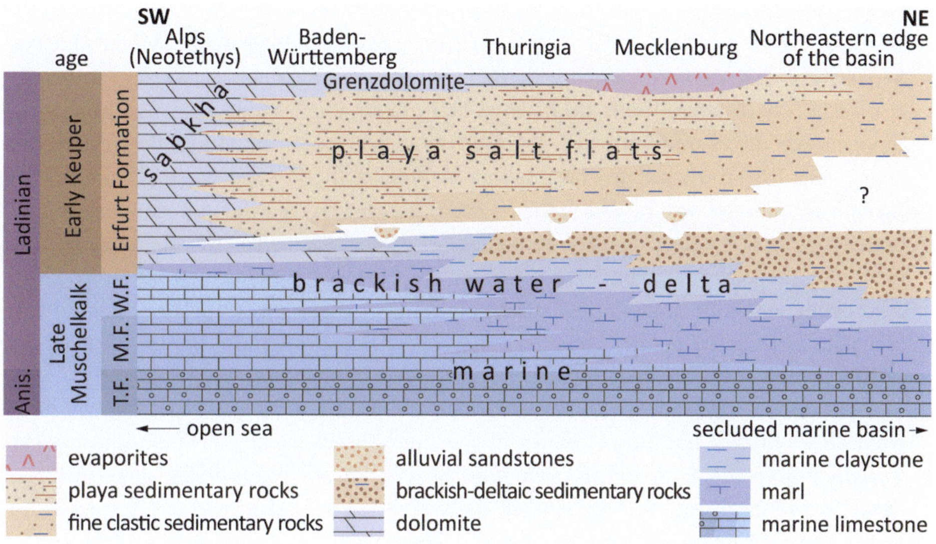

Fig. 11.52 Lithofacial transitions between the Muschelkalk and the Keuper in the Germanic Basin. *T.F.* Trochitenkalk Formation, *M.F.* Meißner Formation, *W.F.* Warburg Formation, *Anis.* Anisium. Modified after Szulc (2000)

Keuper. Renowned are the bonebeds (German: *Grenzbonebeds*) at the boundaries between Muschelkalk and Keuper and between Rhaetian and Lias, the latter marking the transition from Keuper to Jurassic. In Kupferzell, in the Hohenloher Land (northern Baden-Württemberg), the largest ever amphibian to have existed in Earth's history, the *Mastodonsaurus giganteus*, was found in a well-known fossil conservation site together with a number of other vertebrate remains. This was a gigantic salamander-like creature that grew to more than 5 m in length.

During the Middle Keuper, arid climatic conditions again dominated the Germanic Basin. For a limited time, a marine transgression formed a very shallow marine basin, comparable in dimensions to the Muschelkalk Sea, and deposited a fossil-rich boundary dolomite (Aigner and Bachmann 1992).

In addition to playa sediments (Box 12: Playa sediments), predominantly red and greenish-gray marls, clays, and sandstones formed in ephemeral lakes and rivers. Evaporitic sediments such as gypsum and rock salt were also deposited in sabkha deserts. Gypsum-bearing Keuper rocks are also referred to as *Gipskeuper* (Gypsum Keuper). Although salt horizons were deposited across the entire Germanic Basin, a direct marine source for the salt is considered unlikely, particularly in the northern regions. In this case, it is assumed that salts from the Zechstein rocks were dissolved from the underlying bedrock and then reprecipitated. Additional salt may have also been transported along rivers and as aerosols transported by the wind (Trusheim, 1971; Nitsch, 2003). In humid areas, local peat deposits that developed in the Middle Keuper gave rise to thin coals seams following lithification. Thick sands were repeatedly deposited as braided river systems. In the area around Nuremberg, a horizon known as *Bausandstein* (building sandstone) constitutes as well-known building stone material. A similar sandstone of the same age occurs in the Stuttgart area known as *Schilfsandstein* (reed sandstone) from the Stuttgart Formation (Fig. 11.39). Both these sandstones have been used in many distinguished buildings of the region, such as the Ulm cathedral and the Stuttgart main railway station. Most recently, some of the deeper Middle Keuper sandstone beds are also being investigated for use as future geothermal aquifers.

The continental conditions of the Early Keuper continued into the Late Keuper with deposition of predominantly terrestrial deposits in the form of clay, silt, and large quantities of sand that formed in an extensive **alluvial plain** environment. These sediments formed the rocks of the Exter Formation that makes up the whole of the Late Keuper sequence. Three subunits are recognizable, each characterized by different fossil communities: the Postera beds with a shell fauna (*Unionites postera*) of low diversity, the Contorta layers with abundant *Rhaetavicula contorta* shells and the Triletes beds containing spores of the genus *Triletes*. The distinct *Stubensandstein*, which occurs in both the Arnstadt Formation of the Middle Keuper and the Exter Formation of the Late Keuper, was formerly used in powder form for cleaning the wooden floors of rooms from where the name was derived (*Stube*, room). The Stubensandstein was also used in many buildings, especially in the Gothic period as building blocks, but due to the clayey binding material, the stone is not particularly resistant to weathering and in old buildings, it has to be laboriously replaced with rock that is more resistant.

11.5 Triassic/Jurassic Boundary

Similar to the Permian/Triassic boundary, drastic changes in environmental conditions occurred at the Triassic/Jurassic boundary linked to a **mass extinction event** that ranks

Box 16: Tides

We are probably well acquainted with the phenomenon of **tides** from the **low** and **high tidal** range that affects the mud flats of the North Sea coast. Whereas most will be aware that this is related to the gravitation pull of the Moon, the precise way in which this works may not be quite so clear. Why does the time and height of each high tide (and the corresponding low level) vary each day? And how can this be accurately calculated into the future, as listed in a tide calendar?

The Moon orbits around the Earth every 29.5 days, at least to reach the same visual point on Earth. Regardless of this, the Earth rotates in the same direction once every 24 h around its own axis. As a result of these two motions and following each daily spin of the Earth, the Moon moves further in its orbit. Thus we can observe the Moon at a fixed location in exactly the same position as the day before but 50 minutes later. It is this change that explains the continuous shifting of the tidal times. There are always two bulges of water that form around the Earth, one on the side facing the Moon and the other on the opposite side. Therefore, we can observe the high and low tides at intervals of about 12 h and 25 min.

The tidal bulges of water result from the interplay of two forces acting on the Earth system and the water masses on its surface: the **gravitation** and the **centrifugal force**. Whereas the gravitation pull of the Earth keeps the water on the surface, the Moon and the Sun also exert gravitational forces that result in the bulging of the water masses and hence the tidal pull effect. In fact, all astronomical bodies exert such forces, but they are so far away or so small that they do not influence the distribution of water on the surface of the Earth. The second opposing force is caused by the rotation of the Moon and the Earth. Although from the perspective of a viewer on Earth it looks as if the Moon moves in a circle around the planet, the movement, in fact, involves a rotation of both objects whereby the Earth and Moon move around a **common center of mass** that lies between them. This center of mass is not in the middle between the two bodies but lies closer to the Earth, 4700 km from its center because of it being 80 times heavier than the Moon. Therefore, the rotation generates higher centrifugal forces on the side of Earth that is opposite to the Moon. This force is large enough to form a bulge of water slightly smaller than on the side closest to the Moon (Fig. 11.53).

The gravitational force of the Sun also affects the tides by influencing the height of the tidal bulge of water. However, the tidal effect of the Sun is only about 40% of that of the Moon, but both combined lead to significant fluctuations in the height of the tides. When the Sun, Moon, and Earth are in a line, as is the case by a new or full Moon, the tidal pulls are combined and the bulge of water is at its highest, giving rise to a **spring tide**. Unfavorable weather conditions and catastrophic storm tides can develop during spring tides. If, on the other hand, looking from the Earth, the Sun and Moon are at right angles to each other, as is the case with a crescent Moon, the tidal forces are reduced and a **low neap tide** results (Fig. 11.54).

The tidal range is the difference in sea-level between the highstand and lowstand during high and low tides. Regionally, large differences exist that are dependent on the local conditions along the coast and the connection

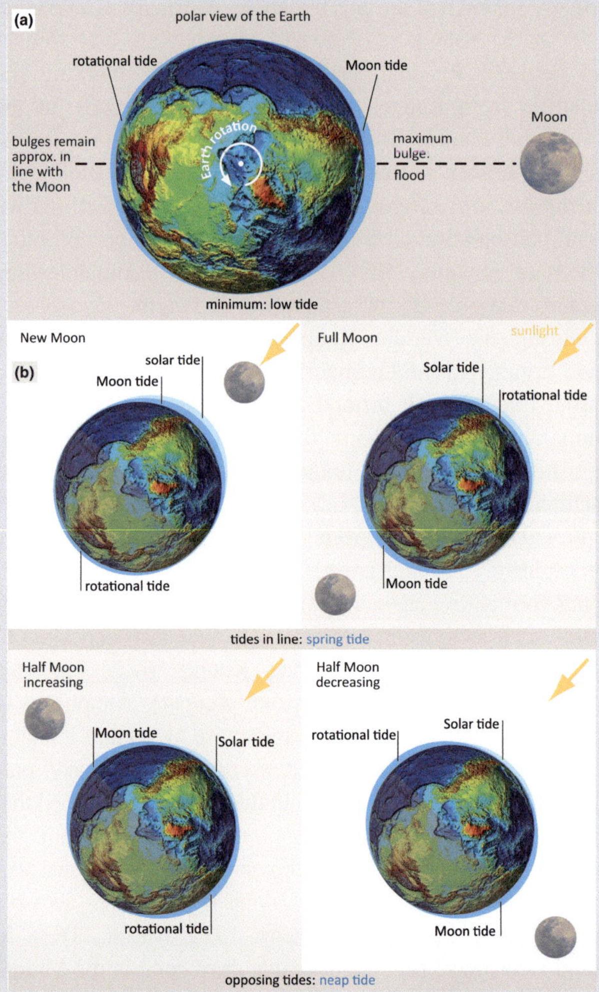

Fig. 11.54 **a** Formation of the tidal motion of water related to the gravitation pull of the Moon and the Earth´s rotation. **b** Tidal cycles, with the formation of spring and neap tides related to the different positions of the Earth, Moon, and Sun relative to each other

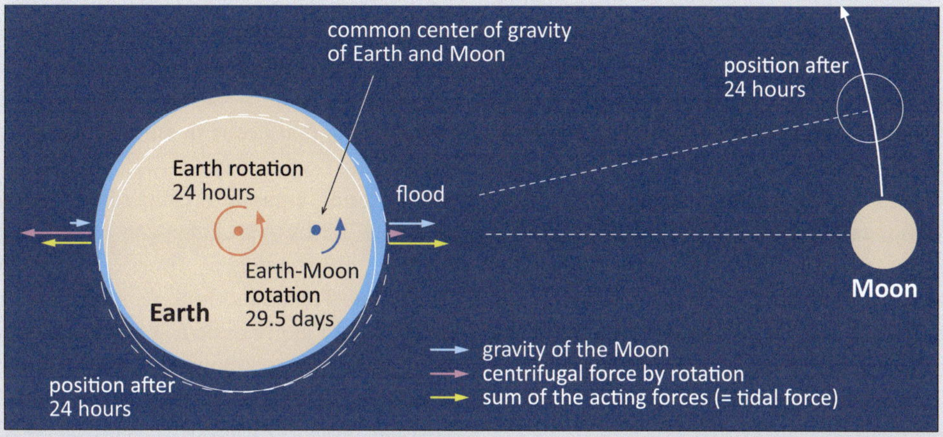

Fig. 11.53 Interaction of tidal forces (gravity of the Moon, centrifugal force) during rotation of the Earth and Moon around a common center of mass

to the open oceans. If the whole Earth was covered by a uniformly thick layer of water, the tides would barely have half a meter difference. However, at funnel-shaped inlets along the seashore, the tide often increases by several meters in height. At the German North Sea coast, the average tidal range is about 2–3 m but may reach more than 6 m in tidally influenced estuaries. In the Bay of Mont-Saint-Michel in Normandy, France, the largest tidal range worldwide occurs at about 15 m. The North Sea alone is too small to produce a tidal system. Instead, it is influenced by the tidal motion of the Atlantic Ocean. In the Baltic Sea, the tidal wave does not enter past the narrow Kattegat, and therefore the tidal range is only between 10 and 30 cm.

among the largest top five episodes of the Phanerozoic (Sepkoski 1996). The most likely cause of this mass extinction event was the intense volcanic activity known to occur within a very short time period of less than 1 Myr (Hesselbo et al. 2002). Other causes such as a meteorite impact (Olsen et al. 2002) or strong sea-level changes may have also played a role (Hallam and Wignall 1999). Covering an area of more than 11 million square kilometers (equivalent to the total area of the United States with 9.6 million square kilometers), the **CAMP province** is probably the largest flood basalt event of all times (CAMP: Central Atlantic Magmatic Province; Hames et al. 2003). This province was formed during the opening of the Central Atlantic and extended from central Brazil, through West Africa, the Iberian Peninsula, and France to the eastern and southern part of North America (Fig. 11.55). With the release of enormous amounts of airborne aerosol particles and volcanic gases, the global climate changed dramatically and resulted in mass extinction.

In the area of the Germanic Basin, the Triassic/Jurassic boundary is characterized by a distinctive facies change, which apart from local variations occurred within a very short time span. The change from the terrestrial-dominated Keuper facies to mostly marine Jurassic conditions is much more pronounced here than in the open, marine depositional area of the Neotethys. The Triassic/Jurassic boundary thus marked a time of rising sea-levels related to a widespread marine transgression. The resultant Jurassic Sea extended beyond the limits of the Muschelkalk Sea and formed an epicontinental shallow marine basin with some archipelagos that remained for a large part of the Jurassic period.

11.6 Jurassic

The Jurassic period saw a continuation of the breakup of the supercontinent of Pangea with the separation of North America from South America, Africa, Europe, and Asia. The oldest documented oceanic crust between Laurasia and Gondwana developed at the transition from Early to Middle Jurassic (about 175 Ma). It is preserved today on both sides of the Central Atlantic along the edges of the ocean. Uplifted areas in southwestern Europe, which in the Keuper still formed a barrier between the Germanic Basin and Neotethys, were gradually submerged during continued crustal subsidence and a global sea-level rise. At the same time, the Nordic Sea advanced further to the south and finally made a connection between the northern sea areas, the epicontinental shallow marine basin of Central Europe and the Neotethys in the south. Major changes also occurred in the catchment area of the Neotethys. Until the Triassic, the NW–SE oriented spreading center of the Neotethys was the dominant tectonic element in Southern Europe. During the Early Jurassic, the onset of the opening of the **Central Atlantic** began to influence the development of Central Europe. The Penninic Ocean opened along a transform fault system within the Neotethys, similar to the Central Atlantic, and was oriented in an NE–SW direction. The **Penninic Ocean** can, therefore, be regarded as the northeasterly branch of the Atlantic spreading ridge system, which was connected to the Atlantic Ocean through the **Ligurian Ocean** (s. below).

In contrast to the arid climatic conditions of the Triassic in Central Europe, the climate during the Jurassic was characterized by subtropical to tropical conditions. Glaciations at the polar caps did not occur during the Jurassic. The climatic conditions, which constituted a greenhouse climate, are not to be compared with the extreme hothouse conditions of the Cretaceous (Sect. 11.7.2). During the Jurassic, dinosaurs dominated animal life on the mainland with some species growing to gigantic sizes. These were in contrast to the mammals that occupied only a few niches and were generally very small, decimeter-sized species. The plant world was dominated by ferns, gingkos, and conifers. In the marine realm, the cephalopods and in particular the ammonites and belemnites became widespread as evident from distinctive fossil assemblages that occur in Jurassic sedimentary sequences worldwide.

The term "Jurassic" (Celtic for forest mountains) was used in 1795 by Alexander von Humboldt (1769–1859) to describe the prominent limestones of the French Jura Mountains. Today, the Jurassic is subdivided into the Early, Middle, and Late Jurassic as marked out on the international stratigraphic table. Following v. Buch's (1839) original color division of a Black, Brown, and White Jurassic for the Swabian Alb, Quenstedt (1843) proposed three basic lithostratigraphic sequences for the Jurassic. These terms were eventually replaced by the English names Lias, Dogger, and Malm, which were introduced by Oppel (1856) and have long been used as the three principal divisions of the European Jurassic strata. The names originated from English quarry workers who used them to describe the Jurassic lithologies: the Lias for Lower Jurassic limestones were used for building stone; the Dogger for Middle Jurassic calcareous or ironstone concretions were used to produce bright, yellowish bricks; and the Malm for the soft Upper Jurassic calcareous loams were also widely used for bricks. In

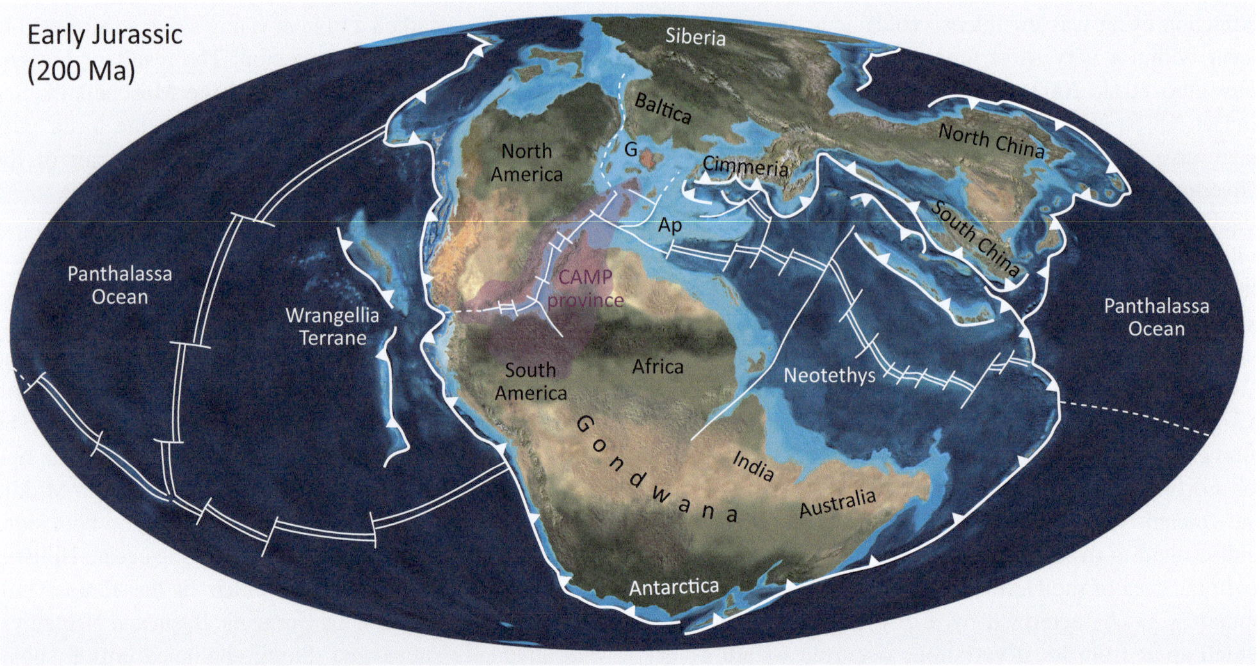

Fig. 11.55 Distribution of continents and oceans at the Triassic/Jurassic boundary (200 Ma). CAMP—Central Atlantic Magmatic Province. The topographic base for the paleogeographic map was provided by Ron Blakey, Flagstaff, Arizona (Blakey 2003). The contour of the CAMP province is modified after McHone and Puffer (2003)

southern Germany, the Jurassic is still divided into the traditional units of the Black, Brown, and White Jurassic.

In Northern Germany and in the Alps, the terms Lias, Dogger, and Malm appear to be preferred in the literature (Menning et al. 2002; Pfiffner 2010). Locally different subdivisions are often used because of the variable and complex fossil content. The most important **index fossils** for the stratigraphic classification of the Jurassic sedimentary record are the **ammonites**; a now-extinct group of marine cephalopods that are related to the living squids (Box 3: Ammonites). Although stratigraphic units can often be correlated over large areas, there are recognizable differences between the areas of basinal sedimentation. The Black Jurassic sequence in the south of Germany does not begin directly at the Triassic/Jurassic boundary but a little later, whereas the beginning of the Lias in the north corresponds well with this boundary. In the detailed classification of the Jurassic, the international stratigraphic table recognizes eleven stages that are named after different type localities in Germany, France, and England, with the exception of the Tithonian, named after *Tithonos* from the Greek mythology (Fig. 11.56). Quenstedt (1843) originally subdivided the three sections of the Jurassic into six subunits that were denoted by the Greek letters α to ζ. Although this classification is no longer used today, it is still commonly listed in the stratigraphic literature when citing older publications.

11.6.1 Early Jurassic

The Early Jurassic is characterized in many places in both north and south of Germany by dark **bituminous shales**. At the base of the sequence, clastic intercalations of sandstones and conglomerates are also found in the eastern border areas (Fig. 11.57). In the south, the sequence starts after some initial claystone with the distinctive **Arietenkalk Formation**. This formation often contains abundant ammonite shells popularly known as *Schneckenpflaster* (snail pavement) in which dozens of large, up to half a meter-sized ammonites of the genus *Arietites* have been found. Beautiful examples of clusters of ammonite shells are exposed in Ofterdingen near Tübingen, in the riverbed of the Steinlach and in the staircase of the Geological Institute (today Department of Earth Sciences) of the University of Tübingen.

During the Jurassic, the South German marine basin was initially connected to the Neotethys via a seaway known as the **Burgundy Gate**. Warm climatic conditions with water temperatures above 20 °C resulted in the development of a rich life fauna with ammonites, nautiloids, fish, crocodiles, and ichthyosaurs. The northern basin was significantly cooler, but even here abundant life forms occupied the marine basin. Despite a shallow water depth of about 150 m, two opposing current systems developed in the Central European Jurassic Sea. The heavier cold water in the north, which also had a slightly higher salt content, flowed along the seafloor bottom towards the south, while the lighter warm water flowed in the upper part of the water column back to the north. This resulted in a prominent **water stratification** that was so stable at times that the deep areas directly above the seabed were isolated from the fresh, oxygenated water supply for extended periods. As the biodegradation of the dead organisms that sank to the seabed consumed the remaining oxygen, this environment rapidly became anoxic (O_2 concentration <0.5 mg/l). As a result, a **hostile organic-rich sludge** formed known as a **sapropel** (ancient Greek *sapros*: rotten; Greek *pelos*: clay, mud). Ground-dwelling organisms such as mussels or sea lilies (crinoids) could no longer exist under the euxinic conditions (Box 17: Euxinic milieu). However, they could form colonies on floating tree logs in the oxygen-rich surface water. One of the most impressive fossils preserved from this environment is the almost 100 m^2 large colony of the gigantic sea lily *Seirocrinus subangularis*. This magnificent specimen was found in the Posidonia Shale in Ohmden, not far from Holzmaden near Kirchheim unter Teck (Fig. 11.58). Today, it can be viewed at the *Urweltmuseum Hauff* (Hauff museum of the primeval world) in Holzmaden in a room specially designed to house the fossil. In earlier times, the crinoids had a frightening effect on people as they were considered to represent a historical deluge. As a result, they were referred to as the *Schwabens Medusenhaupt* (the Swabian head of Medusa).

Unfortunately, the term **Posidonia Shale**, which is still used in current stratigraphic charts for a bituminous-rich mudrock layer of the Early Jurassic (Fig. 11.56) is somewhat misleading. The layer was introduced in the nineteenth century for a frequently occurring shell known at that time as *Posidonia bronni* (today *Bositra buchii*). Earlier, the genus *Posidonia* was used to name certain shell specimens (today attributed to the genus *Posidonomya*) from the much older Carboniferous period, thus adding confusion to the terminology. As the Early Jurassic Posidonia shale has now become entrenched in the literature, the rock name is retained to avoid further confusion with another new name. The name for the Carboniferous formation is now Posidonomya shale (s. Chap. 10.2.2). The German term *Posidonienschiefer* for the Jurassic Posidonia Shale is also misleading since the lithology is not a metamorphic slate or schist in the petrographic sense, but represents a sedimentary shale. Thus, the layered fabric of this rock is not a cleavage

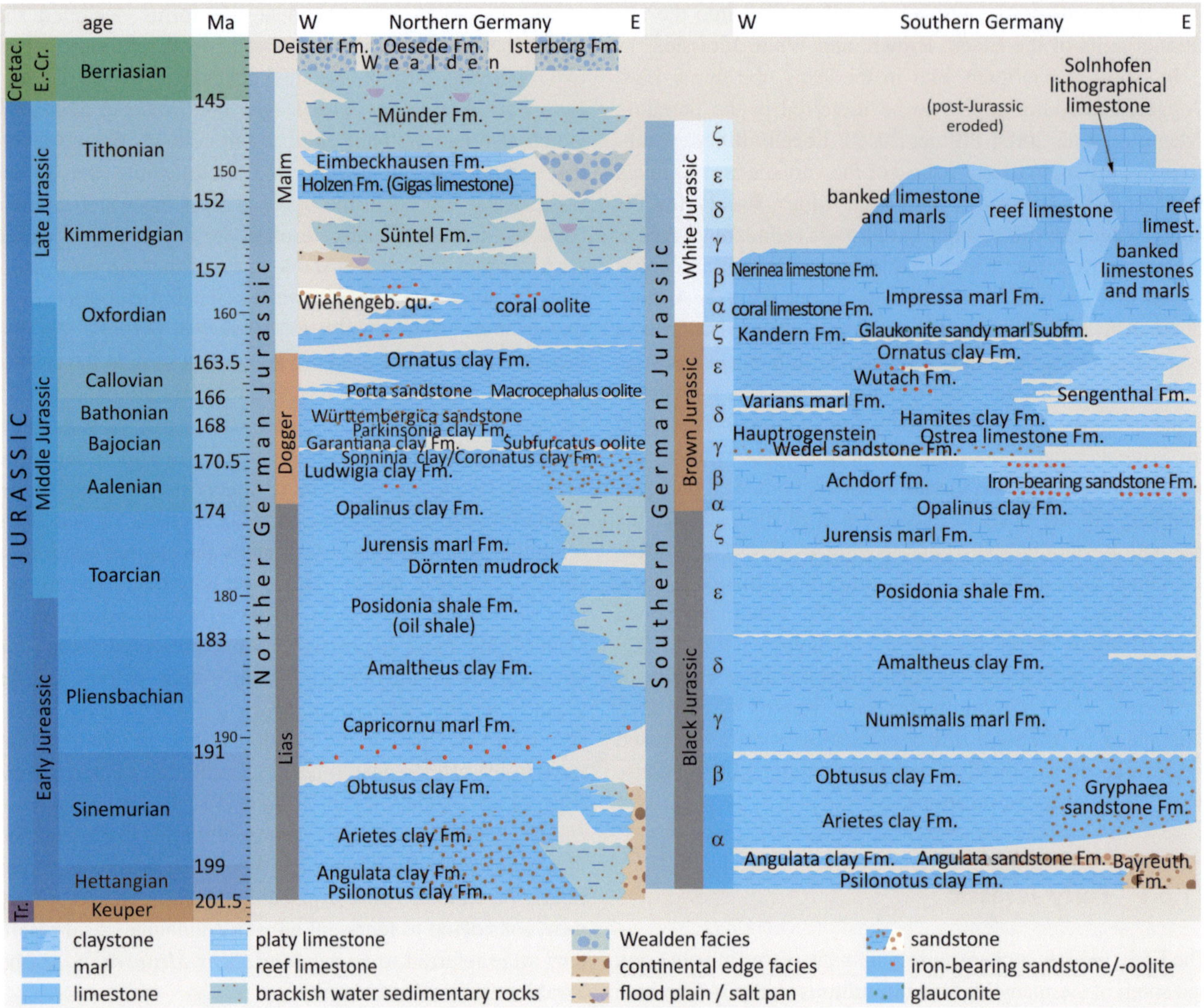

Fig. 11.56 Jurassic stratigraphy of northern and southern Germany. Fm—Formation, Wiehengeb. Qu—Quartzite from the Wiehengebirge. Modified after Menning et al. (2002); and Mönnig (2005), Deutsche Stratigraphische Kommission (2016)

formed by rock deformation and recrystallization. Instead, it represents a finely laminated sedimentary bedding fabric formed during deposition and compaction.

At the bottom of the stagnant Jurassic sea, a large amount of organic remains accumulated that were not degraded by microbes due to the lack of oxygen. In the organic anoxic muds that accumulated on the seafloor bottom, toxic hydrogen sulfide was released, a gas known to be highly unfavorable to most life forms. The sunken carcasses of vertebrates such as fish, crocodiles or ichthyosaurs, but also ammonites and shells and even the remains of soft parts were thus deposited and not destroyed by scavengers or microbe activity. In some cases, even the skin of the animals survived. The black color of mudrock, marl, and limestone usually reflects both the high concentration of organic remnants in the form of an insoluble organic residue known as kerogen, and finely dispersed pyrite (FeS_2). Kerogen is the precursor material for oil and gas that forms by the thermal modification of organic matter during burial and lithification. One famous fossil site is located at Holzmaden, where complete assemblages of ichthyosaur, plesiosaur, and crocodile bones were recovered that are now on display in many natural history museums around the world. Even pregnant females of ichthyosaurs have been found at the stage of giving birth, such as that displayed in the staircase of the Geological Institute of the University of Tübingen. This demonstrates that the ichthyosaurs, like the modern dolphins, gave birth

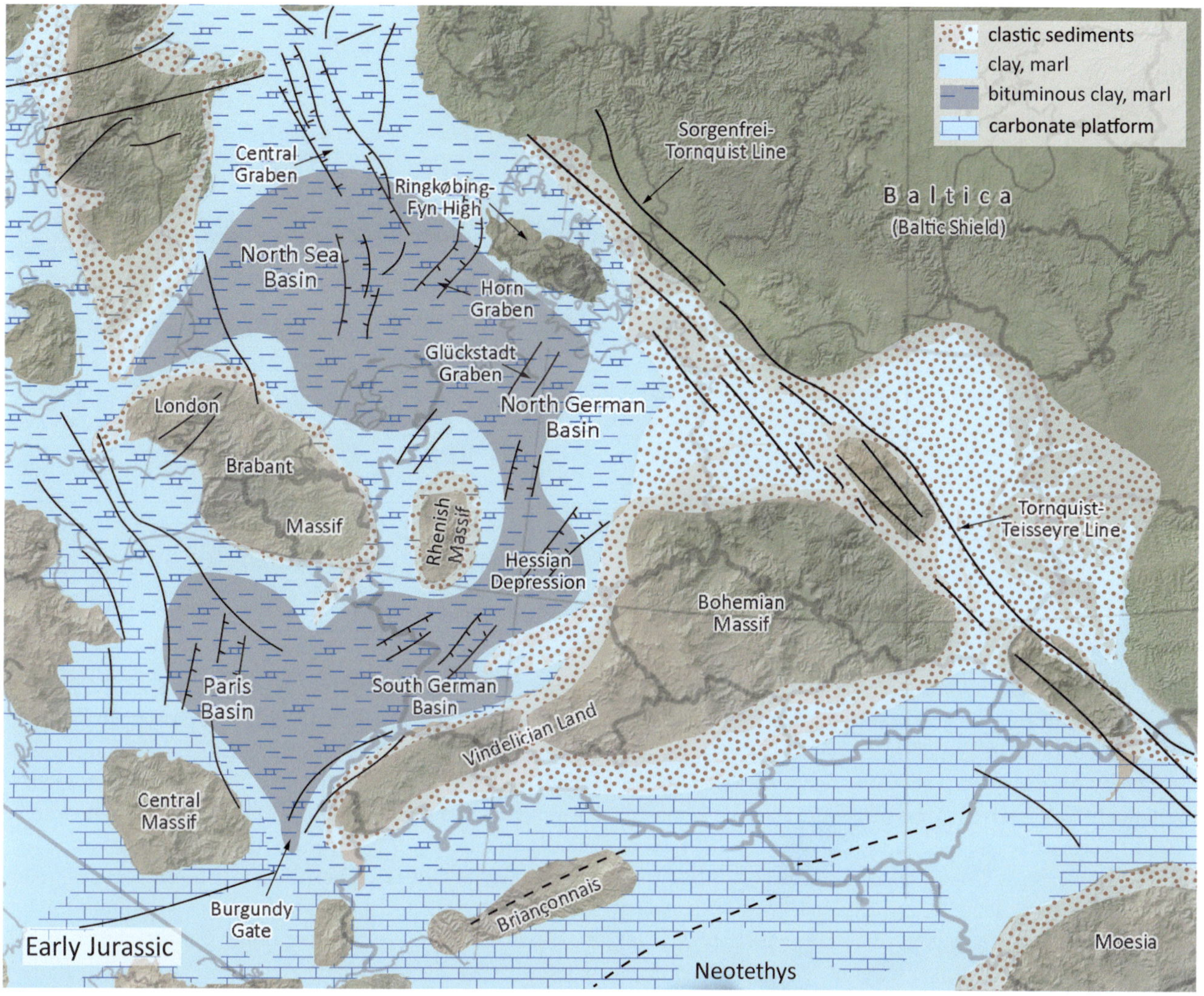

Fig. 11.57 Early Jurassic paleogeographic reconstruction of Central Europe. Topographic base map from Blakey (2011); modified after Ziegler (1990)

directly to a living infant. It is also likely that this particularly memorable fossil from the Black Jurassic reflects a premature birth that was triggered by the stress of a life-threatening situation. One explanation could be a storm event that led to turbulence of the toxic bottom waters, trapping the mother prior to perishing.

In the south of Germany, the Early Jurassic sedimentary sequences reach up to 80 m in thickness, whereas equivalent sequences in Northern Germany and in the North Sea are often thicker and exceed more than 200 m. The bituminous shales of the Early Jurassic were the source rocks for many petroleum reserves that have been found and exploited in the Jurassic graben systems of Northern Germany, the Netherlands, and the North Sea. The main source of organic matter that gave rise to the hydrocarbons was the abundance of microorganisms (plankton), which constantly trickled down like rain to the bottom of the Jurassic sea. It is assumed that large animals such as fish or dinosaurs contributed only a minor part of the material that formed the sapropelic mud. Above 25% of the organic matter in the muddy sediment were converted during diagenesis to mature hydrocarbon oil and natural gas (methane). Some oil-bearing source rocks were initially mined in southern Germany and heated (retorted) to release and recover the oil. In many cases, however, the oil and gas were expelled during burial diagenesis from oil shales, typically at temperatures between 60 and 120 °C, and migrated towards the surface. When entering porous rocks, such as poorly cemented sandstone or some limestones, the oil and gas was trapped within fold, fault or stratigraphic structures known as traps. Today, petroleum resources can be extracted in places by drilling into these geological reservoirs.

Fig. 11.58 *Schwabens Medusenhaupt* (the Swabian head of Medusa), the largest ever prepared block of the gigantic sea lily *Seirocrinus subangularis*, which was fossilized in the state of growing on a large piece of driftwood. The sample was recovered from the Black Jurassic of Ohmden/Holzmaden and is exhibited in the Urweltmuseum Hauff (museum of the primeval world) in Holzmaden (http://www.urweltmuseum.de), who kindly provided the photograph [N 48°38′06″/E 09°31′41″]

Box 17: Euxinian Milieu

The water of the open sea is normally subject to continuous mixing due to the motion of ocean currents. This process is necessary for adequate ventilation of the sea, whereby the amount of O_2 dissolved in the water is maintained at a more or less constant level. There are some areas of the seafloor where very calm stillwaters reside, such as closed basins or in isolated depressions. Here, inadequate water circulation occurs and the water column may become stratified with cold and heavy salt enriched water underlying warmer light water. In this environment, the lower part of the water column becomes stagnant and the existing free O_2 is quickly consumed during decomposition of organic compounds. Eventually, the trickle of dead organism's reaching the bottom of the sea begins to accumulate under the O_2-poor conditions. This accumulation is accompanied by the formation of hydrogen sulfide in the sediment caused by the bacterial reduction of sulfates. Such deposition conditions are known as **euxinian milieu** and characterize the stillwater environment. It is named after the *mare euxinium* (Latin for the Black Sea), where at the base of this silled and landlocked depression, the exchange of water is limited by a narrow connection to the Mediterranean sea via the Bosporus. This stagnation results in O_2-poor and reducing bottom sediment conditions.

Sedimentary deposits formed in an O_2-poor environment, are typically characterized by high organic content. Precipitation of iron sulfides (FeS, FeS_2) make the sediment very dark or black and when struck or ground with a hammer, they release an unpleasant smell of hydrogen sulfide. The high organic content originates from the remains of organisms that were incorporated into the sediment but did not decompose due to the lack of O_2. These black colored fine-grained sediments are known as black shales or bituminous sediments. They are of economic importance as the source rocks for oil, which once released during thermal maturation, accumulate in reservoir rocks prior to industrial extraction. If of appropriate maturity, organic-rich shales may also contain natural gas trapped in the organic matter. This latter type of unconventional hydrocarbon deposit is known as **shale gas**. This source of fossil fuel is currently being exploited in North America from shales buried at depths of between 1500 and 4000 m. It is highly controversial due to the need for intense hydromechanical fracturing of the rock that is of environmental concern. For this reason, the exploitation of shale gas in Germany is not yet possible and the practice of this method is strongly opposed by the public.

Depending on their state of diagenesis or metamorphism, organic-rich mudrock may be classified as a mudstone, shale or slate: the later been strictly a metamorphic term. Well-preserved fossils may occur in bituminous rock and, in some cases, even the structure of soft tissue is retained. One famous black shale fossil site is that of Holzmaden, located at the end of the Swabian Alb. Here, Early Jurassic shales (known as the Black Jurassic) contain numerous completely preserved ichthyosaur skeletons discovered during the construction of Highway 8 in the 1930s. These fossil finds are on display at an extraordinary museum in Holzmaden, which was founded in 1936–1937 by Bernhard Hauff (1866–1950), who was a quarry owner from Holzmaden. Today, it presents one of the most comprehensive collections of Early Jurassic fossils from the Swabian Alb.

There are several periods of Earth's history when sediments were deposited in a euxinic milieu over large areas. The oldest dates back to the Precambrian, but rocks related to this event are not common in Central Europe and only a few local Late Precambrian deposits occur in Germany in the core of the Schwarzburg Anticline of Germany. Examples of Paleozoic anoxic sedimentary rocks are found in the form of the Silurian Bundenbach Slate with a well-preserved diverse fauna of brittle stars, trilobites, and crinoids. The Devonian Kellwasser limestone presents another example of an anoxic deposit, which in this case was related to a global mass extinction event. At the end of the Paleozoic, the organic and metal-rich lithology of the Permian Copper Slate (*Kupferschiefer*) also formed under stagnant anoxic conditions. Now it is well known for its excellent preservation of fossil fish. Similar euxinic conditions occurred in the Mesozoic, such as the mudstone deposits of the Swabian Alb that bear numerous Early Jurassic dinosaur and ammonite fossils. There are also the Early Cretaceous clay deposits of Northern Germany, containing widespread ammonites, belemnites, and other fossil remains. In the Tertiary, lacustrine (lake) euxinic sediments were deposited bearing lignite in the Geiseltal near Halle (Saale) or bituminous hydrocarbons in the shale of Messel.

Such dark mudstones rich in organic carbon often contain very well-preserved fossils. The fossil diversity preserved in this lithology can be over a 1000 times greater than the overlying and underlying rock units in the same sedimentary sequence. This diversity reflects both biological and inorganic processes whereby decomposition processes are significantly reduced when O_2 is limited in the euxinic environment. This includes the reduction in scavengers and burrowing organisms that normally consume remains that may otherwise become fossilized.

Surface exposure of Early Jurassic rocks is, in fact, rare in the north of Germany, as these strata are typically buried beneath thick sequences of younger sedimentary rocks. A number of outcrops are accessible in the Lower Saxon Hills (Niedersächsisches Bergland) at the edge of the North German Basin, where shales of the Lias are still mined. In Mecklenburg-West Pomerania, Lias mudstones located above a salt pillow are also mined in a clay pit near Grimmen, close to Greifswald. At this locality, the only specimen of *Emausaurus ernsti* (named after the University of Greifswald, EMAU), was found in the 1960s. This fossil find was extremely important for determining the phylogeny of the dinosaurs (Fig. 11.46).

The features of fossils preserved in Early and Middle Jurassic rocks (see below) are typically very well preserved, although in most cases these have been reduced to the thickness of a piece of paper. This results from the extreme compaction during burial diagenesis whereby voluminous mud is converted to thin layers of mudstone. The fine-grained clays often lose about 70–90% of their original volume following deposition due to the loss of water that is expelled by dewatering, dehydration, and compaction during subsequent burial diagenesis.

In the Lias of Northern Germany, several layers contain **geodes** in which fossils are well preserved in their three-dimensional form. Geodes occur by secondary carbonate precipitation within the sediment during early diagenesis. As the decomposition of organic matter releases ammonia there is a local increase in the pH value around the decaying organisms. Dissolved $Ca(HCO_3)_2$ is then precipitated as "calcium-carbonate" in these sites, leading to the formation of an ellipsoidal geode that surrounds the fossil and protects it against further compaction and deformation. In this way, the fossil shape can remain largely intact.

11.6.2 Middle Jurassic

In the Middle Jurassic, the marine connection to the Neotethys opened across a broad area as the Vindelician mainland was flooded by the ongoing **transgression**. However, in the north, a continental uplift event occurred despite the global trend of sea-level rise. In the area of the North Sea Central Graben and in the southern Viking Graben, submarine basalts erupted, probably triggered by the rise of a mantle diapir. The crust was uplifted in this area by more than 2 km, causing large parts of the region, including the Ringkøbing–Fyn High, to be elevated above sea-level and eroded (Fig. 11.59). Already at the end of the Middle Jurassic, this land area known as the **Cimbrian Land** subsided again to form a swell that connected the London–Brabant Massif with the Rhenish and Bohemian Massifs.

On the global scale, the final breakup of Pangea occurred in the Middle Jurassic. This resulted in a distinct change in the direction of the crustal stress system in Central Europe. In the Early Jurassic, NE-SW striking graben structures dominated that were already active in the Triassic (e.g., Glückstadt Graben, Horn Graben). In the Middle Jurassic, however, the direction of extensional stresses changed to form NW-SE striking depressions, such as the Lower Saxony Basin and the Lusatian Basin.

A shallow sea extended over Central Europe in the Middle Jurassic with strong water currents developed in the straits between the islands of the London–Brabant, the Rhenish and the Bohemian Massif. The resulting strong wave action close to the coastlines resulted in the formation of shallow water **oolites** (Box 18: Oolite). The high input of iron from the land derived from weathering led to the formation of **iron oolites**. The frequent brown or brownish-red coloration of these rocks has given rise to the term "Brown Jurassic" for this part of the Jurassic sequence. The iron originated from tropical weathering and formation of **laterite soils** on the surrounding mainland. The iron was transported from the laterite soils into the adjacent coastal waters before precipitating around newly forming oolites as limonite (a brown iron ore consisting of a mixture of different iron hydroxides). In some places, the tidal currents of shallow coastal areas led to the reworking and redeposition of the iron oolite sediment, together with fine-grained sand. Some beds of iron oolite are several meters in thickness and were mined as ore with iron contents typically between 20 and 50%. An example of this type of iron deposit is the Lorraine Minette ore (derived from the diminutive of French *la mine*. mine, vein), mined until 1994 around Metz and Thionville in France. In Germany, similar deposits known as the Dogger ore were mined in the Swabian and Franconian Jura Mountains until the first half of the twentieth century. Also of economic importance are iron oolites of the iron sandstone formation lying above the Opalinus clay (Fig. 11.56). In Northern Germany, the Middle Jurassic *Trümmereisenerz* (detrital iron ore) of Salzgitter are also of economic significance. This consists of limonitic iron ore that was partly eroded, reworked, and resedimented during the Early Cretaceous uplift by salt pillows (Sect. 11.7.1).

Box 18: Oolite

An oolite (eggs stone; Greek *oon*: egg; Greek *lithos*: stone) is a sedimentary rock composed of small spherical grains with concentric layers held together by a calcareous or clayey matrix. Ooids are mostly composed of carbonate minerals (calcite or aragonite; $CaCO_3$), but in the presence of sufficient iron, oxyhydroxides of iron may also precipitate most typically in the form of goethite (FeO(OH)) or limonite (FeO (OH)· nH_2O). Their diameter usually ranges between 0.5 and 2 mm. In rare cases, they may reach centimeter sizes, such as in the Rogenstein of the Buntsandstein (Sect. 11.4.1; Fig. 11.40).

Ooliths require intense wave action and warm carbonate or iron supersaturated water for their growth. As a result, they are good indicators of moving water and well-oxygenated conditions. Small particles, such as sand grains or shell fragments that remain buoyant in the moving water, act as the germs for subsequent crystallization with carbonate or hydrated iron oxide precipitating in concentric layers around the grain. The crystals grow in radial–fibrous form that gives them a characteristic appearance in thin section. Depending on the strength of the water movement, ooids gradually sink to the bottom as they become too heavy to remain buoyant. Once on the seafloor, they are rolled back and forth continuously until been buried and compacted as part of the sediment and later solidified. The ooid-bearing sediments are transformed into oolites by the diagenetic processes of lithification.

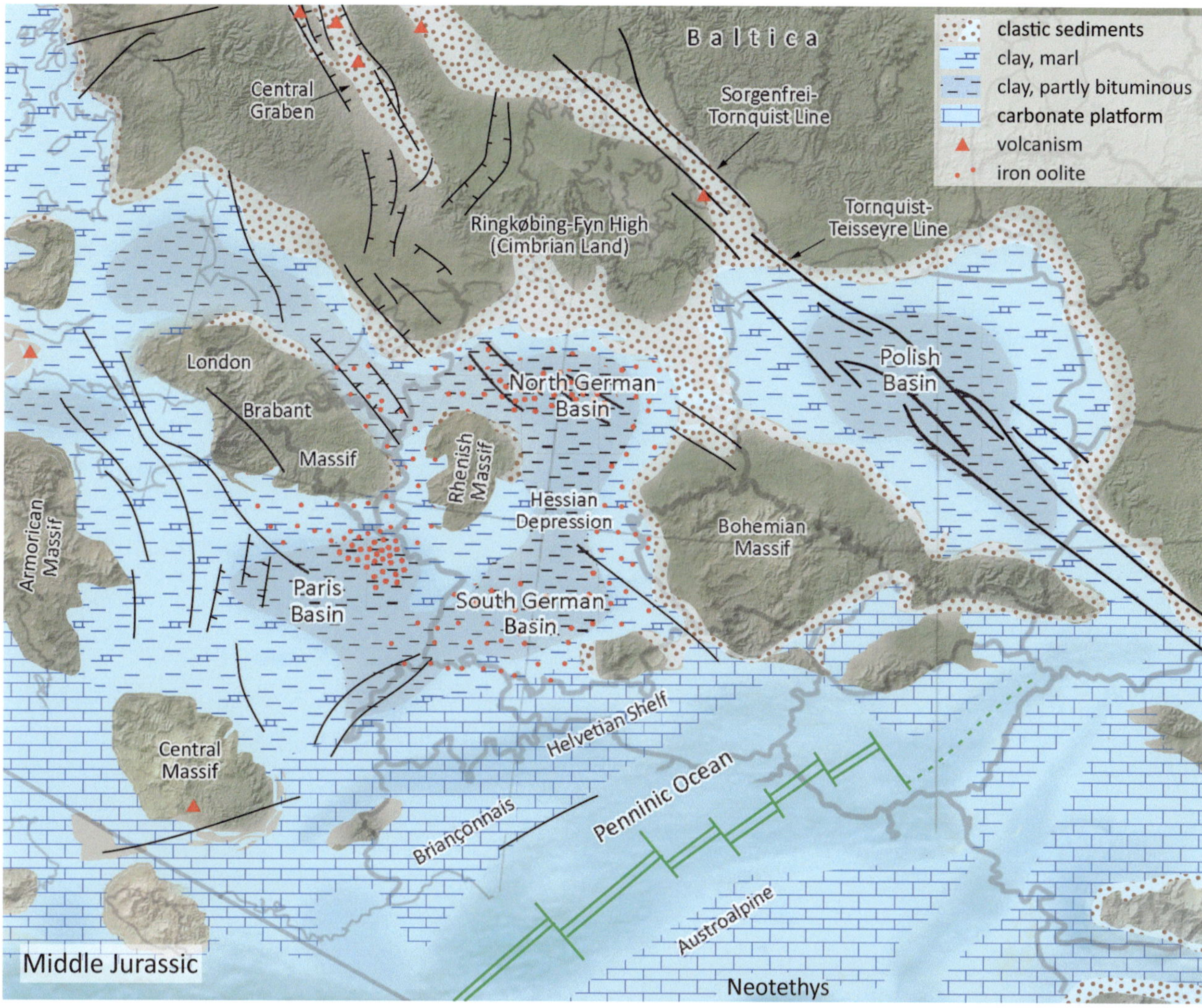

Fig. 11.59 Paleogeographic reconstruction of Central Europe during the Middle Jurassic. Topographic base map from Blakey (2011); modified after Ziegler (1990)

11.6.3 Late Jurassic

During Late Jurassic times, the North German Basin was separated from the Southern German Basin by a continuous land bridge that stretched from the London–Brabant Massif across the Rhenish Massif up to the Bohemian Massif (Fig. 11.60). A marine connection to the Neotethys existed through the seaway of the Saxon Gate, which partly flooded areas of the Bohemian Massif. At the beginning of Late Jurassic sedimentation, mainly clay-rich and partly bituminous sediments were deposited in the north, whereas a **carbonate platform** developed along the passive margin of the Penninic Ocean in the south. The platform was divided by a number of elevated swells that were occupied by microbe sponge reefs (e.g., lime-secreting cyanobacteria) and corals that gave rise to the formation of massive reef facies limestone (Fig. 11.61). The intervening shallow lagoonal areas were characterized by the deposition of finer grained, banked limestones.

The muddy seabed contained a variety of organisms living in the sediment that gave rise to the bulky banked limestones typical of the Late Jurassic of Southern Germany, the Jura Mountains, and the Paris Basin. However, in some tranquil lagoons of the Late Jurassic Sea located in Southern Germany, the rate of evaporation was so high that strong salinization occurred, and consequently, a low-oxygen bottom water zone developed that was not favorable for life

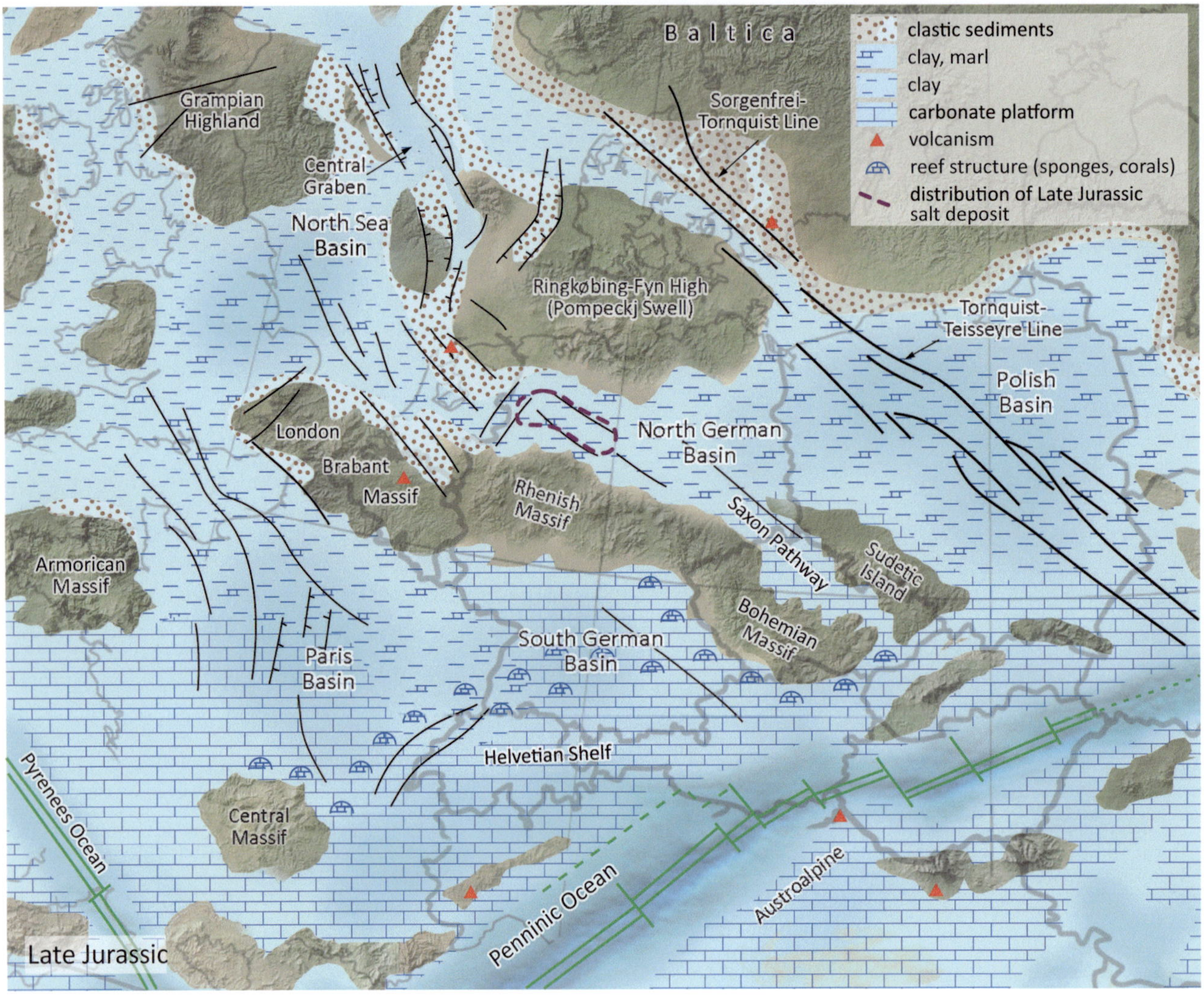

Fig. 11.60 Late Jurassic paleogeographic reconstruction of Central Europe Topographic base map from Blakey (2011); modified after Ziegler (1990) and Schmidt-Kaler (2003)

Fig. 11.61 Sponge reefs of the Altmühltal nature park exposed in the *Zwölf Apostel* rock formation (twelve apostles) near Solnhofen [N 48°53′38″/E 11°01′05″]

(Fig. 11.62). The organic remains that settled on the seafloor decomposed very slowly and some fossils were uniquely preserved. The typical activity of bioturbation and sediment disturbance that takes place following initial deposition did not occur here, thus preserving the finely laminated carbonate-bearing sediment with its rich fossil content. In addition to a large variety of fish, crustaceans, ammonites, bivalves, and crinoids, a complete range of terrestrial and flying animal fossils are found in these deposits due to the close proximity to land. The world-famous **Solnhofen lithographic limestone** of Late Jurassic age (Tithonian; Fig. 11.56) was deposited in such a lagoon on the edge of the South German Basin. This lithology is composed of an extremely fine-grained, laminated limestone of light coloration. Smaller occurrences of a similar facies are the **Nusplingen lithographic limestone** (exposed near Nusplingen) and the Kolbingen limestone with its significantly poorer fossil assemblages.

The well known Solnhofen limestone has been the source of eleven known specimens of the fossil **primeval bird** *Archaeopteryx* (ancient Greek *archaîos*: ancient, very old; ancient Greek *ptéryx*: wing, feather), the oldest known forerunner of today's birds (Fig. 11.46) and a transitional fossil between birds and nonavian dinosaurs. The best-known and most complete specimen of *Archaeopteryx* is exhibited at the Natural History Museum of Berlin. This fossil was found around 1874 near Eichstätt and after triggering an intense debate of Darwin's theory of evolution has become one of the most famous fossils on Earth. Another well-preserved specimen of *Archaeopteryx*, initially described as a small dinosaur (*Compsognathus*), is exhibited in the Jurassic Museum in the Willibaldsburg of Eichstatt. This museum offers a deep and unique insight into the diverse living world of the Late Jurassic (Fig. 11.63).

The Solnhofen limestone has been used as building material for many centuries and is still mined for facing stone and floor tiles. In the nineteenth century, the limestone was used for lithography and hence was referred to as *Lithographenschiefer* (lithographic slate). Its fine grain size, low permeability, low moisture uptake, and general stability are ideal properties for this purpose, especially for printing topographic maps (Fig. 11.64). The lithograph was then the

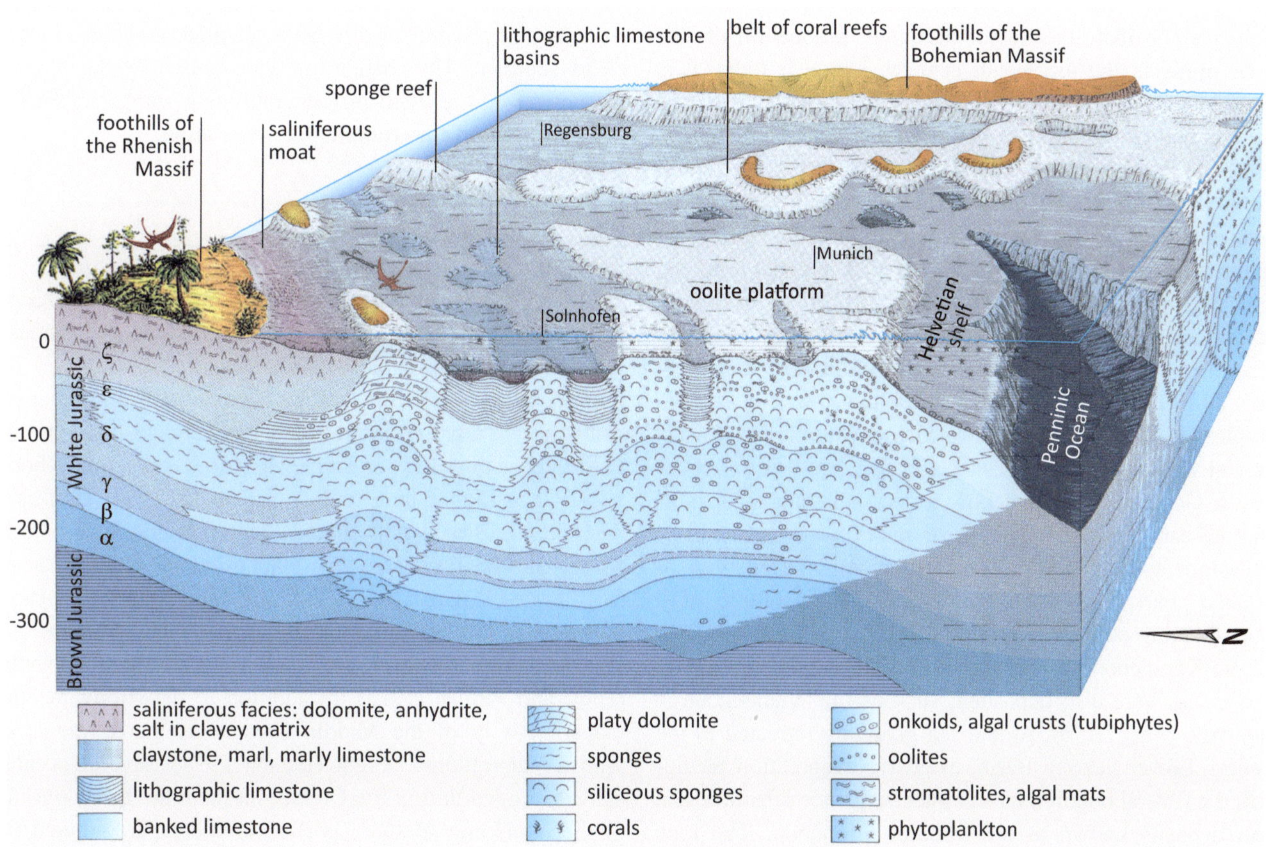

Fig. 11.62 Late Jurassic distribution of lagoonal lithographic limestone lithologies ("*Wannen*", vat) and reef limestones in the Franconian Alb area. Modified after Röper et al. (2000)

Fig. 11.63 *Archaeopteryx lithographica* (Eichstätt specimen) of the Jurassic Museum in the Willibaldsburg, Eichstätt. (Photograph with the kind permission of the Jurassic Museum Eichstätt) [N 48°53′31″/ E 11°10′11″]

only method that allowed for the larger production of colored prints in the nineteenth century, although today it is occasionally used as an artistic technique.

The end of the Jurassic period saw a distinct decline in sea-level and the drying-up of large parts of the South German Basin. A short colder and drier period can be recognized and there is evidence of glaciation at the North Pole around the time of the Jurassic/Cretaceous boundary. At this time the area of the southern basin was no longer covered by water but was dominated by erosion under tropical conditions whereby the carbonate platform underwent karstification (Box 23: Karstification). In some areas, well developed tropical cone karsts developed which during the Cretaceous period were infilled by marine sediments (see below). In the North German Basin, brackish water sediments with bituminous carbonates that gave rise to the Eimbeckhausen platy limestone were formed following the initial deposition of chalky, marly, and oolitic sediments. An example of a coral-bearing oolite (Oxfordian in age) is found in the form of the Korallenoolith Formation. Locally, deltaic and fluvial sediments were also deposited, such as the Wiehengebirge quartzite (Fig. 11.56). As the sea gradually retreated to the central Lower Saxony Basin, extensive evaporation precipitated a several hundred meters thick sequence of sulfate and salt deposits known as the Münder Formation. The Late Jurassic brackish and hypersaline sedimentation continued into the Early Cretaceous with an eventual transition into clastic and fluvial sediments of the overlying Bückeberg Formation. The precise position of the Jurassic/Cretaceous boundary in relation to this facies change, however, is not clearly defined. The change in facies caused by the prominent regression resulted in few continuous marker horizons that can be used for stratigraphic correlation.

11.7 Cretaceous

Following a short pronounced **regression** in Northern Germany around the transition from the Jurassic to Cretaceous, the rest of the Mesozoic was characterized by a series of prominent sea-level rises and the deposition of a number of distinctive **transgressional horizons** (Fig. 11.65). In the Late Cretaceous, the sea-level was for a short time about 170–250 m higher than today (Haq et al. 1987; Haq and Al-Qahtani, 2005). This quickly led to a doubling of the worldwide area covered by epicontinental shallow seas (Fig. 11.65). Apart from the fact that there were no ice caps at the poles during the Cretaceous period, the cause for the unusually **high sea-level** was related to plate tectonic events (Hays and Pitman, 1973). In addition to the strong mantle diaper activity of the Middle Cretaceous that culminated with a "superplume" event (Larson 1991), several oceans started to open during the Cretaceous with the final split up of the remaining southern portion of the supercontinent of Pangea (Gondwana). This led to the opening of the Atlantic and Indian oceans that extended the total length of mid-ocean ridge systems and as a result displaced large parts of the oceanic water content (Seton et al. 2009). Above all,

Fig. 11.64 Lithographic plate of Solnhofen limestone used for printing a topographic map—sheet Hahausen (northwestern edge of the Harz Mountains), which was published in 1878 by the Königlich-Preußische Landesaufnahme. The photograph was kindly provided by the geoscientific collection of the Federal Institute for Geosciences and Natural Resources (BGR) in Berlin; © BGR

the spreading rates at the mid-ocean ridges were much higher in the Middle Cretaceous than at present and reached up to 40 cm per year. In comparison, the fastest spreading rate today is 15 cm per year occurring along the central East Pacific Rise. The high Cretaceous spreading rate produced a large amount of young oceanic lithosphere in a geologically short time. However, since the total area of oceanic lithosphere cannot expand due to the spherical shape of the Earth, a correspondingly large amount of older oceanic crust must be subducted. The simultaneous formation and consumption of oceanic lithosphere led to a gradual reduction in average age. As warmer, young oceanic lithosphere is less dense than cooler and older crust, it is less prone to return back into the asthenosphere. The older lithosphere, in contrast, becomes significantly thicker and denser during cooling and therefore subsides with time to form the deep abyssal plains. If, as in the Cretaceous period, the significantly younger lithosphere is formed, the surface area of the great abyssal plains

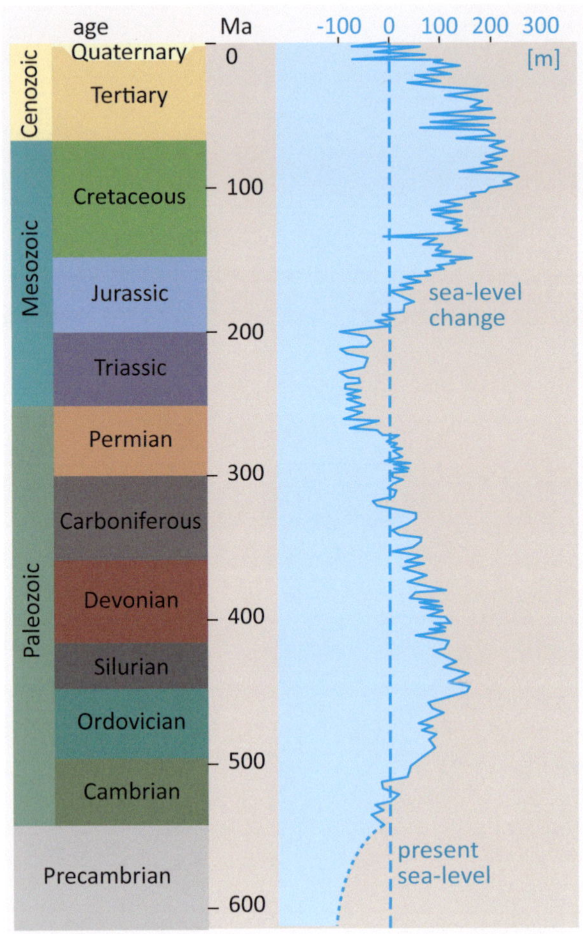

Fig. 11.65 Phanerozoic sea-level changes. 600 Ma until today; modified after Haq et al. (1987), and Haq and Al-Qahtani (2005)

decreases, as does the average water depth of the oceans. However, since the amount of ocean water remains largely the same, the water is forced onto the continental shelves as a result (Fig. 11.66). This is the mechanism behind the widespread transgressions of the Cretaceous period.

11.7.1 Early Cretaceous

During the Early Cretaceous, a continuous mainland area extended from Scotland across the Ardennic–Rhenish–Bohemian Massif to the Eastern Sudetic Basin (Fig. 11.67). This led to different sedimentary facies in the northern and southern sea areas. At the edge of the Rhenish–Bohemian Massif, siliciclastic sediments were deposited that gave rise to sandstone and mudstone lithologies, together with calcareous siltstones, marls, and limestones. Exposures of these continental dominated sequences occur along the northern edge of the Rhenish–Bohemian Massif in the areas of Aachen, the Münsterland Cretaceous Basin, the Lower Saxony Basin and the Saxon Cretaceous Basin. Similar exposures exist at the southern edge of the massif in the area of Regensburg and the Upper Palatinate.

At the transition from Late Jurassic to Early Cretaceous, the Lower Saxony Basin in Northern Germany subsided as part of a series of en echelon basins formed during transtensional tectonics. These include the Sole Pit Basin, the West Netherlands Basin, the Lower Saxony Basin, the Subhercynian Basin, and the Eastern Sudetic Basin (Scheck-Wenderoth et al. 2008; Fig. 11.67). **Transtensional basins** form along strike-slip fault zones due to the extensional motion occurring between the boundaries of the shear zone during lateral displacement. In the cases mentioned, a sinistral shear movement resulted from the large-scale plate motions related to extensional tectonics in the Central Graben in the North Sea and by the opening of the Atlantic (Chap. 12.5; Fig. 12.5). In continental areas, this extension reinforced the upward movement of salt that had already commenced in the Late Jurassic (from around the Oxfordian onward). This motion increased erosion above the ascending diapiric structures giving rise to prominent local unconformities (Fig. 11.34).

In Central Europe, basin formation during the Late Jurassic and Early Cretaceous was related to late Cimmerian tectonic movement during the early phase of the Cimmerian Orogeny. This orogeny formed a fold belt that stretched from the Balkans to Southeast Asia and is today linked to the mountains uplifted during the Alpine Orogeny. The Cimmerian Orogeny marked the collision of the Cimmerian continent with Eurasia (Fig. 11.38; Sengör 1984). In Central Europe, the lateral, strike-slip nature of tectonic movements resulted in the formation of numerous transtensional basins (Fig. 11.67).

In northern Germany, the Early Cretaceous regression led to a regional change in the lithostratigraphy. Following deposition of the Salinar sequences of the Münder Formation (Sect. 11.6.3; Fig. 11.56), sedimentation in the Lower Saxony Basin and the neighboring regions in the east was dominated by lagoonal and fluvial deposits. These strata are described in the older literature as the "**German Wealden**" due to their similarity with the fossil-rich sedimentary rocks of the English Wealden Formation described at the beginning of the nineteenth century in the countryside of Weald, southeastern England. Today the rocks are known under the stratigraphic name Bückeberg Formation, which are now known to have been deposited in the Berriasian (142–137 Ma) and are thus older than the Barrêmian (127–121 Ma) age of the English Wealden Formation.

During the Early Cretaceous, the **Lower Saxony Basin** was occupied by a freshwater lake connected to the open sea via narrow and flat seaways that temporarily dried up

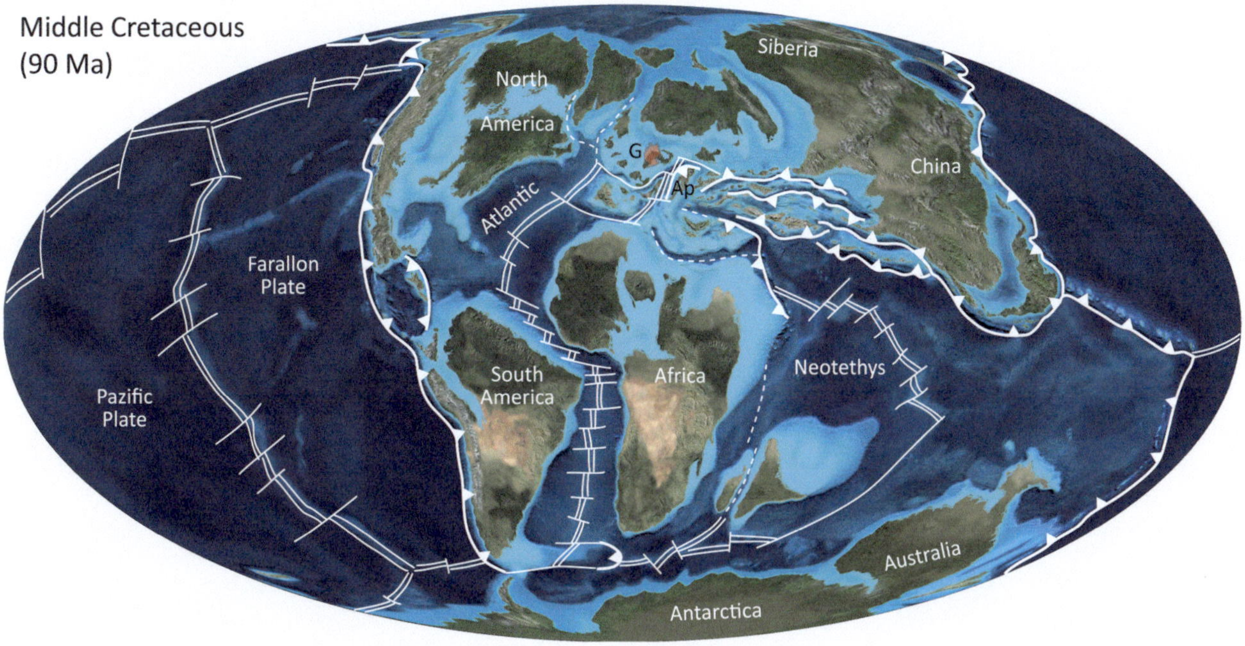

Fig. 11.66 Middle Cretaceous distribution of continents and oceans (90 Ma). Ap—Apulian Plate, G—approximate location of Germany. The topographic base for the paleogeographic map was provided by Ron Blakey, Flagstaff, Arizona (Blakey 2003)

(Fig. 11.68). This basin was limited to the north by the Pompeckj swell where cypress-like trees, conifers, ginkgo species, ferns, and palms thrived in the extensive swamp forests of the warm, subtropical climate. In some places, wetlands formed on the edge of the basin with the accumulation of peat that was later lithified into coal seams. In the area of the Deister hills, near Hannover, up to five coal seams occur, some of which were mined until 1957 as at Barsinghausen. This site is still open today for visitors to explore the historical mining operations (the Klosterstollen of Barsinghausen).

A large number of **dinosaur fossil tracks** have been found in the sedimentary rocks of the Bückeberg Formation in the Rehburg Hills, near Münchehagen. Recently, diverse types of tracks originating from several different species of dinosaur were also discovered in the Obernkirchen sandstones of the Bückeberg, near Minden. Overall, 13 different locations have now been reported and a number of these are open to the public, including some of the more worldwide renowned sites. Theropods (e.g., raptors) and Iguanodon tracks (Fig. 11.69), among other dinosaurs, are so numerous on the bedding surfaces of the Obernkirchen sandstone quarry that the locality is referred to as the "chicken yard" by the paleontologists who work there (Hornung et al. 2012). The sandstone excavated in Obernkirchen is highly valued as a building material due to its resistance to weathering. It has been used in the construction of the Cologne cathedral, the Ulm cathedral, and even the White House, Washington D.C.

At the edge of the Lower Saxony Basin, a **detrital iron ore** (*Trümmereisenerz*) accumulated in the gradually transgressing sea of the Early Cretaceous. This iron ore initially formed the basis of the steel industry in the greater area of Salzgitter, where it is still located. The Volkswagen factory was founded at this site in the 1930s and as a consequence, the industrial city of Wolfsburg was developed around this rural town. The iron in the detrital ores was reworked from sediments of the Early and Middle Jurassic that were eroded and redeposited in the Early Cretaceous as the older Mesozoic strata were uplifted to the surface by rising salt diapirs. As a result, Early and Middle Jurassic geodes composed of siderite were enriched in the surf zone of the Early Cretaceous Sea and then converted into limonite nodules by oxidation. **Limonite** is the mineral giving rise to the brown coloration of the iron ore, which consists of a mixture of various types of hydrous iron oxides that develop during the alteration of hematite or siderite. The limonite nodules accumulated in small sinks known as potholes, where they were mixed with iron ooids to form the detrital iron ore. In places, remnants of resedimented and partly rounded Jurassic ammonites can be found.

In the Salzgitter area, the age of detrital iron ore ranges from Valanginian to Aptian, whereas the ore deposits of Ilsede that lie to the north of Salzgitter are of Late Cretaceous age (Coniacian/Santonian). The latter formed in response to Late Cretaceous compressional tectonics (Sect. 11.7.2), which in combination with salt intrusions, tilted the older iron-bearing strata into the upright position prior to erosion and exposure at the surface.

The Salzgitter deposits have an average iron content of 25–35% and based on archeological findings of crude ore and iron slag have been smelted in the northeastern Harz

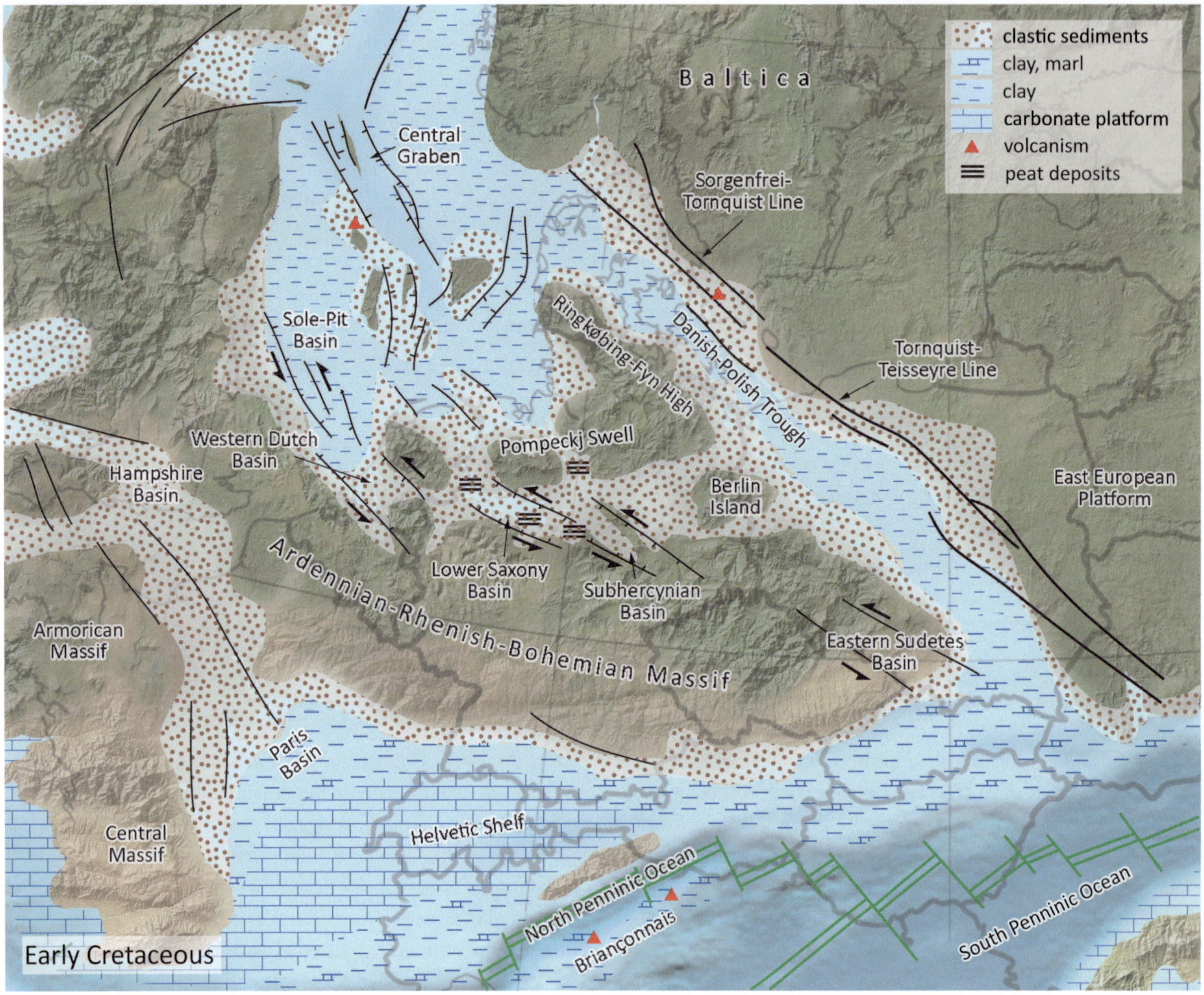

Fig. 11.67 Early Cretaceous paleogeographic reconstruction of Central Europe. Topographic base map from Blakey (2011); modified after Ziegler (1990), and Scheck-Wenderoth et al. (2008)

foreland area for around 2000 years. Mining ceased in 1982 when the Haverlahwiese pit, once the largest iron ore mine in Germany shut down its operation. The iron ore mine of Lengede–Broistedt became well known in 1963 when a mine accident trapped eleven miners at a depth of 60 m. The miners were fortunately rescued against all odds in an event known as "Miracle of Lengede". The iron ore mine of Schacht Konrad located in the urban area of Salzgitter has been closed since 1976. As the mine is notably dry, it is currently been prepared for the final storage of low to medium level radioactive waste that produces negligible heat.

Some of the most striking examples of Early Cretaceous sandstone strata are located at the edge of the Teutoburg Forest, near Detmold, and are known as the **Externsteine**. These layers were uplifted and steeply tilted during Late Cretaceous **compressional tectonics** (Sect. 11.7.2). Due to the very hard siliceous cementation of the sandstone they sometimes resemble quartzite and as a result of differential physical weathering and erosion the partly vertical layers protrude out to form a prominent landscape. Along the fractures within the sandstone, there is a characteristic type of weathering resembling that of a woolsack. The Externsteine are actually assigned to the Osning Sandstone (Osning is an old name for the Teutoburg Forest and the Egge Mountain Range), which was deposited in the Early Cretaceous as a beach and offshore tidal flat facies.

In Northern Germany, the Early Cretaceous continental sediments pass upward into increasingly argillaceous marine deposits. Stillwater conditions had already developed by the Hauterivian and Barremian, with deposition of clayey sediment containing a high organic content. Early Cretaceous fossil-rich clays were mined in numerous clay pits around Hannover, but most of them have been now backfilled or recultivated. Deposition of the dark clays continued into the

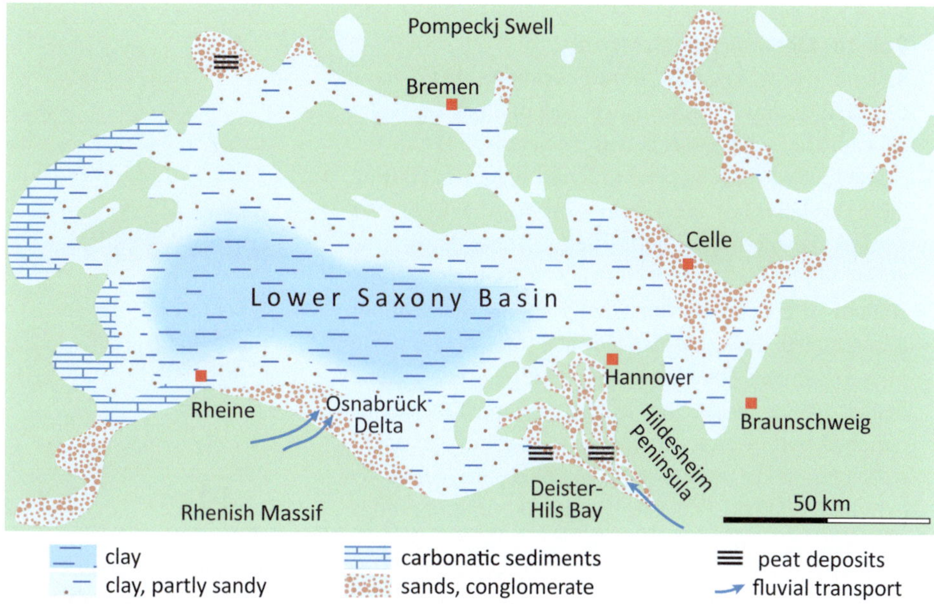

Fig. 11.68 Facies distribution in the Lower Saxony Basin during the Berriasian. Early Cretaceous; modified after Kemper (1976), and Richter (2007)

Fig. 11.69 Dinosaur tracks of the genus *Iguanodon* fossilized in a sandstone bed of the Obernkirchen sandstone [N52°15′54″/E 09° 12′32″]

Albian and after the Early Cretaceous, these were replaced by Cenomanian marls and carbonate sediments. The clay deposits of the Early and Middle Cretaceous are internationally known for their rich ammonite fauna, which in particular contain a variety of heteromorphic ammonites that are often embedded in concretions and retained their original shape (Fig. 11.70). These dark clay deposits also contain layers representing **oceanic anoxic events** that can be recognized and correlated worldwide (Box 19: Oceanic anoxic events).

Along the southern edge of the Rhenish–Bohemian Massif, iron ores of Cenomanian age formed in a similar way to those in the Lower Saxony Basin where the iron was detritally derived from ferruginous sediments of the Middle Jurassic. During the Early Cretaceous, the Late Jurassic carbonate platform of Southern Germany transformed into the mainland and was thus subject to intense karstification with the formation of karst channels, large sinkholes, and elongated polje-like valleys. As the seawater advanced during the Cenomanian transgression, iron ores accumulated in the depressions formed by the karstification with precipitation of siderite ($Fe[CO]_3$) under reducing conditions and goethite (FeO[OH]) under oxidizing conditions. These iron ores were mined until 1974 in the area of Amberg and Sulzbach.

Based on numerous geophysical investigations (geomagnetic, gravimetry, and seismic surveys), the existence of an intrusive body is considered to occur at a crustal depth of 5 km in the Bramsche area, northwest of Osnabrück. This body is known as the *Bramscher Pluton* (or Bramsche Massif) and is assumed to have intruded into Late Carboniferous rocks as a laccolith during the Late Cretaceous (Fig. 11.71). Similar bodies, also detected by geophysical surveys, were discovered in the surroundings of Osnabrück, near Vlotho, Uchte, and Loccum. The high-rank coal seams

Box 19: Oceanic anoxic events

Several major **oceanic anoxic events** (OAE) have occurred during Earth's history. The open ocean water beneath the surface layer was depleted of O_2 to the extent that was unfavorable to living organisms during these relatively short-term events. While the well-oxygenated and light-filled surface layer (the photic zone) was rich in life due to its high nutrient content, the dead remains that sank to the seafloor accumulated due to the absence of decomposing microorganisms that normally oxidize the organic matter. As a result, the seafloor sediments became enriched with a high percentage of organic carbon.

Such anoxic events were probably triggered by extremely intense volcanic eruptions, emitting huge amounts of volcanic gas into the atmosphere. The resultant strong increase in the concentration of CO_2 is assumed to be responsible for past short-lived super-greenhouse climates whereby global temperatures rose dramatically and the oceans were heated up so much that even the Polar Regions reached average temperatures of 27 °C. Because of such rising temperature, the rates of erosion subsequently increased, thus supplying a higher concentration of nutrients to the rivers and seas. It is the increased supply of nutrients to the oceans that then led to enhanced bio-production and greater amounts of organic carbon been preserved in the sedimentary record during such anoxic events. As the carbon sink started to remove large quantities of CO_2 from the atmosphere, the greenhouse effect was gradually reversed. Eventually, the average temperature then decreased and a normal state of equilibrium was reestablished. This explanation is supported by a general correlation between the occurrence of anoxic events and the formation of large igneous provinces (LIP) as the probable sites for the rapid extrusion of large volumes of volcanic gas. As Mesozoic examples, anoxic events of the Toarcian, the Early Aptian and that of the Cenomanian–Turonian boundary correspond well with the age of major igneous eruptions, namely, the Karoo-Ferrar Basalt (183 Ma), the Ontong Java Plateau (125–120 Ma) and the Caribbean igneous province (95–88 Ma).

characteristic of the coal mining area of Ibbenbüren are attributed by some researchers to the intrusion of a nearby laccolith, which heated the surrounding rocks to an elevated temperature higher than the normal diagenetic conditions of the Mesozoic strata.

Fig. 11.70 *Crioceratites sp.*, a heteromorphic (partly unrolled) ammonite of the Early Cretaceous preserved in Barrêmian clay deposits. The fossil was found in a clay pit of the former brick factory of Moorberg, near Sarstedt, which has since been backfilled [N 52°14′48″/E 09°52′57″]

Other explanations have been proposed for the thermal modification of the rocks that are today found at relatively shallow depths in the Osnabrück area (Brink 2013). A likely cause of the increased coalification of the **Ibbenbüren coal** is that it was buried to greater depths within the Lower Saxony Basin during the Early Cretaceous and subsequently uplifted. A number of transtensional basins formed at that time (see above; Fig. 11.67), with strong local subsidence and deep burial of the coal-bearing strata of the Late Carboniferous that reached depths where the regional geotherm was high enough (typically >200 °C) for **anthracite** formation (Box 14: Coal, oil, natural gas). The rocks were then exhumed toward the surface during Late Cretaceous uplift and erosion, where the high-rank coalified rocks are found today at relatively shallow crustal depths. This type of crustal extension (or transtension) and subsequent compression (or transpression) and uplift is referred to as **inversion tectonics** (Cooper and Williams 1989) due to the opposite motions involved (see below, for example, Fig. 11.80).

Anthracite coal has been mined in the Late Carboniferous layers of the Ibbenbürener Bergplatte up until 2018. Due to its high rank of coalification, it constitutes one of the most valuable coals produced in Germany. Whereas in the Ruhr area mainly high-volatile flame and gas flame coals (*Flammkohle*, *Gasflammkohle*) were mined, the Ibbenbüren mines produced higher rank forge and nonbaking coals (*Esskohle*, *Magerkohle*) while recovering the highest rank

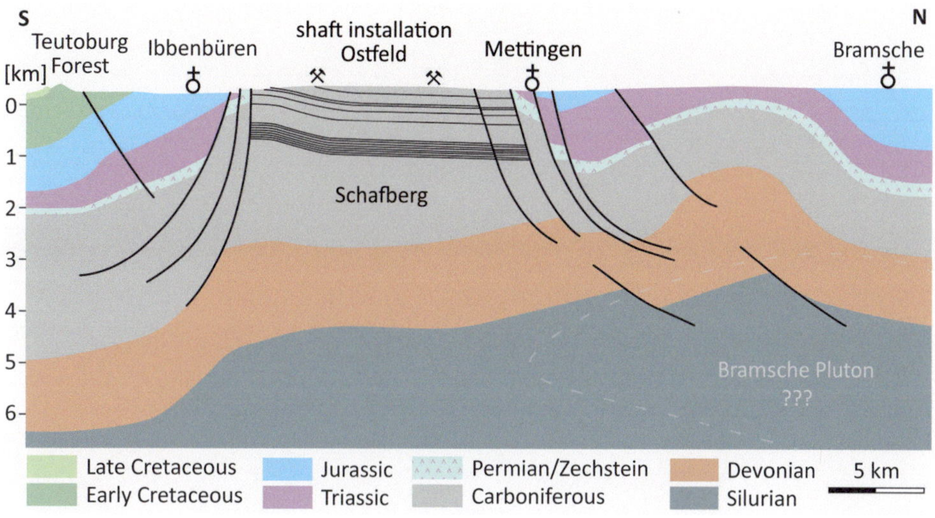

Fig. 11.71 Geological cross-section through the horst structures of the Schafberg area near Osnabrück, which forms part of the Ibbenbürener Bergplatte. The high-rank coalification of the Late Carboniferous coal seams mined in the *Ostfeld* (East field) mining shaft was for a long time considered to result from the intrusion of a pluton (Bramscher Pluton). However, new investigations favor deep burial of the coal during Mesozoic transtension. Modified after RAG Anthrazit Ibbenbüren GmbH (2002)

anthracite coals from greater depths. Gas and fat coals (*Gaskohle*, *Fettkohle*) are located in younger deposits of the Earliest Cretaceous known as Wealden coal. These occur close to the surface and are higher in rank than comparable coals from other regions. Based on the degree of coalification, these coals are classified as subbituminous (*Glanzbraunkohle*; Stadler and Teichmüller 1971).

In the area of the Ibbenbürener Bergplatte, mobilization of hydrothermal fluids and precipitation of ore bodies occurred along permeable fault systems. At the border of the Schafberg region, which is part of the Ibbenbürener Bergplatte, the Zechstein rocks were enriched in predominantly sulfidic nonferrous metal ores that were mined until the 1920s. The Ibbenbürener Bergplatte is one of several Carboniferous blocks that were uplifted as horst structures during compressive movements of the Late Cretaceous (see below). In addition to the Schafberg, these blocks include the Piesberg, north of Osnabrück, where Late Carboniferous sandstones are mined in one of the largest quarries of Europe and the Hüggel, where metal ores were mined until the 1940s.

11.7.2 Late Cretaceous

At the beginning of the Late Cretaceous, the sea had spread across most of northern Germany and the mainland regions of the Late Jurassic. Northern Central Europe was again submerged by a marine transgression that was in full flood during the Aptian. These Nordic Seas extended from Scandinavia and northern Scotland to the Rhenish–Bohemian Massif that remained as a source area for clastic sediments on the southern edge of the basin (Fig. 11.72). This massif separated the shallow cool (boreal) Nordic Seas from the warmer southern Cretaceous Sea that was connected to the Penninic Ocean and merged into the Helvetic shelf of the Alpine depositional area.

The Late Cretaceous witnessed a distinct change in the plate tectonic framework. The North Atlantic opened at the same time as activity ceased in the Alpine oceans (Penninic Ocean, Ligurian Ocean, Pyrenees Ocean, Valais Ocean) and the spreading center of the Bay of Biscay (Torsvik et al. 2002). While during the Early Cretaceous, tectonic activity in Central Europe was still dominated by crustal extension, in the Late Cretaceous compressive movements prevailed. Normal faults that were initiated during graben formation underwent reactivation and inversion as reverse or thrust faults. The cause of the compression was probably not related to the early stage of the Alpine–Carpathian arc collision with the European Plate, as supposed by Ziegler et al. (1995) and others, but was more likely related to the large-scale change in the plate movement of Africa in relation to Europe (Chap. 12.5; Fig. 12.5).

During the Early Cretaceous, the southeast movement of Africa away from Europe resulted in a phase of oblique extensional tectonics and formation of transtensional basins. The motion, however, changed toward the northeast in the early Late Cretaceous with a switch to compressional deformation. As the **Pyrenees Ocean** had already closed, these stresses were distributed across the continental lithosphere (Kley and Voigt 2008). A narrow residual ocean did still exist between the gradually developing Alpine Orogen and the European Plate, which prevented the propagation of stress across this area. The compressional NNW-directed forces of the Alpine Orogen were mechanically not

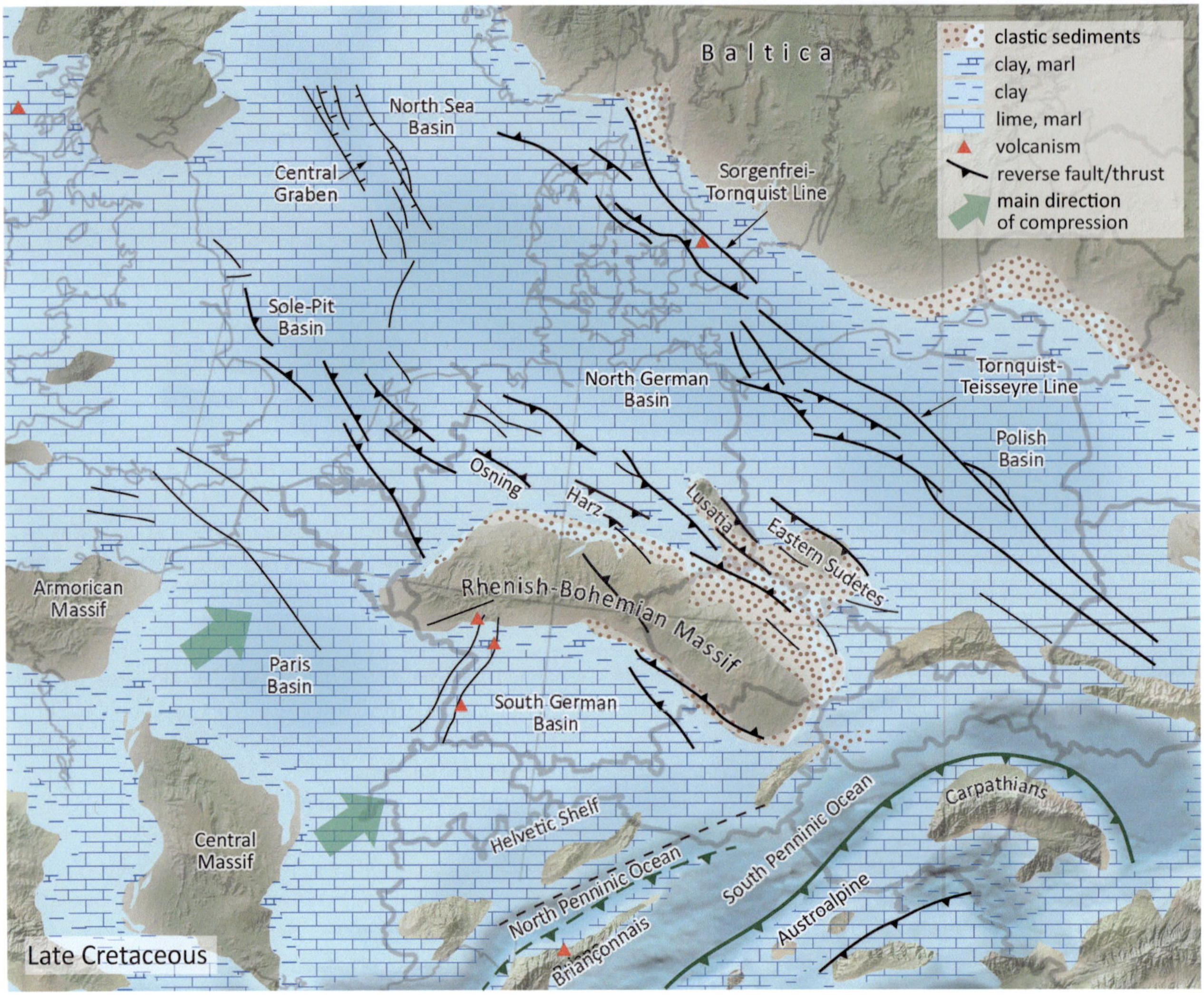

Fig. 11.72 Late Cretaceous paleogeographic reconstruction of Central Europe. Topographic base map from Blakey (2011); modified after Ziegler (1990) and Vejbæk et al. (2010)

responsible for the NW–SE (to NNW–SSE) trending reverse and thrust faults formed in the intraplate region. These resulted from the northeasterly directed compression related to the **movement of the African Plate** that initiated **folding and thrusting in the Pyrenees** (Fig. 11.73).

A diverse set of structures developed during Late Cretaceous compressional tectonics across Central Europe (Fig. 11.74). One of the best examples is the **Northern Harz Boundary Fault** (Fig. 11.75) along which Santonian to Campanian uplift of the Harz occurred to form a **half-horst structure** with a vertical displacement of ca. 6000–7000 m (Franzke et al. 2004; Voigt et al. 2006; von Eynatten et al. 2008). Other well-documented examples of Late Cretaceous activity include the **Osning Thrust Fault** of the Teutoburg Forest, fault structures in the Kyffhäuser, the Thuringian Forest, the **Gardelegen Fault** (Fig. 11.74) and the **Lusatian Thrust Fault**. Many other related structures occur in an area that extends from southern Sweden to the South of France, Spain, and into North Africa (Fig. 11.73). During uplift of the Harz Mountains and other prominent blocks, the Mesozoic sedimentary cover was simultaneously eroded while other areas, such as the Subhercynian Basin, underwent subsidence where the eroded products accumulated during deposition. As older strata, such as the widespread Triassic sequence, unconformably underlie the region, this indicates there was widespread sedimentation prior to the uplift and Late Cretaceous block faulting.

The Northern Harz Boundary Fault is a fine example of the interrelationship between tectonics and sedimentation. In the ***Aufrichtungszone*** (zone of steepening) along the northern edge of the Harz Mountains, an area referred to as the “classic square mile of German geology” occurs. Here, the Mesozoic strata are steeply tilted into the upright position and in places overturned (Fig. 11.75). Beginning in the

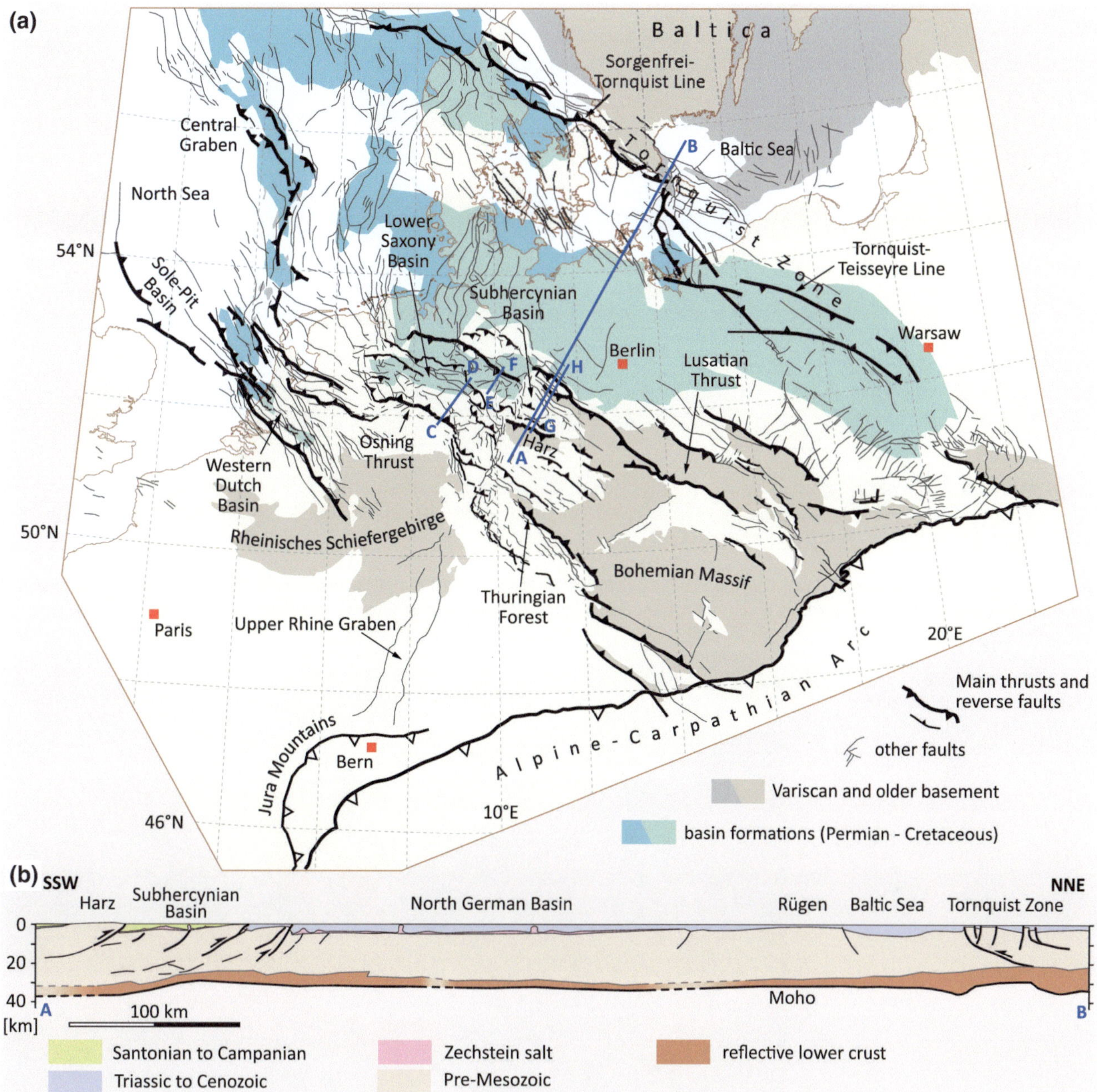

Fig. 11.73 **a** Fault systems and uplifted basement in Central Europe; **b** section (A–B; for the location of the section see Fig. 11.73a) through the continental crust from the Harz Mountains to the Tornquist zone based on deep seismic investigations of the DEKOPP BASIN research group (1999). Sections C–D, E–F, and G–H in Fig. 11.74. Modified after Kley and Voigt (2008)

Coniacian, a change from extensional to compressive tectonics occurred when a number of normal faults related to the formation of graben structures were reactivated as reverse faults (Fig. 11.76) during inversion tectonics (see above, Sect. 11.7.1). The steep tilting of the Santonian to Campanian sedimentary rocks caused by thrust movements was followed by several **unconformities** within the strata, with local sedimentation continuing during inversion. The notable discordance between the Early Cretaceous rocks and the underlying Triassic and Jurassic deposits was caused by the migration of the Zechstein salt that had already begun in Late Jurassic and Early Cretaceous times. The salt was folded and steeply tilted together with the Mesozoic strata beneath the Coniacian/Santonian unconformity during fault inversion. Erosion of the steepening strata of the *Aufrichtungszone* during compression resulted in another angular unconformity (Fig. 11.75). Whereas the youngest tilted strata are of Santonian age and lie discordantly upon the inclined older deposits, there is only a minor time gap between the two sides of the unconformity, indicative of a limited time of compressive deformation. Regionally, this

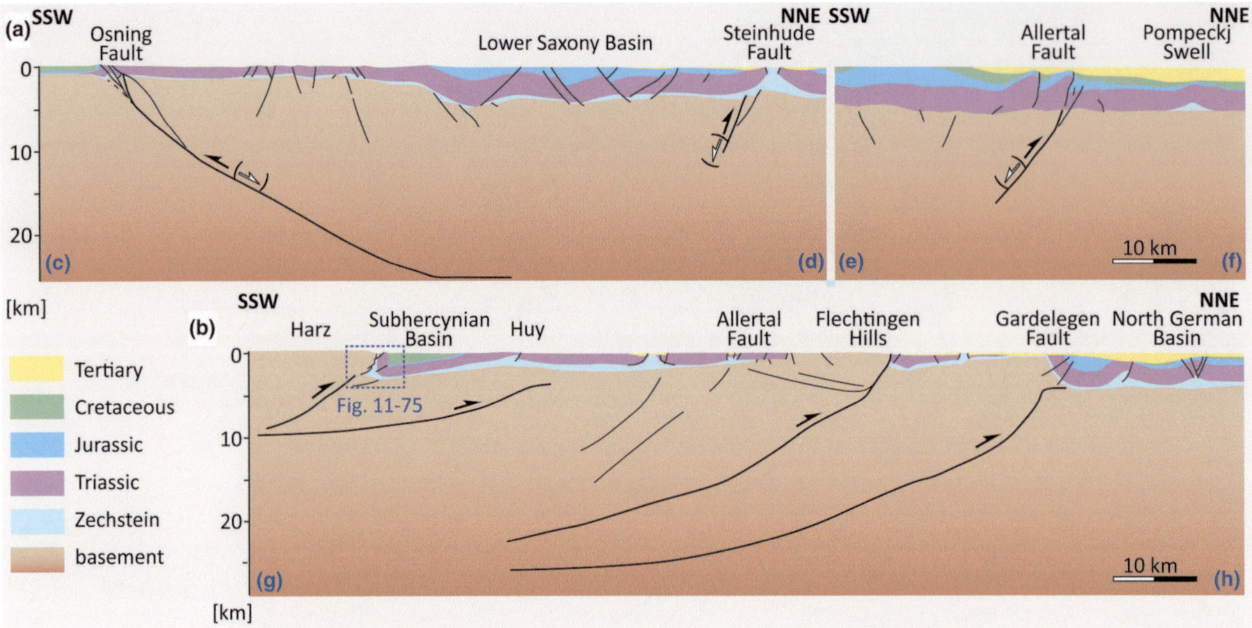

Fig. 11.74 Cretaceous reverse and thrust faults in the Central European Basin formed during inversion tectonics (normal faults reactivated as reverse faults) in the Lower Saxony Basin, and b uplifted basement blocks linked to thrust faulting Modified after Kley et al. (2008)

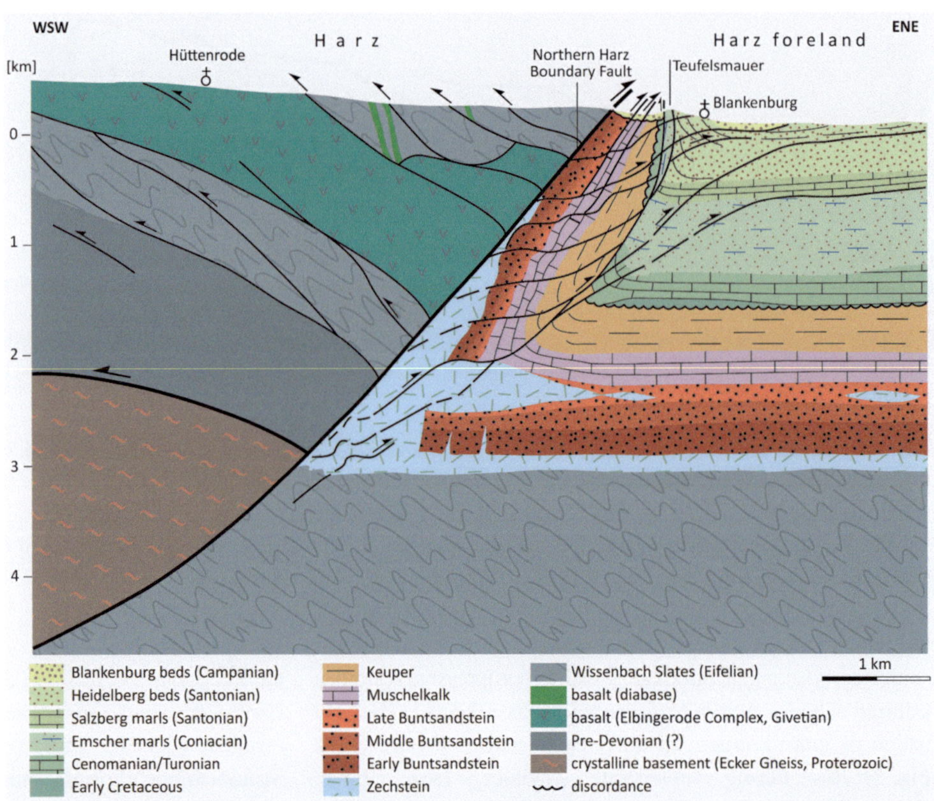

Fig. 11.75 Geological cross-section across the Northern Harz Boundary Fault, near Blankenburg. Modified after Franzke et al. (2004)

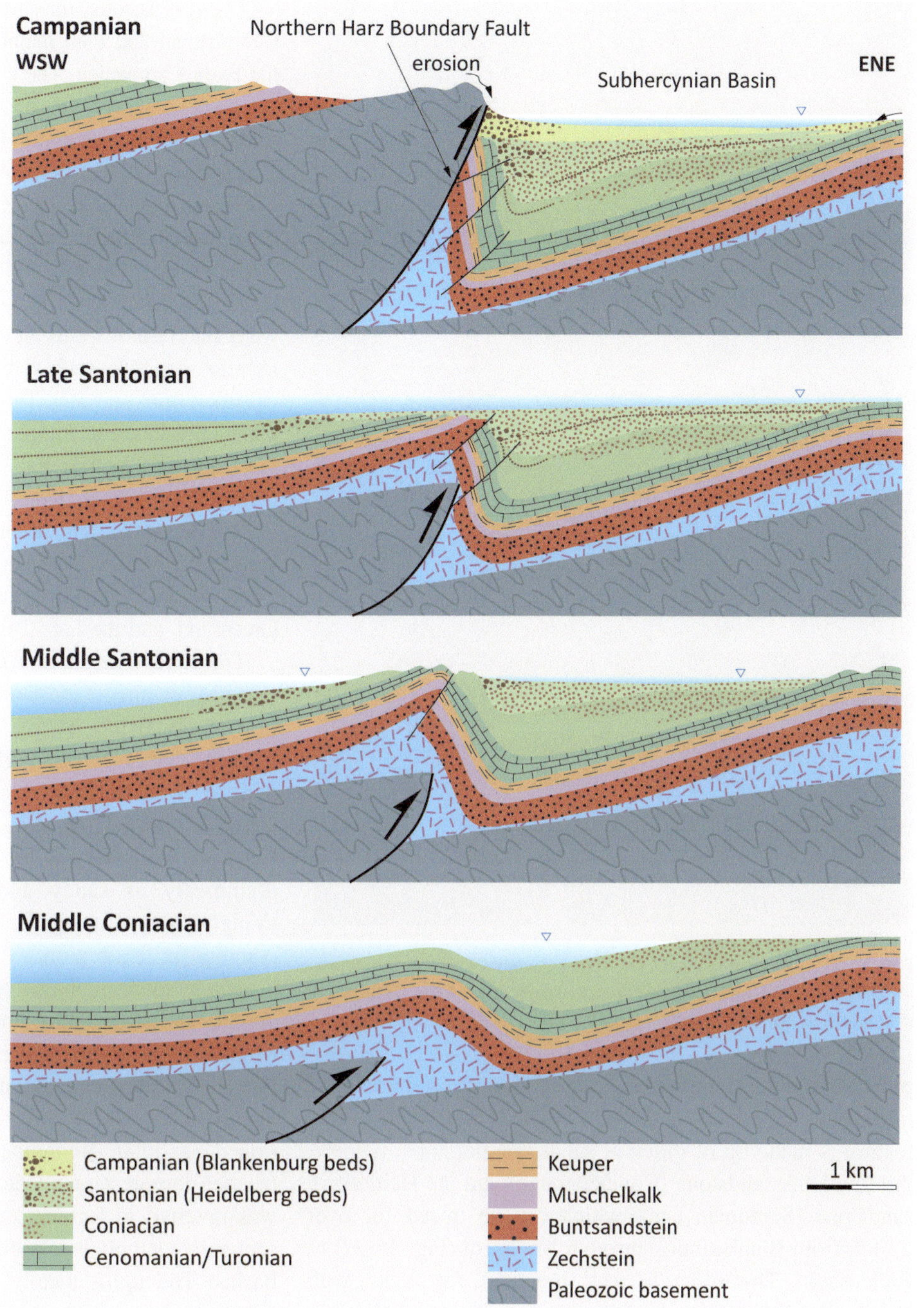

Fig. 11.76 Middle Coniacian to the Campanian tectonic and sedimentary evolution of the Northern Harz Boundary Fault in the area of Blankenburg. Modified after Voigt et al. (2006)

event is placed between the Middle Coniacian to the Early Campanian, giving a 2–4 million year time span for the uplift of the Harz Mountains and thrusting along other NW–SE oriented thrust faults.

During uplift of the Harz Mountains, sediments of a different depositional environment were deposited along the northern edge that gave rise to thick accumulations of sandstone. The deposition occurred as small-scale deltas formed in response to the uplifting half-horst structure of the Harz, which shed clastic sediment northward during erosion. Examples of such deltaic sequences are found in Blankenburg, at Burg Regenstein or at the so-called Hamburger Wappen: a rock formation resembling the emblem of Hamburg. Over a broader area, the deltaic sandstones pass into thinner layers of only a few meters thickness. These beds are more resistant to weathering due to a silica cement that precipitated from circulating formation waters. The rock is sometimes referred to as quartzite, although not petrologically correct because a quartzite represents a type of a metasedimentary rock. The resultant hardness of this rock is the reason the vertical sandstone beds of the *Aufrichtungszone* are particularly resistant to weathering and form a

Fig. 11.77 Devils Wall between Neinstedt and Weddersleben showing steeply tilted and silicified quartz sandstone of the Heidelberg Formation of Santonian age [N 51°45′27″/E 11°04′58″]

prominent natural mural and an impressive geographical monument: known as the **Devil's Wall** (*Teufelsmauer*). This feature is made up of different sandstone horizons, namely, the Involutus sandstone (Coniacian age) and the Heidelberg sandstone (Santonian age), which can be traced for over 20 km from Blankenburg, through Neinstedt (Fig. 11.77) to Ballenstedt. The route is well prepared for visits with appropriate information boards provided at several sites.

The Northern Harz Boundary Fault was again reactivated during the Tertiary, together with other fault systems in the area of the low mountain ranges, where blocks were uplifted by several hundred meters. A few occurrences of marginal marine and fluvial Tertiary sediments are located in the Harz Mountains (Fig. 10.37; gravels, sands, and silts from the Oligocene; König 2008). They were deposited on gradation plains at an average altitude of 500–600 m, with subsequent uplift occurring since the Oligocene by a further ca. 500 m during half-horst reactivation. This uplift of the Harz Mountains continues to this day.

The Cretaceous transgression reached its climax in the Cenomanian and then maintained its sea-level high stand until the Early Campanian. The transgression is frequently marked at the base by sandstones of Albian to Cenomanian age. These rocks, commonly known in the literature as greensands, contain a high proportion of **glauconite** which gives them a characteristic green color (e.g., the Rüthen Greensand found within the Cenomanian Essen Greensand Formation; Fig. 11.78). Glauconite is a clay mineral (phyllosilicate) that forms under reducing conditions during early diagenesis not far below the sediment–water interface. It forms either by the direct alteration of detrital iron-rich minerals, such as biotite and Fe-illite or by direct precipitation from seawater. This characteristic green clay mineral is commonly found in filling the cavities of small microfossils or replacing fecal pellets and forms mostly under marine conditions. Glauconite is, therefore, a relatively good indicator of marine deposition and occurs throughout the Late Cretaceous. Particularly notable rocks containing abundant glauconite are the Cenomanian Essen Greensand, the Middle Turonian Bochum Greensand, and the Late Turonian/Coniacian Soest Greensand.

The sandstones of the ***Elbsandsteingebirge*** (Elbe Sandstone Mountains), near Dresden (Fig. 11.79), were deposited in a marginal bay of the Cretaceous sea during the Cenomanian and Coniacian (Fig. 11.72). Weathered detritus derived from the basement and overlying Mesozoic strata were transported from the northeasterly adjacent Prignitz–Lusatian Wall (Fig. 11.80) into the bay and deposited predominantly as sandy and partly sandy-clayey sediment (Voigt 2009). During the Turonian, greensands were also common in this region.

The tectonic evolution of the Prignitz–Lusatian Wall was similar in nature and timing as the uplift of the Harz Mountains and subsidence of the Subhercynian Basin. In this case, the areas of subsidence were that of the East Brandenburg–North Sudetic basin and the frontally positioned Saxonian–Bohemian Cretaceous Basin. During Cretaceous compressional tectonics, the Prignitz–Lusatian Basin was inverted to form a prominent horst structure that subsequently supplied clastic sediment to the surrounding basins. The uplift locally reached more than 4000 m and exhumed a number of magmatic bodies to the surface, including the Lusatian granodiorite.

The Elbsandsteingebirge, which is part of the National Park of Saxon Switzerland (*Nationalpark Sächsische Schweiz*) founded in 1990, is characterized by thickly bedded and horizontally layered sandstones that are separated into large rectangular blocks by a distinct fracture system (Fig. 11.79). The high rates of erosion occurring along the fractures have given rise to often bizarre and rugged landforms in the shape of steep towers, mesas, and caves. These features resulted from various interacting weathering processes such as frost damage, salt bursting, sand

Fig. 11.78 Quarry in Rüthen Green Sandstone of Cenomanian age (Essen Greensand Formation). The enlargement shows worm traces in glauconite-bearing sandstone. Quarry Kirsch, Rüthen on the river Möhne, Mildental (Milde Valley) [N 51°30′03″/E 08°25′ 19″]

Fig. 11.79 Fractured sandstones of the Coniacian exposed in the Schrammsteine of the Elbsandsteingebirge. Horizontal layers and vertical fractures arranged at a right angle to each other result in the cube-shaped weathering forming high rock towers, steep walls, and ledges [N 50°54′40″/E 14°12′01″]

polishing, and karstification. Solution weathering and even cave formation occurred in the sandstones of the Elbsandsteingebirge, even though these rocks were not typically affected by karst processes. At the edge of the Elbsandsteingebirge, the granites of the Lusatian Massif have been thrust northeastward onto the sandstone along the Lausitz Thrust and in part were later steeply tilted (Figs. 11.73, 11.80 and 11.81).

Above the glauconite-bearing sandstones and dolomites, which are most abundant along the margins of the Münsterland Cretaceous Basin, lie a thick-bedded series of limestone and calcareous marl, known as the **Plänerkalke**. This series was deposited throughout Northern Germany and often contains shells of the genus *Inoceramus*. These are typically very large, some even reaching up to 3 m in length and partly used as index fossils for the Late Cretaceous. Gigantic ammonites also occurred at this time. The world's largest ammonite known so far has a diameter of about 1.8 m. This specimen was found in 1895 in the Münsterland Cretaceous Basin in a quarry near Seppenrade, southwest of Münster (*Parapuzosia seppenradensis*, Early Campanian; Kennedy and Kaplan 1995).

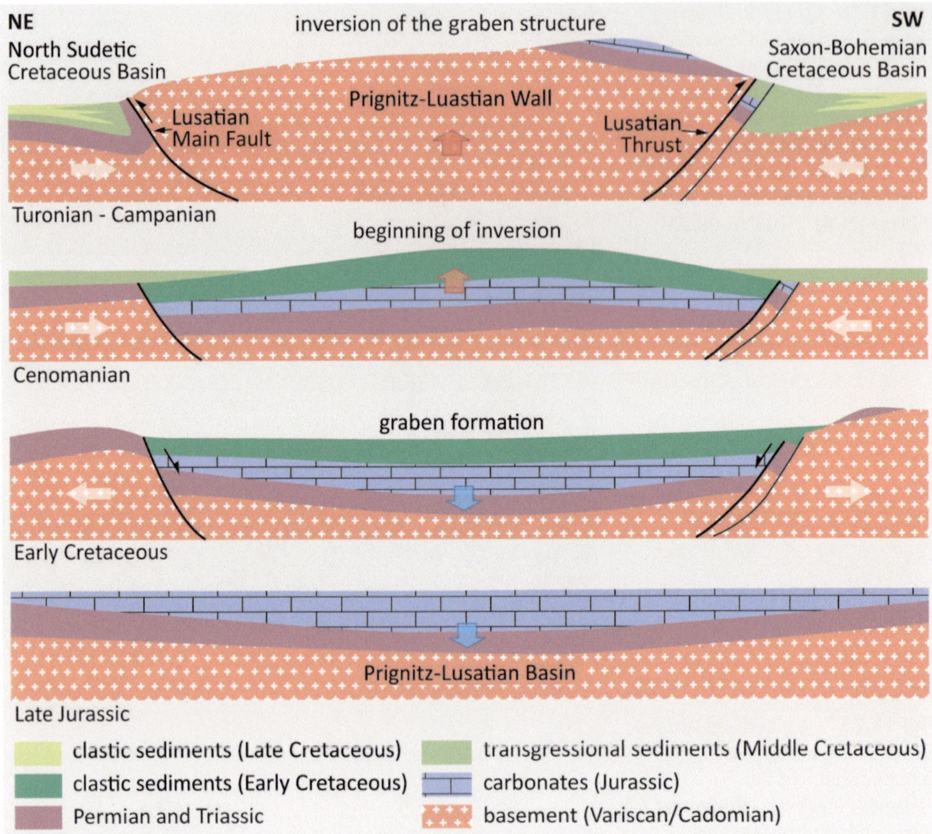

Fig. 11.80 Evolution of the Prignitz–Lusatian Basin showing the Late Cretaceous tectonic inversion of the Prignitz–Lusatian Modified after Voigt (2009)

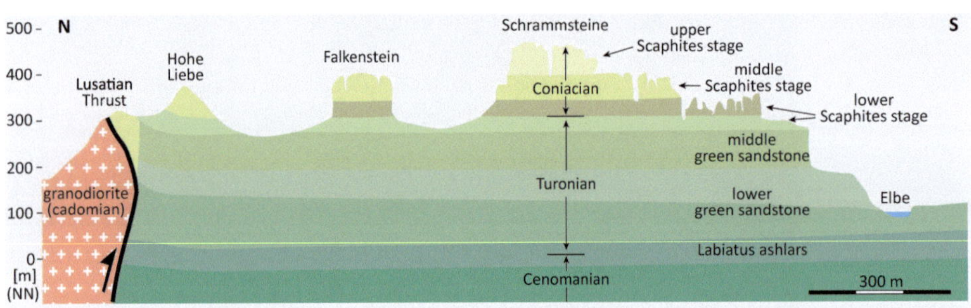

Fig. 11.81 Geological cross-section through the Elbsandsteingebirge crossing the area of the Schrammsteine from the Elbe to the Lusatian Thrust Modified after Richter (1999); scale not exaggerated

Late Cretaceous sedimentary rocks are widespread throughout Northern Germany, but they are typically buried below a thick cover of younger Tertiary and Quaternary strata. Only at the southern edge of the Cretaceous basin do the light-colored Cretaceous chalks come close to the surface where they are covered by a thin layer of sediment. Open pit mining is possible, with large limestone quarries located at Hoever and Misburg, east of Hannover. These Campanian limestones are generally mined for cement production. The limestones exposed in the chalk cliffs of Dover (United Kingdom) are also of Late Cretaceous age (Turonian to Campanian). Near Lägerdorf, northwest of Hamburg, limestones of Coniacian to Campanian age are mined. These deposits were uplifted from several kilometers depth to the surface by the rising Zechstein salt.

During the Late Campanian and Maastrichtian, a regression started that continued into the Early Tertiary where large parts of Germany dried out, all except the northernmost and southernmost parts. Thick beds of light-colored limestone of Maastrichtian age occur at the edge of the Baltic shield, these carbonate rocks range into even younger ages. The almost pure Maastrichtian **chalk** that forms the famous Rügen Island white cliffs and those of Møn (Denmark) is eponymous for the stratigraphic name of the Cretaceous (*Kreide*; Fig. 11.82). The chalk deposits on the islands of Rügen and Møn have been deformed by later glacial

Fig. 11.82 Rügen chalk (Maastrichtian) exposed beneath the Königsstuhl on the Jasmund peninsula of Rügen. Sedimentary layers bearing flintstones were folded during the glacial deformation of the last ice age [N 54°34′50″/E 13°38′41″]

tectonics that has led to complex imbrication and thrusting (Chap. 16; Fig. 16.14).

On Rügen Island, the chalk has been mined for cement production in more than 30 quarries since the early twentieth century. Today, the chalk is only recovered from a single site at a quarry near Sassnitz. Historically, the chalk is known as the *Rügener Heilkreide* (Rügen healing chalk) due to its supposed healing powers. The full history of the Rügen healing chalk and its mining as a whitening substance can be discovered in the highly recommended Gummanz chalk museum, near Sassnitz.

The chalk of northern Germany was deposited in a cool, boreal shallow sea, where the abundance of planktonic **calcareous algae** accumulated to form thick sediment. The water depth is considered to have been between 50 and 200 meters. The calcareous algae of the order *Coccolithophorida* form calcareous scales called **coccoliths**. These coccoliths are only a few microns in size but occurred in such vast quantities on the seafloor and with time formed thick layers of sediment. It has been estimated, for example, that a cubic centimeter of Rügen chalk contains about 800 million coccoliths.

The soft limestone of the chalk often contains a layered stratification marked by the occurrence of **flintstones** (or—synonymously—firestones). Flintstone is composed of SiO_2 in the form of cryptocrystalline quartz, the crystallinity of which is only visible under the microscope at very high magnification. As the SiO_2 precipitates during diagenesis, the chalk is progressively cemented and its density increases. The subrounded nodules of flint form as a chemical imbalance develops around dead organisms, initiating a decrease

Fig. 11.83 Flintstone containing a silicified sea urchin from the Cretaceous chalk of Rügen [N 54°31′43″/E 13°40′30″]

in pH. This favors the precipitation of SiO_2 which once started, acts as a nucleus for further growth of the nodule. Such concretions formed in large quantities in the chalk. Due to their hardness and the soft nature of the marly limestone, the flints are easily weathered out and concentrated as beach pebbles. Sometimes fossils such as sea urchins, sponges or belemnites are preserved within the flintstones and the original calcareous shells are completely silicified (Fig. 11.83). In many places, Stone Age prehistoric findings indicate that the flints (also known as **silex)** were a valuable resource in the early days of humankind and used to produce tools (hand axes, arrowheads, knives, etc., Figure 2.1). The flintstone deposits of the Cretaceous chalk on Rügen Island are well known for this purpose. These stones were transported several kilometers along the beaches probably during storm events and redeposited in adjacent promontories. The several square kilometer large flintstone fields of Mukran were formed in this way.

11.8 Cretaceous/Tertiary Boundary

Much like the boundaries of the Permian/Triassic and Triassic/Jurassic, the Cretaceous/Paleogene boundary was also characterized by a distinct mass extinction event during which a broad range of animal and plant groups died out, including the ammonites, belemnites, and famously, the dinosaurs. Only one evolutionary branch of the dinosaurs, the birds (see Fig. 11.46), survived the catastrophic climate change and still populate the world today. Although still debated, the principal cause of the mass extinction is generally considered to have resulted from one of the most prominent events of Earth's history: a major **meteorite impact** that occurred at the Cretaceous/Paleogene boundary. By chance, Alvarez et al. (1980) discovered a significantly elevated concentration of the element iridium (Ir) in sedimentary rocks at the Cretaceous/Tertiary (C–T) boundary. Iridium is a member of the platinum group elements common to meteorites. On Earth, these elements are normally found in extremely low concentrations. Since this iridium anomaly was also detected in other outcrops worldwide where the C–T boundary is exposed without disturbance, Alvarez et al. (1980) considered this a clear indication of a meteorite impact strong enough to distribute the iridium-enriched dust globally.

Exposure of the C–T boundary is not known in Germany. However, only about 100 km northwest of Rügen Island, located on the Danish Island of Seeland, are the cliffs of the peninsula of Stevns (Stevns Klint). Here, the **fish clay** (Fischton) is exposed where the first iridium anomaly was discovered by Luis and Walter Alvarez. Today, this outcrop remains one of the most important locations characterizing the global change that occurred across the C–T boundary.

The marked environmental change at the C-T boundary led to a pronounced transformation of the flora and fauna occurring in a relatively short period of only about one million years. The large meteorite impact crater, known as the Chicxulub Crater, which was discovered in the early 1990s (Hildebrand et al. 1991) on the Yucatan Peninsula in Mexico, is currently proposed as the main cause of the climate change. There are, however, critics of the impact theory and others point out links between the extinction and one of the largest volcanic events of Earth's history: the formation of the Deccan Traps in India. This volcanism probably lasted a few hundred thousand years but was so large that even if it was not the main cause, it is likely to have contributed to the global environmental change associated with the C–T boundary.

12 The Evolution of the Alps

Although only a small part of the European Alps lies in Germany, the geological history of Central Europe following the Variscan Orogeny is integrally linked with the evolution of the Alpine Orogeny. This chapter presents a concise overview of the key plate tectonic movements related to this orogenic development. It is a history that ranges from the rifting phase of the Neotethys, the first collisional phase in the Jurassic, the origin of the Penninic Ocean, to the final collision between the Adriatic-Apulian Plate and Eurasia continent.

The Permo-Mesozoic and Cenozoic geology of the Alpine region has in parts a significantly different history than from the rest of Central Europe. However, the European edge of the Neotethys, and later of the Penninic Ocean (opening in the Jurassic; Sect. 12.4), formed in a similar way. Today, the Alps represent a high mountain range that was elevated during the collision of the Adriatic-Apulian and European plates that prior to the Jurassic were separate continental fragments. The tectonic structure of the Alps, which is subdivided into the Helvetic (European Plate), Penninic (Penninic Ocean) and Eastern (Austroalpine) and Southern Alps (Adriatic-Apulian Plate), plays testimony to this collision.

12.1 Overview of the Tectonic Structure of the Alps

12.1.1 Helvetic

The southern European continental margin is commonly referred to as the Helvetic zone (Latin *Helvetia*: Switzerland), a unit which corresponds to the *Dauphinois* in the French Alps. During the Triassic, the Helvetics were still located in a continental setting and formed part of the Germanic facies area. In the Jurassic, the area became part of a passive continental margin located on the northern edge of the Penninic Ocean, where it was affected by higher rates of subsidence. At this time, there was a clear difference between the Austroalpine and Southern Alpine units (Sect. 12.1.3). The European continental margin was almost entirely covered by a shallow **shelf sea** that deepened slightly towards the south together with the deposition of Jurassic and Cretaceous calcareous marly sediments that reached considerable thicknesses. Local facies transitions associated with the shallowing of the epicontinental waters occurred toward the north.

The Variscan basement of the European continental margin is exposed in several autochthonous massifs in the Central and Western Alps. These include the Aar and the Aiguilles-Rouge/Mont-Blanc massifs and the Belledonne, Pelvoux, and Mercantour (Argentera) massifs (Fig. 12.1). The Variscan massifs are covered by Mesozoic sedimentary sequences that were deposited on top of them prior to their uplift. Both units were then imbricated and displaced far to the north and northwest as part of the Helvetic nappes during the collision between the European and Adriatic-Apulian plates. During the last phase of the Alpine Orogeny, Late Miocene folding of the Jura Mountains occurred. This externally positioned mountain range is characterized by regular fold trains with long anticlines and synclines that formed above a shallow decollement zone composed of a mechanically weak evaporate layer. At the transition from the Helvetic zone to the Penninic Ocean in the south, a unit known as the **Ultrahelvetic zone** occurs. The rocks of this unit were formed mostly as deep-water sediment (deep oceanic red clay and pelagic deep-sea carbonates) deposited on the Tethyan continental slope.

12.1.2 Penninic

The Penninic zone (Latin *mons penninus*: Great Saint Bernard) consists of rocks deposited in the areas of the **South Penninic** and **North Penninic oceans**. These areas were underlain mostly by oceanic crust but contained an elevated swell facies of the **Briançonnais zone** (named after the landscape around

M. Meschede and L. N. Warr, *The Geology of Germany*, Regional Geology Reviews,
https://doi.org/10.1007/978-3-319-76102-2_12

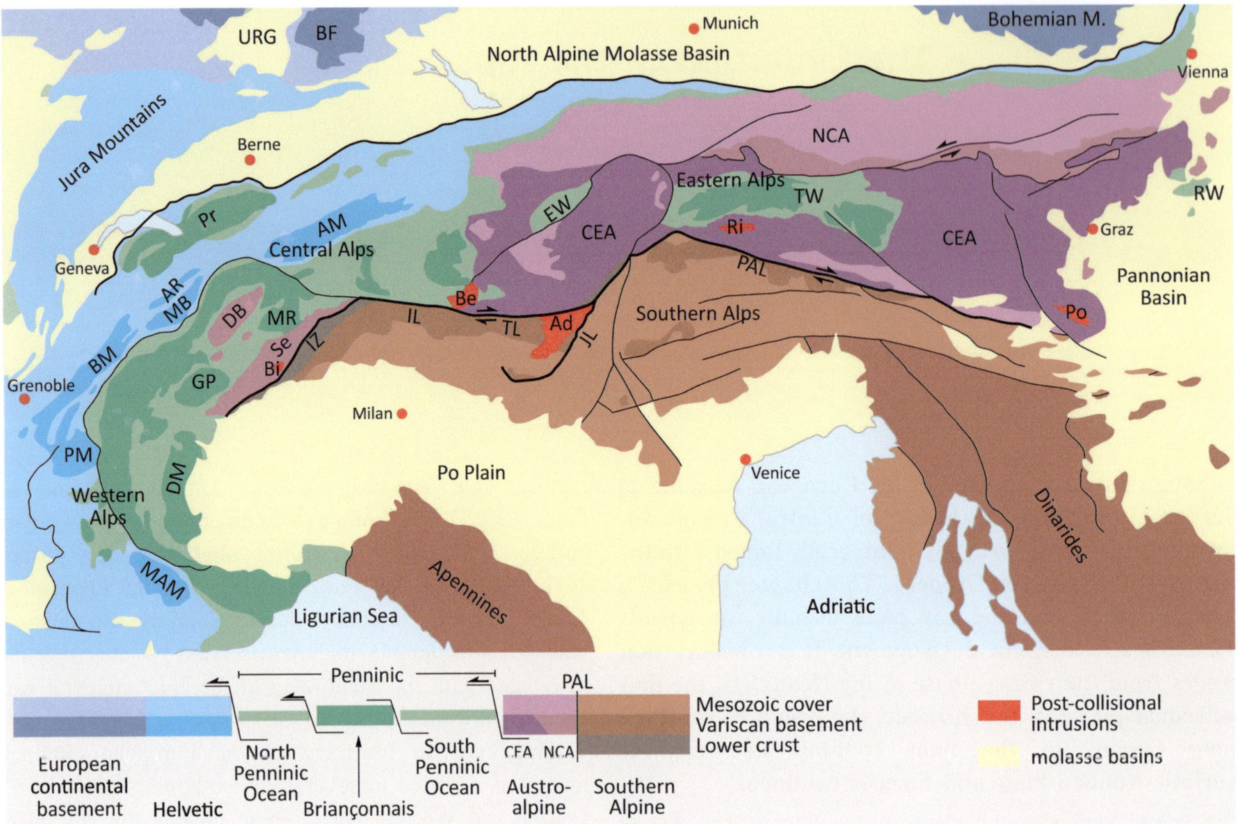

Fig. 12.1 Overview of the tectonic structure of the Alps with a schematic representation of the original tectonic positions of the Alpine units. Ad Adamello, AM Aar Massif, AR Aiguilles-Rouge Massif, Be Bergell, Bi Biella, BM Belledonne Massif, Bohemian M. Bohemian Massif, DB Dent Blanche, DM Dora-Maira Massif, EW Engadine Window, GP Gran Paradiso, IL Insubric Line, IZ Ivrea Zone, JL Judicarie Line, MB Mont Blanc, MAM Mercantour-(Argentera-)Massif, MR Monte Rosa, NCA Northern Calcareous Alps, URG Upper Rhine Graben, PAL Periadriatic Lineament, PM Pelvoux Massif, Po Pohorje, Pr Prealps, RW Rechnitz Window, Ri Rieserferner, Se Sesia Nappe, BF Black Forest, TW Tauern Window, TL Tonale Line, CEA central Eastern Alpine. Modified after Bousquet et al. (2012)

Briançon in the Western French Alps) that represent a continental terrane fragment. The oceanic areas were characterized by up to several kilometer thick sequences of argillaceous sediment that was lithified to form the **Graubünden Slate Formation**. These rocks are also referred to as *Glanzschiefer* (lustrous schist) and *Schistes lustrés* in French. The formation is named after its widespread occurrence in the Swiss canton of Grisons. Clastic deep-water sediments with calcareous sands, quartz sands and clays were deposited in the deep basins of the ocean and later metamorphosed during the Alpine event into calcareous phyllite and schist. Moreover, ophiolitic rocks with metamorphosed basalts, gabbros and serpentinites occur. These are often associated with mudstones or siliceous limestones containing radiolarians. The swell facies of the Briançonnais are regarded as being part of the Penninic zone, but they were probably deposited on Variscan deformed continental crust. The overlying rocks of the Mesozoic were strongly reduced in thickness with a high degree of variability. The Penninic is stacked into several nappes that today overlie the Helvetics, partly as klippen (i.e., a separated part of a nappe structure that has been isolated from the original nappe sheet by erosion). The pre-Alps (Préalpes Romandes) represent the largest klippe and lies to the northwest of the Central Alps (Fig. 12.1).

12.1.3 Austroalpine and Southern Alpine Units

South of the Central Alps, the Southern Alpine unit is connected directly to the Penninic unit and is partly divided by the **Ivrea zone**: a deeply exhumed section of the lower crust, with granulitic rocks showing a transition to mantle rocks, now exposed at the surface. In the Eastern Alps, the Penninic is largely hidden beneath the Austroalpine nappe unit following tectonic overthrusting towards the north during the Alpine Orogeny. The Austroalpine is subdivided into two areas: (1) the **Northern Calcareous Alps** composed mostly of unmetamorphosed Mesozoic rocks, which is partly underlain at the base by deformed rocks of the Variscan graywacke zone, and (2) the **central Austroalpine** consisting mostly of

Variscan and Alpine deformed basement rocks, overlain by small areas of Mesozoic rock. The central Austroalpine unit is comprised of a stack of tectonic nappes (Fig. 12.1) with a metamorphic grade in some parts reaching the eclogite facies.

In the Eastern Alps, the Penninic is exposed only on the northern edge of the Austroalpine unit and in some tectonic windows (Engadine, Tauern and Rechnitz windows; Fig. 12.1). The Austroalpine and Southern Alpine tectonic units both originated from the Adriatic-Apulian Plate. These geological units do not correspond with the geographical terms "Eastern Alps" and "Southern Alps". The Austroalpine represents a stack of nappes that lies on the Penninic rocks, whereas the Southern Alpine autochthonous unit structurally remained as part of the Adriatic-Apulian Plate. In the latter unit, a number of Variscan deformed basement complexes are exposed at the surface, including the Brixen quartz phyllite, the Ivrea zone and the crystalline rocks of the Maritime Alps in the southwestern part of the Alps. Both the Austroalpine and Southern Alpine tectonic units are characterized by thick Triassic carbonate platforms and the deposition of carbonate sediments continued in the Jurassic in many places. The carbonate platforms were strongly influenced by tectonic movements during the opening of the Penninic Ocean.

In the Central Alps, the **Dent-Blanche** and the **Sesia nappes** are assigned to the Austroalpine unit (Fig. 12.1). Here, the deformed Variscan basement of the Adriatic-Apulian Plate lies within a tectonic nappe on top of the Penninic. For instance, the Matterhorn belongs to the Dent-Blanche Nappe, where deformed gneisses were thrust on top of much younger ophiolitic rocks of the Penninic unit (with Jurassic to Early Cretaceous rocks) derived from the Zermatt- Saas-Fee zone.

The Austroalpine is today separated from the Southern Alpine by a prominent large-scale fault system known as the **Periadriatic Lineament** (or Periadriatic Seam). This structure is more than 700 km long and constitutes one of the most important tectonic lineaments of the Alps. It is morphologically evident by major E-W and NE-SW trending valleys. The NE-SW oriented central part of the lineament is represented by the **Giudicarie Line**, to the west follows the **Tonale Line**, and finally, this links to the **Insubric Line** (Fig. 12.1).

12.2 Development of the Alpine Region During the Permian

Similar to the area around Halle an der Saale or in the Saar-Nahe region (Sect. 11.2.1), Early Permian rhyodacitic eruptions also occurred in the Alpine region to form the **Etschtal Volcanic Complex** and the volcanic rocks of the Monte Cinto caldera on the island of Corsica (see above). This volcanism resulted from a distinct phase of extensional tectonics linked to the formation of rift and graben systems. The volcanic rocks can be regarded as forming part of a late phase of Variscan magmatism, but they were probably also related to the beginning of the breakup of Pangea.

During the Permian, the sedimentary rocks of **Alpine Verrucano** were deposited in the Alpine region. These strata were originally known as "Verrucano" named after occurrences at Monte Verruca in the Apennines. However, the latter deposits turned out to be of Triassic age whereas the Alpine strata formed in the Permian. Hence, to avoid confusion the latter is referred to as the Alpine Verrucano. These are typical Permian continental deposits containing poorly rounded boulders of the reworked Variscan basement and Permian volcanic rocks set in a fine sandy-clayey matrix. The formations also contain fanglomerates deposited in hot arid climatic conditions and locally occur together with basic to acidic volcanic rocks.

The Alpine Verrucano of the Central Alps has been variably metamorphosed and in places reached lower greenschist facies conditions. Well known is the exposure at the *Lochsite* (Swiss German for "hole side"), near Glarus, in Switzerland, where the metamorphosed Alpine Verrucano unit is found thrust onto weakly metamorphosed Tertiary flysch deposits or Cretaceous limestone. This structure, known as the Glarus Thrust, was added to the World Heritage List of UNESCO in 2008 as part of the Tectonic Arena Sardona.

Continental sedimentary rocks in the Southern Alps comprised of sandstones and conglomerates occur in the Val Gardena Formation. These were deposited following a transgressional series comprised of fanglomerates and conglomerates (the Waidbruck conglomerate) that discordantly overlie the eroded Variscan basement and the Etschtal Volcanic Complex. The first marine ingressions occurred as shallow coastal waters and lagoons with deposition of sediments that gave rise to the limestones and dolomites of the Bellerophon Formation (named after a genus of snails that became extinct in the Triassic). Repeated evaporation in the shallow lagoon areas also led to the formation of gypsum layers within the Bellerophon Formation.

The Permian sedimentary record of the Northern Calcareous Alps also contains the Alpine Verrucano deposits, with bodies of largely rhyolitic volcanics occurring in the northwest. The Late Permian rocks of the ***Haselgebirge*** (Hazel Mountains) to the southeast, temporally correlate with the North German Zechstein series. The Haselgebirge sequence consists of salt clays, anhydrite, gypsum, and rock salt and reaches thicknesses of several 100 m. Salt deposits in these layers were already mined around Hallstatt in the Salzkammergut during the Bronze and Iron Age (Hallstatt culture, ca. 1200–450 B.C.). Location names such as Hallstatt, Hallein, Hall in Tyrol, Bad Reichenhall or Salzburg testify to the importance of salt production and its trade in former times. The words "Hall" and "Hazel" are derived from a Celtic word for salt.

12.3 The Alpine Triassic

Continental sedimentation in the area that later became the Alpine Mountain Range prevailed until the Early Triassic. The Buntsandstein exposed in the western part of the Northern Calcareous Alps (Austroalpine) and the Southern Alps was deposited mainly as fluvial sediment similar to the Buntsandstein of the Germanic Triassic. Towards the south and east, the Alpine Buntsandstein merges into the predominantly marine Werfen beds comprised of sandstones, mudstones and calcareous sedimentary rocks that represent deposition in shallow marine basins. In the Southern Alpine area, a marine influence became apparent in the Late Permian with deposits of the Bellerophon beds and in the southern Austroalpine by evaporitic formations of the Alpine Hazel Mountains. To the west and north, however, the coarse clastic sediments of the Alpine Verrucano dominated. During the early Middle Triassic (Early and Middle Anisian), the Alpine region was characterized by shallow marine carbonates.

From the Late Anisian onward, the spreading of the Neotethys Ocean, which commenced in the Permian, continued into the northeastern peripheral regions of Pangea with the formation of an ocean and the development of a passive continental margin. At that time, the future Adriatic-Apulian Plate (Fig. 12.2) still belonged to Pangea. Platform and basin carbonates of several kilometers thickness were deposited in the Triassic during ongoing subsidence on the passive margin of the later Adriatic-Apulian Plate. This is in contrast to the Germanic Triassic where relatively thin limestone deposits accumulated in a shallow sea environment (see Sect. 11.4). Organisms with carbonate shells were abundant in the warm waters of the western Neotethys, which at the time of the Triassic was located at low latitudes. This led to the formation of substantial quantities of carbonate that formed the kilometer-thick **limestone and dolomite platforms** of the Triassic. Examples are the Wetterstein platform, which reached about 1 km in thickness, and the Dachstein platform that was twice as thick. Overall, up to 3 km of sediment accumulated on the continental margin. The reef growth continued during prolonged periods of subsidence whereby a shallow marine and partly lagoonal sedimentation environment prevailed for a long time. Similar depositional conditions are found today along Australia's Great Barrier Reef and in the Bahamas of the northern Caribbean. The older reefs (Wetterstein platform in Fig. 12.3) were largely built by sponges and calcareous algae, whereas the reefs of the upper Late Triassic were dominated by corals (most frequently: *Thecosmilia*).

The areas of the Northern Calcareous Alps and the Southern Alps were connected during the Triassic and Jurassic, and probably remained so until the Middle Cretaceous. As a result, these areas went through the same sedimentary evolution during this time. The Werfen beds were deposited at the beginning of the Triassic and are presently found in both regions at the base of the thick carbonate sequence. These areas were also subject to extensional tectonics related to the opening of the Neotethys. This resulted in **horst and graben structures** that produced similar geomorphological features. In the upper Middle Triassic, the distribution of carbonate platforms and deeper basins was strongly controlled by the topography. In the Southern Alps, the older shallow-water carbonates are known as the Schlern and Cassian dolomite, whereas in the Eastern Alps these deposits are referred to as the Wetterstein platform. Only in the Late Triassic during the upper Carnian, did the widespread Main dolomite/Dachstein limestone platform develop as the older intervening basins between the platforms were filled. The Main dolomite in the Southern Alps is known as *Dolomia principale*. Basins located within the older platform cycle were filled in the Middle Carnian by clastic sediments of the Raibler and Lunzer beds. This cycle of sedimentation was characterized by multiple regressions and transgressions of the sea, giving rise to a wide range of lithologies, namely limestones, dolomites, marls, mudstones, and evaporitic horizons.

In the eastern Northern Calcareous Alps, the Raibler beds are substituted by the Lunzer beds. These formations, interpreted as delta and shallow marine or brackish sediments, resemble the Keuper of the Germanic Basin. Locally coals occur that were even mined in some places in the nineteenth and early twentieth century.

A number of sedimentary facies can be recognized in the Triassic sequence of the Northern Calcareous Alps. The Main dolomite facies (formerly known as the Bavarian-North Tyrol facies) of Late Permian to earlier Middle Triassic age is a very similar facies to that of the Germanic Triassic. However, from the Late Anisian onward this facies changed to open marine conditions during the development of the Neotethys. At the Ladinian/Carnian boundary, the first **platform carbonates** developed (Wetterstein limestone and dolomite) with both reef and lagoon facies. As further to the south other Wetterstein platforms developed, a rather complex pattern of platforms and basins evolved in the period from the Ladinian to Carnian. The Partnach beds represent basinal sedimentation between the platforms, as do the clastic sedimentary rocks of the northern Alpine Raibler or Lunzer beds. Following infilling of the basins between the platforms, the Main dolomite was formed in the upper part of the Late Triassic. During the Rhaetian, carbonate production decreased due to the renewed input of clastic sediment, with the formation of a deep lagoon known as the Kössen Basin (Fig. 12.3).

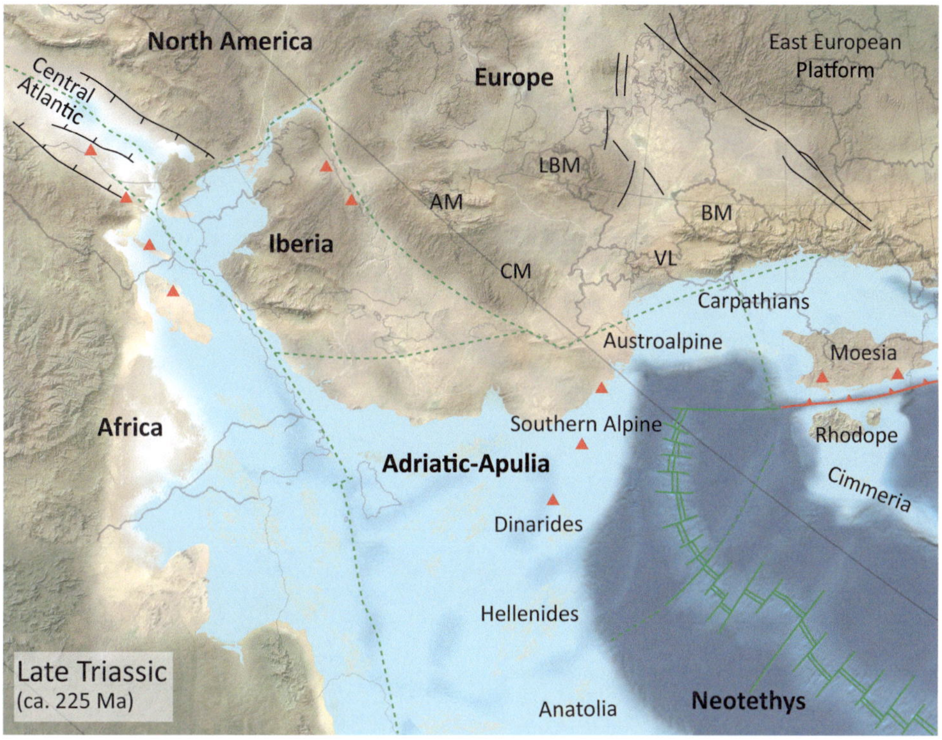

Fig. 12.2 Late Triassic plate tectonic reconstruction of the northern region of the Neotethys. AM Armorican Massif, BM Bohemian Massif, LBM London-Brabant Massif, VL Vindelician Land, CM Central Massif. Topographic base map from Blakey (2011), modified after Frisch (1979), Dercourt et al. (2000), and Ziegler (1990)

The Dachstein limestone facies (formerly known as the Berchtesgaden facies) developed during Middle Triassic to Carnian times and is similar in appearance to the Main dolomite facies. Only in the upper part of the Late Triassic did differences become apparent in the Main dolomite/Dachstein limestone platform. The Dachstein limestone was characterized by a **lagoonal environment** (Lofer facies) with **reefs**. The transitional area from reef to basin resulted in the deposition of the Hallstatt facies that represents the deeper part of the passive continental margin (Fig. 12.3). Here, from the Late Anisian to the Norian, open marine Hallstatt limestones formed. This was followed by marly sedimentation of the Zlambach beds during the Rhaetian.

Tectonic movements intensified in the Alpine area in the course of the Late Triassic during the collision of the easterly lying Cimmerian continental fragments that commenced in the Middle Triassic (Fig. 11.38). Extensional movements and accompanying volcanism marked the opening of the Penninic Ocean in the Early Jurassic. With the mass extinction event at the Triassic/Jurassic boundary (Sect. 11.5 ; Fig. 10.12), carbonate production in the Alpine area came to a virtual standstill. Early Jurassic crustal extension in the Penninic area led to horst and graben structures, which influenced deposition in the main dolomite facies area. As the continental crust rifted apart and the Penninic Ocean began to open, the Austrolpine and South Alpine areas were located along the southern passive margin of the newly formed ocean.

12.4 The Alpine Jurassic

The Jurassic witnessed extensive tectonic changes in the Alpine region with the opening of the **Penninic Ocean** from Middle Jurassic times onward (Fig. 12.4). With the opening of this ocean, the Adriatic spur developed in the Cretaceous to form an independent plate tectonic unit known as the Adriatic-Apulian Plate (Frisch 1979). The Penninic Ocean is in some publications referred to as the "Piemont-Ligurian Ocean" or "Piedmont Ocean". Others refer to it as the "Alpine Tethys" (e.g., Handy et al. 2010). The latter denotes the southern, larger section of the Penninic Ocean, which in the Eastern Alps is called the **South Penninic Ocean**. Another ocean opened to the north at the beginning of the Late Jurassic, referred to as the **North Penninic Ocean** in the Eastern Alps region. This ocean formed as a prolongation of the **Valais Ocean** in Switzerland and represented a continuation of the Pyrenees Ocean separating Iberia from the rest of Europe (Sect. 12.5; Fig. 12.5). The term "Alpine Tethys" is purposely not used here to emphasize its difference from the older Neotethys that was regionally aligned in

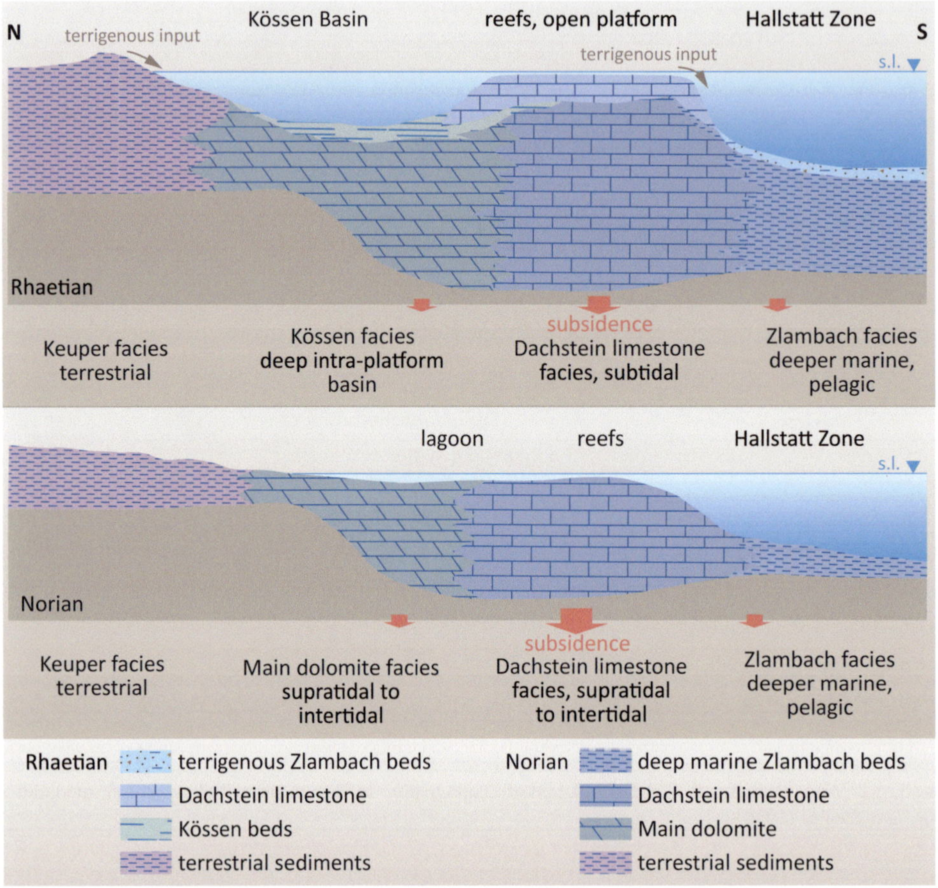

Fig. 12.3 Facies development in the Northern Calcareous Alp during the Norian and Rhaetian. Simplified after Gawlick (2000)

a different direction. Spreading of the South Penninic Ocean accompanied the opening of the Central Atlantic that was connected via a large transform fault. Therefore, in recent studies this ocean has been referred to as the "**Alpine Atlantic**" (e.g., Missoni and Gawlick 2011). From Jurassic times onward, essentially three separate tectonic units can be recognized in the area of the younger Alps (Fig. 12.4): the Southern European continental margin including the Helvetic zone, the Penninic Ocean with its northern (only from the Cretaceous) and southern part (Penninic), and the shelf areas of the Adriatic spur (Austroalpine and Southern Alpine units).

At the northwestern edge of the Adriatic spur, the Triassic carbonate platform broke apart at the beginning of the Jurassic. At this time, a horst and graben structure developed with deposition of the basinal sediments of the **Graubünden Slate** (Tauern Window), the Allgäu beds and the *Fleckenmergel* (spotted marls) in the Northern Calcareous Alps. A condensed sedimentation sequence formed on swell areas characterized by the deposition of the reddish **Adnet limestones**. The condensed horizons of this formation were deposited during very low rates of sedimentation. The **Klaus limestone** and **radiolarite** of the Ruhpolding Radiolarite Group formed during the Middle and Late Jurassic. The radiolarites developed at the edge of the Neotethys and were partly formed by turbiditic currents and deposited below the calcite compensation depth (CCD), which at that time was slightly higher than today because of the reduced carbonate production. The CCD indicates the depth at which sinking particles of calcite are completely dissolved in the water.

The opening of the Penninic Ocean can be subdivided into three stages: (1) Earliest Jurassic extension of continental crust with formation of graben structures, leading to (2) Late Lias development of the South Penninic Ocean with formation of oceanic crust, and (3) Latest Jurassic opening of the North Penninic Ocean and the Valais Ocean in the north. While the areas in the surroundings of the Penninic Ocean underwent crustal extension, the edge of the Neotethys was characterized by convergence and the transport of thrusted nappes. The extensional movements of the Austroalpine, in particular at the southern edge of the Penninic Ocean, led to slumping and breccia formation in many places along submarine slopes. Such slumps were most likely triggered by seismic activity. The sedimentation regime changed towards a calcareous facies during the Late Jurassic with increasing intercalations of coarser turbidites.

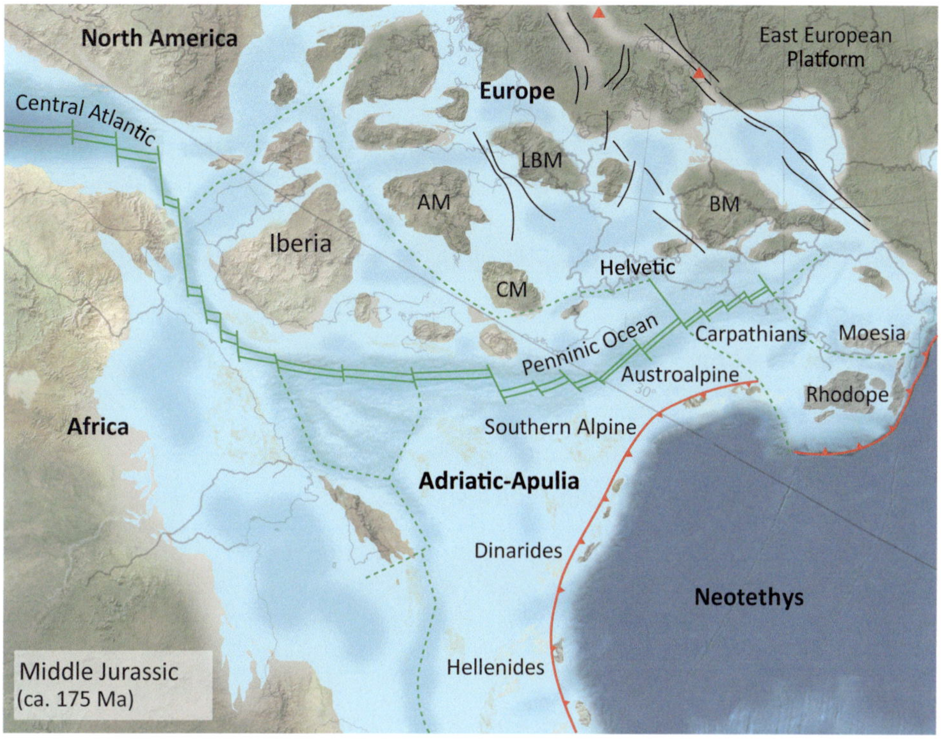

Fig. 12.4 Middle Jurassic plate tectonic reconstruction of the surrounding areas of the Mediterranean region. AM Armorican Massif, BM Bohemian Massif, LBM London-Brabant Massif, CM Central Massif. Topographic base map from Blakey (2011), modified after Frisch (1979), Dercourt et al. (2000), Handy et al. (2010), and Missoni and Gawlick (2011)

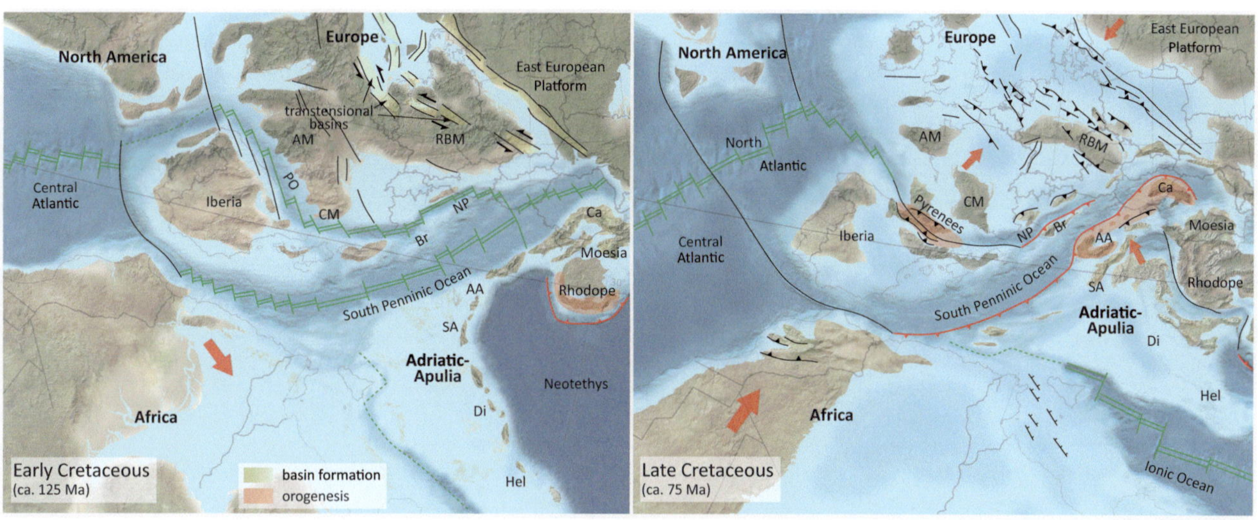

Fig. 12.5 Early and Late Cretaceous plate tectonic reconstruction of the surrounding areas of the Mediterranean region. The changing directions of the African and European plates led to inversion tectonics in Central Europe. AM Armorican Massif, Br Briançonnais, Di Dinarides, Hel Hellenides, Ca Carpathians, NP North Penninic Ocean, AA Austroalpine, PO Pyrenees Ocean, RBM Rhenish-Bohemian Massif, SA Southern Alpine, CM Central Massif. Modified after Kley and Voigt (2008) and Handy et al. (2010), topographic base map from Blakey (2011)

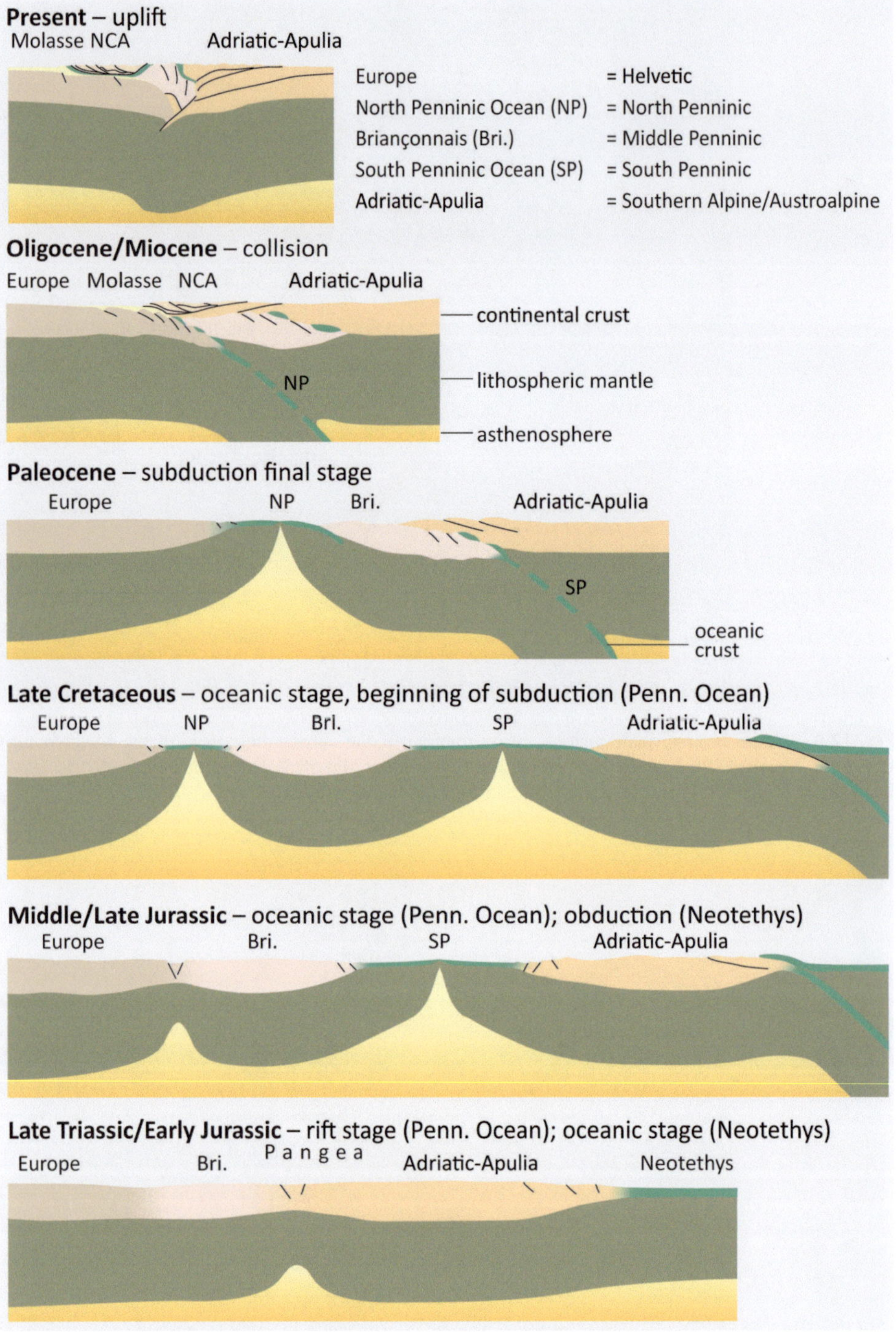

Fig. 12.6 Evolution of the Alpine region in the sense of the supercontinent Wilson Cycle; modified after Frisch and Meschede (2013) and Frisch and Gawlick (2003)

In the Southern Alps, the late Early Jurassic sedimentary deposits consist of characteristic reddish nodular limestone of the **Rosso Ammonitico**, which is largely equivalent to the Adnet and Klaus limestones of the Austroalpine unit. The numerous ammonite fossils and the reddish color caused by hematite-bearing iron ore have given this rock its name. This lithology originated as a pelagic carbonate sediment in waters deeper than 200 m and, similar to the Adnet limestones, represents a condensed sedimentary sequence. The red limestone of the Rosso Ammonitico have been used since Roman times as a building material and decorative stone and has found widespread usage in historic buildings throughout Europe.

With the opening of the Penninic Ocean, intraoceanic subduction began in the Neotethys and at the end of the Middle Jurassic resulted in the first nappe displacements at the southeastern edge of the Adriatic spur (Frisch and Gawlick 2003; Sect. 12.6; Fig. 12.6). During this collisional phase, the area of the Hallstatt facies and Dachstein limestones were imbricated into thrust nappes. The sedimentary successions of the Hallstatt nappes contained synorogenic units such as mélange zones, olistostromes and turbidity current deposits (turbidites) reflecting the subduction activity at that time (Frisch and Gawlick 2003; Fig. 12.4).

During the Middle Jurassic, ophiolites originating in the Neotethys were obducted from the southeast onto the Austroalpine and transferred to the overriding plate as a large block. They are preserved in the Western Carpathians and in the Balkans, whereas limited remains can be found in the Hallstatt nappes of the Austroalpine. A complicated tectonic evolution followed, as demonstrated by the numerous names given to the oceans from which the ophiolites were derived. Depending on the region of occurrence they are called Meliata, Maliac, Hallstatt, Dinarides, Vardar or the Pindos Ocean (e.g., Schmid et al. 2004, 2008; Handy et al. 2010; and others), but all of these units probably belonged to the northernmost part of the Neotethys.

12.5 The Alpine Region in the Cretaceous and Early Tertiary

During the Earliest Cretaceous, the southeastern edge of the Adriatic-Apulian Plate was characterized by only small after-movements of the nappe thrust system. The displacements of the Jurassic deformation were largely complete and the uplift of the mountain range that began in the Latest Jurassic had reached its peak. Early Cretaceous sedimentation in foreland molasse basins was characterized by the abundance of eroded material that was shed off from adjacent mountains and transported into the foredeep depressions. However, sedimentation in the Austroalpine and Southern Alpine areas developed differently during the Cretaceous, where it was characterized by deeper water deposits. In the Southern Alps, these conditions continued in some places up to the Miocene. However, in the Austroalpine, Western Carpathians and the Pannonian region, a new orogenic phase occurred during late Early Cretaceous, characterized by high-pressure metamorphism and large-scale lateral displacements. The cause of this orogeny is still largely unknown but was probably related to continued subduction of the Neotethys as well as the northward shift of intracontinental compression and/or major lateral movements. The orogeny again resulted in nappe thrusting and deformation that gave rise to an early stage of uplift and erosion. In the Turonian, the basal Gosau beds (named after the village of Gosau at the Dachstein) were unconformably deposited on top of the folded layers of the older Mesozoic strata of the Austroalpine. There occur terrestrial sedimentary rocks consisting of conglomerates, sandstones, and local coal deposits all of which represent deposition in small, restricted basinal environments.

Following completion of the Middle Cretaceous orogenic phase, sedimentation took place under calmer conditions with the deposition of the Gosau beds lasting into the Paleogene. Shallow marine sediments were deposited from the Late Cretaceous to the Paleocene. These were followed by deep marine carbonates and locally by thick deep marine, mostly turbiditic sediments, the sedimentation of which began during the Campanian in some places. Extensional tectonics in the Austroalpine area at this time resulted in the deepening of the sedimentary basins.

The opening of the South Penninic Ocean continued into the Early Cretaceous. At this time, the North Penninic Ocean opened north of the Briançonnais to form a continental block between the two parts of the Penninic Ocean. The opening of the North and South Penninic Oceans were related to the **opening of the Atlantic Ocean**. During the Early Cretaceous, a spreading and transform fault system developed through the Bay of Biscay, the Pyrenees Ocean and into the North Penninic Ocean (Fig. 12.5). This fault zone became inactive in the Late Cretaceous as the spreading axis of the Atlantic Ocean shifted northward to its present position. Thick layers of clastic deep-water sediments (the Graubünden Slate) were deposited in the North Penninic Ocean. These pass upward into the Prättigau flysch that was sedimented during progressive narrowing of the oceanic basin. The Graubünden Slates and the Prättigau flysch occur in the Western and Central Alps. In the Eastern Alps, the Penninic series are characterized in the Late Cretaceous by sedimentation of the Rhenodanubian flysch. It should be noted that the Graubünden Slates in the North Penninic Ocean are of Early Cretaceous age, wheareas in the South Penninic Ocean they represent Jurassic deposits.

The various synorogenic deep-sea trench basins of the Penninic Ocean that lay in front of the advancing Adriatic-Apulian Plate became the catchment basins for Late Cretaceous to Eocene turbiditic sediments. These rocks consist mainly of sandy and argillaceous lithologies and hard sandstone beds which, at the beginning of the nineteenth century, were designated as “”**flysch** in local dialects of the Alpine region. This term is now internationally accepted as a descriptive term for this type of sedimentary deposit. Flysch forms from turbidity currents that develop at an active continental margin as unconsolidated sedimentary material slides down-slope as a mixture of sediment and water. Within a short time (i.e., within a few hours or days) up to several meter thick sedimentary packages can be deposited, the individual turbidite layers having typical thicknesses of a few decimeters or centimeters. These turbidite beds have a characteristic structure with graded stratification (Box 12.20: Turbidites). The intervening shales or marls represent longer periods of background sedimentation in the deep-sea environment. Flysch sedimentation in the Alps accompanied the entire subduction process and, in places, lasted until the Eocene. In the western Eastern Alps, this type of sedimentation probably ended in the Late Cretaceous. Flysch deposits are located tectonically below the Austroalpine nappes that were thrust over the Penninic. Together with the overlying nappes of the Austroalpine, the Rhenodanubian

Box 20: Turbidites

Interbedded strata of distinguishable layers of muddy, silty (partly marly) and sandy lithologies in fining-upward sequences are widespread in the geological record. The sandy layers are often of turbiditic origin, generated by underwater **turbidity currents** (or suspension currents) that flow downslope. This sedimentation process, recognized in the 1960s, explains how many features known in shallow water sediments are formed in the deep-sea environment. During this process, large amounts of sediment are transported within a short time from the shallow shelf edge into the depths by turbidity currents. Triggered by instabilities, such as those caused by earthquakes, sedimentation caused by a turbidity current is geologically speaking a rapidly occurring process. Within a few hours, meter thick sand layers can form. Turbidity currents flowing downslope on the bottom of the sea have been known to reach speeds of up to 100 km/h, whereby only small slope angles of 2–3° are sufficient to initiate a turbidite flow.

In spatial terms, turbidity currents form debris cones at the foot of a slope and are known as **turbidite fans**. They develop in a similar way as avalanches. The dimension of the fans typically ranges from a few tens of meters in lakes to up to several hundred kilometers when formed along continental slopes. The further the distance they are deposited from the origin of the sedimentary material, the more and more fine-grained are the individual turbidite beds. In the case of short transport distances, the sandstone beds contain coarser-grained material and form part of units termed **proximal turbidites**. After traveling long distances, the beds contain finer particles and are known as **distal turbidites**. In addition to the lateral changes in grain size, turbidites exhibit very characteristic **graded bedding**, whereby the grain size decreases upwards towards the top of the turbidite sequence. This grading results from the broad spectrum of particle sizes that are transported in the suspension current. In the beginning it is mainly the coarse grains that settle and sediment on the seafloor, while the finer fractions are transported further within the suspension. As the speed of the current declines and the energy decreases, fewer coarse grains remain in suspension both upwards and towards the toe of the turbidity current. The result is a stratification that is divided into numerous grain-sized sorted layers. This type

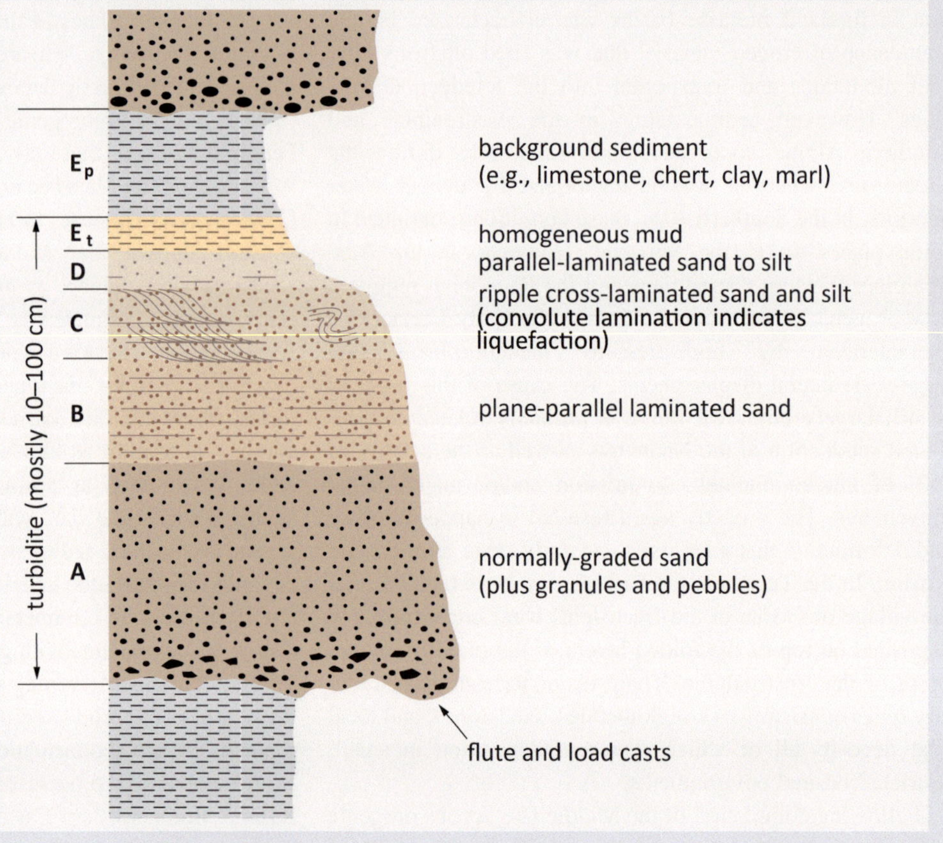

Fig. 12.7 Subdivision of a turbidite into the five typical units (A to E) of a Bouma sequence. Its deposition results from a suspension current. Modified after Kuhlemann et al. (2009)

of stratification was first described by Arnold H. Bouma (1932–2011; Bouma 1962) and now named accordingly as the **Bouma sequence** (Fig. 12.7).

A complete Bouma sequence begins with an often significantly graded layer (A horizon) characterized by the occurrence of coarse components, largely in the range of sand sizes. This layer typically has an erosional base and contains rip-up clasts of mudrock. The sandy B horizon is the lower, often finely parallel laminated layer, which varies from fine- to medium-grained sandstone. In the sandy-silty C-horizon, oblique layering and flame sedimentary structures that resemble folds occur caused by the escape of water (*convolute bedding*). The upper finely laminated layer of silt and sometimes clay-sized (D horizon) follows above. The lower part of the E horizon (or E_T horizon; T for turbiditic) preserves the finest grain fractions of the suspension current as it ceases to flow, whereas the upper part of the E horizon (or E_P horizon; P for pelagic) represents the slow but continuous deep-water background sedimentation that occurs between turbidite flows. Dependent on the distance from its source and the capacity of the turbidity current to transport its load, the Bouma sequence is not always fully developed. Generally, the further away from its location of origin, the more likely the lower horizons of the Bouma sequence are missing.

At the beginning of avalanche-like transport, turbidity currents have a high erosional force that can carve deep valleys (*canyons*) into the upper part of the turbidite fan. Such canyons may be several hundreds of meters wide and on some continent slopes are known to reach several kilometers in depth with the development of meandering systems. The erosive power of turbidity currents commonly forms **sole marks** at the base of the coarser turbidite beds. These may occur as groove casts, flow casts, scour marks or load casts.

Examples of turbiditic rocks in Germany are found in the Late Devonian/Early Carboniferous basinal facies of the Variscan where they occur as immature sandstones known as **graywacke**: a term that is no longer recommended (Vinx 2011). These deposits are now considered to present Variscan flysch, which occurs in the Rheinisches Schiefergebirge, in the Harz Mountains and the Thuringian Slate Mountains. Alpine turbidite sequences that formed as basin fill in the Penninic Ocean are mainly represented by the Rhenodanubian flysch and Graubünden Slates. The term **flysch** is derived from a local dialect of Swiss German (spoken: fleesh) and refers to loamy rocks that are often profoundly weathered and have little internal cohesion. They tend to flow downslope and result in long rolling hills and rounded hilltops.

flysch was uplifted and eroded at the end of the Eocene and Oligocene. From this material, the molasse accumulated in the Tertiary, infilling the externally located foreland basin that lay in front of the Alps (Sect. 13.3).

The plate tectonic conditions of the Alpine region underwent a significant change from the Middle Cretaceous onward. The African Plate switched its direction of movement in a way that the opening of the Alpine oceans (South Penninic Ocean, Pyrenees Ocean, and North Penninic Ocean) ceased and subduction of the European Plate beneath the Adriatic-Apulian or African Plate began (Fig. 12.5). The timing of the onset of subduction is still unclear but general opinion places it in the Late Cretaceous (Santonian; Putis et al. 2008).

The Cretaceous period of the European continental margin (Helvetian shelf; Fig. 11.72) is characterized mainly by limestone, marl, and clay with minor amounts of sandstone. These strata are found today in the Helvetic sequences that are best preserved in the Western Alps. The relatively undisturbed sequence of Helvetic flysch was deposited up until Early Tertiary times before being deformed in the late stages of the Alpine Orogeny from the Oligocene onward.

12.6 The Tectonic Evolution of the Alps

In terms of the supercontinent **Wilson Cycle**, two such cycles can be recognized in the tectonic evolution of the Alpine Orogeny (Fig. 5.5). The first cycle started with the birth of the Neotethys, which at the turn of the Permian to Triassic age was telescopically opened to the northwest with the formation of a spreading center. During this development, the oceanic stage of a supercontinent cycle was reached but did not go to completion. While at the northern end of the Neotethys a first collision occurred, a new rift stage began in the Jurassic that was no longer related to this ocean but instead represented a continuation of the Atlantic spreading ridge system. The resultant South Penninic Ocean started the second cycle that initially opened nearly perpendicular to the spreading system of the Neotethys and later paralleled the North Penninic Ocean. Although subduction at the edge of the Neotethys started in the Middle Jurassic, it was not until the Late Cretaceous that the two Penninic Oceans were subducted.

It is evident that ocean spreading persisted while subduction along their edges has already begun (Fig. 12.6). In

Box 21: Isostatic adjustment and mountain uplift

Isostatic adjustment

The collision of two continental lithospheric plates results in crust thickening by the stacking of tectonic nappes and finally the uplift of a mountain range. In the process, the two thick crustal packages are stacked, interlinked and strongly deformed. Folds and tectonic nappes with numerous faults develop. At depth, the rocks are transformed into metamorphic rocks by high pressures and temperatures. Rocks also melt and rise upward as magma from which igneous plutonic rocks evolve at depth or volcanic rocks crystallize at the surface. The strong thickening of the continental lithosphere that results from collision pushes the asthenosphere down to greater depth. Simultaneously, the upper surface of the collisional zone rises upward above the other crustal areas of normal thickness. This is comparable with two wooden blocks floating in the water when stacked on top of each other and connected. Together they immerse deeper into the water but at the same time, they will rise higher above the water surface. This process is known as **isostatic adjustment** and the continental crust behaves in very much the same way.

A simple experiment illustrates the effect of isostatic adjustment. Consider a normal thickness of continental lithosphere as a column with a 1 cm^2 base area and a height of exactly 200 km (Fig. 12.8). An imaginary plane at 200 km depth is defined as the reference horizon for compensating movements (the theoretic compensation plane). Based on average density values for the lithosphere (mantle: 3.3 g/cm^3; continental crust: 2.8 g/cm^3) and asthenosphere (3.25 g/cm^3) the column is divided into 100 km of mobile asthenosphere, 70 km of rigid lithospheric mantle and 30 km of continental crust (Fig. 11.54a, left column). It has a total weight of 64,000 kg:

$(100\ \text{km} \cdot 1\ \text{cm}^2 \cdot 3.25\ \text{g/cm}^3) + (70\ \text{km} \cdot 1\ \text{cm}^2 \cdot 3.3\ \text{g/cm}^3) + (30\ \text{km} \cdot 1\ \text{cm}^2 \cdot 2.8\ \text{g/cm}^3) = 32{,}500\ \text{kg} + 23{,}100\ \text{kg} + 8400\ \text{kg} = 64{,}000\ \text{kg}$

If the column was changed by increasing the crustal thickness to 60 km, corresponding to a collision of two continental plates, the weight of the column with its constant height of 200 km would be reduced by 1350 kg (Fig. 12.8a, middle column). This amount is compensated from below by movement of the mobile asthenosphere so that the weight acting on the theoretic compensation plane at 200 km depth is again 64,000 kg. In order to achieve this, 1350 kg of

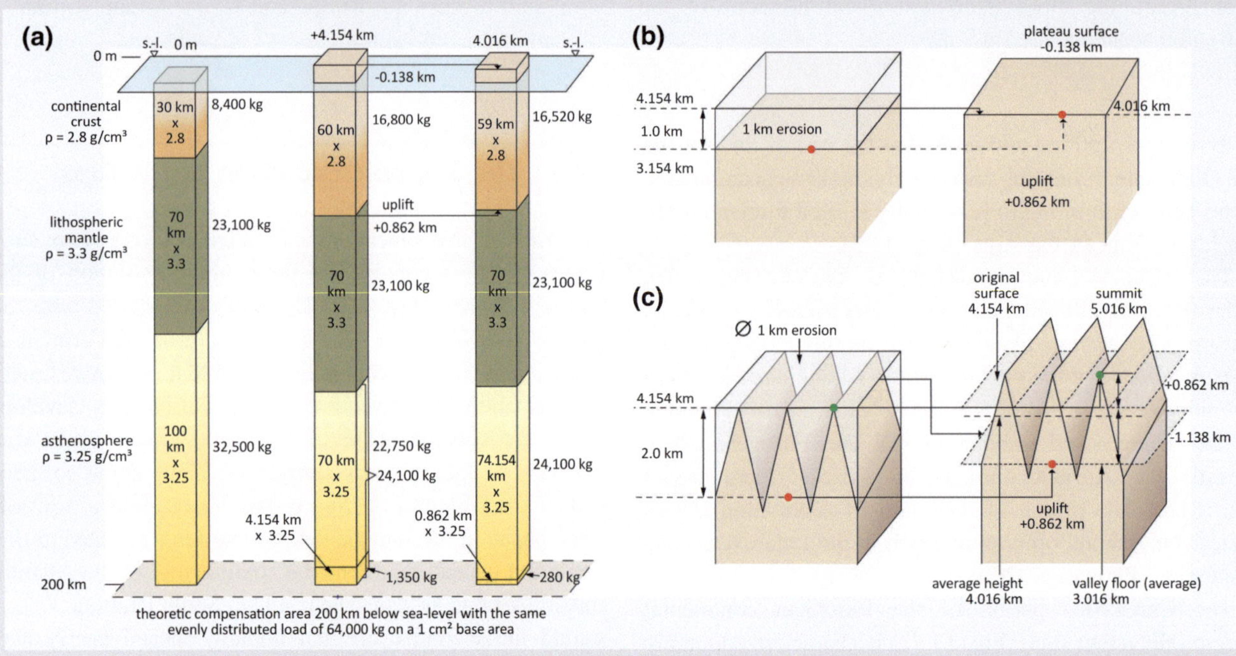

Fig. 12.8 **a** Effects of mountain building (crustal thickening) and discharge of rocks (erosion) on the state of isostatic equilibrium (floating balance). A theoretical compensation area is assumed in the asthenosphere (here at 200 km depth) at which the pressure per unit area (here in kg/cm^2) is everywhere the same. The thickening of the crust (here from 30 to 60 km) leads to the uplift. Although erosion leads to a lowering of the Earth's surface relief, it results in further uplift of the lithosphere. **b**, **c** effects of erosion and relief on uplift and shaping of the topographic surface. While the average height is lowered by erosion, the incision of valleys leads to a rise in the height of the summits. With a uniform erosion of 1000 m (**b**), together with the same erosion volume forming valleys and peaks (**c**), the surface or the average height in this theoretical example is lowered by 138 m and the rocks and thus the summit ascends to 862 m. Modified from Frisch and Meschede (2013)

asthenospheric material has to be added to the 200 km column, which corresponds to a column of 4.154 km height. This additional height is the amount by which the thickened crust moves upwards due to isostatic adjustment.

Mountain uplift and erosion

The highest mountains on Earth, such as the Himalayas, are significantly more elevated than the ca. 4 km height outlined in the example above. These mountainous peaks reach twice the height amounting to nearly 9 km above sea-level (Mount Everest, 8848 m ASL). Such heights are not explained by isostatic adjustment alone. Instead, an important factor, the erosion of the mountains has to be considered. This might at first sound paradoxical because erosion is the process that removes rock material by chemical and physical weathering, and as such is expected to lower the height of mountains. However, the rock material is not evenly transported out from the mountains. While valleys are deeply incised, mountain peaks retain more height. This process is best illustrated by another experiment. If we removed by erosion 1 km of rock material from the column above but consisting of a double crustal thickness of 60 km and a lithospheric mantle of 70 km, the final balance of 64,000 kg would have to be reset again by isostatic adjustment (Fig. 12.8a, right column; Fig. 12.8b).

The 1 km of continental crust removed from the top with a weight of 280 kg is replaced below by replenishing asthenospheric material. This corresponds to a slightly smaller column of 862 meters of asthenosphere because of the higher specific weight of the material and therefore it is the exact height by which the lithosphere is uplifted. Therefore despite the erosion, and even though it becomes thinner, the column responds by uplift. However, the column will be in total 138 m lower down beneath the theoretic compensation plane than before, since the 1000 m removed by erosion is accompanied by only 862 m of isostatic uplift. If the erosion is not uniform everywhere but cuts localized valleys (Fig. 12.8c), the lithosphere will still rise by the same 862 m, but the eroded mass remains identical. The tops of the mountains are thus uplifted accordingly by 862 m if it is assumed that they retain their original vertical dimension. At the same time, the valleys are incised. If the valleys have an average depth of 2 km, the valley floor will be on average 1138 m lower than the surface before erosion (2000 m valley erosion minus 862 m isostatic uplift). For this reason, the erosion and the formation of a strong relief is an important prerequisite for the uplift of a high mountain range.

the Early Tertiary, the collision of the Adriatic-Apulian and the European plates occurred following the closure of the South Penninic Ocean. The closure of the North Penninic Ocean followed a little later in the Eocene and as subduction ended, it was followed by the building of today's Alpine mountain chains. During the Oligocene, the subducted oceanic lithosphere broke off from the slab and resulted in the rise of mostly tonalitic melt along the Periadriatic Lineament (e.g., the intrusions of the Bergell pluton and the Rieserferner; Fig. 12.1). A tonalite is a felsic type of rock composed mainly of quartz, plagioclase, hornblende, and biotite, but without alkali feldspar. Following slab breakoff of the subducted oceanic lithosphere, the **uplift of the Alps** commenced due to the isostatic adjustment that continues to this day (Box 12.21: Isostatic adjustment and mountain uplift).

Nappe thrusting started in the Alps during the Middle Jurassic and initially affected only the southeastern area of the Austroalpine, which was divided by thrusts into several subunits. In the Austroalpine, representing the highest structural level of the Alps, the Lower Austroalpine is located at the bottom, followed by the Upper Austroalpine. The latter unit is further subdivided into a lower (previously also designated as the Middle Austroalpine) and a higher part (Fig. 12.9) that are spatially separated. The higher part of the Upper Austroalpine with its Permo-Mesozoic sedimentary strata is primarily found in the Northern Calcareous Alps and the Drau Mountain Range, south of the Tauern Window. Some deposits are also found on top of the crystalline rocks of the Lower Austroalpine. The lower part of the Upper Austroalpine (also called Central Austroalpine) contains predominantly Variscan and Alpine deformed rocks that occur mainly south of the Northern Calcareous Alps, up to the Periadriatic Lineament, and include the graywacke zone. The main stage of nappe thrusting in the Austroalpine carried on until the end of the Late Cretaceous. After a period of inactivity in the Paleocene, the Eocene collision finally resulted in the displacement of the entire nappe pile onto the Penninic.

The deposition of the Gosau Group took place after the mountain-building "phase" of the Middle Cretaceous during submergence in an open sea, with various islands developed across the region of the Austroalpine. There are different opinions as to the origin of the **Gosau basins**. One view is that the extensional basins, which formed on top of the Austroalpine unit at the end of the Late Cretaceous, were

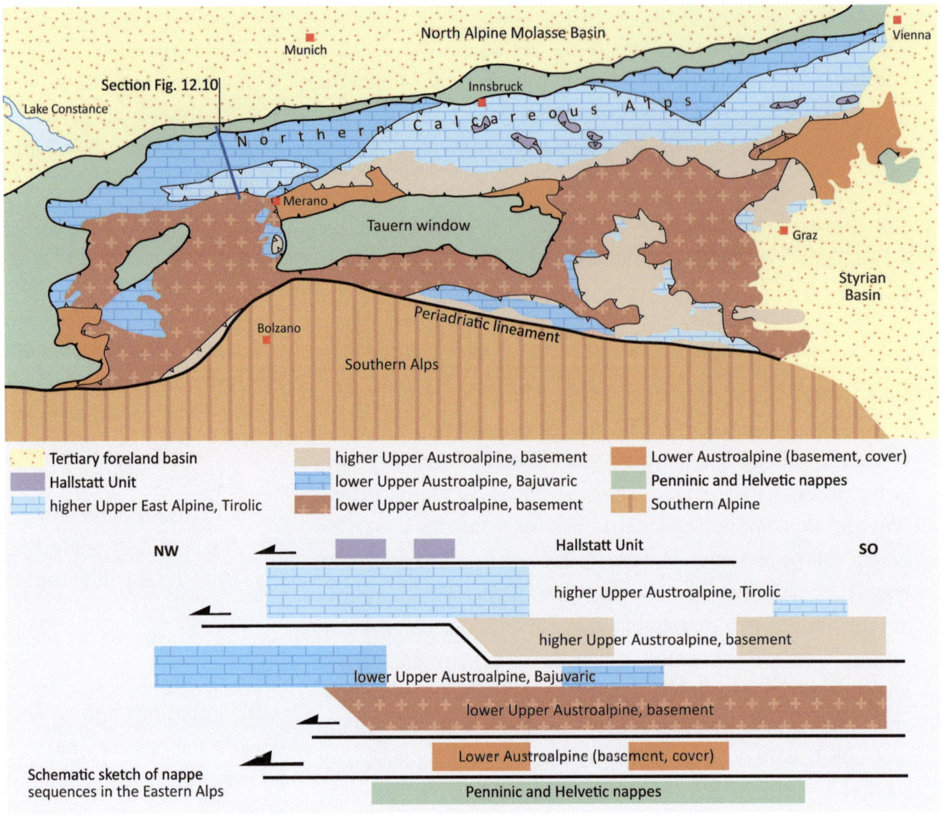

Fig. 12.9 Tectonic map and a schematic sketch of overthrusts in the Eastern Alpine region. Modified after Frisch and Gawlick (2003)

related to the retreat of the subduction zone (**subduction rollback)** to the northwest. Another interpretation considers the extension to have resulted from a change in the subduction zone that led to **subduction erosion**, inducing subsidence and extension in the overriding plate (Wagreich 2003). Later during the Eocene, the Gosau basinal strata were thrust, together with the older rocks of the Northern Calcareous Alps, onto the Penninic and finally all the units were together subsequently displaced onto the Helvetic zone.

In the Austroalpine of the Northern Calcareous Alps, a fold-and-thrust belt formed with deformation of varying intensity resulting from the different mechanical strength of the rock units (Linzer et al. 1995). The competent, thick-bedded limestones of the Berchtesgaden facies were more difficult to fold (Fig. 12.3) and formed large, slightly deformed horizons most of which are still relatively flat-lying. In contrast, clay-rich and evaporitic horizons such as the Partnach Formation, the Raibler Formation, and the Permian Haselgebirge were often strongly deformed and served as shear horizons for nappe thrusting (Fig. 12.10).

Alpine metamorphism affected only the lower units of the Austroalpine during the stacking of nappes in the Cretaceous. Whereas some parts of the Northern Calcareous Alps

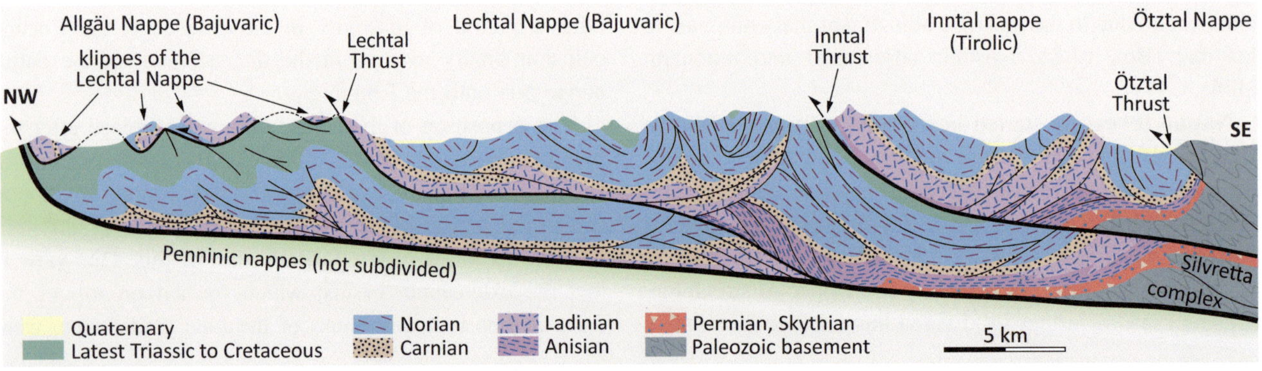

Fig. 12.10 Geological cross-section through the western part of the Northern Calcareous Alps. Location of the cross-section shown in Fig. 12.7. Modified after Eisbacher (1990)

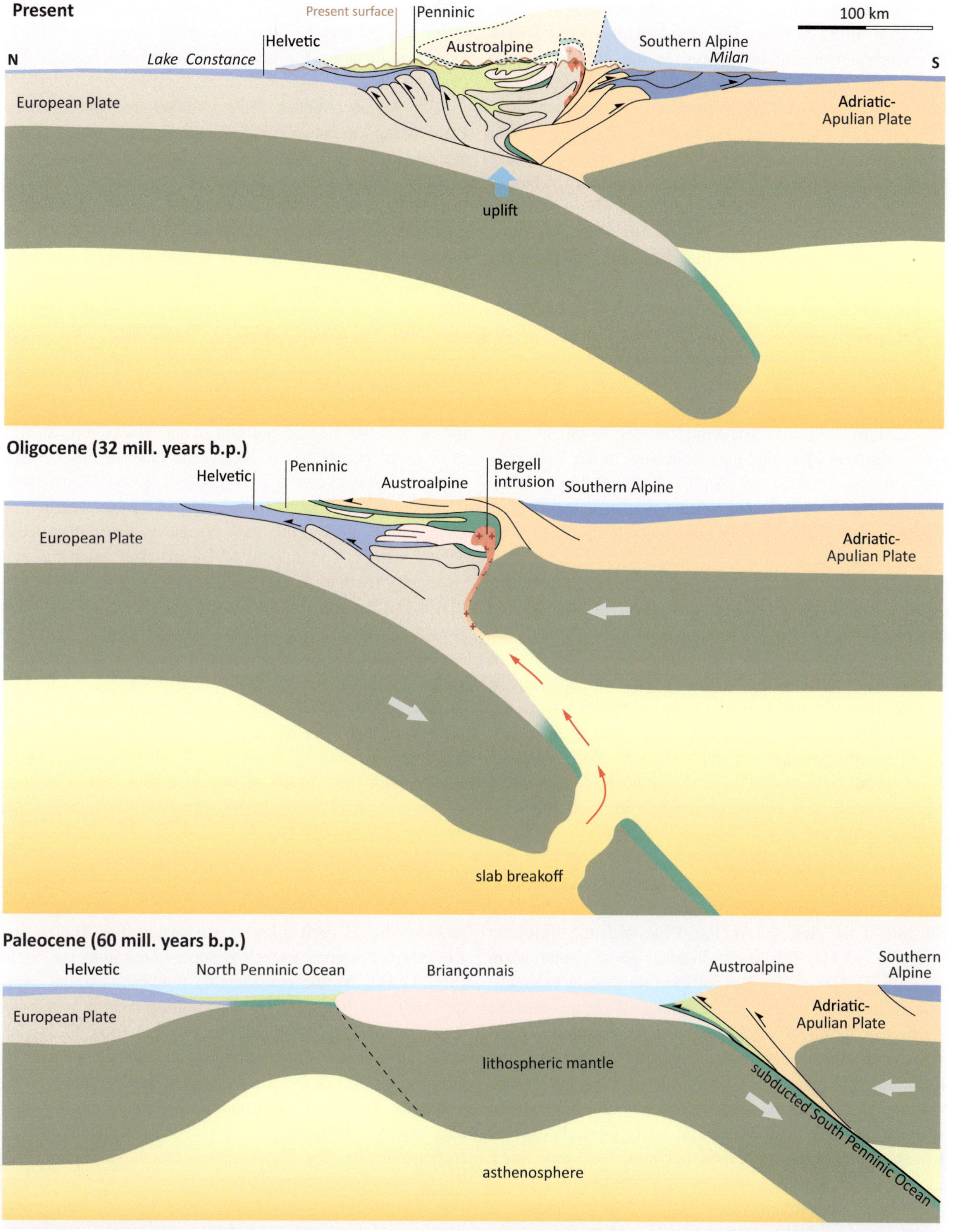

Fig. 12.11 Nappe formation and folding in the Alps resulting from the collision between the Adriatic-Apulian Plate and the European Plate. Schematic sketch showing the evolution of the Alpine Orogeny along a cross-section through the eastern Central Alps. Modified after Schmid et al. (1996, 2004), Froitzheim et al. (2003)

remained unmetamorphosed, towards the western edge and at the base of the Northern Calcareous Alps, an increase in metamorphic overprint is evident. In the deeper tectonic levels, even metamorphic rocks of eclogite facies occur. Such rocks are also found in the Variscan deformed basement, which was overprinted a second time by the Alpine metamorphism. The Southern Alpine rocks were much less affected by thrusting and were split into a number of blocks during the **main Eocene compressional phase** prior to thrusting onto each other (Fig. 13.16). However, there was no nappe thrusting to the extent that occurred in the Northern Calcareous Alps.

The Austroalpine unit is tectonically located on top of the Penninic nappes. These units were exposed in the Eastern Alps within a narrow zone lying at the northern edge of the Alps and in three major tectonic windows (from west to east: Engadine Window, Tauern Window, and Rechnitz Window; Fig. 12.1). In the Tauern Window, Variscan basement rocks of the European Plate that may be related to the Briançonnais were overprinted by Alpine metamorphism. They belong to the Venediger Nappe and represent the deepest tectonic level in the Tauern Window. In the Glockner Nappe, oceanic crust of the Penninic Ocean and its sedimentary cover is partially preserved as in the Graubünden Slates. Some authors assume that parts of the bedded sequence of the Tauern Window must belong to the Helvetic unit (e.g., Tollmann 1977; Pfiffner 2010). The Tauern Window is considered in this sense to represent a Penninic-Helvetic double window. In the Western Alps, the Penninic nappes extend over a large area whereby the lowest nappes are comprised of rock units from the North Penninic Ocean. These pass upward into the continental blocks of the Briançonnias, which in turn are overlain by rock units representing the South Penninic Ocean.

During the Eocene, the South Penninic Ocean was already fully subducted and only the narrow North Penninic Ocean remained until it was also consumed prior to the final **collision** of the Adriatic Apulian Plate with the European Plate (Fig. 12.11). This last collisional phase continued in part until the Oligocene. The collision resulted in extensive folding and thrust stacking of the Penninic nappes and the synorogenic flysch sequences that were previously deposited in the remaining oceanic basins prior to the Eocene. The Austroalpine nappe pile, which suffered Late Cretaceous thrusting, was further displaced as a structural package far to the north to overlie the Helvetic unit. In the Central Alps, the Dent-Blanche Nappe and the Sesia Nappe were also thrust onto the Penninic rocks.

The lowermost structural level of the Alps is represented by the Helvetic nappes. These units are composed of the Permo-Mesozoic sedimentary cover of the European continental margin that was stripped off from the basement and thrust northwards in the Early Miocene during the late phase of the Alpine Orogeny. The Helvetic nappes are widespread in the Central and Western Alps but are only exposed as a narrow strip along the northern edge of the Eastern Alps (Fig. 12.1).

The Adriatic-Apulian Plate intruded into the European Plate during the collision like a battering ram, which in geological terminology is known as an **indenter**. Caused by the collision, far-reaching horizontal movements occurred throughout, whereby the Eastern Alps were compressed and displaced eastward along large lateral faults. This process, described by Ratschbacher et al. (1991) as **lateral extrusion**, led to an N-S shortening of the Eastern Alps and simultaneous stretching in the E-W direction by an average of 30–40%. The crust was dissected into several tectonic blocks and formerly continuous nappe sheets were divided into segments that were often displaced by many kilometers from each other (Fig. 12.12). The shortening can be restored like a jigsaw puzzle and the original position of the blocks reconstructed prior to the penetration of the Alpine indenter that occurred about 30 million years ago (Frisch et al. 1998). The locations of the large lateral faults that developed during extrusion are today geomorphologically recognizable by long, often deeply carved valleys. The amount of lateral offset along the fault zones may reach up to 100 km, such as along the SEMP Line (Salzachtal-Ennstal-Mariazell-Puchberg) or along the Periadriatic lineament (Fig. 12.12). The Periadriatic Lineament represents a fault line that was reactivated during extrusion, with the uplift of the Austroalpine nappes to the north of the eastern part of the lineament by several kilometers: a displacement that occurred together with far-reaching lateral movements.

The Oligocene **uplift of the Alps** that started at about 32 Ma and resulted from slab breakoff (Box 12.22: Slab breakoff) is an event that is well documented by the simultaneous intrusion of plutons. The age of the intrusions that have been radiometrically dated indicates they, ascended at about the same time along the Periadriatic Lineament. These include the Bergell pluton, the Adamello, Biella, and Rieserferner plutons, as well as some minor intrusive bodies (Fig. 12.11). As the considerably thickened Alpine nappe pile suddenly lost its downward pulling weight as the slab broke off, isostatic rebound was initiated (Box 12.21: Isostatic adjustment and mountain uplift). This force drove the Oligocene uplift of the Alps: an event that continues today. Currently, the rate of maximum uplift in the Central Alps is still about 1 mm/year, an isostatic adjustment that is responsible for the high Alpine mountain chain today. Therefore, the highest mountain in the Alps of Mont Blanc, which is 4810 m above sea-level, is in fact relatively young, despite not being composed of the hardest rocks. The current geomorphology of the Alps resulted mainly from the erosive action of glaciers that were widespread during the last glaciation periods.

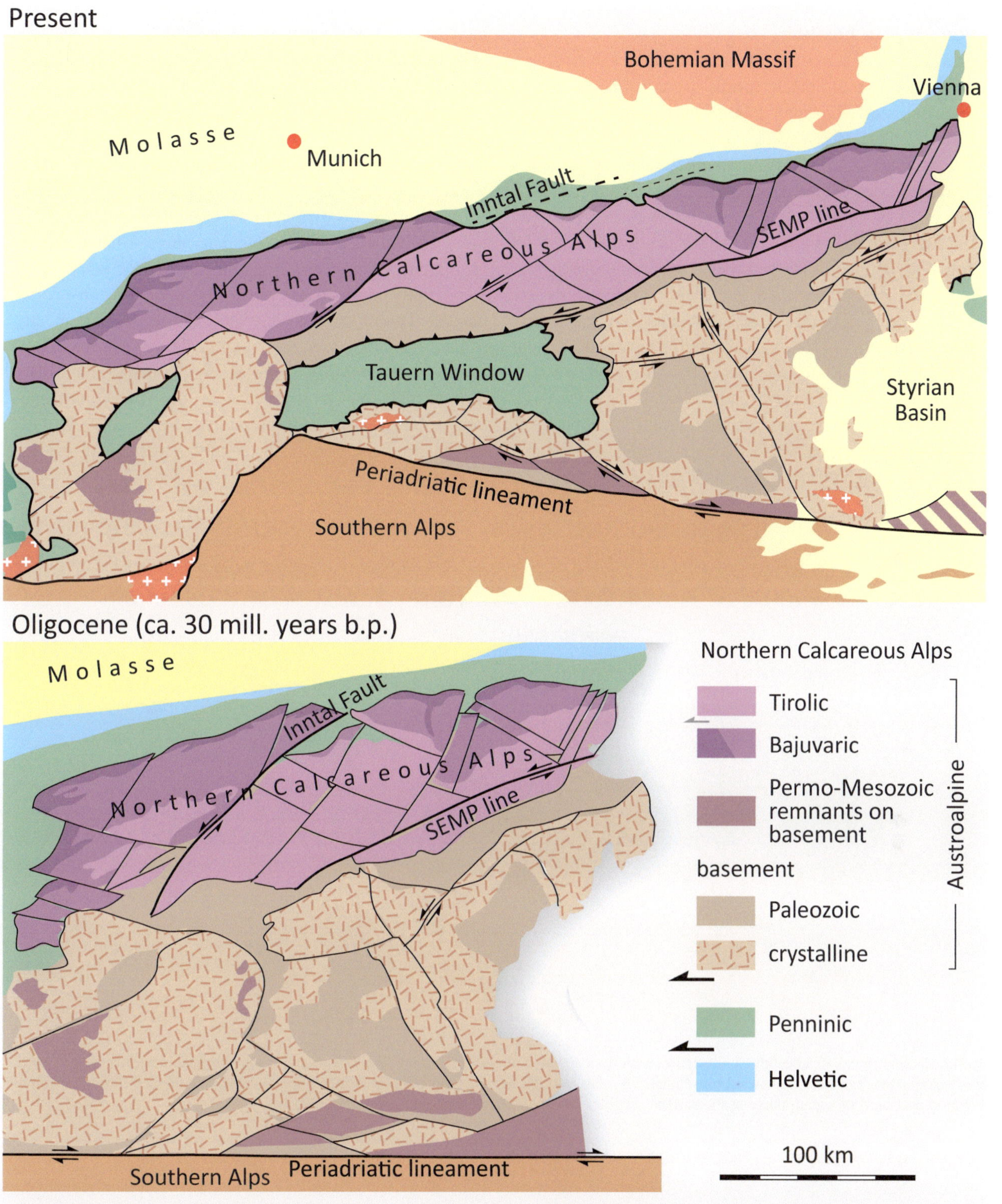

Fig. 12.12 Tectonic map of the Eastern Alps and its subdivision in tectonic blocks. The lower map displays the reconstructed position of the blocks relative to each other in the Oligocene prior to displacement. Modified after Frisch et al. (1998)

Box 22: Slab breakoff

The closure of an ocean by subduction at a convergent plate boundary ultimately results in the collision of continental plates. The subduction of the downgoing oceanic plate continues for a while following the collision with the continental crust, as the underthrusted continental margin is dragged down by the subduction zone. Initially, this gives a counterweight against the buoyant lift caused by the relatively light continental crust thickened by collision. The colliding continents are wedged together, deformed and transported as tectonic nappes. As the lightweight thickened crustal package moves rapidly down to depth, the forces of slab pull become insufficient to pull it further down. The uplift forces of isostatic adjustment are eventually large enough to cause subduction to cease. In contrast, the oceanic lithosphere of the subducted ocean reaches a depth where the basaltic rocks are metamorphosed to much denser **eclogite**. This increases the downward motion and subducts the oceanic lithosphere even deeper (Fig. 12.13). Eventually, it tears off from the continental lithosphere and sinks down into the mantle in a process known as **slab breakoff**. As the previously attached thickened continental crust loses its heavy counterweight, isostatic uplift forces then dominate leading the crust to rise back upward. The thickened continental crust is then uplifted, together with the surface topography of the high mountain range. Slab breakoff also initiates deep-rooted fracture zones in the lithosphere where magma can ascend rapidly from the asthenospheric mantle. **Tonalites**, with a granite-like composition, are formed from the ascending magma. These lack the alkali feldspars of common granites and contain amphibole (hornblende), mica (biotite) and in some cases pyroxene (augite) as dark mineral components. A bright variety of tonalite containing less dark mafic components is known as **trondhjemite**.

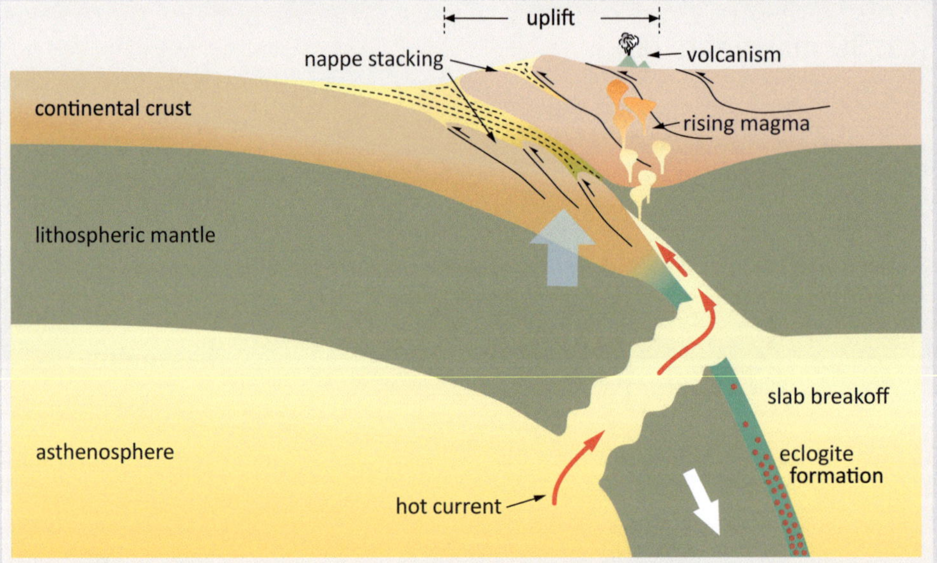

Fig. 12.13 Slab breakoff affecting subducted oceanic lithosphere following collision. Basaltic rocks of the lithosphere are metamorphosed to dense eclogite that enhances downward movement. The slab breakoff supports the buoyant uplift of the thickened continental lithosphere and the formation of a high mountain range at the surface. Slightly modified after Frisch and Meschede (2013)

Tertiary Basins 13

The Tertiary geological history of Germany was mainly characterized by basin deposition in restricted tectonic settings, either as graben structures or as the molasse-filled foreland foredeep of the Alps. The most prominent basin structure of the Tertiary is the N–S trending Upper Rhine Graben that represents one of the most striking geomorphological elements of southern Germany today. Other graben structures are the Lower Rhine Valley, the Eger Graben and some smaller grabens such as the Hessian Depression. During this time, thick Tertiary lignite deposits formed in coastal environments in the areas of basinal sedimentation. Another major type of Tertiary deposit is the molasse. These clastic sediments represent alternations of marine and freshwater deposits that were derived from the uplifting Alpine mountain chain that lay to the south. These sediments accumulated in a foreland basin area of down flexure and subsidence that lay in front of the advancing fold-and-thrust belt.

There has been some discussion over the last years as to the suitability of the term Tertiary as a designation for the period between the Cretaceous and Quaternary. The naming of these periods originates from the early days of geological science in the eighteenth century, when the Italian geologist Giovanni Arduino (1714–1795) divided up geological time into "Primitive" (the Paleozoic), "Secondary" (the Mesozoic), and "Tertiary". In its international usage today, the **Tertiary** period has now been replaced by the Paleogene and Neogene periods, and the younger **Quaternary** is retained as the last period of the Cenozoic era. Due to its long-term usage in the geological literature, however, the name, Tertiary, is unlikely to disappear and is retained in this book. There has also been a lot of uncertainty as to the timing of the end of the Tertiary and the beginning of the Quaternary, and a number of working definitions have been used. In the past, the internationally accepted boundary was based on marine deposits in the south of Italy and therefore commonly placed at 1.8 Ma while in Central Europe, the boundary was traditionally placed at 2.59 Ma. More recently, it has been reconfirmed by the International Commission on Stratigraphy that the boundary between the Piacenzian and Gelasian is to be placed at the latter geochronological position, giving the Tertiary (the Paleogene and Neogene combined) an age range of 66–2.59 Ma.

Tertiary deposits are widespread throughout Germany and are located close to the surface in almost half of the entire country. In most cases, and particularly in the north, these deposits are covered by up to several hundred meter thick sequences of Quaternary sediment and therefore hidden from direct access. The Tertiary strata were formed in marine and continental regions, with a predominance of clayey and sandy sediment as well as rare limestone. Typical of the Tertiary was the development of a series of sedimentary basins that resulted from extensional tectonics. The most striking example is the Upper Rhine Graben, but the Hessian Depression, the Lower Rhine Bay, and the Eger Graben region also form distinctive structures. In front of the Alps, a molasse basin evolved during the Tertiary, while the north of Germany was mostly covered by an epicontinental sea.

Tertiary **brown coal deposits** are widespread across Northern Central Europe and represent deposition typically in bay environments along the southern coastline of the continental shelf sea. During the Eocene, dense swamp forests and moors developed whereby the accumulation of the plant debris was excluded from oxygen and converted to brown coal during progressive burial. Tertiary volcanic events of considerable dimension (Chap. 14) also occurred in many places. Other events include the Middle Miocene asteroid strike in the area of the Swabian-Franconian Alb, where traces of the impact crater are still visible today (Chap. 15).

The global plate tectonic evolution during the Tertiary was characterized by the continued breakup of Pangea. The southern supercontinent of Gondwana was dissected into four large plates (Fig. 7.1). The subduction of Neotethys along its northern border resulted in the convergence and collision of Africa and Europe, including the smaller plates that lie between them. The Teritary geological history of Central Europe was very much influenced by the collision of

M. Meschede and L. N. Warr, *The Geology of Germany*, Regional Geology Reviews,
https://doi.org/10.1007/978-3-319-76102-2_13

the Adriatic-Apulian Plate with the European Plate, resulting in nappe stacking and folding, and finally uplift of the Alpine Mountain Ranges (the Alps, Carpathians, Apennines, Dinarides, Hellenides, Pyrenees and Betic Cordillera) across the European realm in the Late Tertiary.

In contrast, in the northern part of Central Europe, a shallow marginal sea extended from Poland to southern England with deposition of predominantly sandy and clayey sediment. The largest sediment thicknesses accumulated in the North Sea, where structures such as the Viking Graben, formed during the Jurassic were related to the opening of the Atlantic Ocean. These depocenters were progressively infilled with sediment during the Paleocene times onward (Fig. 13.1). The Tertiary sequences in the Viking Graben area reached thicknesses of more than 3000 m and contain numerous large petroleum reservoirs that remain of significant economic importance as Europe´s main oil reserve.

In the Paleogene, formerly known as the Early Tertiary, the global mean temperature was much warmer than today whereby the highest temperatures were reached at the beginning of the Eocene with a yearly average of about 30 °C. This event is known as the **Paleocene/Eocene Thermal Maximum** (PETM), as marked by a global increase in temperature related to the higher concentration of greenhouse gases in the atmosphere about 55 million years ago. The cause of the temperature rise is still unclear, but it is commonly considered to result from the large amounts of methane gas released during intense volcanic eruption. Only at the end of the Tertiary did it become colder, prior to leading into the Quaternary glaciation phases of the ice age (Chap. 16).

In Europe, subtropical forests became widespread and extended up to Scandinavia for several millions of years in the Tertiary. During the Late Eocene, large parts of Scandinavia and the present Baltic Sea were covered by **amber forests**. The trees produced large amounts of **resin** that was particularly resistant to weathering and as a result, is found today across the Baltic region in the form of amber (Fig. 13.2). One of the most common variety is succinite

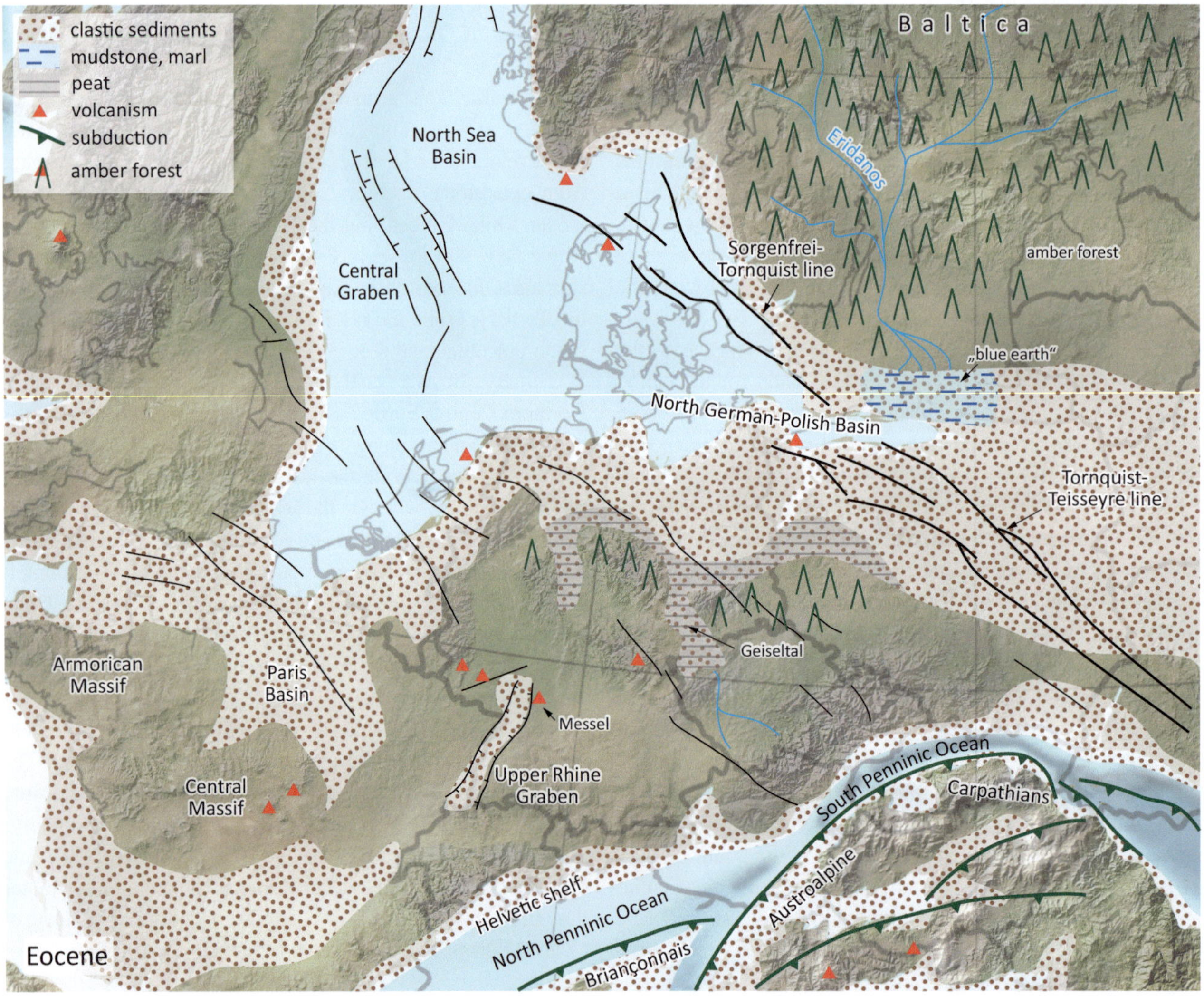

Fig. 13.1 Eocene paleogeographic reconstruction of Central Europe. Topographic base map from Blakey (2011); modified after Ziegler (1990)

formed by the "amber pine" (Langenhein 2003). Amber is strictly speaking not a mineral or rock, but a solid, amorphous (non-crystalline) organic type of resin. The golden yellow resin is well known for its excellent preservation of insects and it is a valued gemstone of the jewelry industry.

The Eocene amber forests extended in a broad belt from Scandinavia to the Urals and covered part of the land lying to the south of the North German Basin (Fig. 13.1). The main deposits of Baltic amber are found today along the southeastern edge of the present Baltic Sea, occurring in a belt extending from Lithuania to Samland (the Kaliningrad region of Russia) and Poland to Gdansk and beyond. This section of the Baltic Coast is also known as the **Amber Coast**. At the time of the deposition of the amber, this coastal section lay on the northern edge of an Eocene sea that extended over much of the present-day Central European lowland. An extensive river system, known as Eridanos

Fig. 13.2 Amber recovered from Quaternary sediments, Greifswald (with the kind permission of the Staatliche Landessammlung Mecklenburg-Vorpommern, Greifswald, Germany)

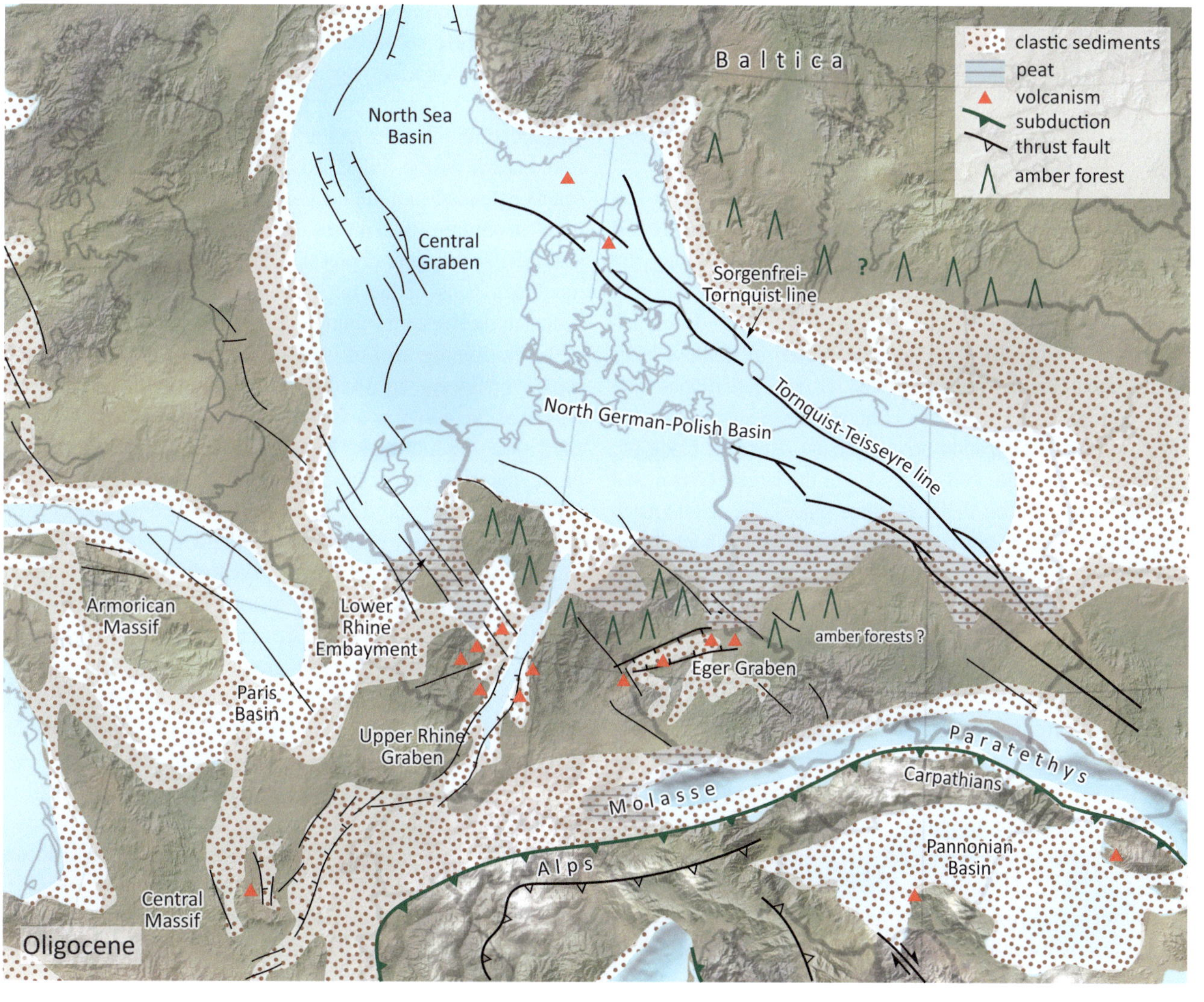

Fig. 13.3 Oligocene paleogeographic reconstruction of Central Europe. Topographic base map from Blakey (2011); modified after Ziegler (1990)

(in the Greek mythology, a large river at the end of the world; Fig. 13.1), which flowed into the Eocene sea from the north, transported the ambers to areas of deposition that were not far from their places of origin.

In the Late Eocene, about 35 million years ago, the so-called Blue Earth deposit originated from the deposition of clayey mica-rich fine sands (Fig. 13.1). Abundant quantities of amber accumulated in these deposits (Schlüter 1990). Some researchers consider that most of the amber found in the Baltic region originated from the north of the depositional area in the Eocene amber forests, and was later redeposited in numerous locations across the Baltic region. In the area of Bitterfeld and in northern Saxony, separate amber deposits are considered to represent the spread of Late Eocene and Early Miocene amber forests (Fuhrmann 2005). During the Quaternary glaciations, extensive re-deposition of amber-containing sediment occurred in melt-water rivers that transported ambers to the southern Baltic and up into the North Sea. Due to their low density (1.05–1.15 g/cm^3), ambers are particularly susceptible to mobilization during storms when they are torn up from the sea bottom by strong water movements and washed up onto the shoreline.

13.1 Tertiary Brown Coal Deposits

Tertiary brown coal seams are widespread across Germany and formed at different times in coastal environments along the edge of the Tertiary sea that flooded large areas of northern Germany and Poland (Figs. 13.1 and 13.3). They developed from thick deposits of peat that accumulated in shallow lakes and lagoonal swamp forests as well as emerged moors. The peat deposits were covered by clastic sediments that excluded oxygen as the coalification process began (Box 14: Coal, oil, natural gas). These deposits form one of the largest brown coal deposits developed on Earth. In some places, the seams reach thicknesses of more than one hundred meters, where subsidence in the surroundings of salt domes often played a role in influencing seam thickness.

The brown coal of Germany is mined largely in three areas: (1) the **Lusatian mining area** south of Cottbus, which continues eastward to the vicinity of Łódź in Poland, (2) the **Central German mining area** near Halle and Leipzig, including also the Helmstedt mining area, and (3) in the **Rhenish mining area** in the Bay of Cologne. Other smaller deposits, which are no longer mined today, exist in the Hessian Depression south of Kassel, in the Eger Graben and in Eastern Bavaria near Schwandorf. With ca. 40 billion tons of commercially exploitable brown coal deposit reserves, Germany possesses more than 10% of the remaining global brown coal resources. This corresponds to an energy equivalent of all known North Sea oil deposits. However, the mining of brown coal has a large impact on the environment of these districts, with huge areas of open pit mining leading to the repeated relocation of entire villages and, as a result, large conflicts with society. The village of Garzweiler, which lent its name to one of the largest open-pit mines in the Rhenish mining area, was relocated in the 1980s and is now used by the opponents of open-pit mining as a synonym for the negative effects of open-pit mining.

The brown coals of the Rhenish mining area formed during the Miocene in the catchment area of the **Lower Rhine Bay**, which originated as an Oligocene graben structure and since then has steadily subsided by up to a kilometer. Although the degree of tectonic subsidence increased during this time, no Pliocene brown coal was formed. The Tertiary sedimentary basin was split apart into many separate blocks, each with different subsidence histories. These are known as the Rur Block, Erft Block, Cologne Block, Venlo Block, and Krefeld Block (Fig. 8.6). The Erft Block suffered most subsidence, with brown coal seams occurring on its northeastern flank where they are covered by up to 600 m of younger sediment. The brown coal-bearing strata in the Lower Rhine Bay are subdivided into three sections: (i) a lower seam group of Late Oligocene to Miocene age, containing the sheeted coals (*Blätterkohle*) of Rott, near Hennef, with rather unique Late Oligocene fossil plants and animals, (ii) the main seam group, also known as the Ville beds of Early Miocene age. These seams extended into the central part of the Lower Rhine Bay, with maximum thicknesses of up to 70 m, and (iii) the youngest brown coal deposits of Late Miocene age containing the Inden beds of the upper seam group. These are mainly found in the western mining area of the Rur Block.

In the mining area of central Germany, the brown coal is mostly of Eocene age, with isolated occurrences of Paleocene coal that are too small for mining (Fig. 13.4). The Tertiary shelf sea of northern Europe repeatedly advanced into embayments to the south, such as the area around Leipzig (Figs. 13.1, 13.3 and 13.5). During the Eocene, one of these bays stretched from Helmstedt via Magdeburg to the south of Leipzig in what is known as the **Weißelster Basin**. The sediments were initially deltaic or estuarine with limnic-fluviatile deposition but by the Middle Eocene, the formation of brown coal predominated in swampy areas. The area of the ***Geiseltal*** (Geisel Valley), southwest of Halle (Saale) was the site of an extensive moor that formed in proximity to the coast and today it is known as a world-famous fossil site. The brown coal seams at this location reach a total thickness of up to 120 m. Among the fossils, which typically occur in the interlayers between the brown coal, are well preserved ancient horses, turtles, amphibians and many other vertebrates found in addition to various insects and plants. Even the soft parts, such as hair feathers and even chlorophyll in the case of plants, are partly preserved. Overall, the Geiseltal sequence covers a time of about 6 Myr (50–44 Ma). Only the lowermost part, the so-called sub-coal of the Geiseltal seam corresponds to the

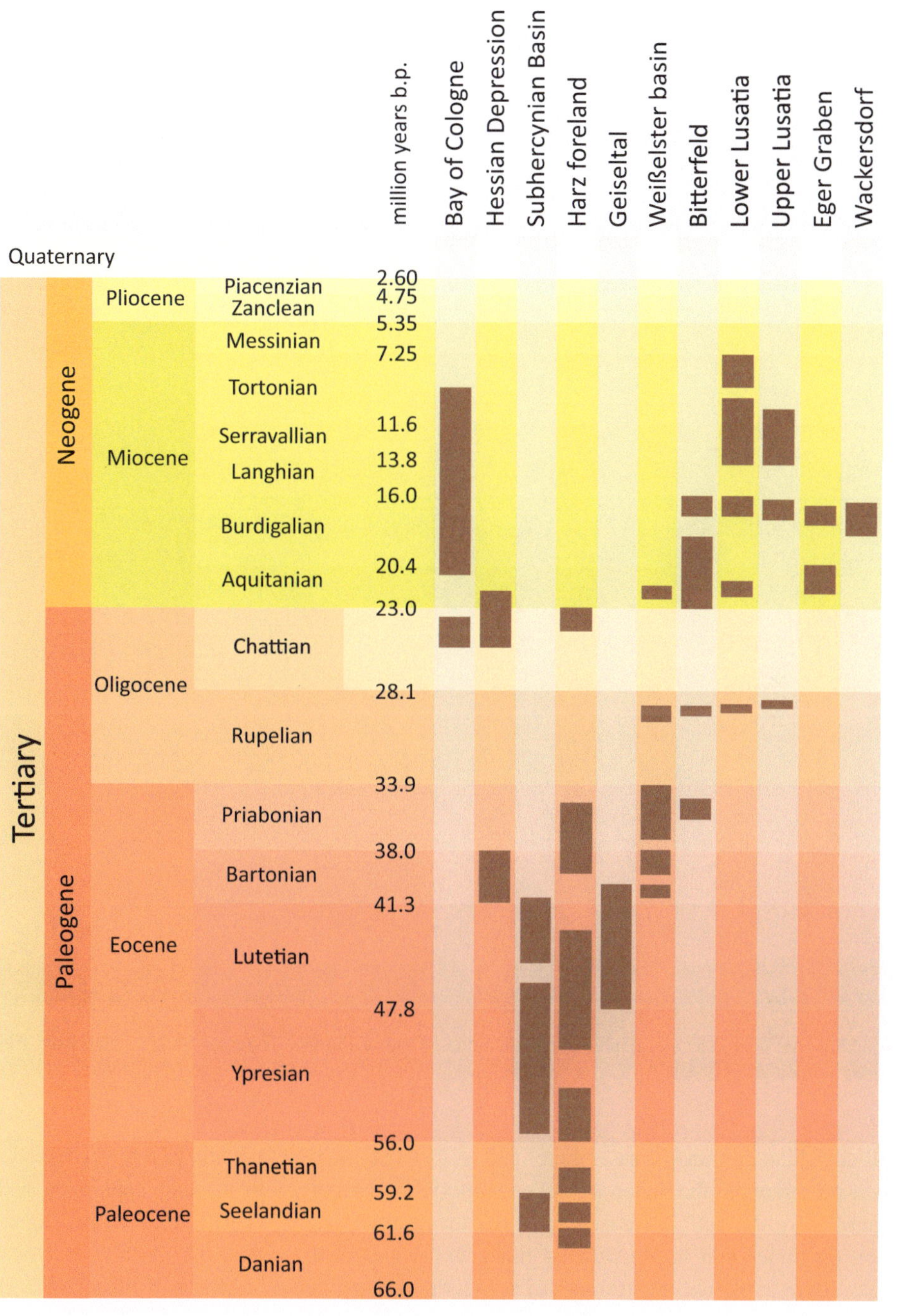

Fig. 13.4 The geological age of brown coals in Germany Simplified after Henningsen and Katzung (2006), Deutsche Stratigraphische Kommission (2016)

Messel Formation (Chap. 14.3), which was deposited over a much shorter time.

A regression occurred at the end of the Eocene as marked by a hiatus in many places. The overlying Oligocene sands and silts, which gradually merge into the more than 100 m thick Rupelian clays (*Rupelton*), contain fossils such as shark teeth, foraminifera and trace fossils that reflect the marine nature of these sediments. During the Oligocene, brown coals originated as paralic deposits mainly in peripheral marine regions as the wetlands shifted slightly to the north of the Bitterfeld area. These brown coal layers, collectively known as the upper seam complex, are mostly overlain by marine sands. Smaller deposits of Oligocene brown coal were also formed south and east of the Kyffhäuser Mountains.

Tertiary sedimentation ended during the Early Miocene in the Central German mining area, while in the easterly adjacent Lusatian mining region, sedimentation started only in the Late Oligocene with brown coals formed mainly in the Miocene (Figs. 13.4 and 13.5). In both the Central German

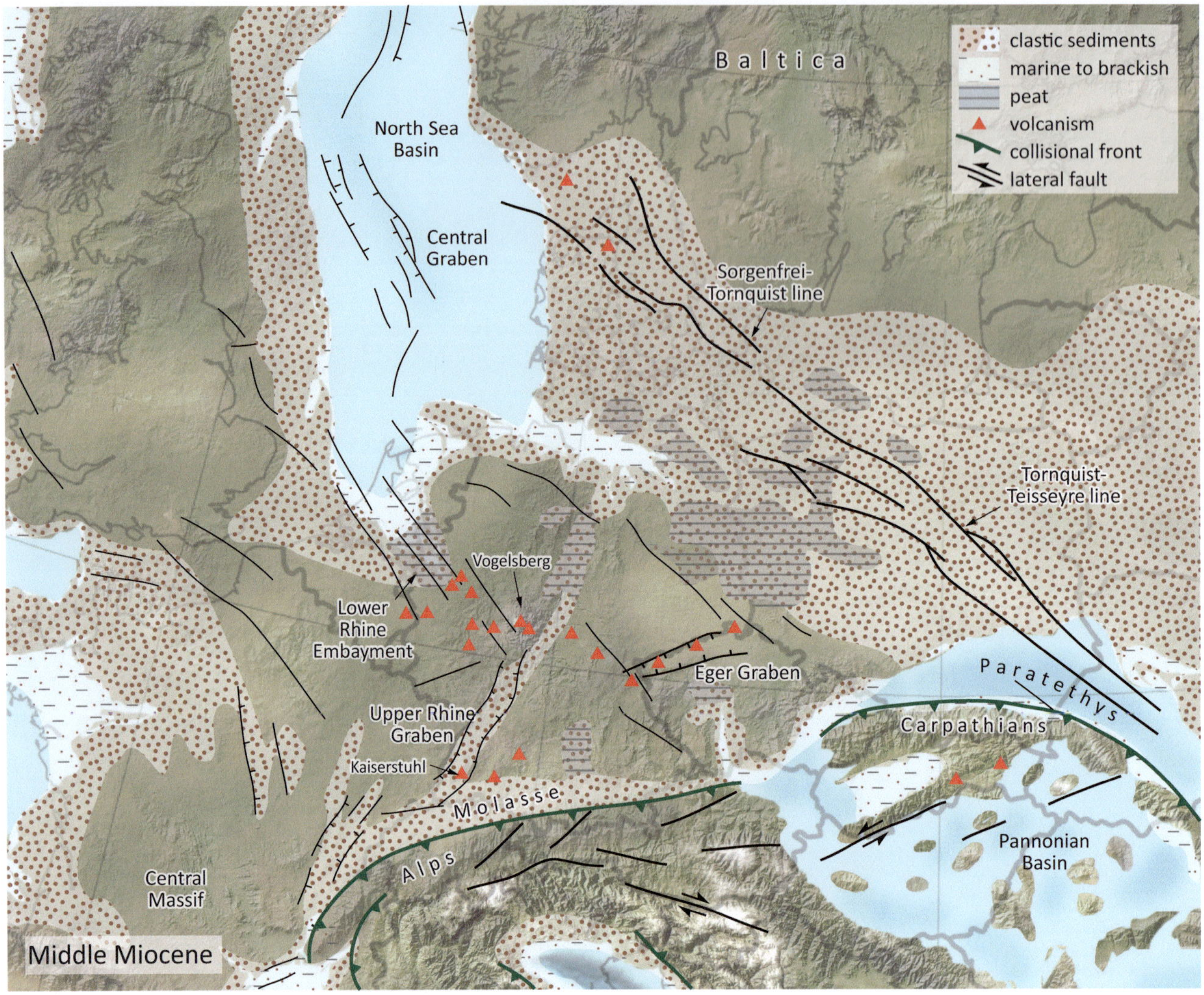

Fig. 13.5 Miocene paleogeographic reconstruction of Central Europe. Topographic base map from Blakey (2011); modified after Vinken et al. (1988) and Ziegler (1990)

and Lusatian mining areas, there are still large active open-pit mines. In the region of Halle—Bitterfeld—Leipzig, a new "lake district" has arisen from the many abandoned open pit mines that have gradually filled with water. What looks like simple and environmentally friendly water-filled open pits, represents in some cases a major threat. The often very deep mining pits have steep slopes and as the groundwater rises, the soil becomes saturated as it gradually fills with water. As this happens, the loose sediment may lose its coherent strength and causes slope failure. This occurred in a major incident in 2009, in the village of Nachterstedt, when a 350 m wide strip containing a road and one and a half houses slipped down the slope.

A special type of Tertiary coal occurs near Kassel, in the area of the **Hoher Meissner**. Here, 11.5 million years ago a basaltic lava flow covered the brown coal that had accumulated in a small depression and heated it strongly in the process (Fig. 13.6). This heating enhanced the coalification so strongly that the brown coal was from transformed in places to anthracite. Containing a higher concentration of carbon, hard coals are energetically of much higher quality than brown coal and therefore even underground mining was used at this site to recover this valuable commodity. Mining began more than 400 years ago at the Hoher Meissner and lasted until 1929 with underground extraction and until 1970 as an open pit mine.

Further to the south, the Molasse Sea increased its extent during the Tertiary. In a bay at the northern edge of this sea, known as in the **Bodenwöhr Bay**, Miocene brown coal deposits formed. These were mined in the open pit of the Wackersdorf area for nearly 100 years until 1982.

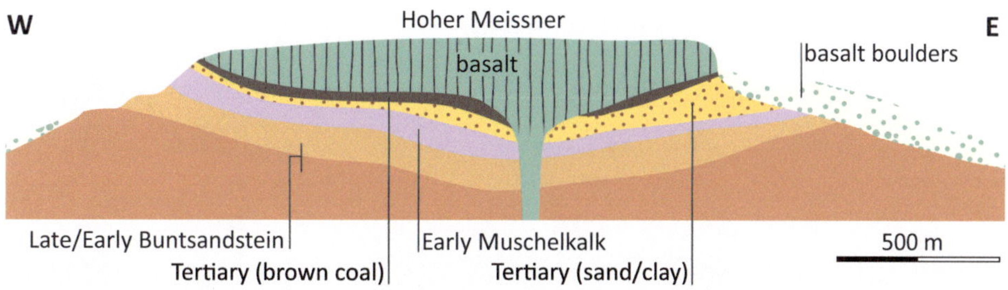

Fig. 13.6 Cross-section through the northern part of the basalt flow at the Hoher Meissner. Modified after Finkenwirth (1978)

13.2 The Upper Rhine Graben

The Upper Rhine Graben of southwest Germany is one of the world's most distinctive continental rift systems. It has long been a subject appearing in geoscience examinations and, together with the much larger East African graben system, is taken as a **type locality** for the geological concept of **graben** formation. In his classic work *The Face of the Earth* (1883), Eduard Suess (1831–1914) defined a graben as an elongated extension zone bounded by inwardly dipping faults or fault zones. In its plate tectonic context, rifting and formation marks the beginning of a supercontinent cycle (Chap. 5). If the lithosphere is stretched by extensional forces or is uplifted above a rising mantle diapir, hot and ductile asthenospheric material ascends closer to the surface than normally possible beneath thick continental crust (Box 21: Isostatic adjustment and mountain uplift; Fig. 13.7). As a result, a bulge forms above the thinned zone of the lithospheric mantle. However, since the upper portions of the continental crust respond in a brittle fashion and fractures during the bending that accompanies stretching, the graben structure sinks in the apex region of the bulge (Fig. 13.8).

The Upper Rhine Graben is an NNE–SSW striking rift structure with a length of about 300 km and an average width of 40 km (Fig. 13.9). In the south, it is bound by the folded rocks of the Jura Mountains which overprinted the southern edge of the graben during the Late Alpine deformation. The graben system merges to the west into a diffuse lateral fault system (Rhine-Bresse Transform Zone; Laubscher 1970). This connects the Upper Rhine Graben with the Bresse Graben in the south, as part of a tectonic continuation of the ECRI system (European Cenozoic Rift System; Chap. 8). This diffuse fault system is considered today to represent an inactive transform fault system (Frisch and Meschede 2013). In the north, the Upper Rhine Graben borders the WSW–ENE extending southern boundary fault of the Rhenish Massif. In its continuation northward, it follows a number of smaller depressions (Wetterau depression, Horloff Graben) and the Miocene volcanic area of the Vogelsberg (Sect. 14.2).

The pre-rift stage of the Upper Rhine Graben coincided with the end of the Penninic Ocean subduction and collision of the Adriatic-Apulian Plate with Eurasia that led to the formation of the Alps (Sect. 12.6). The forces that acted on the European mainland continental crust reactivated old fault systems, including the Early Carboniferous Upper Rhine shear zone system (Illies 1965; Edel et al. 2007). Around the Paleocene/Eocene boundary, a dome-like structure began to develop along the suture, marking the course of the later graben system. This up-doming was somewhat stronger in the south than in the north. Seismic and gravity data in the

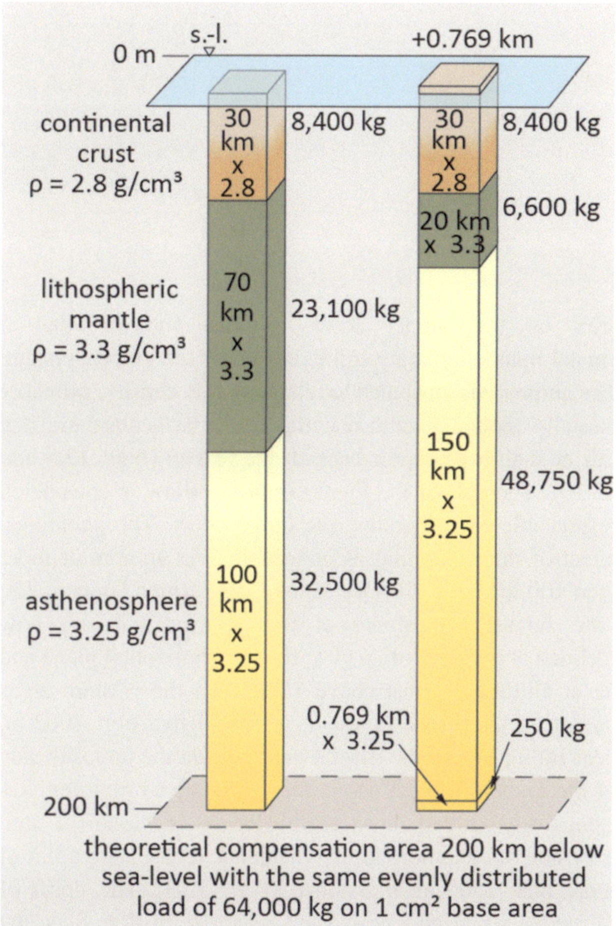

Fig. 13.7 Isostatic adjustment of an ascending mantle diapir composed of hot asthenospheric material. A theoretical compensation plane in the asthenosphere is assumed (here at 200 km depth) at which the pressure per unit area (here in kg/cm^2) is universally the same. The thinning of the lithospheric mantle (here from 70 to 20 km) leads to uplift since the asthenosphere with $\rho = 3.25$ g/cm^3 is slightly lighter than the lithospheric mantle with $\rho = 3.3$ g/cm^3. The resulting deficiency in mass is replaced from below by the inflowing asthenosphere

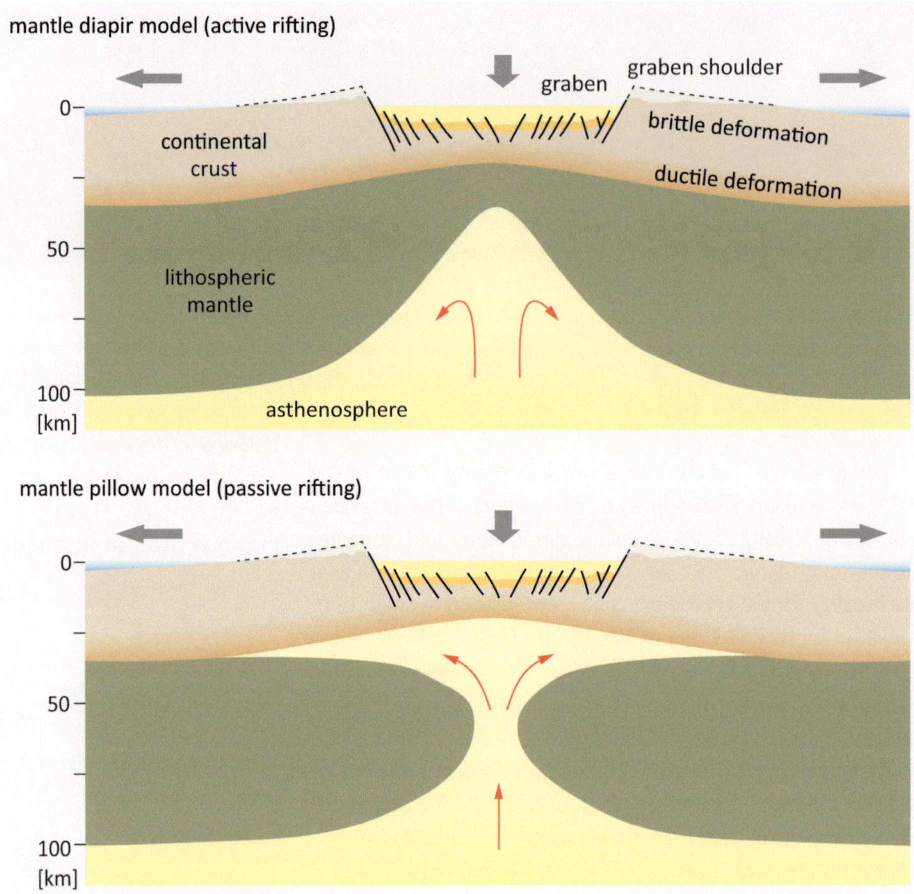

Fig. 13.8 Graben structures form either by active rifting related to an ascending mantle diapir or passive rifting caused by extension and drift of the continental plate along a fracture zone. A bulge caused by isostatic adjustment evolves with a graben structure developed in its apex region (Fig. 13.7)

area of the Upper Rhine Graben indicate that a crustal-mantle anomaly still exists under the graben system. The anomaly is attributed to relatively low density, probably partially molten mantle material of the asthenosphere that lies at shallower depths beneath the graben (Figs. 13.8 and 13.10). The rise of the lighter **mantle pillow** is considered responsible for generating the bulge effect. The geological effect of the up-doming is noticeable over an area of more than 100 km on both sides of the Upper Rhine Graben. The Late Jurassic limestones of the Swabian Alb are now inclined at an angle of up to 4° from the horizontal plane and lie at altitudes of just above 1000 m in the western areas bordering the Black Forest (e.g., the Plettenberg, 1002 m, near Balingen). In the Black Forest and on the opposite side of the graben in the Vosges, the Mesozoic overburden was extensively eroded down to the Paleozoic basement with exposure of metamorphic and magmatic rocks. Here uplift of more than 2500 m is assumed (Fig. 13.13). The uplift of Mesozoic strata and subsequent selective erosion since the Oligocene led to the formation of the cuesta landscape characterizing the Mesozoic rocks of south Germany (Box 7: Cuesta landscape). Moreover, it resulted in the strong karstification of calcareous deposits (Box 23: Karstification).

The fact that the Upper Rhine Graben is restricted in its southerly and northerly extent is related to the rheological properties of the subsurface and its tectonic location (Greek *rhei*: flow; rheology deals with the deformation and flow behavior of matter). To the south, the graben is limited by the northern edge of the Alps where the stresses that built-up during the collision between the Eurasian and Adriatic-Apulian plates resulted in an extensional regime lying perpendicular to the direction of compression. This led to the rifting of an NNE–SSW striking subsiding graben system on the European foreland.

The basement lying beneath the Upper Rhine Graben system consists of crystalline rocks dominated by metamorphic and felsic igneous units. In contrast, the equivalent crustal level rocks of the Rheinisches Schiefergebirge to the north are composed mainly of softer shale and graywacke lithologies, with igneous and metamorphic rocks occurring only at greater depths. During the compression of the Alpine Orogeny, the Upper Rhine Graben responded by

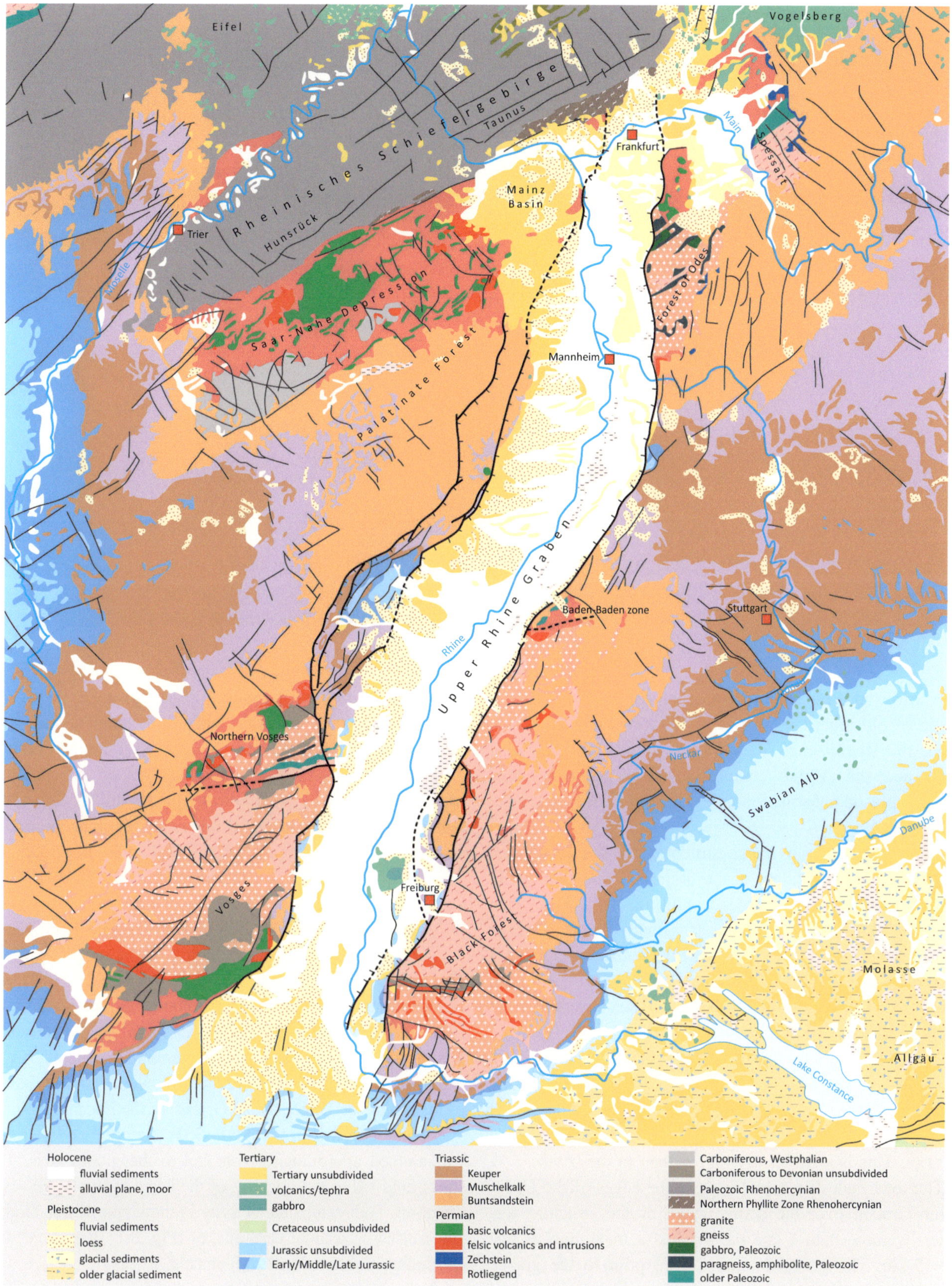

Fig. 13.9 Geological overview map of the Upper Rhine Graben in southwest Germany. Simplified after the 1: 1,000,000 geological map of the Federal Republic of Germany, BGR (1993)

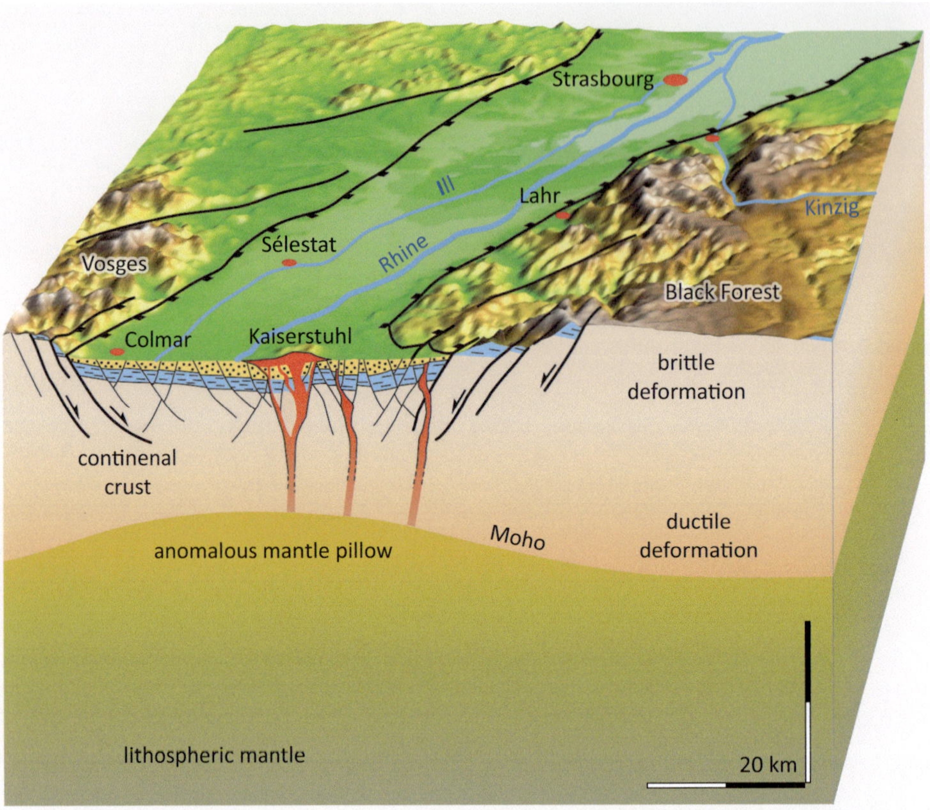

Fig. 13.10 Block diagram of the Upper Rhine Graben near the Kaiserstuhl. The upper part of the continental crust is characterized by normal faulting, whereas the deeper crust is dominated by ductile deformation. After Frisch and Meschede (2013)

E–W directed extensional brittle deformation and reactivation of old inherited basement structures. Further to the north, the Rheinisches Schiefergebirge was uplifted as a block and the Alpine compression was accommodated more by plastic deformation that extended to deeper crustal areas.

In the south of the graben structure, the total fault displacement along the edges of the basin is about 5000 m. In the northern part, it also exceeds 4000 m. The offset along the eastern side of the graben is actually up to 1000 m greater than on the western side where the graben underwent less subsidence. The bulge along the shoulders of the graben is highest in the southern part of the Black Forest and the Vosges, with uplift of these areas by more than 2500 m. In contrast, the northern shoulders were uplifted by no more than 1000 m. There are also differences in subsidence along the axis of the graben structure. The largest amount of subsidence occurred in the north, where it reached more than 3000 m. This is significantly higher than in the south.

The total fault displacement along the edges of the Upper Rhine Graben over the entire length of the structure was generally more than 4 km and reached a maximum of just over 5 km. The peneplain that existed prior to the graben subsidence, but is now long eroded, would have stretched across the Vosges and Black Forest region at a level of about 2500 m above sea-level (Fig. 13.13). The highest peaks today in the Black Forest lie at 1493 m (Feldberg) and in the Vosges at 1424 m (Grand Ballon). Therefore, erosion has since removed at least 1000 m and locally even more than 2000 m of overburden since the beginning of graben subsidence: a process referred to as **denudation**. The upward doming of this region, which reached several kilometers in some areas, was accompanied by extension of the graben in the horizontal direction. During rifting of the upper crust, the graben flanks drifted apart by about 4–7 km. The extensional displacement is today reduced due to subsequent transtensional lateral movements that led to a shortening of the graben's width. This lateral motion resulted from the change in the Central European stress field that rotated counterclockwise after the Late Tertiary (about 6–7 million years ago). Today, the main compressive stress direction is oriented NW–SE, with extension in the NE–SW direction. However, at the time of formation of the Upper Rhine Graben in the Eocene, the main stress direction was in the NNE–SSW direction and therefore was aligned approximately parallel to the graben axis. This stress field led to the opening of the basin in the WNW–ESE direction. Today, tectonic activity in the Upper Rhine Graben still causes occasional earthquakes (Fig. 8.1) associated with continuing transtensional movements.

At the beginning of the Cenozoic, Mesozoic sedimentary deposits covered most, if not all, of the Upper Rhine Graben and its shoulders. The age of rocks exposed at the surface at

Box 23: Karstification

In principle, all areas composed of rock vulnerable to solution weathering may undergo karstification. A major factor controlling this process is the amount of rainfall. The term **karst** incorporates the features produced by **solution weathering** common to limestone, anhydrite, and gypsum, but such features may also develop in silicate rocks weathered under certain climatic conditions (Box 10: Silicate weathering). Karst structures, named after the Dinaric landscape *Kras* in northern Italy and Slovenia, typically form at various altitudes ranging from coastlines to high mountainous regions. Significant water-rock interaction is a necessity for such structures to form and they are most commonly developed in massive limestone formations.

The solubility of calcite ($CaCO_3$) is not particularly high, with only 14 mg of $CaCO_3$ dissolving in one liter of pure water. However, this value increases many times when CO_2 is dissolved in rainwater. In the presence of dissolved CO_2, up to 500 mg of $CaCO_3$ can be solved. This is because CO_2 forms **carbonic acid** when dissolved in water, which reacts with $CaCO_3$ to Ca bicarbonate:

$$CaCO_3 + H_2O + CO_2 \rightleftarrows Ca(HCO_3)_2$$

The amount of dissolved calcite therefore strongly depends on the amount of dissolved CO_2. However, when water is heated it quickly loses its CO_2 so that the opposite effect occurs and $CaCO_3$ precipitates. As the reactions of dissolution and precipitation are reversible, the activity of such solution processes often leads to huge cavities combined with mineral precipitations and the formation of thick **travertine** layers in karst areas. Whether the dissolution or precipitation of calcite occurs depends on three main factors: **temperature, pressure and the concentration of dissolved CO_2**. As warmer water holds less CO_2, the solubility of $CaCO_3$ decreases with increasing temperature. In contrast, the solubility of $CaCO_3$ increases with pressure. Therefore, as rainwater penetrates rock crevices and percolates downward, it will dissolve more and more $CaCO_3$ as the water pressure increases (corresponding to the height of the water column above). If the local pressure drops as the water enters an underground cavity, calcite will precipitate to form **dripstones** that hang from the cave ceiling (**stalactites**; Figs. 13.11 and 13.12). As the falling drops still contain dissolved Ca and CO_2 that escapes upon striking the floor by the mechanical effect of the impact, calcite also

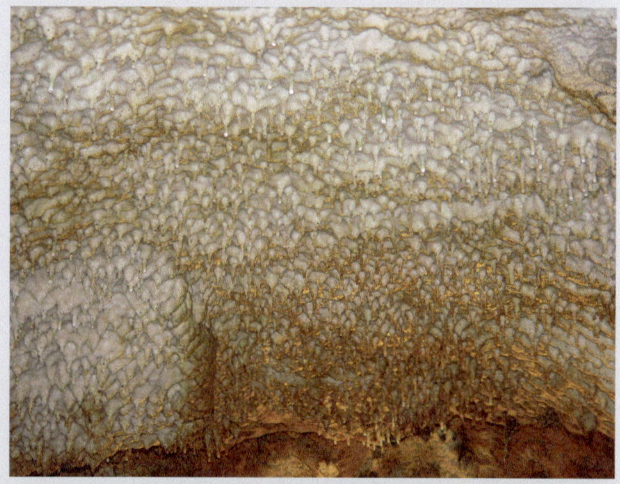

Fig. 13.11 Cave ceiling with stalactites in the Bärenhöhle (German: bear's cave) near Erpfingen in the Swabian Alb. [N 48°22′14″/E 09°12′ 54″]

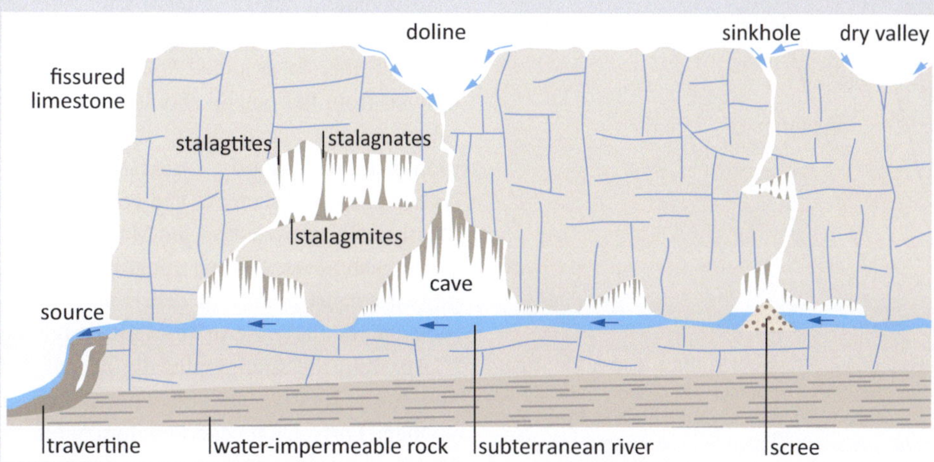

Fig. 13.12 Illustration of possible karstification features developed in fissured limestone

precipitates on the cave floor as dripstones (**stalagmites**). If a stalactite and a stalagmite grow together, they form a pillar known as a **stalagnate** (Fig. 13.12).

If karst water charged with dissolved $CaCO_3$ flows out at the surface, it often leads to the precipitation of **travertine terraces**. This process occurs especially in summer as the water warms up when leaving the cave system. This may also happen when the water cascades down a waterfall, resulting in loss of dissolved CO_2 that initiates precipitation of calcite. Impressive examples of travertine precipitation in waterfalls can be found in the Swabian Alb, for instance at the Urach waterfall, near Bad Urach, where a stream falls nearly 40 m. Another example is seen in the Echaz Valley, near Reutlingen, directly below the Honauer Steige (German for a steeply climbing road near Honau) where a cave was formed in the travertine limestone at the steep scarp (Olga cave, N 48°24′49″/E 09°15′39″). The Olga cave is a rare case of a **primary cavity** formed simultaneously with the surrounding rock. The cavity does not represent a karst cave developed by dissolution of the rock but it is overgrown with a **travertine apron**.

Secondary karst caves result from the continuous dissolution of limestone in underground water systems. As rainwater with its dissolved CO_2 migrates along fractures and bedding planes through carbonate rock it forms a complex branching cave system which, in massive limestone formations, can develop huge voids reaching several tens of meters height. An example is the 30 m high Mörikedom in the cave system of the *Blautopf* (blue pot), near Blaubeuren. In the Swabian Alb alone around 2500 caves are known, some of which are open to the public for visiting. When the roof of a cave becomes fragile and collapses, sinkholes known as **dolines** may form at the surface (Fig. 13.12). These often occur suddenly without warning and can cause serious damage in the karst areas.

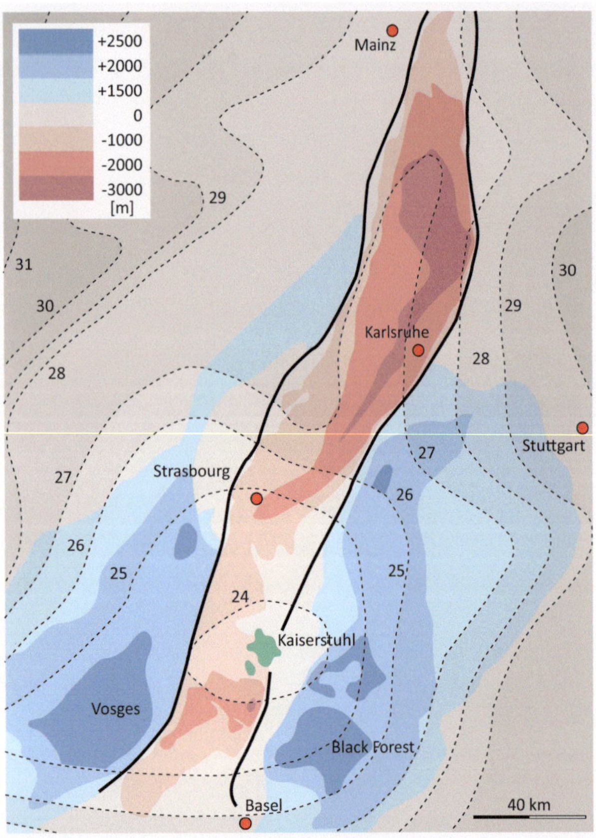

Fig. 13.13 Distribution of uplift and subsidence across the Upper Rhine Graben including the graben shoulders. Red zones indicate the subsided areas, blue zones the uplifted areas of the erosional peneplain (in meters above present sea-level). Dotted lines indicate the depth to the base of the crust (Moho; in km below sea-level; after Frisch and Meschede 2013)

that time was older in the south than the north. The Mesozoic strata are hidden deep within the graben beneath the Tertiary sediments and have been encountered in numerous drill holes. In the south, the Mesozoic sequence ends with Late Jurassic limestones whereas, in the north, Permian sedimentary rocks are found. Within the areas of up-doming, erosion began along the graben flanks simultaneous with subsidence and sedimentation in the graben. Today, remnants of the Mesozoic overburden on top of the graben shoulders are preserved in the middle part of the rift structure, along its western edge in the northern section, and in step faults along the graben edge (Fig. 13.9). In the area of the Black Forest, the Vosges and the Odenwald, Tertiary erosion exposed metamorphic and igneous rocks of the Paleozoic basement.

Altogether, about 20,000 km^3 of Tertiary sediment accumulated in the Upper Rhine Graben (Roll 1979) and most of this is assumed to represent erosional products derived from the graben shoulders. The edges of the graben often contain coarse clastic sediment in the form of conglomerates and poorly sorted sandstones, which suggests short transport distances. The composition of the sediments mostly reflects erosion of the Mesozoic overburden from the graben shoulders. Eocene conglomerates in the south of the graben consist mainly of debris derived from Late Jurassic rocks, particularly the more extensively eroded upper layers. Clasts from the stratigraphically deeper beds of the Middle and Early Jurassic are, in contrast, less abundant. In the younger Early Oligocene conglomerates, this changed with Middle Jurassic clasts prevailing due to the deeper erosion of the source area. The size and shape of the clasts also indicate a steep and strong relief existed at the graben edges, with the rapid deposition of torrent-like fluviatile deposits related to the strong vertical (>1000 m) relief. Finer sediments were deposited towards the center of the graben with sand, silt and mud, as well as carbonates and marls, together with the local occurrence of evaporate deposits (salts).

Subsidence in the south of the Upper Rhine Graben began during the Early Eocene, whereas in the north it was initiated somewhat later in the Late Eocene. At first, small local depressions spread across the graben area, which linked up to the Hessian Depression in the north (Fig. 13.14). Multiple marine ingressions occurred in the Middle Eocene with the deposition of widespread brackish water sediments that covered most of the Upper Rhine Graben. As these sediments contain no nummulites, which would indicate a Mediterranean origin, it is assumed that the ingressions were Atlantic waters that reached the graben via the Paris Basin. A lake developed in the Late Eocene under warm subtropical to tropical conditions and spread across the entire area of the Upper Rhine Graben. Following this stage, the sea advanced several times from both the north and the south into the elongated depression of the graben. During these advances, the basin was temporarily connected to open waters in the north via the Hessian Depression and the Wetterau. To the south, connections existed through the Molasse Sea that lay in front of the rising Alps (Figs. 13.1, 13.3 and 13.5). The flooding of the Rhine Graben led to the deposition of clays and marls, some of which formed suitable petroleum source rocks due to their high organic content (Box 14: Coal, oil, natural gas). As a result of the rapid and continuous subsidence of the sediments that were buried to several kilometers depth in the rift, they were heated to appropriate temperatures for the organic kerogen to transform to oil and natural gas. Hydrocarbon reservoirs have been widely exploited in the Upper Rhine Graben and some fields are operational today.

Another form of energy can be obtained from the hot Variscan granites buried beneath the graben that are blanketed by clastic sedimentary rocks of low thermal conductivity. In a demonstration project of hot-dry-rock geothermal energy, 1.5 MW of electricity was generated from the fractured Soultz-sous-Forêts granite located just across the German border in France. By drilling three boreholes and pumping down warm saline water to depths of about 3.5–5 km, hot water of ca. 175 °C was returned to the surface and used to generate electricity as a renewable energy. The project, however, was discontinued in 2008 due to the fear of inducing earthquake activity. Studies of the mineralization history of the Soultz-sous-Forêts granite have revealed at least four hydrothermal fluid-flow events affected the granite since its intrusion (Schleicher et al. 2006; Bartier et al. 2008). These European-wide events crystallized authigenic illite (K-mica) and K-feldspar in the pores of rocks throughout the Mesozoic and Tertiary and were probably linked to extensional tectonics during the opening of the Atlantic and Upper Rhine Graben rifting.

The **main stage of rifting** in the Upper Rhine Graben commenced in the Early Oligocene during the continuing northward movement of the Alps (Dezes et al. 2004). At this time, the thickest sedimentary deposit of the Tertiary period was formed with more than 2000 m of Oligocene strata. Overall, the subsidence of the Upper Rhine Graben reflects large-scale tectonic movements in Central and Southern Europe. These were associated with the Alpine Orogeny and the beginning of the Mediterranean drift of the Corsica-Sardinia arc (Figs. 13.3 and 13.5) away from the European continent.

During the Oligocene and Miocene, the sea advanced no less than five times into the Upper Rhine Graben again both from the north and the south (Martini 1990). Connections to the Nordic Sea existed through the Hessian Depression, the area of the not yet uplifted Rheinisches Schiefergebirge and the Lower Rhine Bay. Since the sea repeatedly dried out during the intervening periods of regression, **evaporate deposits** formed typical saline sequences of Eocene and Early Oligocene age. The last beds of the sequence are usually composed of 1–3 m thick potash salt layers. Today, these are buried at depths between 400 and 1100 m where they were mined underground until 2002, for example, near Mulhouse in the region of Alsace. In some places, the salt deposits were so thick that they rose as diapirs due to their low density and high mobility. Salts have a lower density (2.2 g/cm^3) than other sedimentary rocks (2.4–2.5 g/cm^3) and due to their crystal and dislocation structure can move by ductile flow when subjected to sufficient differential stress. When covered by several hundred meters of thick denser rock, they initially form salt pillows and later develop into salt diapirs (see Sect. 11.2.2; Fig. 11.32).

A significant change in geodynamic development that affected the sedimentary sequence of the Upper Rhine Graben occurred in the Early Miocene about 18 million years ago (Grimm et al. 2011). The southernmost part of the Upper Rhine Graben was included in Alpine thrusting and was consequently uplifted and partly eroded during formation of the so-called Rhine dome. The formation of the dome formed a watershed in the area of the Kaiserstuhl as sedimentation ceased in the southern half of the Upper Rhine Graben. Up to this time, this area had shared a common Tertiary sedimentary history with the northern Alpine Molasse Basin. Despite the up-doming in the south of the rift system, deposition continued in the northern part of the graben with initially brackish and lacustrine sediments, followed later by fluvial and continental strata that continued until the Middle Miocene.

In the Late Miocene, a precursor of the Upper Rhine Valley formed with local deposition of the Dinotheria sands. The Rhine River, which was at this time not yet sourced

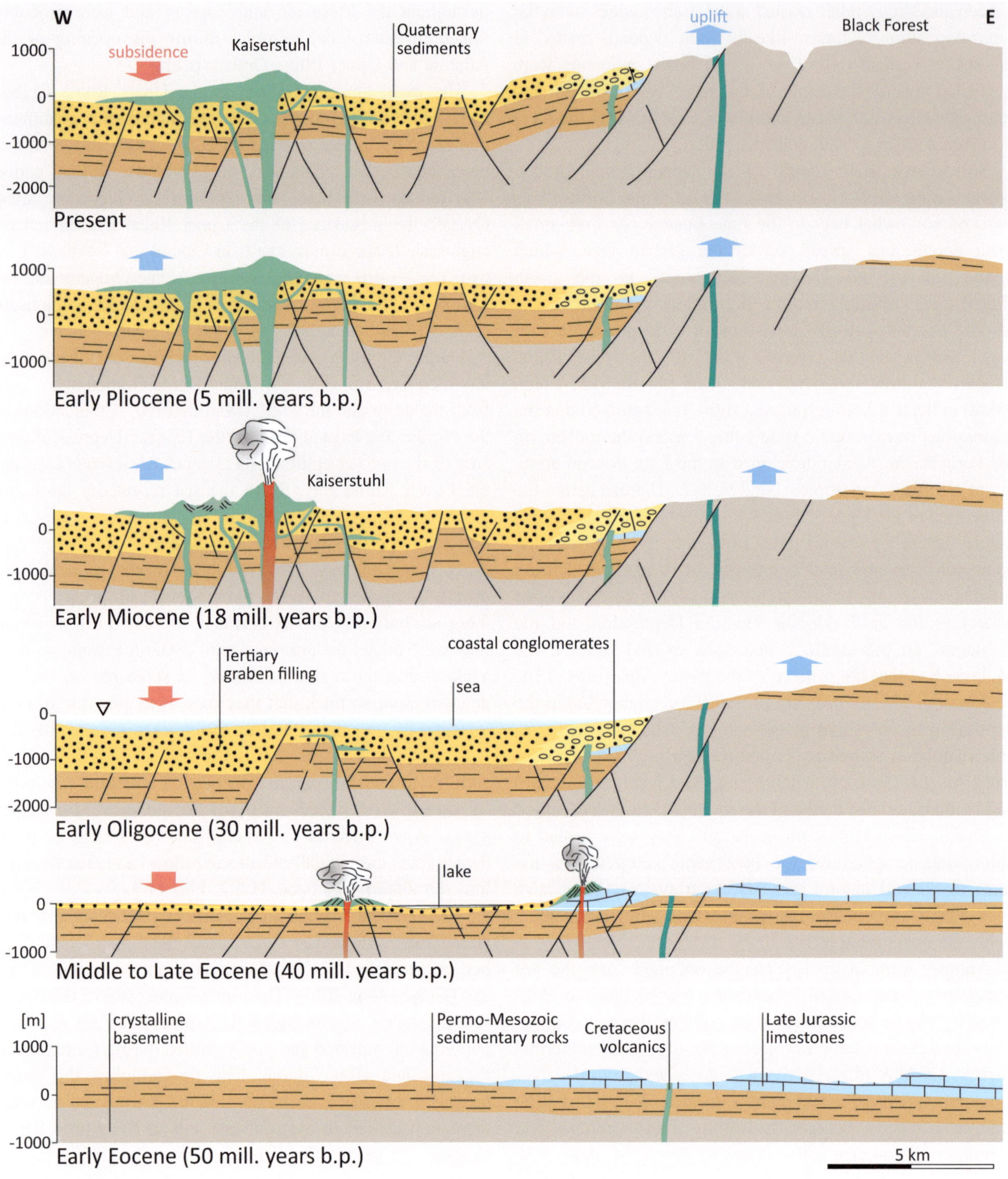

Fig. 13.14 Eocene to present evolution of the southern Upper Rhine Graben in the area of the Kaiserstuhl. Modified after Schreiner (1977)

from the Alps, ran from the Kaiserstuhl to the Mainz Basin (see below). This valley also called the "Kaiserstuhl Rhine", flowed northward into the North Sea. Widespread deposition of fluvial and continental sediments also occurred in the Pliocene. The drainage systems were shifted during uplift of the southerly lying folded Jura Mountains during the Late Pliocene. The connection to the Bresse Graben towards the west was then closed, causing the Aare to flow into the Upper Rhine Graben and, a short time later, the Alpine Rhine became connected by this path.

The youngest sediments of the Upper Rhine Graben are represented by dune sands and air blown loess that were deposited during the glacial periods of the Pleistocene. At no time, however, did glaciers reach the area of the Upper Rhine Graben.

Today, the Upper Rhine Valley is characterized by fluvial sediments deposited by the Rhine River that constantly changed its course by meandering across the flat valley floor. The switching of the path of the river was even noted in historically documented times (Fig. 13.15). Many long meanders were shortened during regulation of the river in the nineteenth century by the engineer Johann Gottfried Tulla (1770–1828) from Baden. The purpose was to make the river continuously navigable—with far-reaching consequences. Although the flood levels dropped by an average of one meter, the increased erosive power of the river that arose due to its higher flow rate was underestimated. Locally, the Rhine deepened its riverbed by 10 m, causing wet floodplains to dry up and the groundwater table to drop. This rendered some areas no longer suitable for agricultural use. In some cases, the remains of the old river branches of the Rhine were preserved along with its original vegetation. An example is the Taubergießen, near Freiburg, which is now a protected area.

Another Tertiary depression was the **Mainz Basin**, located at the northwestern edge of the Upper Rhine Graben. According to the strict definition, it does not represent a basin in itself but formed as a fracture zone at the edge of the graben system (Fig. 13.9). The term basin, however, has long been used in the literature and is retained in its regional context. The Mainz Basin was initiated as part of the Permo-Carboniferous Saar-Nahe Basin (Sect. 11.2.1), the rocks of which were folded at the end of the Variscan Orogeny. Following the deposition of Jurassic platform sediments, the area was subject to erosion from Early Cretaceous to Eocene times. Eocene and younger sediments of a similar nature to those found in the northern part of the Upper Rhine Graben were then deposited. However, the thickness of these sediments was significantly less than in the Rhine Graben and produced a sequence only a few hundred meters thick. Toward the end of the sedimentation, a lagoon facies gave rise to the deposition of limnic brackish and chalky marly sediments with occasional brown peat layers that are now preserved as coals.

13.3 The Northern Alpine Foreland Basin—The Molasse

During the Tertiary, a wide basin spread out in front of the Alps and collected the detritus from the ascending Alpine Mountain Range. The **northern Alpine foreland basin**, known as the **Molasse Basin**, formed as a continuation of the deep-sea trench sedimentation associated with subduction of the Penninic Ocean. In its final phase, the trench deposits of the **Rhenodanubian flysch** and, further to the west, the flysch series of Switzerland formed (Fig. 11.72). The flysch series, whose youngest deposits extend into the Early Eocene, are in a tectonic sense regarded to be part of the Penninic. These deposits accumulated during the closure of the ocean between the European and the Adriatic-Apulian plates (Sect. 12.6).

The subduction of the South Penninic Ocean ended in the Eastern Alps during the Early Eocene. To the west, the North Penninic Ocean closed a little later in the Eocene by subduction of the Eurasian Plate beneath the Adriatic-Apulian Plate and their subsequent collision. In the process, the deep sea trench migrated toward the Helvetic sequence on the European continental margin and became the precursor to the foreland basin of the Alpine orogen. Following initial flysch-like sedimentation, the basin was filled from the Oligocene onwards by large volumes of river-transported gravel, sand and finally silt. This style of **molasse** sedimentation that developed in the Northern Alpine foreland basin is considered worldwide as the type locality for foreland (or foredeep) basin development in front of a collisional orogen. The term generally refers to sediments deposited in a flexural subsiding basin formed in front of the thickening lithospheric load of the ascending mountain range.

The term "molasse" was introduced by the French naturalist Horace-Bénédict de Saussure (1740–1799) but its origin is not entirely clear. As many millstones in the region were made from fine-grained molasse sandstones, it is either based on the Latin word *mola*, meaning millstone or on the French word *molle*, meaning soft. A basin also formed in the hinterland area of the Alpine orogen where thick clastic sedimentary sequences accumulated. The younger Po Basin is mostly of Quaternary age (Fig. 13.16) and represents the southern foreland basin of the Alps. Molasse sedimentary strata of several kilometers thickness have accumulated in this deposition center and today the basin continues to subside at a rate of up to 1 mm/year.

The sediments of the Molasse Basin are composed largely of terrestrial, freshwater and shallow marine deposits. At the beginning of molasse sedimentation, a deep-sea basin developed that was gradually infilled. The clasts deposited in the Molasse Basin were transported by large river systems sourced from the rising mountains and ranged from coarse pebble, sand, silt to clay-sizes. Large deltas developed at the edge of the mountain range with conglomerates and sands predominating. At greater distances towards the basin center, mainly fine sands and silts with lake sediments were deposited. Small carbonate platforms also developed with the deposition of marl and calcareous sediment. Along the northern edge of the North Alpine Molasse, basin talus fans formed that received gravel material from the northerly-located Mesozoic rock series of the Swabian and

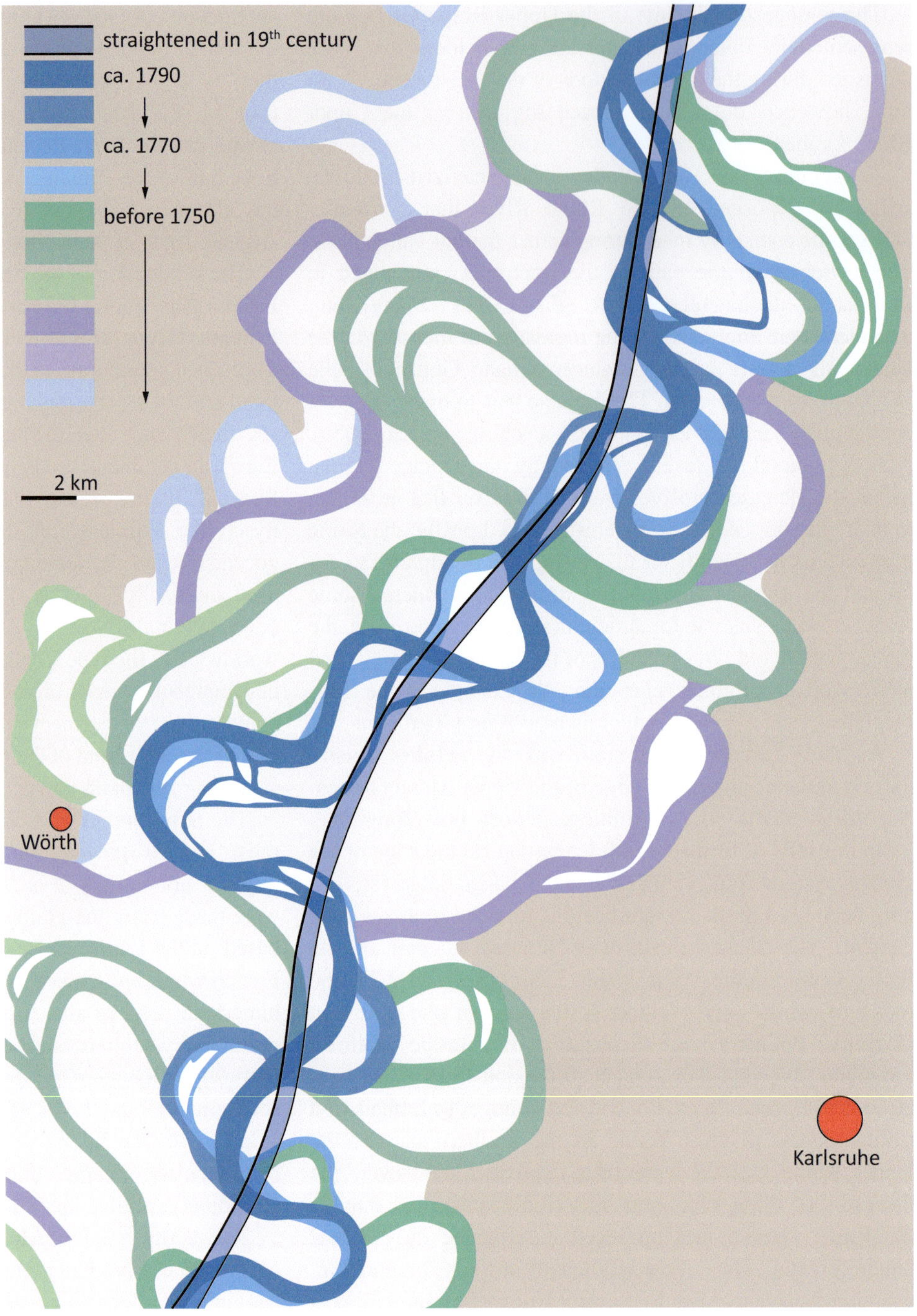

Fig. 13.15 Meanders in the Upper Rhine Valley around Karlsruhe showing old branches of the river that were straightened in the 19th century. Modified after Thürach (1912)

Franconian Alb. Further east, the alluvial fans collected debris derived mainly from the Bohemian crystalline.

Four major stages of development are recognized in the sedimentation history of the North Alpine Molasse related to widespread marine transgressions and regressions. Marine molasse (*Meeresmolasse*) was replaced upwards in the sequence by freshwater molasse (*Süßwassermolasse*) characterized by terrestrial sediment. Lithostratigraphically, the Molasse Group is divided into four formations: (i) the **Lower Marine Molasse** (LMM) of Oligocene age, (ii) the **Lower Freshwater Molasse** (LFM) of the Late Oligocene and Early Miocene, (iii) the **Upper Marine Molasse** (UMM) of Early Miocene age, and (iv) the **Upper Freshwater Molasse** (UFM) of the Middle and Late Miocene (Fig. 13.17; note the possible confusion of the acronym UMM, with different meanings in the German and English language, see Table 13.1). At the end of the Early Miocene, a brackish stage of sedimentation known as the **Brackish Water Molasse** (*Brackwassermolasse*; BM) occurred in the middle part of the molasse stratigraphy and marks the transition from the UMM to the UFM units. In the eastern part of the Molasse Basin, marine conditions prevailed longer than

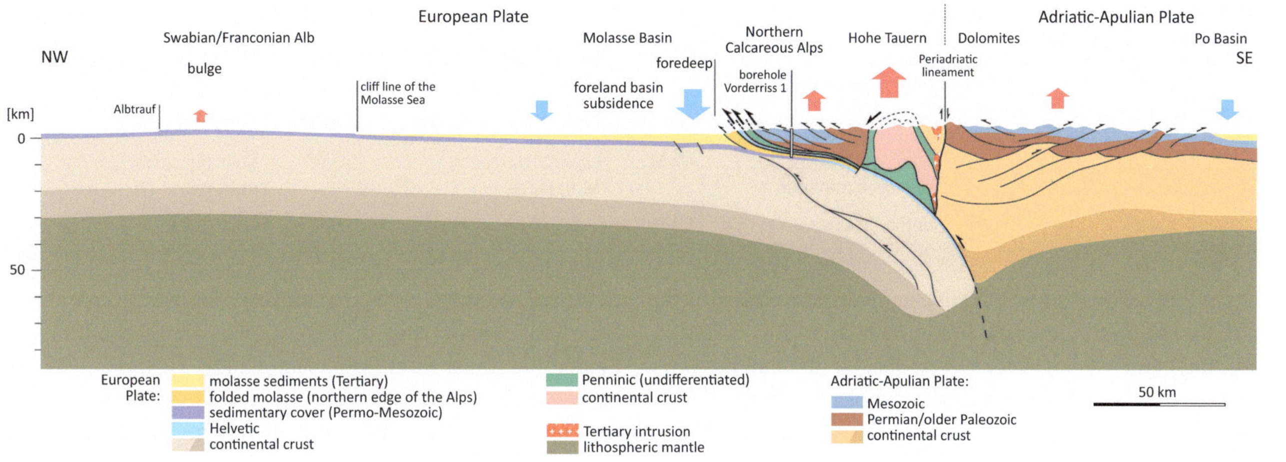

Fig. 13.16 Cross-section through the Eastern Alps from the Po Basin to the Swabian Alb. Modified after Schmidt et al. (2004) and Pfiffner (2010)

in the west. The facies distribution within the foreland basin and the amount of sedimentary material deposited were very much related to the tectonic evolution of the Alpine Orogeny. The various phases of tectonic activity characterized by folding and thrusting, and the northward propagation of the orogenic belt, were characterized by increased sediment supply, which in turn influenced the change from marine to terrestrial deposits (Kuhlemann and Kempf 2002). The sedimentary development of the North Alpine Molasse Basin was also influenced by external global factors, such as eustatic variations in sea-level (Haq et al. 1988; Fig. 13.17).

The collision between the Adriatic-Apulian and Eurasian plates resulted in major thickening of the crust by the stacking together of the two continental lithospheric plates

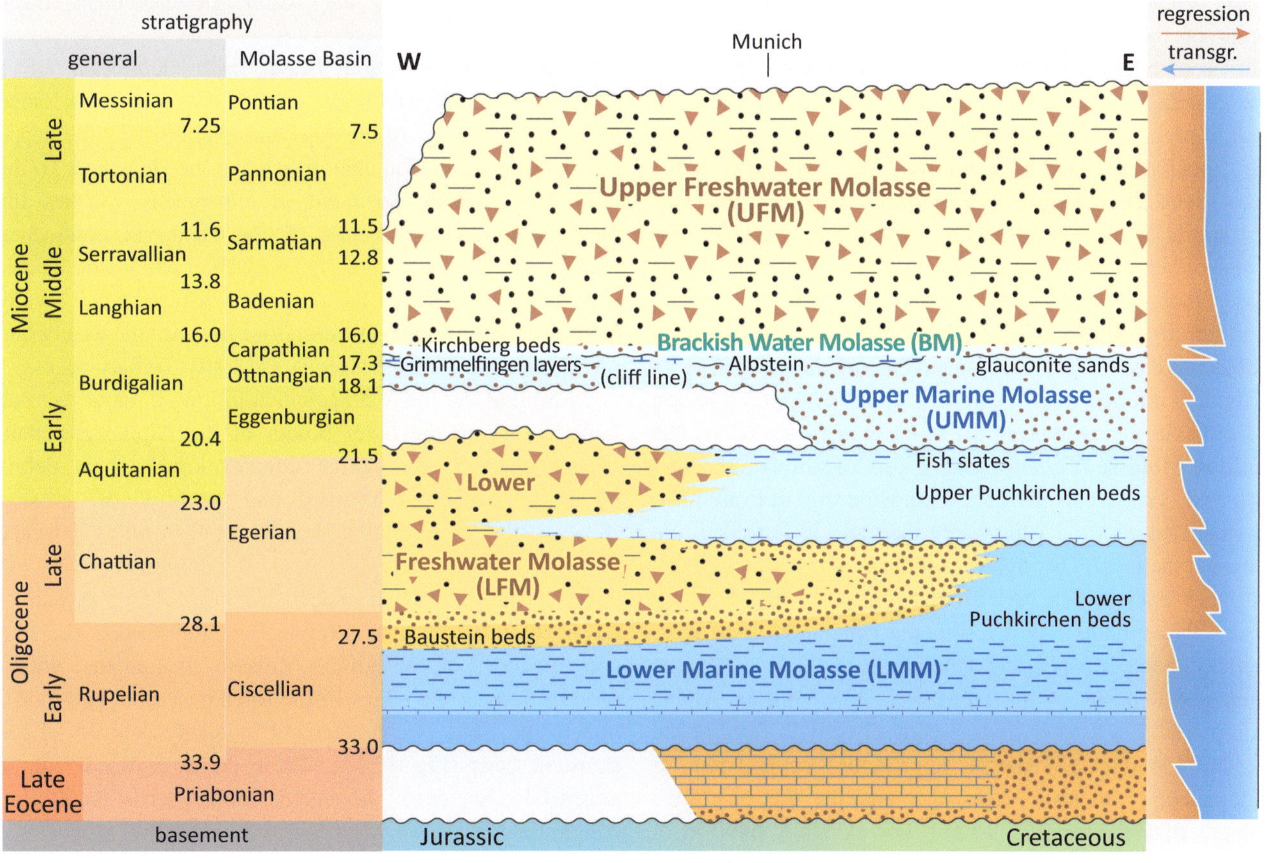

Fig. 13.17 Stratigraphy of the Alpine foreland molasse with a schematic representation of the facies changes from west to east. LMM Lower Marine Molasse, LFM Lower Freshwater Molasse, UMM Upper Marine Molasse, BM Brackish Water Molasse, UFM Upper Freshwater Molasse Modified from Lemcke (1988), Kuhlemann and Kempf (2002), Menning et al. (2002), and the Deutsche Stratigraphische Kommission (2016); sea-level curve after Haq et al. (1988)

Table 13.1 German and English subdivisions used for the North Alpine Molasse

German		English		Age
Obere Süßwassermolasse	OSM	Upper freshwater Molasse	UFM	Middle and Late miocene
Obere Meeresmolasse	OMM	Upper Marine Molasse	UMM	Early miocene
Untere Süßwassermolasse	USM	Lower freshwater Molasse	LFM	Late oligocene—early miocene
Untere Meeresmolasse	UMM	Lower marine Molasse	LMM	Oligocene

(Fig. 13.16). The load of the advancing Alpine Orogen lowered the northerly adjacent lithosphere in a down flexure, resulting in an asymmetrical wedge-shaped foreland basin fill characterized by continuous sedimentation. The basinal strata were thickest at the edge of the advancing orogen with sedimentary sequences approaching 5000 m in vertical extent. Approximately 150–200 km north of the orogenic thrust front, and behind the inflexion point of lithospheric **down-flexure**, a saddle-shaped **bulge** (or forebulge) formed from the rebounding upward-flexure. This led to the uplift of the Jurassic limestone platform of the Swabian and Franconian Alb by several hundred meters as it was tilted by a few degrees to the southeast. Due to the additional up-doming associated with the formation of the Upper Rhine Graben (Sect. 13.2), the tilt of the Jurassic limestone platform in the western part was even more extensive at about 3°–4°.

The roots of the European continent continue far to the south beneath the North Alpine Molasse Basin and the nappe units of the Alps (Fig. 13.16). This is known from the numerous boreholes that have been made in the region. In the eastern part of the basin, the basement consists of crystalline rocks of the Bohemian Massif overlain by sedimentary rocks of the Late Carboniferous and Rotliegend. In the west, these rocks are followed by Early and Late Cretaceous strata and in the central part of the basin by Late Jurassic limestones exposed in the Swabian and Franconian Alb. The older molasse strata in the south were deformed by the northernmost thrust of the Alps, where they were again subject to erosion and resedimented within the same basin in a process referred to as "cannibalization". The incorporation of molasse into Alpine thrust units is documented by the **Folded Molasse** (*Faltenmolasse*), also known as Subalpine Molasse. This occurs in a ca. 20 km wide strip in front of the northern edge of the Alps where the molasse is folded into broad synclines. The tighter anticlinal structures suffered stronger deformation and in many places were detached along thrust faults. As a result, the synclines often occur without their corresponding anticlines (Fig. 13.18). Even the youngest deposits of the UFM were tilted during Alpine compression in a zone of steepening directly in front of the Folded Molasse. Thicknesses approaching several tens of kilometers of molasse strata also underlie the overthrust nappes of the Alps (Fig. 13.16). For example, 6400 m of molasse sediment was encountered in the Vorderriss 1 borehole, which was drilled in the 1970s about 20 km behind the thrust front in the Isar Valley. However, the base of the molasse was not encountered. Geophysical surveys indicate with reasonable certainty that the base of the sequence would have been encountered at a depth of about 7 km if drilling had continued. These surveys also indicate that the oldest overthrust molasse strata are located internally about 40–50 km behind the northern thrust front.

The sedimentary history of the Northern Alpine Molasse trough was marked by various stages of marine regressions and transgressions characterized by alternating sequences of marine and freshwater molasse (Figs. 13.17 and 13.19, Table 13.1). These units can be partially correlated with marine transgressions observed in the Upper Rhine Graben. During the Oligocene, there existed a marine connection between the Rhine Graben, the Northern Alpine foreland basin, the southern maritime regions and the North German/Polish Basin (Sect. 13.2).

Molasse sedimentation began with deposition of a sequence of marine sediments (LMM) that was partly folded at the northern edge of the Alps and overthrust by the nappe stacks of the Rhenodanubian flysch and the Helvetics. To the north, the molasse sediment unconformably overlies the platform sedimentary rocks of the European continental margin represented by limestones of the Late Malm. To the east, the Molasse Basin lies on Early Cretaceous and locally **bean ore formations** (*Bohnerz*), together with the weathering products of the Late Cretaceous and Early Tertiary rocks.

In front of the Inn Valley, which follows a major tectonic fault line along the outer border of the Alpine mountain chain, a Tertiary submarine fan developed where debris accumulated from the surrounding area of erosion (Fig. 13.19a). East of the valley, the shallow hilly landscape of the Northern Calcareous Alps was karstified during the Late Eocene and buried under siliciclastic debris as part of the developing Northern Alpine Foreland Basin. The first conglomerates of the Alpine Molasse accumulated within **alluvial fans** at the edges of the foreland basin. The transition to the LFM is characterized by sandy deposits of the Baustein beds (Fig. 13.17). These merge upwards into a terrestrial sequence of mudstones, sandstones, and conglomerates.

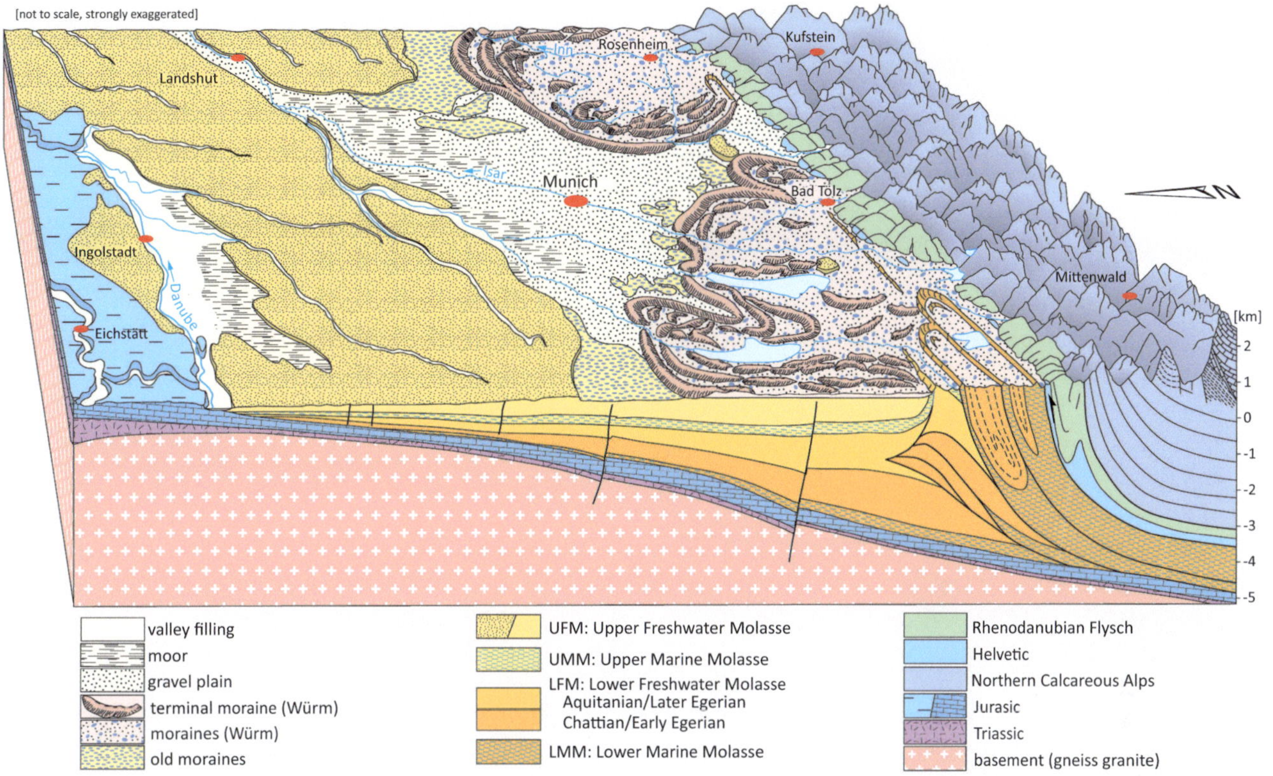

Fig. 13.18 Block diagram of the Northern Alpine Foreland Basin in the area of Munich. Modified after Wellnhofer (1983) and Glaser et al. (2008)

A regressive phase followed during Late Oligocene to Miocene times (Fig. 13.19b). The associated increase in the supply of clastic sedimentary material for the first time filled up the Northern Alpine Molasse trough and silted up the western and central parts of the basin (LFM). The total thickness of the terrestrial sediments of the LFM reached up to 2500 m near the edge of the Alps, whereas the underlying marine sediments of the LMM are in some areas only about 300 m thick. The regression was most prominent in the western part of the Alps and corresponded with the beginning of rapid uplift of the Alpine orogen. It also correlates with one of the strongest global sea-level drops in the entire Tertiary (Haq et al. 1988), which occurred in the Late Oligocene about 28.5 million years ago. This event marked the beginning of a glaciation phase in Antarctica (Fig. 13.17; Berggren et al. 1995), an event that certainly had an impact on the molasse sedimentation. However, the regression phase that affected the Northern Alpine Molasse did commence a little earlier than the glaciation and was most likely initiated by the final phase of the Alpine collision.

In the Western Alps and along the Periadriatic Lineament, an increase in volcanic and magmatic activity occurred about 32–30 million years ago (Sect. 12.6). This is reflected in the composition of the older molasse conglomerates along the northern rim of the Alps, with clasts of andesitic volcanics derived from an eroded volcanic chain that lay along the Periadriatic Lineament that no longer exists today (Frisch et al. 1998).

Several river systems drained into the Molasse Basin from the gradually ascending Alps and from the Jurassic limestone platform in the north. During the LFM, the northern edge of the North Alpine Molasse Basin was also fed by rivers from the Bohemian crystalline massif (Fig. 13.19b). At that time the main outflow was to the east, where a marine sedimentary environment persisted (Paratethys; Fig. 13.3). The shoreline existed in a location east of Munich and remained stable for a period of about 9 Myr. To the east, the **Puchkirchen Trough** opened where deep marine depositional conditions prevailed with water depths exceeding 1000 m. This topographic contrast between the western terrestrial-dominated part and the eastern marine-part of the North Alpine Molasse Basin reflected the differences in the geological evolution of the Western and Eastern Alps. While the Western Alps and the western part of the Eastern Alps had already developed a strong relief with a mountainous character by the end of the Oligocene, the eastern part of the Eastern Alps remained relatively flat, with low hills composed of gravels and sands that were deposited up to the Pliocene. Here, uplift was initiated only a little later in the Miocene.

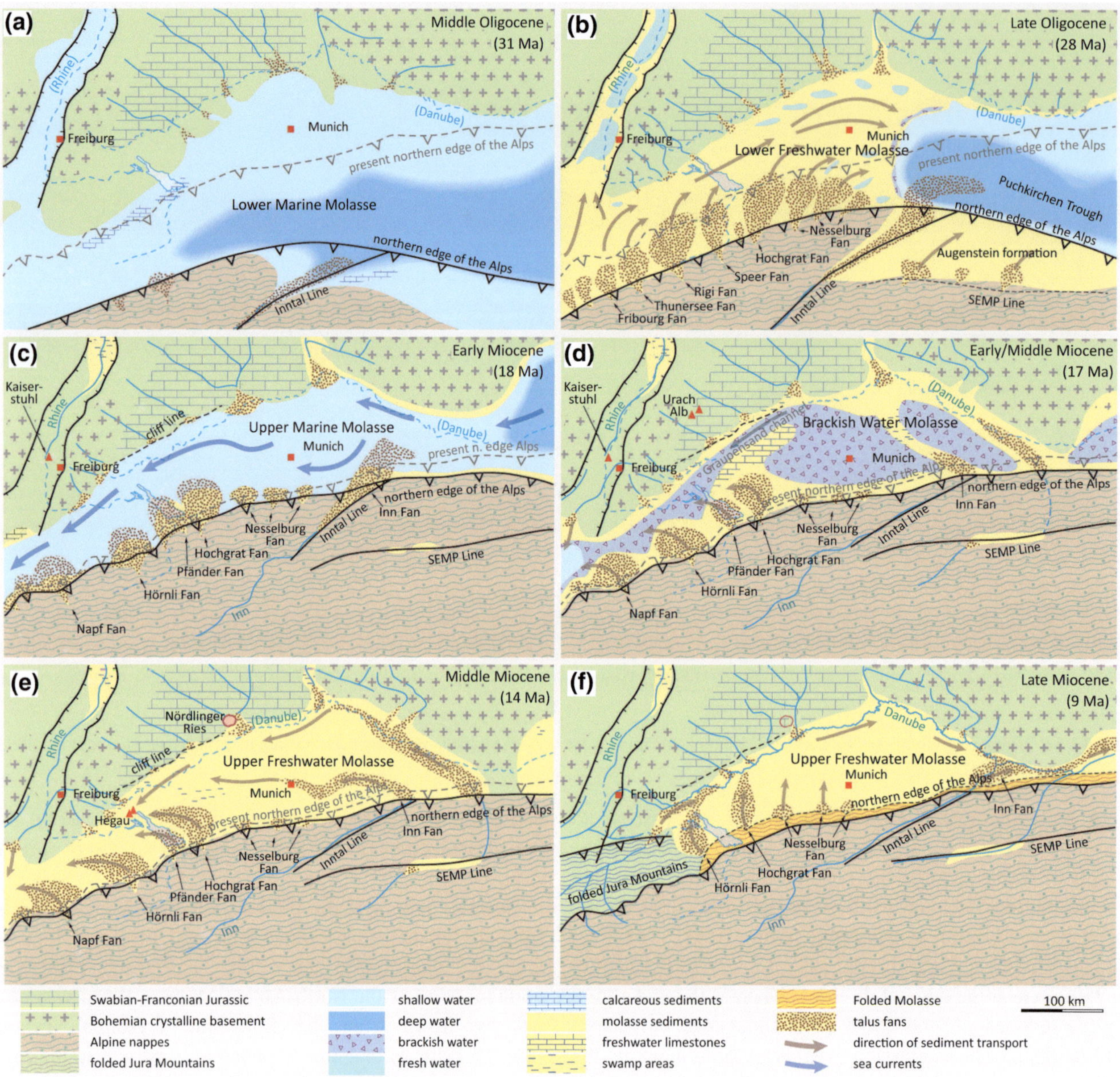

Fig. 13.19 Oligocene to Miocene stages of molasse evolution. See text for further explanation. Modified after Kuhlemann and Kempf (2002), Bayerisches Geologisches Landesamt (1996)

The Early Oligocene **Augenstein Formation** of the central and eastern Northern Calcareous Alps consists of terrestrial sediment that covered large areas of the karstified hills with gravel deposits. Although the conditions of sedimentation were similar, this area did not strictly belong to the Northern Alpine Foreland Basin. The fluvial conglomerates, including the bean ore and reworked tropical lateritic soils, are today uplifted as part of the Northern Calcareous Alps in parts more than 2000 m above sea-level (Frisch et al. 2001). In a few remaining areas, the sediments of the Augenstein Formation are exposed, for instance, on top of the Dachstein plateau or in the northeasterly adjacent mountain range of the Totes Gebirge.

In the Eastern Alps, the eastward **extrusion** of the Austroalpine units and the **exhumation of the Tauern Window** began about 21 million years ago in the Early Miocene. The large lateral displacements are considered to result from the continuing **indentation** of the Adriatic- Apulian Plate into the European continent and the evasive movement of the Alpine orogen to the east into the Pannonian basin (Frisch et al. 2000; Sect. 12.6). The Early Miocene saw a renewed transgressive advance of the sea (OMM; Fig. 13.19c) related to increased subsidence and narrowing of the Northern Alpine Foreland Basin during continued nappe stacking of the Alpine orogen. The reduced supply of sediment may have been related to the eventual **collapse of the orogen** by gravitational sliding.

The flow directions of water currents in the Molasse Basin were influenced by the development of antiforms formed during the UMM, with the main drainage directed to the west and concentrated mainly along the northern edge of the basin. It is assumed that these conditions led to the so-called ***Graupensandrinne***, which formed as the Brackish Molasse (BM; see below) erosively cut down into the previously deposited molasse sediments. A marine connection to the Upper Rhine Graben did not exist at that time.

In the north, the Molasse Sea responsible for deposition of the UMM surfed against the Late Jurassic limestones of the Swabian and Franconian Alb. The **cliff line** is still visible and traceable at numerous locations along a line running from Tuttlingen in the Black Forest, through Temmenhausen near Ulm, to Donauwörth in Bavaria (Fig. 13.19c). A good example of an Early Miocene **wave-cut cliff** riddled with paddock, which marks the paleo-surf zone, occurs at a well-known and meanwhile protected exposure (Fig. 13.20) in the village of Heldenfingen, north of Ulm. This former cliff reached 100 m height in some places and, in Heldenfingen, is now located at an altitude of 600 m above sea-level. Further to the east, it can be found at 500 m altitude and on the western Alb at 900 m above sea-level. Since the sea-level during the Oligocene/Miocene was still at least 100 m higher than today, the present altitude indicates that the region has been uplifted by 400–800 m since the

Fig. 13.20 The Heldenfingen cliff in the village of Heldenfingen. A paleo-Tertiary surf zone is exposed in the form of a wave-cut section developed in Late Jurassic limestone that is riddled with notches and paddock shaped holes. This erosional feature is of Early Miocene (Burdigalian to Ottnangian) age and was formed when waves crashed against the steep cliff from the south. [N 48°36′23″/E 10°04′25″]

Miocene epoch. The cliff line, therefore, provides something like a fossil spirit level. Since the Early Miocene, the area of the Swabian-Franconian Alb was not only uplifted but also simultaneously tilted by 400 m causing layers in the southwest to dip somewhat steeper than in the northeast. Just south of the cliff line, shallow coastal marine sediments were deposited consisting of conglomerates, coarse sands, and broken shell carbonates. These deposits include the fossil-rich beds of the Erminger *Turritellenplatte* (turritella plate), exposed near Ulm, and the *Randen Grobkalk* (coarse limestone), which represents a mixed facies with sandy and carbonate components. An example of the latter occurs in the region of the Hegau and has a maximum thickness of 10 m.

At the end of the Early Miocene, the UMM passed into a brackish water stage characterized by diverse depositional conditions. This stage is also referred to independently as the Brackish Water Molasse (BM; Fig. 13.17). During this stage, the *Graupensandrinne* developed at first as an erosive deposit. A trench-like structure formed on the northwestern edge of the Molasse Basin, within which a 10–20 km wide river system transported sand and gravel derived from the Bohemian Massif (Fig. 13.19d). The sedimentary fill of the Graupensandrinne consists of the fine gravel to sandy Grimmelfingen beds, also called Grimmelfinger Graupensande. These are interpreted as estuarine deposits, with the clayey to marly Kirchberg beds representing the last transgressional advance of the UMM from the west (Sach and Heizmann 2001). Brackish lakes developed at this time in the Northern Alpine Foreland Basin and were gradually filled with clastic sediments. These were very sensitive to short-term and minor sea-level changes. On an area of higher topography lying south of the Graupensandrinne, the Albstein Swell formed with the development of a largely unfossiliferous Albstein nodular limestone (*Knollenkalk*). However, the genesis of this limestone is still controversial. Some consider it to represent a purely terrestrial calcrete formation developed during dry periods by the precipitation of carbonate within soil horizons (caliche). Others consider there is evidence that it was formed by algae in a freshwater lake as a limnic carbonate deposit.

The transition from the BM stage to the UFM was characterized by the approach of the Northern Alpine Thrust Front to the southern spur of the Bohemian Massif (Fig. 13.19e). The Northern Alpine Foreland Basin was slightly uplifted by the interaction of two continent masses in this area and, as a result, a general westerly flow direction within the Molasse Basin was established. The collision also led to an increase in the eastward lateral extrusion of the Alpine orogen and, as a consequence, to the decrease in sediment supply from the Central and Eastern Alps. With increasing lateral strike-slip movement, the degree of thrust motion was also reduced. This led in turn to a reduction in subsidence within the Northern Alpine Foreland Basin. Despite the decrease in the sediment supply from the rising mountain range and the enhancement of eastward lateral movements, the Northern Alpine Foreland Basin was overfilled in the Miocene and by the Middle Miocene became completely silted up.

The UFM of the Late Miocene was characterized by a west-directed drainage system. The numerous alluvial fans that developed along the southern edge of the Molasse Basin preserved a variety of erosional products from the rising Alps in the form of diverse pebble compositions. For example, distinctive cobbles originating from the Tauern Window can be recognized by characteristic rock types unique to the area, making it possible to reconstruct the uplift history of the source areas during the Miocene. Such reconstructions (Frisch et al. 1998; Kuhlemann et al. 2001) indicate that the north-directed drainage system of the Alps extended much further to the south at the beginning of the Miocene than it does today. Thus, the alluvial fans contain cobbles of Oligocene volcanic rocks from source areas that have been completely eroded by Alpine exhumation. Such volcanism originated above plutons intruded at depth and today is only partially exposed at the surface south of the main Alpine divide. In the Miocene, the river drainage of these areas was maintained in a northerly direction.

14.6 million years ago, the **Nördlinger Ries** and its "little brother", the Steinheim Basin (Sect. 15) were created by an asteroid impact. The impact caused large amounts of ejecta to be relocated over distances of several hundred kilometers. The **Brock horizon** occurring within the UFM (Sach 1997) is considered to represent such ejecta. It contains numerous fragments ranging from several cm to dm size of Malm limestone, some with shatter cone structures, Malm-derived fossils, and bean ore deposits. Due to the precise radiometric dating of the Ries event, the Brock horizon represents a good **marker horizon** in the sedimentary record of the Molasse Basin.

The westward draining system of the Molasse Basin continued until the Late Miocene. However, about 10 million years ago the flow direction changed with the alluvial fan of the Inn River switching to flow toward the east (Fig. 13.19f). At this time, the Danube River made its way along the northern edge of the Northern Alpine Foreland Basin. The reversal of the drainage direction resulted from the Jura Mountains folding to the west of the basin with the deformation extending nearly as far as Lake Constance. The western part of the Northern Alpine Foreland Basin was from that time subject to erosion and uplift caused by the Jura fold-and-thrust deformation. This uplift blocked the westward drainage of rivers causing them to find another way to the east. The end of the molasse sedimentation was marked by the uplift of the Jura Mountains, which as a belt, moved diachronously towards the east with deposition ending about 8–9 million years ago (Kuhlemann and Kempf 2002). The youngest deposits of the Northern Alpine Foreland Basin are closely associated with the Quaternary glacial history of the Alps (Fig. 13.18). They are represented by terrace deposits that developed during the glaciation period when the rivers transported less material.

A special feature of the North Alpine Molasse is the gravel sediment that consolidated into conglomerates known as ***Nagelfluh*** (or gompholite). Nagelfluh consists of fluvial gravels that formed during the terrestrial periods of the molasse sedimentation and formed during the Quaternary. Fine examples of these rocks occur at the Mönchsberg in Salzburg, Austria. Highly calcareous groundwater completely filled the large pores between the gravel components and precipitated a carbonate cement that lithified the conglomerate. In the Oberallgäu region, this concrete-like rock is referred to as *Herrgottsbeton* (Lord's concrete). The German term *Nagelfluh* has its origin in the appearance of its weathered surface: the *Fluh* (an old word for rock face) referring to conglomeratic components which protrude like nail heads. In the mountain range of the Nagelfluhkette in the area of the Oberallgäu, thick layers of Nagelfluh are exposed for instance in the Nagelfluhkette nature park at the Hochgrat (1834 m). These rocks belong to the Folded Molasse (Fig. 13.18) and consist of conglomerates deposited in the Hochgrat alluvial fan (Fig. 13.17). They were steeply tilted during overthrusting along the northern edge of the Alps and today form prominent mountain ridges. Exposures of Nagelfluh also occur along the northern edge and western part of the Molasse Basin, where conglomerates containing components from northerly-located areas of erosion were cemented together.

The Pliocene and Pleistocene uplift related to crustal bulging in the region of the Upper Rhine Graben caused far-reaching changes to the drainage system of the Molasse Basin. These changes started with a shift of the Rhine River during the Pliocene so that it was sourced from the Alps (Sect. 13.2). Today the **Aare** flows as a tributary into the **Rhine** near Waldshut-Tiengen. However, at the beginning of the Pliocene, it was still connected to the **Danube** and drained as the Aare-Danube River eastward along the northern edge of the Molasse Basin to the Black Sea. During the Middle Pliocene, the situation changed. For a short time, the Aare passed through the Doubs into the Mediterranean between the Black Forest and the folded Jura Mountains. As the southern Upper Rhine Graben finally subsided in the upper Pliocene, the Aare then connected to the Rhine. Initially, it was the source river of the Rhine but later became a tributary after the connection to the Alpine Rhine was formed through Lake Constance.

Changes in the drainage system of this region continue into recent times, an example being the **Wutach** canyon. Even during the last ice age, the so-called **Feldberg-Danube**

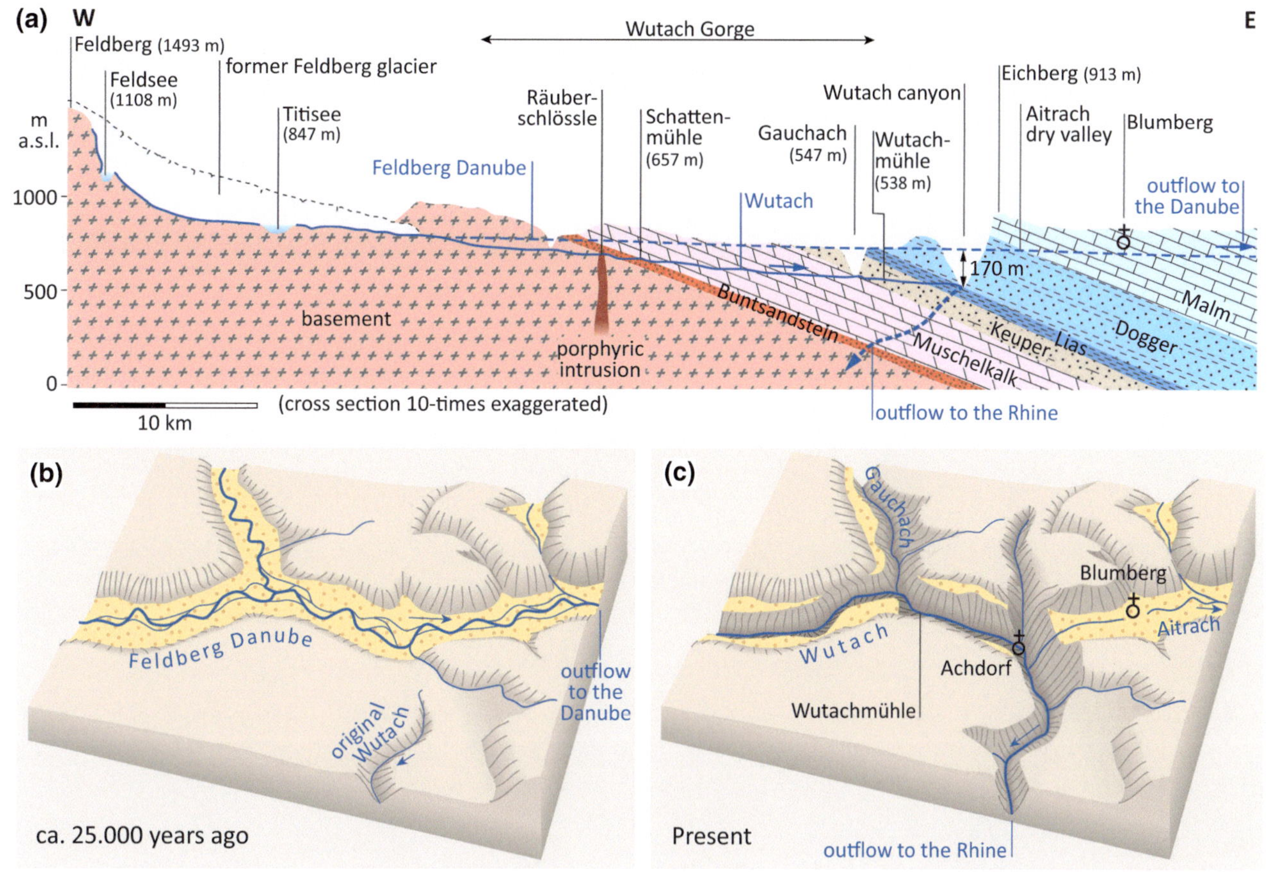

Fig. 13.21 **a** Profile section along the Wutach River from the Feldberg to the junction with the Rhine. The Aitrach dry valley represented the continuation of the Feldberg-Danube until 25,000 years ago. **b** Course of the Feldberg-Danube before the Wutach River became active. **c** Present stage of the deeply incised Wutachschlucht. Profile section modified after BNL (2004)

River was a wide braided river system that was fed by a mountain glacier located in the Feldberg of the Black Forest and flowed into the Danube (Fig. 13.21). Following the peak of the last glaciation period (last glacial maximum, LGM) about 22,000 years ago, the river was deflected towards the Rhine due to regressive erosion. Since the base level of the Rhine is significantly lower than the Danube, a new river course was incised between the two main river systems to form the impressive Wutach Canyon (*Wutachschlucht*). The ancient Feldberg-Danube River cut deep into the Mesozoic strata and down into the Paleozoic basement to form broad valleys above the Wutach Canyon, which had much stronger erosive power than the small volume of water flowing along the valley today.

14 Tertiary and Quaternary Volcanism

Tertiary and Quaternary volcanic events occurred in several regions in Germany, some of them continuing into the Holocene and not completely extinct to this day. One of the most recent eruptions was that of the Laacher See volcano of the Eifel area occurring just ca. 13,000 years ago. In addition to the two active centers of Eifel volcanism, there existed a number of other Tertiary volcanic fields often related to tectonic structures. These include the volcanic rocks of the Kaiserstuhl, located in the Upper Rhine Graben, and the volcanics of the Eger Graben area.

There is plenty of evidence to be found across Germany of volcanic activity from the more recent periods of geological history: the uppermost Cretaceous, the Tertiary and the Quaternary. In particular, young volcanic rocks are found in a broad zone extending from west to east that includes the Eifel and the Eger Graben. Other smaller areas with similar rocks are located in the southern area of Baden-Württemberg (Fig. 14.1). The most recent volcanic formations of the Eifel were formed during the eruptive explosion of the Ulmener Maar located in the eastern Eifel region. With an age of about 11,000 years, this event represents the youngest proven volcanic eruption in central Europe (Zolitschka et al. 1995). Tertiary volcanics also occur in areas of the Westerwald, the Vogelsberg, the Hessian Depression, the Rhön and the Heldburger Gangschar. Other notable features in the south of Germany are the Kaiserstuhl volcanics, the Urach tuff vents, and the sculptured volcanic pipes of the Hegau that belong geologically to young volcanic formations. Further east, the volcanics of the Eger Graben region occur, most of them are located in the Czech Republic and in Lusatia. These types of volcanic formations that are scattered across Germany are in fact quite diverse and vary from extensive areas of basaltic volcanic rocks occurring as thick basalt flows to highly explosive volcanics with maar formations and widespread pyroclastic deposits.

14.1 Volcanism in the Eifel

On a spring day in May, about 13,000 years ago, one of the largest volcanic eruptions occurred in the northern Eifel area, the effects of which extended far beyond the present borders of Germany. Although the Laacher See, with its almost circular shape, is often referred to as the largest maar of the Eifel and equated with the typical Eifel maars, it is in fact not a maar but a lake developed in a collapsed crater (**caldera**) that formed at the center of a large volcano. In contrast, a maar forms during a relatively small, highly explosive eruption that occurs when the water encounters the magma (Box 24: Maars). During the eruption of the **Laacher See volcano**, a large amount of loose volcanic ash was produced and transported by the wind over hundreds of kilometers in a northeasterly direction. This material reached areas of the present-day Baltic Sea, Southern Sweden and, to the south, also the Po plain (Fig. 14.2; Schmincke 2008, 2013). The volcano is estimated to have erupted about one and a half times more material than that produced during the eruption of Mount Pinatubo in 1991 on the Philippines island of Luzon: one of the largest volcanic events of the twentieth century. The Laacher See volcano also released six times more ash than the 1980 eruption of Mount St. Helens in the US. According to the volcanic explosivity index (VEI) both the Pinatubo and the Laacher See volcano have a VEI value of 6 on a scale ranging from 1 to 8.

M. Meschede and L. N. Warr, *The Geology of Germany*, Regional Geology Reviews,
https://doi.org/10.1007/978-3-319-76102-2_14

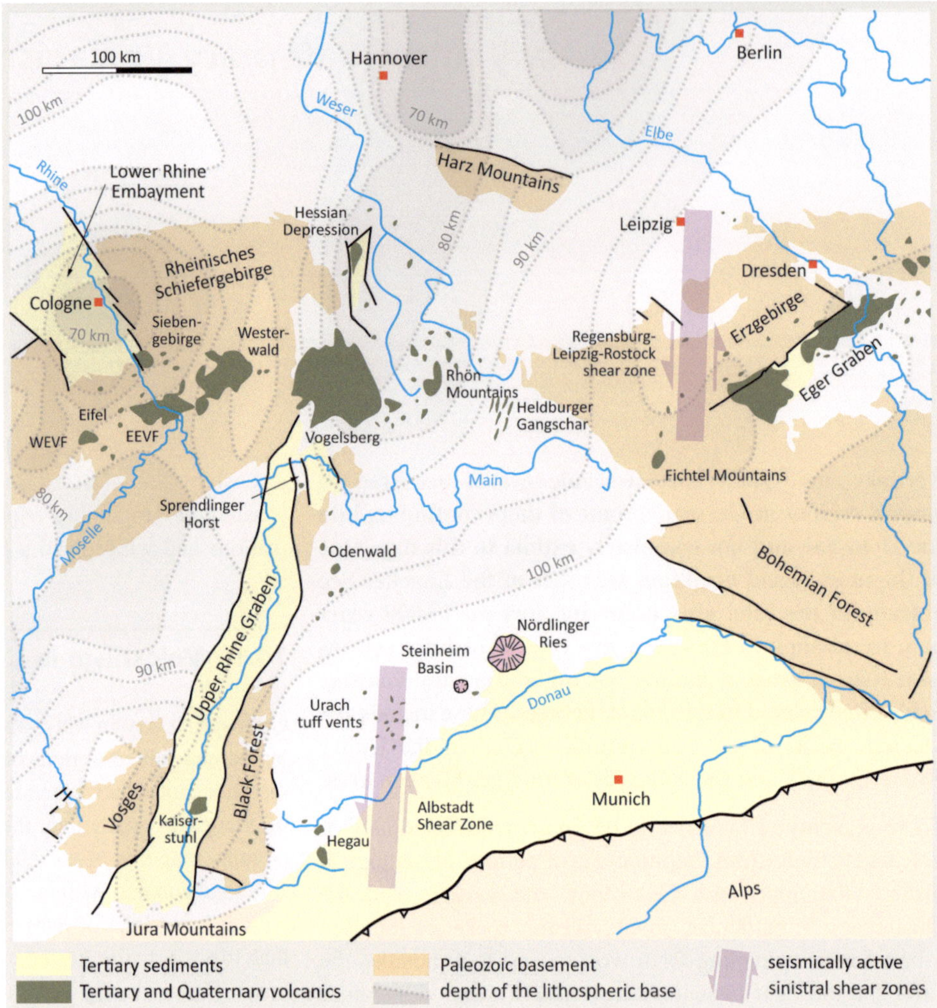

Fig. 14.1 Tertiary and Quaternary volcanic areas in Germany and its surroundings. Modified after Henningsen and Katzung (2006); base of lithosphere after Plomerová and Babuška (2010); location of shear zones after Bankwitz et al. (2003)

The eruption of the Laacher See volcano is considered to have lasted only a few days, but during this extremely explosive event, an enormous amount of ash was ejected, probably reaching an altitude of about 40 km. Such eruptions are called **Plinian eruptions**, named after Pliny the Younger who observed and described the eruption of Mount Vesuvius in 79 A.D.

Although the Laacher See is not strictly a maar, the eruption was nonetheless accompanied by **phreatomagmatic eruptions** caused by contact between the rising magma and the groundwater. The eruption led to massive deposits of pumice tuff and welded pyroclastic slag formed from the glowing clouds, the so-called fallout. The ejected hot ash clouds of the Laacher See volcano were so powerful that the Rhine Valley in the area near Koblenz was sealed off by a dam that formed from an early eruption phase (Park and Schmincke 2009). During a second eruption, a **pyroclastic flow** then blocked a valley northwest of Andernach, just about 9 km from the crater center and about 40 km southeast of Bonn (Fig. 14.3). This resulted in the formation of the Brohl dam, named after the *Brohltal* (Brohl Valley). For a few weeks or months, a natural reservoir formed about half the size of Lake Constance with a depth more than 25 m. This lake filled the valley all the way up to the region of the present city of Mannheim and deposited volcaniclastic sediment which is still preserved close to Koblenz and Neuwied. These are the so-called *Bimsflöße* (**pumice rafts)** composed exclusively of pumice that floats on water in its dry state. These fragments are enclosed in the normal tuff and welded pyroclastic deposits, indicating the water of the Rhine River was temporarily dammed into a lake where pumice rafts accumulated.

In the immediate vicinity of the Laacher See, the tuff deposits reached a thickness of more than 20 m. In the

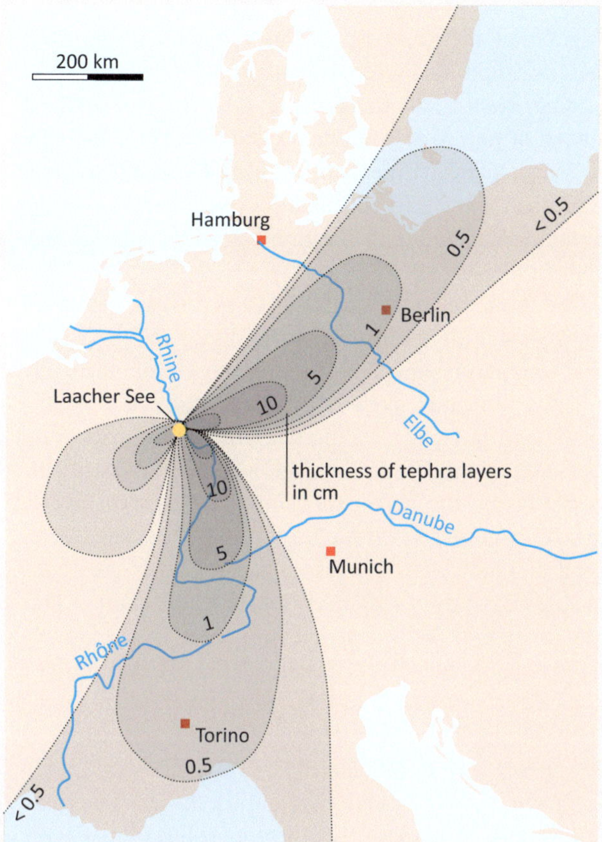

Fig. 14.2 Distribution of Laacher See tephra in Central Europe. After Schmincke (2008, 2013)

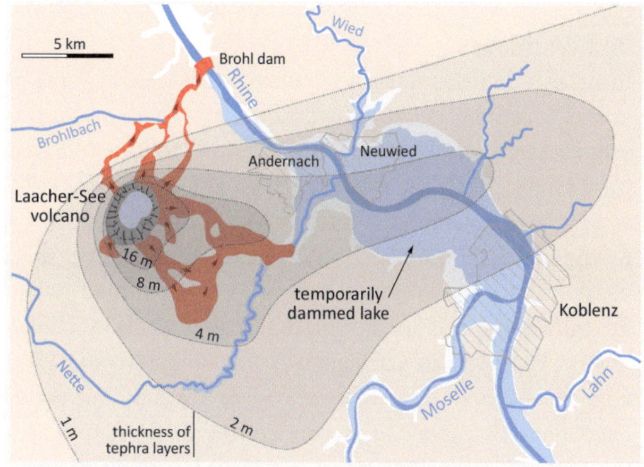

Fig. 14.3 Extent of a lake in the Rhine Valley after a dam formed at the end of the Brohl Valley by large pyroclastic flows from the Laacher See volcano eruption. The level of the lake was about 25 m above the Rhine. Modified after Schmincke (2009), and Park and Schmincke (2009)

impressive Wingertsberg wall, sedimentary structures occur within the ash layers reflective of very turbulent and destructive depositional conditions (Fig. 14.6). Wave ripples of several meters size indicate ash flow velocities exceeded 100 km/h. Such deposits, which are extremely destructive in close proximity of a volcano, were observed and studied in detail during the eruptions of the Mt. Unzen volcano complex in Japan. In 1991, a hot pyroclastic flow (several hundred degrees) ran down the slopes of the volcano at speeds of over 200 km/h burying everything in its path. This was not, however, as destructive as the collapse of a lava dome in 1792, which triggered a tsunami wave that killed over 14,000 people. Similar pyroclastic deposits, partly also formed under the influence of water, are located at distances of several tens of kilometers around the Laacher See. A number of pyroclastic flows with a mixture of water, pumice and ash traveled their way down through the Brohl Valley, a few kilometers north of the Laacher See (see above), before solidifying. The result was a highly porous and soft but stable rock material that was mined as a mortar during Roman times and it is still widely used today due to its excellent hydraulic properties for producing specialized underwater setting mortar (Fig. 14.7). The **trass** caves in the Brohl Valley are remnants of centuries-old mining activity and testify to the mechanical stability of the trass that reaches a total thickness of more than 50 m in some places.

Pumice and **tephra layers** with a thickness of around 1 m can be detected at distances of 50 km away from the Laacher See volcano and testify to the intensity of the eruption. At a distance of 200 km (to the northeast, Fig. 14.2), the thickness of the ash layers are still 5 cm, and at a distance of more than 700 km on the island of Bornholm, a half centimeter thick ash layer has been discovered.

The age of the Laacher See volcano eruption can be dated with high precision using a variety of methods. The most recent volcaniclastic layers deposited by the ash clouds of the Laacher See are geochronologically well constrained and represent excellent marker horizons. In particular, the method of **dendrochronology** (from Greek *dendron*: tree, *chronos*: time and *logos*: teaching) has played a decisive role in dating such events. In the case of the Laacher See, the connection to the dendrochronological time scale is not yet fully complete but new research is expected to provide more clarity soon (Kaiser et al. 2010). The growth of tree rings depends on the variations in the seasonal weather of each year, resulting in annual rings of different thicknesses.

When the weather does not change regularly for a number of years, a sequence of tree rings with similar thickness represents a stable climatic time segment. This sequence

Box 24: Maars

In the Eifel area, numerous remains of a special form of volcanic activity occur known as maars. These structures originated largely by phreatomagmatic eruptions (derived from Greek *phreatos*: fountain) only a short time ago: the youngest of them, the Ulmener Maar, is just a little bit more than 10,000 years old. Maars, which owe their name to the Latin *mare* (the sea), are formed when groundwater meets the hot molten rock at shallow crustal levels. This type of volcanism is not unique to the Eifel area but occurs worldwide whenever magma and water encounter each other in the underground. In many cases the remains of ancient maars can also be found in older rock series, for example, in the area around Bad Urach of the Swabian Alb, where a large number of maar explosions (Sect. 14.5) were formed in the Miocene about 16–17 million years ago. There is also the maar deposit of the Messel Pit, where a unique Eocene fauna and flora was preserved (Sect. 14.3).

The explosive effects of the rapid evaporation of water in a confined environment, which causes a maar explosion at the depth, is often used publicly as a demonstration on firefighting. For this purpose, a container of oil (representing the risk of a burning fryer) is heated to almost boiling point at about 200–220°C and then set on fire. If one attempts to extinguish the burning oil with water, then the opposite will be achieved. A violent explosion occurs by the instantaneous evaporation of water. This is the precise reaction that occurs when hot liquid magma encounters groundwater. During evaporation, the water expands its volume in a very short time by 1680 times, equivalent to one liter of water yielding 1680 L of water vapor. Due to the confined volume of pore spaces in the rock at a depth of 2–3 km, the **steam pressure** generated during the encounter can be so great that it exceeds the lithospheric pressure of the overburden, thus giving rise to a volcanic explosion that ruptures the surface.

The formation of maars has been in the past a point of controversial discussion. However, today the process of maar formation is much better understood from the study of active volcanic systems, such as those around the Pacific and in particular, the two phreatomagmatic vent structures of the Ukinrek maars in the northwest of the Alaska Peninsula. Maar explosions are caused mainly by surface waters and groundwater that circulate down along fractures into the bedrock. In some cases, the fractures extend to very deep levels where molten lava may ascend. If the water and lava meet at a relatively shallow depth, the effects described above can result in a violent explosion (Fig. 14.4). A mixture of shattered wall rock, molten rock, and water vapor is discharged at the surface via a tubular pipe. Tuffs consisting predominantly of rock pulverized during the explosion are formed with only relatively low proportions of magmatic rock remaining. The tuffs typically build a wall around the edge of the maar where they are deposited as a thick layer. Such explosions are rather short-lived events and usually last only a few days or weeks. They only remain active until the fluid overpressure declines to more

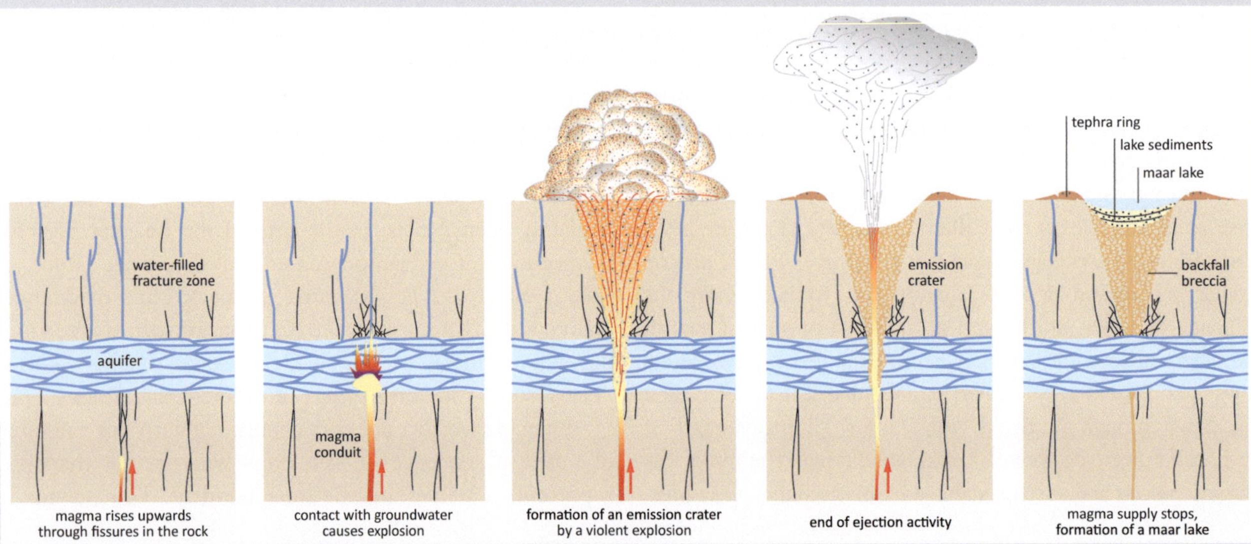

Fig. 14.4 Evolution of a maar when groundwater and ascending magma come into contact

normal levels. Thereafter, the pipes are sealed by the backfall breccia (Fig. 14.5) and are usually not activated a second time. Subsequent explosions generally follow new pathways to the surface along with preexisting fracture networks and, therefore, emerge at a new location close to the already existing maar (e.g., Fig. 14.10). Once a maar explosion has occurred, the amount of water infiltration is ultimately responsible for the size of a maar. Smaller subsequent explosions involving lesser quantities of water typically take place at depths of 30–100 m. However, a large influx of water into the volcanic vent can lead to explosions at several hundred meters to kilometers depth in the vicinity of rivers and lakes or during the influx of water-rich stream.

Once dormant, the ejection funnel created by the explosion is partly filled with pulverized rock material (Fig. 14.4) and the crater forms a maar lake. The diameter of such maars often varies between 50 m and up to 2 km, whereby the Eifel maars usually do not exceed a few hundred meters. The lake Laacher See, which is often referred to as a maar, is much larger than the true Eifel maars as it was not caused by a phreatomagmatic eruption. It instead represents a volcanic caldera (Sect. 14.1).

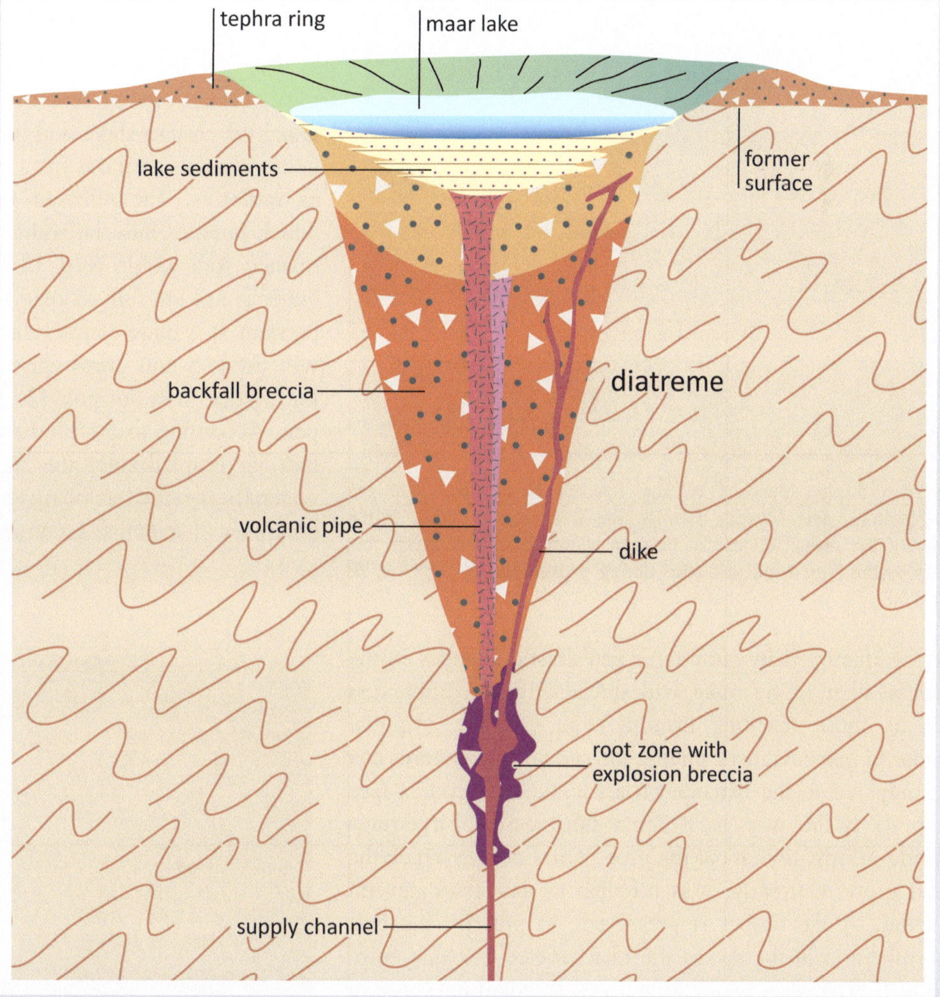

Fig. 14.5 Cross-section through a maar funnel after the termination of volcanism and development of a maar lake. Modified after Lorenz (2000)

Fig. 14.6 Tephra deposits of the Laacher See volcano at the Wingertsberg, near Mendig, forming the Wingertsberg Wall. In the center of the image, dunes can be seen within the tephra layers that testify to the high wind velocities during deposition. [N 50°23′31″/E 07°16′30″]

may be disrupted by short-term and drastic changes in the local weather or by more widespread climatic differences such as those resulting from a large volcanic eruption. Similar to paleomagnetic dating, the annual ring patterns can be compared to the barcode on a commercial product (see Chap. 4). In this way, the different thicknesses of tree rings provide information about the time period during which the wood grew. Currently, it is possible to date back several millennia if the timber is preserved. At present, the best determination of the age of the Laacher See volcanic eruption is 12,916 years before today (as a standard reference to 1950, Baales et al. 2002) or 10.966 years BC (Jöris and Weninger 2000). The approximate time of eruption was also probably in May, as determined by means of the plant pollen found in the ash layers. The weeds (sedge) had already faded, while bird cherries belonging to the rose family were not yet ripe.

Fig. 14.7 Trass caves in the Brohl Valley, north of the Laacher See. [N 50°27′37″/E 07°17′48″]

The activity of the Laacher See volcano has not completely ceased as evident from the release of gas bubbles along the eastern shore and in the open part of the lake. Gas still nowadays extrudes from outlets in the lake floor known as **mofettes**. The released gas is carbon dioxide (CO_2), which bubbles into the water with an intensity of a freshly opened soda bottle (Fig. 14.8). This activity indicates the crater is not yet fully cooled and a hot volcanic hearth lays beneath. It is quite conceivable that the Laacher See volcano will awaken and erupt again. However, based on current knowledge, it is not possible to predict when this will happen. According to an international agreement, any volcano that has erupted since the end of the last glacial period is generally regarded as being active. From this point of view, a number of scientists recommend continuous monitoring of

Fig. 14.8 CO_2 mofettes (gas release from the lake floor) at the edge of the Laacher See. [N 50°24′46″/E 07°17′08″]

the volcanic fields in the Eifel area to detect any possible future threat.

The rather exaggerated disaster movie "*Vulkan*" (Volcano) based on the science fiction novel by Ulrich Schreiber (2006) and broadcast by the private TV station RTL, played out the scenario of a volcanic eruption and the disaster it could cause. Such a scenario is certainly not impossible in the Eifel, but at present, there is no indication of any build-up leading to a new volcanic eruption. According to Schmincke (2009), the only serious statement that can be made with regard to the Eifel volcanism is that a volcanic eruption will occur in the future but the time, place and type of eruption cannot be predicted. When volcanic sites are monitored and the available data completely documented and carefully analyzed, the imminent danger of an eruption can be usually recognized decades in advance (Schmincke et al. 2011).

The Laacher See volcano has at least two related volcanic centers: the **Rieden volcano** (about 400,000 years old) and the **Wehr volcano** (about 220,000 years old) both of which lie only a few kilometers east of the Laacher See. These are not to be seen as direct predecessors of the Laacher See volcano, whose history of eruption began some 30,000 years ago. Based on isotopic age dating of individual minerals (zircon and pyrochlore; Schmitt et al. 2011), it has been shown that the magma chamber beneath the volcano developed at this time at a shallow depth. About 4000 years before the eruption, the chamber gained in size but it is not really known why the eruption then waited several thousand years. Such eruptions are probably triggered by the buildup of tectonic stress in the Earth's crust, the accumulation of volcanic gases that increase fluid pressure, or by the addition of new magma sourced from depth.

The Laacher See volcano is located in an area where volcanic events have occurred regularly since the Eocene. Volcanic centers can be found at more than 400 places in the Eifel with the occurrence of effluents, ejected volcanic material and maars (Fig. 14.9). Maars are typical for the Eifel area, whose volcanic origin was recognized already at the beginning of the 19th century (Box 24: Maars). The origin of these volcanic structures was central to the dispute between the Neptunists who interpreted all rocks as sediments deposited in the waters of the oceans (including the famous German writer Johann Wolfgang von Goethe) and the Plutonists who considered rocks to be originally derived from volcanic-related forces.

The older Tertiary volcanic formations in the Eifel are concentrated in the High Eifel (*Hocheifel*). Younger vol-

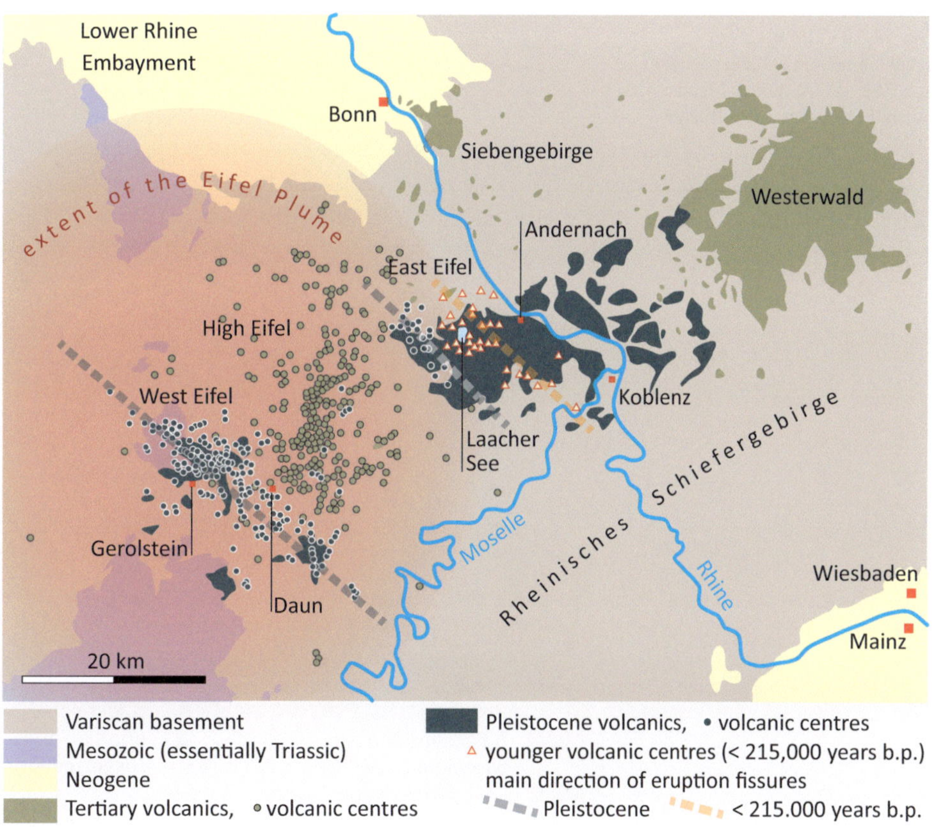

Fig. 14.9 Quaternary und Tertiary eruption centers in the volcanic areas of the Rhine-Moselle region. Western Eifel, eastern Eifel, Siebengebirge and Westerwald; modified after Schmincke (2008); extent of the Eifel plume after Bräuer et al. (2013)

canism occurred in the West Eifel (*West Eifel volcanic field*, WEVF) and the East Eifel (*East Eifel volcanic field*, EEVF) and erupted along NW-SE striking fault zones during the Quaternary about 650,000 years ago (Fig. 14.1). Since volcanic eruptions in the Eifel usually produced easily eroded tuff deposits, the High Eifel is characterized by volcanic chimneys carved out of the landscape forming roundish hilltops. The Tertiary tuff deposits have long since been eroded away but the alkali basalts remain together with andesites and trachytes dated between 44 and 35 Ma. A striking mountain consisting of Tertiary basalts is the *Hohe Acht,* at 747 m above sea-level. This represents the highest mountain in the Eifel, together with the neighboring *Raßberg* (667 m) and the *Nürburg* (678 m), near Adenau.

The volcanic remnants of the High Eifel differ considerably in their chemical composition from the Quaternary Eifel volcanic rocks, which are not genetically related despite their close geographical proximity. Volcanic craters are still preserved in the East and West Eifel, where the most recent eruption took place just 11,000 years ago. The youngest maars, whose ages have not yet been accurately determined (Ulmener Maar, Pulvermaar), may well be younger than the Laacher See eruption as the attempts at age dating spread between 6500 and 12,400 years. Ages around 10,000 to 11,000 years are considered the most likely.

At Schalkenmehren and Gemünden, there is a group of water-filled and dry maars occurring in a distance of little more than a kilometer (Fig. 14.10). The Gemünden maar, the Weinfeld maar and the double maar of Schalkenmehren, comprising of a western water-filled and an eastern dry maar, are arranged like a string of pearls along a suspected **feeder channel** in the ground where the magma ascended. As the groundwater seeped downwards into the fracture zone, it led to the maar explosions as the water came into contact with the hot ascending magma (Box 24: Maars). The alignment of the feeder channel in an NW–SE direction is in alignment with the Pleistocene stress field (Fig. 8.3), whose main direction of extension was NE-SW and perpendicular to the feeder channels (Fig. 14.10).

In addition to the well known Eifel volcanic maars and the Laacher See, other forms of volcanic structures and deposits occur in the same region. **Lava flows** composed of alkali basalt are widespread and often extend over several kilometers. These are well developed in the Nette Valley, near Mayen, and at the Ettringer Lay between Mayen and Mendig. The basalt flow exposed here belongs to the volcano group of the Bellerberg, which erupted 200,000 years ago and spread lava flows over the landscape. The basalts have been mined since ancient times with some of the oldest basalt lava mines extending back more than 7000 years. The

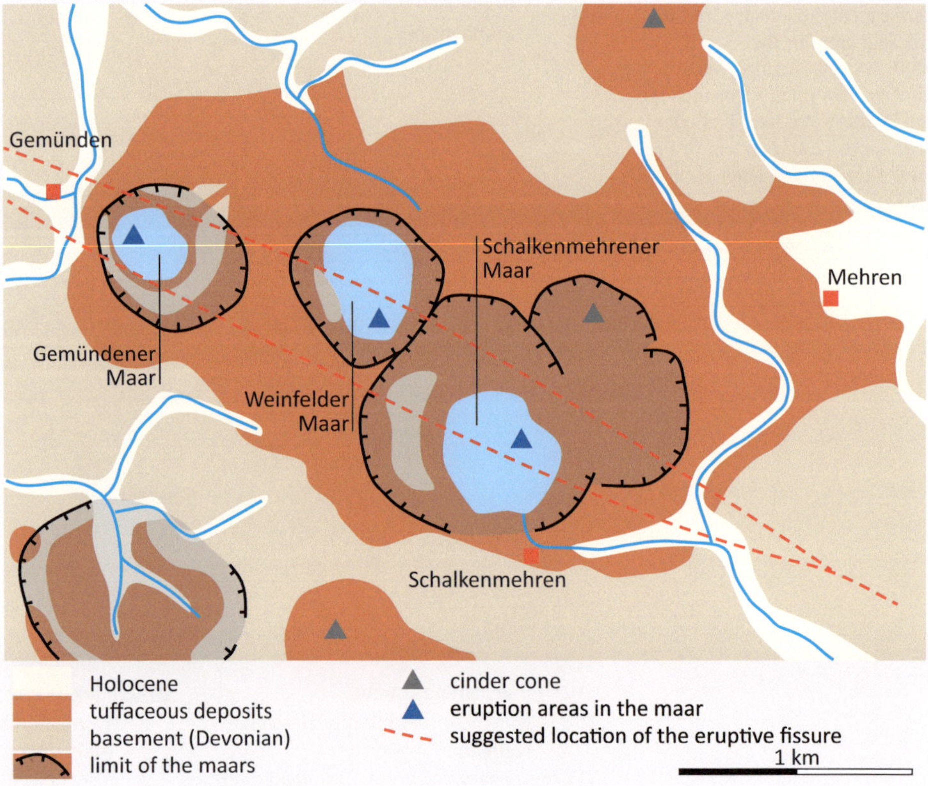

Fig. 14.10 The maars of Schalkenmehren and their relation to a ESE-WSW directed feeder channel. Modified after Büchel and Krawczyk (1986)

rock was used as a millstone for grinding cereals. Since the Middle Ages, the basalt was also mined underground creating large cavities. These were subsequently used as storage cellars, in particular for the fermentation and storage of beer in a cool environment. Today the cavities of the cooling cellars are no longer in use but instead are inhabited by one of Germany's largest bat populations.

The chemical composition of the basaltic rocks of the Eifel volcanism indicates that they are intracontinental basalts probably originating from a hot spot (Fig. 14.9). This hot spot is known as the **Eifel Plume** and has been extensively studied over the last two decades. It mainly extruded alkali basalts and other volcanic rocks with minerals that are referred to as feldspathoids (foids) due to their deficiency of SiO_2 in the rocks chemical composition (Fig. 3.12). The most common rock types are nephelinites (mostly composed of the mineral nepheline: a foid), basanites, tephrites, and phonolites. The lithosphere beneath the Eifel, comprised of the upper mantle and continental crust, has a total thickness of 70 km, whereas the crust alone is about 30 km thick at this location. This is considerably thinner than the normal lithospheric thickness of west and east Europe, which is normally about 90–110 km thick (Fig. 14.1). In areas of extension, the ascending mantle material of the asthenosphere partially melts during decompression and ascends along magma pathways to the surface where volcanism occurs.

Based on seismotomographic investigations, a three-dimensional image of the Eifel plume has been developed (Fig. 14.11, modified according to Ritter 2004). It is constructed from the fluctuations in the travel-times of seismic waves that pass through the Earth after the signals are filtered to image different temperature domains. High seismic velocities indicate relatively cold material, whereas low velocities indicate hot material assumed to represent the underlying hot spot where molten material ascends. Such large bodies with low seismic velocities are interpreted as **mantle diapirs** which, similar to the Eifel region, are found beneath many active volcanic regions. A clear connection, therefore, exists between ascending mantle diapirs and volcanic events at the surface.

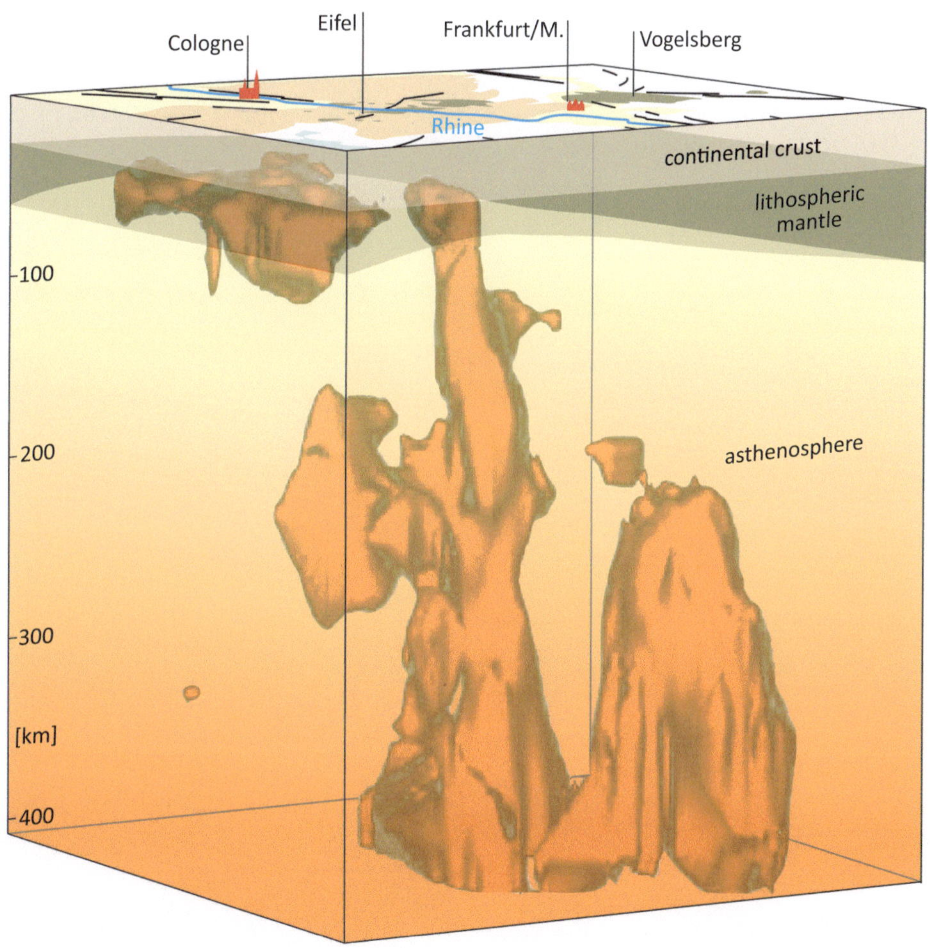

Fig. 14.11 The mantle (the Eifel-Plume) beneath the volcanic fields of the Eifel based on seismotomographic investigations. Modified after Ritter (2004)

14.2 Westerwald, Siebengebirge, Vogelsberg, Rhön, and Heldburger Gangschar

Relatively recent deposits of basalt occur in the **Westerwald** with ages of about 300,000–800,000 years. These are genetically related to the volcanism of the East Eifel volcanic field (EEVF) (Fig. 14.1). However, most of the large basaltic and trachytic volcanic deposits of the Westerwald are older and developed in the Late Oligocene to Early Miocene (28–22 Ma) and the Late Miocene (8–5 million years ago). The latter differs from the volcanic rocks of the Eifel not only in age but also in chemistry and correlate with the volcanic rocks of the neighboring **Siebengebirge** (28–22 Ma). A striking example of the older volcanism of the Westerwald is the ca. 25 Ma old basalt cone of the Druidenstein in the northern Rhineland-Palatinate, located a few kilometers east of Betzdorf.

Both the volcanism of the Westerwald and the Siebengebirge (Fig. 14.1) formed during extensional movements associated with initiation of the Tertiary graben systems. The volcanoes of the Siebengebirge are aligned along NW–SE striking fault zones that belong to the graben faults at the northeastern edge of the Lower Rhine Bay. The extensional movements caused subsidence in the Neuwied Basin and resulted in the ascent of magma from great depths typical of an intracontinental crustal setting.

After a period of plinian eruptions, trachytes and latites typical of the Siebengebirge intruded into previously deposited tuffs. Particularly impressive is the **trachytic cryptodome** at the Drachenfels, near Bonn, and the neighboring **latitic lava dome** of the Wolkenburg (Fig. 14.12). These volcanic structures were later exposed by the progressive erosion by the Rhine River. Today, the trachytes of the Drachenfels are exposed in outcrops below the castle ruins. These rocks are particularly known for their several

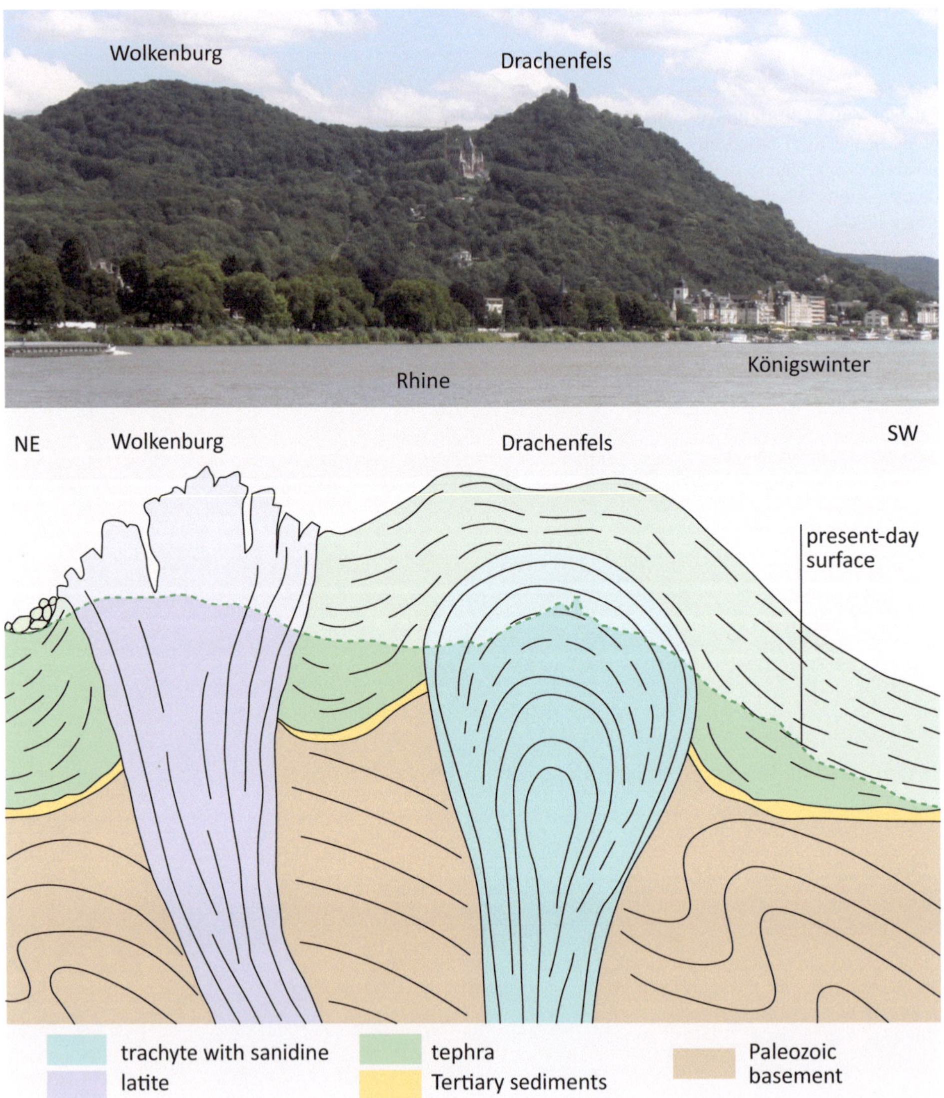

Fig. 14.12 Cross-section through the trachytic cryptodome of the Drachenfels and the latitic chimney of the Wolkenburg in the Siebengebirge, near Bonn. Modified after Wunderlich (1968)

centimeter-sized sanidine inclusions (a high-temperature form of alkali feldspar), which are often developed as twins. In the 13th century, the trachytes of the Drachenfels were used for the construction of the Cologne Cathedral because the quarries were conveniently located in direct proximity to the Rhine and the material could be the easily transported by boat. Unfortunately, the rock is very prone to weathering and, therefore, basalt from the Vogelsberg area is still used for necessary repair work.

Although the **Vogelsberg** is the largest coherent volcanic structure in Central Europe (Fig. 14.1) with a size of 2100 km^2 (about the size of the Saarland), it is not a stratovolcano as suggested by various authors and drawn by Cloos (1950) in his structural map of the German low mountains. It is instead composed of many layers of predominantly basaltic and trachytic lava. These findings are mainly the result of a research project carried out in 2007 at the Rehberg in the Vogelsberg area. The current shape of the Vogelsberg Mountain, with its branches extending all the way to Frankfurt am Main, is reminiscent of a shield volcano. However, its form resulted from the complex interplay of uplift and erosion rather than representing a large eruptive center with a central crater. The Vogelsberg volcanic complex, which includes the volcanic areas of the Hessian Depression and the Knüllwald, was active during the Miocene with volcanism beginning about 20 million years ago. The volcanic activity reached its maximum between 15 and 18 Ma and ended at about 10 Ma. Following the initial extrusion of trachytes and phonolites, basalts (alkali basalts and tholeiites) were formed. The volcanic rocks were then partly converted into red clays by **laterite weathering** under the then subtropical conditions. Such deposits are preserved at many sites in the Vogelsberg area. In some places, bauxite (Al ore) and iron ores were also formed at this time and were mined from the Middle Ages up to the 20th century.

The **Hessian Depression** forms the northern continuation of the Upper Rhine Graben having developed around the same time (Fig. 14.1). Volcanism started in this area later in the Miocene and took place between 11 and 15 million years ago. It partly coincided with the Vogelsberg volcanism, although the main eruptive phase in the Hessian Depression did begin earlier. The volcanic rocks are mainly composed of basalts and foid-bearing basic varieties as, for example, those developed at the Hoher Meissner, near Kassel, which erupted 11.5 million years ago. Basaltic tuffs dominate the Habichtswald, also near Kassel. The volcanic activity in the Rhön, located toward the southeast, occurred over a longer time from 25 to 11 Ma, with a main eruptive phase between 22 and 18 Ma. The composition of the magma was diverse with the eruption of basalt, basanite, nephelinite, and trachyte (Fig. 3.12). A similar diversity is also developed at the Milseburg in the western **Rhön**, where phonolithic and trachytic magmas dominate. This contrasts with the hill range of the High Rhön, with basanites occurring at the highest elevation marked by the Wasserkuppe. In the Grabfeld area, some 40 km to the east of the Rhön, NNE-SSW trending dikes are located at the **Heldburger Gangschar**. These parallel the direction of the graben system of the Upper Rhine Graben (Fig. 14.1). The dikes are mainly derived from basaltic to phonolithic magma. Overall, it is estimated that several hundred meters of overburden may have been removed from these Tertiary volcanic areas.

14.3 Small Chimneys in the Odenwald and the Messel Pit

The oldest volcanic rocks connected to Tertiary graben formation in Germany are located at the northeastern edge of the Upper Rhine Graben in the Odenwald area. The volcano of the **Katzenbuckel** began its activity already at the end of the Cretaceous period at about 68 Ma and erupted again at 40 Ma. It was presumably an explosive maar volcano with funnel breccia-containing rocks from the overlying layers that in the meantime had been eroded away. The volcanic remnants of the volcano are composed of foid-bearing basalt and funnel breccia located in the midst of the Buntsandstein plateau of the Odenwald. As the breccia also contains fragments of Late Triassic and Jurassic rocks, it is therefore assumed that the land surface at the time of the eruption was 500–600 m higher than it is today. Two other basaltic volcanoes occur in the vicinity centered at Rossberg (52 Ma) and the Otzberg (35-20 Ma).

The formation of the renowned oil shale pit of **Messel** was also related to Tertiary volcanic activity. This oil pit is located on top of the **Sprendlinger Horst**, in the northern extension of the Odenwald, near Darmstadt (Fig. 14.13). The depositional conditions at the site were particularly favorable for the preservation of a large number of Eocene fossils and provides a unique insight into the flora and fauna of that time. Bituminous shale, iron ore, and brown coal have been mined in the Messel pit for more than 100 years until 1970. The oil shale was heated to produce petroleum products in a process known as retorting. In the 1980s the mine was to be used as a dump, but this was stopped by citizens' initiatives, demonstrations and finally due to a formal error by the authorities in 1990. Only a short time later, in 1995, the former oil shale pit was included in the list of UNESCO World Natural Heritage sites, being a protected location since.

In the Middle Eocene, about 49 million years ago, maar eruptions occurred in the Messel region, similar to the more recent Eifel volcanism. Water that met ascending magma in the upper crust was presumably related to active extension and subsidence in the Upper Rhine Graben area, which began at that time. The resulting phreatomagmatic eruptions

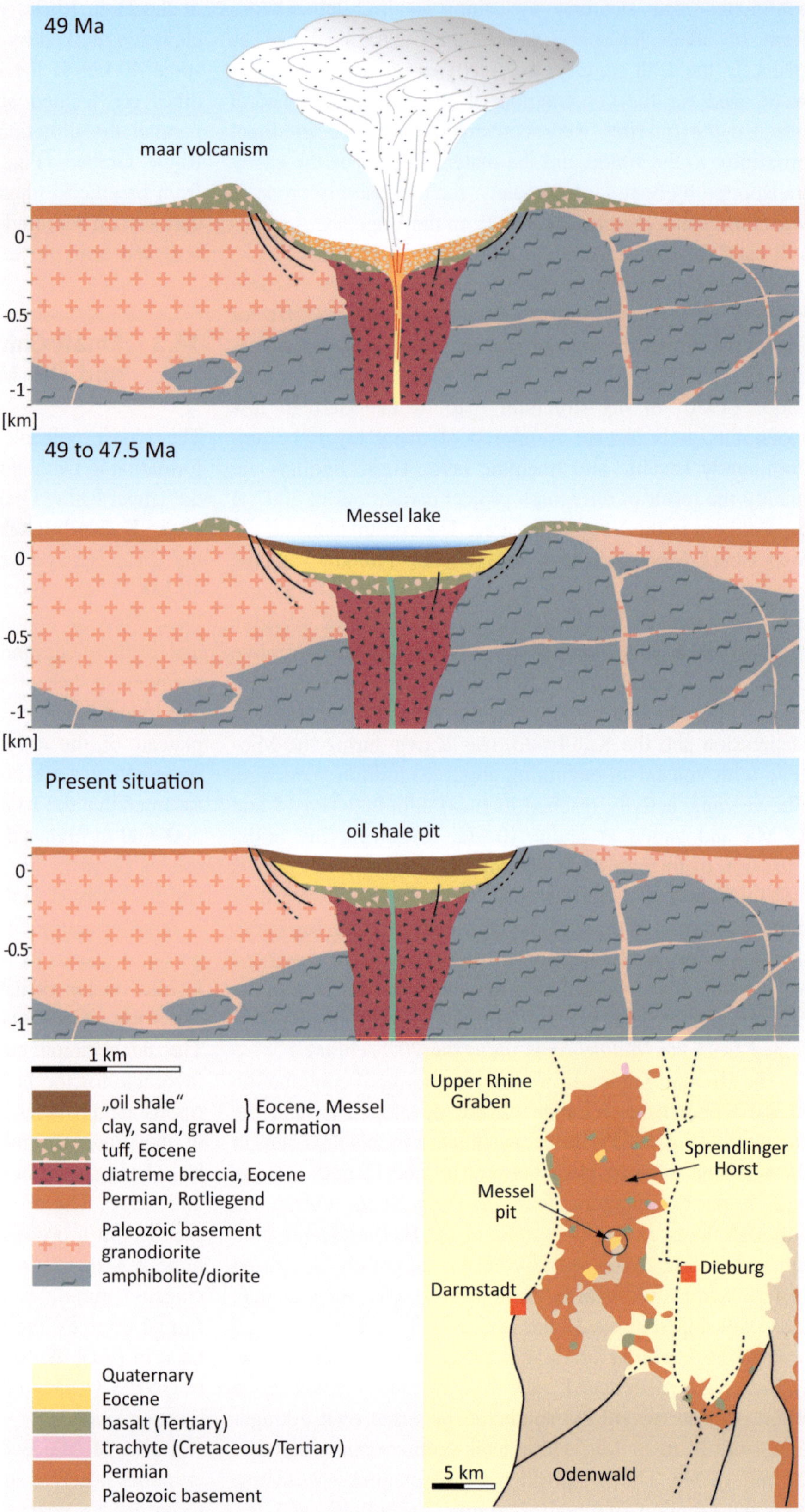

Fig. 14.13 Formation of the oil shale pit at Messel during Tertiary maar volcanism and the subsequent development of a lake in the maar crater. The lake bottom developed an anoxic zone where organic residues were preserved. Modified after Harms et al. (1999a), and Buness et al. (2005); map section after Harms et al. (1999b)

produced a number of maars in the area. In Messel, a particularly deep maar formed, where for about 1.5 Myr a lake developed with O_2-poor water characterizing the deeper water layers due to the lack of ventilation (Fig. 14.13). At the time of the maar formation, tropical climate conditions prevailed in Central Europe, allowing a rich fauna and flora to thrive. Dead organisms washed into the lake during periods of flooding did not decay. Instead, they were preserved as a bituminous shale layer with a total thickness of up to 150 meters. The **oil shale** is a finely laminated alternation of algal laminates and layers of shale with a total organic content of about 30% (Goth 1990). The paper-thin fine layering of the sediment formed during deposition has remained intact since bioturbation by burrowing organisms was absent. In this way, a large number of organisms that are generally not preserved under normal circumstances, have been handed down to posterity and today reveal to us a unique picture of the living Eocene world (Box 17: Euxinian environment). The fossils are partly so well preserved that even the dazzling colors of the insect carapaces are still visible. The most famous fossils of the Messel pit are the primeval horses (*Propalaeotherium*). These are dog-sized relatives of the present horse, which although not representing their immediate ancestors, developed shortly afterward as an extinct side-branch. There are also countless remains of crocodiles, turtles, birds, fish, frogs, insects and diverse plant assemblages preserved. Even a completely preserved skeleton of a primate species (*Darwinius masillae*) was recently found (Franzen et al. 2009). In short, the Messel pit turned out to be a real stroke of luck for paleontology.

14.4 Kaiserstuhl

The **Kaiserstuhl**, located northwest of Freiburg in Breisgau, was formed in a relatively short time in the Miocene between 18 and 16 Ma. It resulted from the largest and the last volcanic event in the Upper Rhine Graben, forming a classic stratovolcano probably more than 1000 m high. Today, however, only the strongly eroded volcanic remnants remain. The composition of the magmatic rocks produced all belong to the foid-bearing rocks of the alkaline series (SiO_2 undersaturated; Fig. 3.12). These differ significantly from the Tertiary volcanic complexes of the Middle-German volcanic belt extending from the Eifel in the west to the Eger Graben in the east. Pyroclastic rocks, mostly of tephritic composition, dominate the surficial volcanic rocks of the Kaiserstuhl. Lava flows with basanites also occur with olivine nephelinite. These rocks include limburgite, named after the Limburg near the Kaiserstuhl. Although the rock looks like a basalt in appearance, it normally contains no feldspar. The subvolcanic rocks of the Kaiserstuhl, exposed after erosion of several hundred meters of overburden, contain some special features. These rocks consist mainly of essexites (the subvolcanic equivalents of phonotephrites and tephrophonolites, Fig. 3.12) and theralites (nepheline bearing gabbros), both of which are often cut by dikes of similar composition. These volcanics also contain enclaves (xenoliths) derived from great depths in the upper mantle of the Earth. Overall, the subvolcanics of the Kaiserstuhl are characterized by a large variety of rare rock types. In the center of the Kaiserstuhl, a 1 km^2 large carbonatite complex occurs, representing the youngest part of the volcanic sequence. **Carbonatites** are magmatic rocks that are composed of 70–90% carbonate minerals (calcite, dolomite, and siderite) rather than silicate phases. These rock types are not common and occur largely in rift systems and hot spots. Today, carbonatite complexes are most abundant in the East African graben system. Another example is the Fen complex in southern Norway, located in the area of the Oslo Graben. As the Fen carbonatite complex is about 540 Ma in age and formed at the beginning of the Cambrian Period, it has no genetic relationship with the young rifting that gave rise to the graben.

Large areas of the Kaiserstuhl are today covered by loess sediment that formed during the last glacial period. The silt-sized loess was deposited as airborne aeolian sediment in a vegetation-poor and dry periglacial climate. The loess accumulated in the wind shadow of the Kaiserstuhl, forming a layer with thicknesses of up to 40 m. The deposits form extremely fertile soil of great importance for the viticulture of the Kaiserstuhl.

14.5 Tuff Chimneys of Bad Urach, Hegau

In the surroundings of Bad Urach, more than 350 small volcanic structures occur in an area of 40 × 40 km, all of which are between 17 and 16 Ma old. They were described by Cloos (1941) and referred to as the "Swabian volcano". In many places, volcanic rocks and ashes can be observed in outcrops, but some of the eruption sites are detectable only by geophysical methods. The volcanic events were dominated by explosive volcanism producing cinder cones and maars. For example, a maar crater formed 17 million years ago in the **Randecker Maar**, near Kirchheim/Teck, and only half of it is exposed today, for the other half has been eroded away. Lacustrine sediments accumulated in the crater lake where the fauna and flora were fossilized to preserve the life forms of the Early to Middle Miocene. Lava was extruded only in isolated cases as the magma typically solidified in the vent. One of the few occurrences of melilitic basalt (melilite is a rarely occurring silicate mineral mainly found in strongly mafic and SiO_2-undersaturated volcanic rocks) is exposed at the surface at the Sternberg, near Gomadingen, and is thought to represent part of the solidified fillings of a lava lake (Kröchert et al. 2009). The relatively large volume of volcanic rock at the Sternberg is considered to influence the source of a local spring, the *Lauterquelle*, located at the foot of the hill. Unlike other sources of karst in the region, the spring is considered to have a permanent source as it is mainly fed from the water-storing basaltic rocks that act as a hydrological reservoir. The water is much softer than in the surrounding area due to exchange reactions with the White Jurassic limestone on the east side of the Sternberg. The Mg originally dissolved from the basalt was exchanged by Ca, causing millimeter-sized authigenic (=newly formed in the rock) dolomite crystals to grow in the limestones. When the limestone is weathered, the dolomite crystals remain insoluble, while calcite is dissolved away. This results in the formation of dolomite sand with notably sharp-edged grains. This was mined in earlier times and used as a scouring powder to sand down-floors. Traces of the excavation remain today as disused pits.

The edge of a volcanic vent is well exposed in a roadcut at the Neuffener Steige (Fig. 14.14). Funnel breccias are also exposed in the hills of the Jusi, right on the edge of the Swabian Alb, near Metzingen, and at the Calwerbühl, near Dettingen/Erms. The funnel breccia of these explosive volcanoes contains rocks from the Mesozoic cover (Jurassic and Triassic) and the underlying metamorphic rocks and granites of the lower continental crystalline basement. In the Swabian Alb, the volcanic vents are sometimes visible as sinks where water reservoirs are located due to the water-storage properties of the vent fillings. Some place names with the endings -hilben or -hülben point to this fact, since the old German terms *Hüle*, *Hilbe* or *Hülbe* refer to ponds in the area of the Swabian Alb. However, in the foothills of the Swabian Alb, known as the *Albvorland*, volcanic vents are formed that are more resistant to erosion and thus form roundish hilltops, sometimes with circular elevations (e.g., Calwerbühl near Dettingen/Erms, Georgenberg near Reutlingen, Floriansberg near Metzingen, and Limburg near Kirchheim/Teck).

Fig. 14.14 Contact between White Jurassic limestone and a volcanic vent filling of Tertiary age at the Neuffener Steige, Swabian Alb, between Reutlingen and Kirchheim/Teck. [N 48°32′17″/E 09°24′25″]

In the Middle Miocene, between 14 and 6 Ma, intense volcanic activity in the **Hegau** region, on the western end of Lake Constance, produced several volcanoes characterized by large quantities of tephra (Fig. 14.1). Volcanic vents of predominantly phonolithic and later melilitic composition intruded into the several hundred meters thick tephra layers. During the last glacial period, the covering tephra was then eroded and the more resistant vents were left behind, thus shaping the current landscape of the Hegau. Remains of this volcanic activity can be found in the area around Singen, such as the Hohentwiel, Hohenkrähen, Hohenstoffeln or Höwenegg, together with eight to ten other vents preserved elsewhere. The formation of the Hegau volcanic vents was probably related to the extensional tectonics of the Upper Rhine Graben.

14.6 Eger Graben Area, Fichtel Mountains, Vogtland, and Lusatia

The **Eger Graben area** represents a rift zone that has been active since the beginning of the Tertiary. It is characterized by distinct zones of subsidence in the central Eger Graben in Bohemia, the Upper Palatinate, and the Fichtel Mountains. The oldest volcanic rocks occurring in this area formed some 65 million years ago in the Czech Osecna complex. The main phase of volcanism occurred in the Late Oligocene with volcanic centers in the Czech Duppauer Mountains (Fig. 14.15), as well as funnel breccias, domes, and dikes in the Bohemian Highlands. Like the Vogelsberg (see Sect. 14.2), the Duppauer Mountains are also regarded as a volcanic center consisting of a complex series of smaller eruptive centers, the remains of which can be seen today. Volcanic rocks, often with impressive columnar basalt (Box 2: Basalt), can be observed at the following locations: (i) Scheibenberg (21.5 Ma; Fig. 14.16), (ii) Hirtstein (24 Ma; Fig. 14.17) in the upper Erzgebirge, (iii) Hoher Parkstein (24 Ma), near Weiden in the Upper Palatinate,

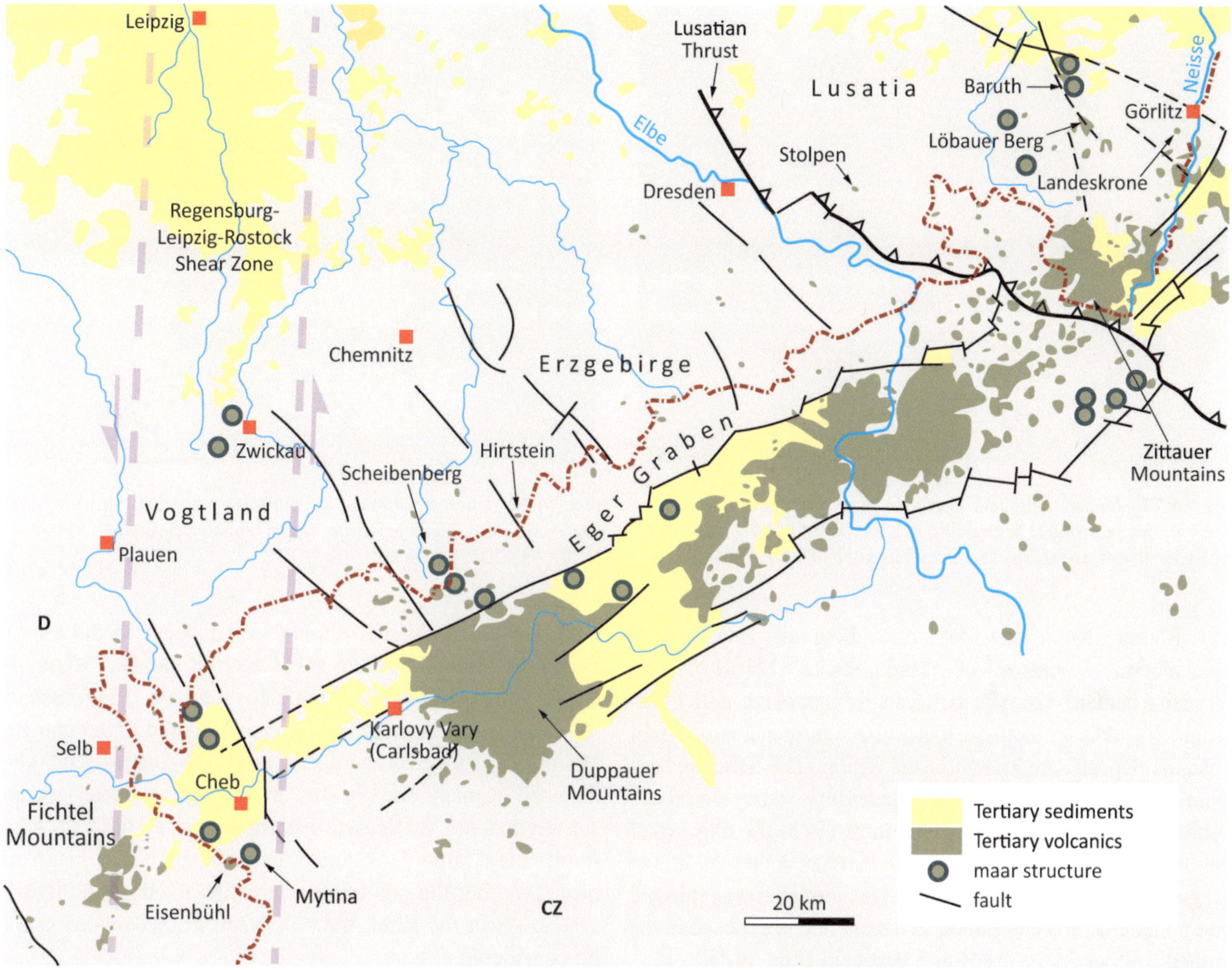

Fig. 14.15 Occurrence of Tertiary magmatic rocks in the area of the Eger Graben. Modified after Pfeiffer and Suhr (2008); location of the shear zone and maar structures after Nickschick et al. (2014)

Fig. 14.16 Vertical columnar basalt within a lava flow at the Scheibenberg, southwest of Annaberg in the Erzgebirge. [N 50°32′22″/E 12°55′25″]

Fig. 14.17 Pennate columnar basalt developed within a cryptodome (subvolcanic intrusion at several tens to hundreds of meters depth) at the Hirtstein, east of Annaberg in the Erzgebirge. [N 50°32′11″/E 13°11′36″]

Fig. 14.18 Curved columnar structures in a basaltic intrusion at the Parkstein, near Weiden in the Upper Palatinate, Fichtel Mountains. [N 49°43′42″/E 12°04′12″]

(iv) Rauher Kulm (21 Ma), near Bayreuth, (v) Großer Teichelberg, southeast of Marktredwitz (24 Ma), and (vi) Hirschentanz (26 Ma) between Konnersreuth and Pechbrunn. In addition to these locations, a further 15–20 small volcanic deposits are known in the region. The Scheibenberg represents a cooled lava flow and therefore consists entirely of vertical columnar basalt. A basaltic intrusion is also exposed at the Hoher Parkstein, where the columnar joints mark the irregular and curved cooling front. This intrusion was injected into a diatreme breccia formed as a maar and is exposed in the immediate vicinity of the Hoher Parkstein (Fig. 14.18).

In the eighteenth century, a dispute arose between the Neptunists and Plutonists over the origin of the basalt columns at the Scheibenberg. Abraham Gottlob Werner (1749–1817) was the founder of Neptunism, a school of thought also supported by Johann Wolfgang von Goethe (1794–1832). They believed that all rocks (including basalt) were formed one after the other by sedimentation in water during the successive flooding of a large ocean. The horizontal superposition of the basalts on top of sedimentary rocks exposed at the Scheibenberg was put forward as a major argument for this case. In contrast, the plutonists, led by James Hutton (1726–1797), believed such basalt lava flow were of volcanic origin and originated from the "central fire" in the interior of the Earth. This theory prevailed at the end of the controversy and marked the start of modern geology as we know it today.

Volcanism in the Eger Graben continued into the Quaternary. The Eisenbühl, near Neualbenreuth, is the youngest volcano in the Eger Graben area with an age of about 300,000 years. Another structure, known as the Mytina maar was recently discovered not far from Eisenbühl and dated as 288,000 years old (Mrlina et al. 2007).

Sources of CO_2-rich mineral water occur in the famous spas of Karlovy Vary in the Czech Republic, Mariánské Lázne (Marienbad), and Františkovy Lázně (Franzensbad), as well as the Saxonian health resorts of Bad Elster and Bad Brambach. Numerous mofettes and CO_2-sources have been detected at many sites in this area, which indicate volcanic activity has not yet ceased. Recent studies of the CO_2 gases imply they largely originate from independent magmatic processes, but the potential hazards involved are difficult to assess. As in the Eifel, the volcanism in these areas cannot be considered extinct.

A zone of magmatic-volcanic activity, whose importance has not yet been recognized, can be traced from Regensburg, through Cheb and Leipzig, to Rostock (Regensburg-Leipzig-Rostock shear zone, RLRZ; Bankwitz et al. 2003; Figs. 14.1 and 14.15). This zone is comprised of magmatic and subvolcanic rocks of the Delitzsch Massif that formed in the Late Cretaceous but is still seismically active today (Chap. 8, Fig. 8.1). Numerous dikes and some diatremes (maar structures) contain some unusual carbonatite and kimberlitic rocks (Seifert et al. 2000; Krüger et al. 2013). Other diatremes of probable Late Cretaceous to Pleistocene age are located in the intersection of the RLRZ with the Eger Graben (near Zwickau) and Cheb in the Czech Republic (e.g., Mytina maar, see above; Fig. 14.15). This zone probably marks the location of deep-seated faults that reach into the asthenosphere.

A number of other volcanic features occur as an eastern continuation of the Eger Graben and to the north of this area (Fig. 14.15). These represent eroded remnants of basalt or phonolithic lava and vent structures isolated by the effects of weathering. Examples include the Lausche and the Hochwald in the Zittau Mountains, the Finkenkoppe (Pěnkavčí

Fig. 14.19 Vertical columnar basalt at the Stolpener Burgberg in Lusatia, near Dresden. [N 51°02′53″/E 14°05′02″]

vrch) on the Czech side of the border, the Kaltenberg (Studenec), the Tannenberg (Jedlová), and the particularly striking cone-shaped Kleis (Klíč).

The landmark mountain of Görlitz, known as the Landeskrone, represents a volcano composed largely of volcanic slag that formed 34 million years ago as a solidified lava lake. The basaltic rock of the Löbauer Mountain, near Löbau, represents the largest volcanic deposit of Upper Lusatia. Due to the slow cooling of the lava, its medium-crystal size formed a dolerite instead of the typical fine-grained basalt. A series of volcanic explosions occurred at Baruth, near Bautzen, 28 million years ago. The volcanoes formed here are comparable to the Quaternary maar explosions of the Eifel and the resultant basalt flows produced a lava lake. This basalt was mined until the year 2000.

In the Lusatian area, east of Dresden, the eruption of a volcano occurred 25 million years ago, the rocks of which are now exposed on the Stolpener Burgberg (Fig. 14.19). According to the modern classification of igneous rocks, the basalt of Stolpen is an olivine-augite-nephelinite. The regular upright position of the basalt columns indicates solidification in a lava lake. This location is renowned for being worldwide regarded as the **type locality for basalt** (Box 2: Basalt). The eminent naturalist Georgius Agricola (1494–1555) introduced the term "basalt" in 1546 to describe the columnar basalt at the Stolpener Burgberg. The word probably originated from Egyptian, but was used in both Greek and Latin as a term for "very hard stone". The place name Stolpen also has something to do with the occurrence of the basalt because the Old Slavic word *stolpno* means nothing else than pillar. At the end of the 18th century, several famous scientists visited the village during the course of the historical dispute between the Neptunists and Plutonists (Scheibenberg, see above).

15 Asteroid Craters

An asteroid struck the southern Swabian Alb in the Middle Miocene resulting in two craters: the Nördlinger Ries and the Steinheim Basin. The consequences of this impact and the rocks produced by the shock metamorphism are described in this chapter.

In the Miocene, 14.6 million years ago, the area around Nördlingen was struck by an event that was to change the landscape in southern Germany for a long time. A 1.5 km large asteroid moving at a speed of between 50,000 and 150,000 km/h (about 15–40 km/s) hit the earth and impacted the Earth's crust to a depth of well over one kilometer. The strike formed the **Nördlinger Ries** with an **impact crater** of 24 km in diameter. Even parts of the crystalline basement and the overlying 400 m of the sedimentary cover was partly vaporized or ejected into the atmosphere. Within a radius of 100 km around the impact center, everything was flattened to the ground and all life destroyed. The effects on the landscape extended to several hundred kilometers away from the impact site.

Contemporaneous to the Nördlinger Ries, the **Steinheim Basin** was formed at a distance of 40 km away, which resulted from another asteroid impact (Fig. 15.1). It is now assumed that this was a small satellite asteroid of about 150 m large, whose impact accompanied the larger Ries body. Due to its smaller size, the effects of this asteroid were not as extreme as those caused by the Ries asteroid. Its collision crater is about 3.5 km in diameter and, unlike the Ries structure, the Steinheim Basin has a central elevation point characteristic of many impacts. With an age of 14.6 Ma (Buchner et al. 2010), the two asteroid craters represent some of the youngest well-preserved impact structures on Earth. The rocks and structures formed during the impact can be studied in a number of exposures.

During the asteroid strike at Ries (Fig. 15.2), the Earth's crust, including the asteroid, vaporized at the impact site due to the extremely high pressure and temperatures reached (up to 30,000 °C). This was accompanied by shock melting of the underlying rock caused by intense frictional heating. In 1973, a research borehole was drilled into the Ries crater (Fig. 15.3). The traces of metals, in particular chromium, nickel, and iron, which were discovered at a depth of 600 m, are presumed to represent remnants of the stone meteorite responsible for the Nördlinger Ries impact.

During the initial phase of impact, large quantities of rock material were ejected from the crater. Blocks up to kilometers in size were displaced and rock fragments derived from sedimentary rocks and the largely granitic basement were compressed together into a breccia known as the **Bunte Trümmerbrekzie** (colored debris breccia) of the Nördlinger Ries. The color varies because of the diverse mixture of different rocks including the granite basement, together with Triassic gray limestones, red clay, sandstones, and dark claystones from the Black and Brown Jurassic as well as the light colored limestones of the White Jurassic. Large Jurassic limestone blocks, known as **Reuterian Blocks**, can be found even at distances of 70 km from the crater. These blocks were named after the geologist Lothar Reuter, who mapped them in 1926 as ejected material from the Nördlinger Ries.

The Brock horizon, comprised of Ries-ejection material, is a distinct, widely distributed horizon representing a useful time marker in the Upper Freshwater Molasse (UFM) (Sect. 13.3). As the ejection cone migrated outward (Fig. 15.2), the ejecting masses were dragged across limestone surfaces of the White Jurassic, leaving behind abundant scratch marks (Fig. 15.3). These can be observed on preserved surfaces of the limestone and are arranged in a

M. Meschede and L. N. Warr, *The Geology of Germany*, Regional Geology Reviews,
https://doi.org/10.1007/978-3-319-76102-2_15

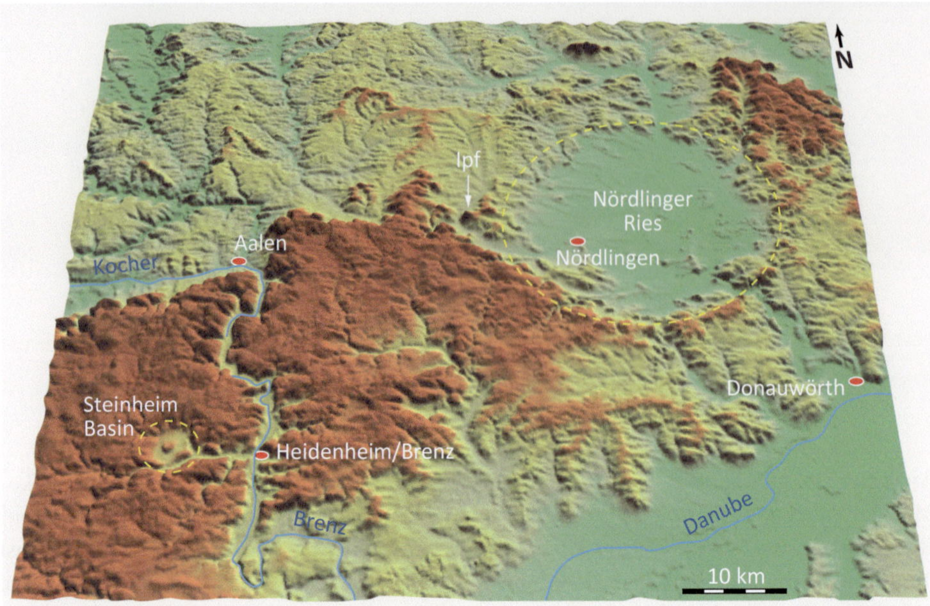

Fig. 15.1 Nördlinger Ries and the Steinheim Basin. The circular asteroid craters in the Swabian Alb of southern Germany are clearly visible in a digital terrain model (about 5 times exaggerated in vertical scale). The central elevation in the Steinheim Basin is also clearly visible

radial pattern around the center of the crater. Impressive examples reminiscent of glacier striations are exposed in a quarry, near Gundelsheim, east of the Nördlinger Ries (Fig. 15.4).

The molten rock ejected from the innermost part of the crater immediately after impact was carried large distances away, much farther than the Reuterian Blocks. The still liquid material was transported at high speed largely toward the east and solidified during its flight to form glass particles (Fig. 15.5). These glass particles, known as tectites, formed a forest-green colored variety called **moldavite**. This type of tectite was named after their location in the upper Moldavia of South Bohemia and Moravia and lie at a distance of up to 450 km from the impact crater (Fig. 15.6). The age of the moldavites, as determined by the ^{40}argon-^{39}argon dating method, falls in the range of 14.6–14.8 Ma (Di Vincenzo and Skála 2009) and corresponds very well with the age of the Ries event. The scattered strew field of tectites is located to the ENE of the impact site. This suggests that the meteorite impact was not perpendicular to the earth but probably reached the surface at an angle of 30°–40° to the horizontal.

Another special rock of the Nördlinger Ries impact is called **suevite**. This material consists largely of crushed rock, molten **impact glass**, modified **high-pressure minerals** and **diaplectic glass** (Fig. 15.7). It was named according to its place of origin in Swabia (lat. *suevia*) and its type locality is the Nördlinger Ries. The term is now used worldwide for rocks formed by asteroid impact with their similar compositions reflecting a mixture of high-pressure minerals and diaplectic glass. In contrast to "normal" rock glass formed by melting, diaplectic glass is produced without passing through a liquid phase. The lattice structure of crystals is destroyed by the extremely high pressures during impact and transformed directly into an amorphous state.

Due to the high pressures and temperatures at the impact site, quartz (density 2.65 g/cm^3) was transformed into the rare high-pressure polymorphs of **coesite** (3.01 g/cm^3) and **stishovite** (4.3 g/cm^3). Stishovite is not a typical mineral on Earth and does not form during orogenic processes even at the highest possible metamorphic conditions. Therefore, the occurrence of stishovite in the suevite of the Nördlinger Ries is a clear indication of an impact event. Moreover, the suevite also contains small micrometer-sized diamonds formed during high-pressure shock metamorphism. It is the discovery of these high-pressure minerals that led to the acceptance of the Nördlinger Ries as an impact crater in the 1960s.

Prior to the 1960s, the Nördlinger Ries was considered to represent a volcanic formation known as the Ries volcano. A series of volcanic formations of similar age occur in the Bad Urach area of the Swabian Alb and in the Albvorland. The nature of the rocks, especially the suevites are very similar in appearance to volcanic tuff or ignimbrite. However, as early as the beginning of the twentieth century, the

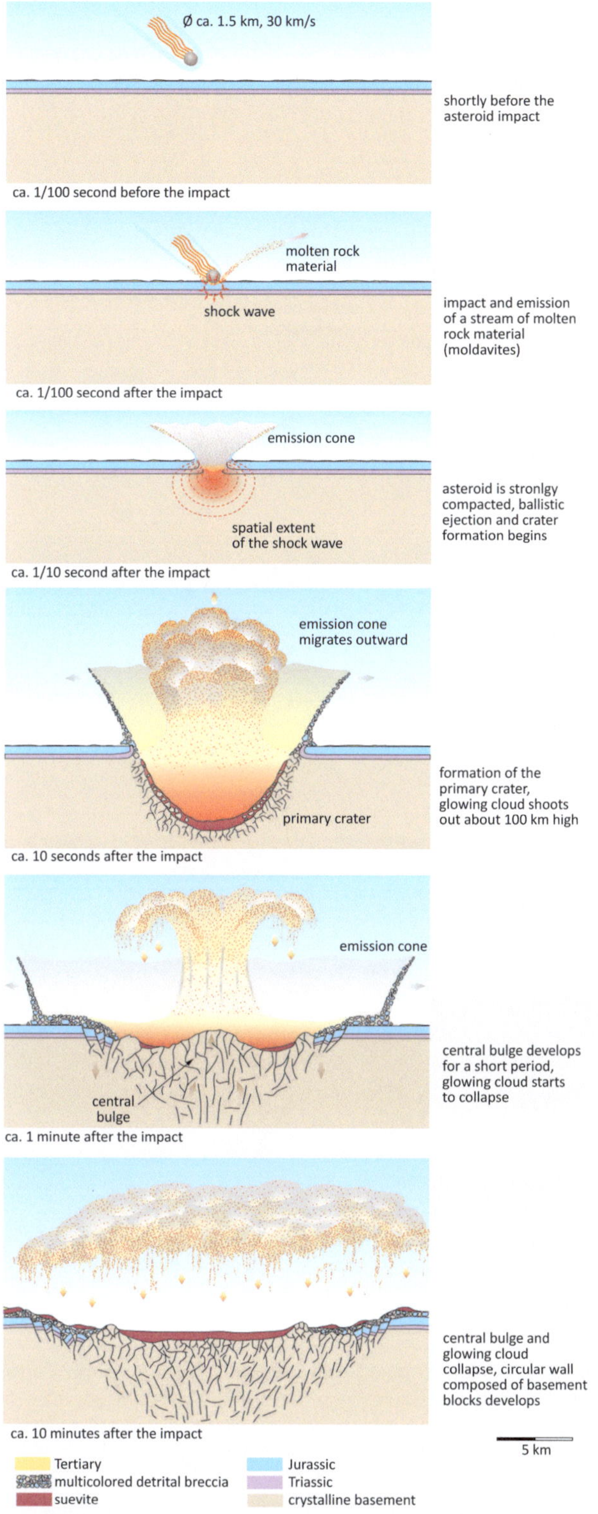

Fig. 15.2 Reconstruction of the first ten minutes of the asteroid impact in the Nördlinger Ries. Modified after the website of the Geopark Ries: www.geopark-ries.de/index.php/de/entstehung_rieskrater

possibility of an asteroid impact was also considered, but in the years that followed the theory was strongly opposed. Finally, Shoemaker and Chao (1961) provided convincing evidence for an asteroid crater by recognizing the high-pressure modifications of quartz at Nördlinger Ries. At the end of the 1960s, when the astronauts from the American Apollo program (NASA missions Apollo 14 and 17) visited the Nördlinger Ries, they were prepared for their moon mission by the Tübingen mineralogist and Ries researcher Wolf Juergen von Engelhardt (1910–2008), who presented them rock samples arising from asteroid impacts.

Following the ejection of suevite, the crater of the Nördlinger Ries was filled with a thickness of up to 400 m of debris, which today is covered by a thick sedimentary cover. Outside of the Ries crater, suevite overlies the colorful brecciated debris. During weathering, the vitreous constituents were physically removed from the groundmass of the suevite and are occasionally found on the ground at Ries or in its vicinity. Due to their shape, which resembles pieces of Swabian pancakes, they are called *Flädle* in the local language. Geochemical investigations have shown that the glass fragments were formed from the claystones in the area during the asteroid impact. The groundmass of the suevite is composed mainly of thermally altered sedimentary rock (Baier 2009).

A number of striking structures developed near both craters due to the immense pressure wave produced by the asteroid impacts. In places, shatter cones formed where the energy was not enough to vaporize or melt the rock. These were already described as early as 1905 in fine-grained limestones of the Steinheim Basin (Branco and Fraas 1905). The so-called ***Strahlenkalk*** **(radiant limestone)** was at the time of its discovery difficult to explain (see below, Fig. 15.11). Today, radiant limestones provide clear indications of a larger impact structure. Particularly impressive are the so-called Ries belemnites and Ries tektites formed by the intense pressure wave. Belemnite rostrae (part of the inner skeleton of a belemnite, often also called thunderbolts) or single rounded pebbles were fractured during impact and afterward, the spaces were filled with diagenetic mineral cement (Fig. 15.8).

Following ejection, the crater walls became unstable and slid inward forming the crater structure of today, with a diameter of about 24 km. An inner crystalline ring wall developed composed largely of strongly brecciated basement rock that rebounded into the central impact area. This effect formed the central hill structure found in the neighboring Steinheim Basin in the same way as a central elevated area is observed in high-velocity images of a droplet falling into a liquid. Following stabilization of the crater structure, with a depth of several hundred meters, it was then filled with water

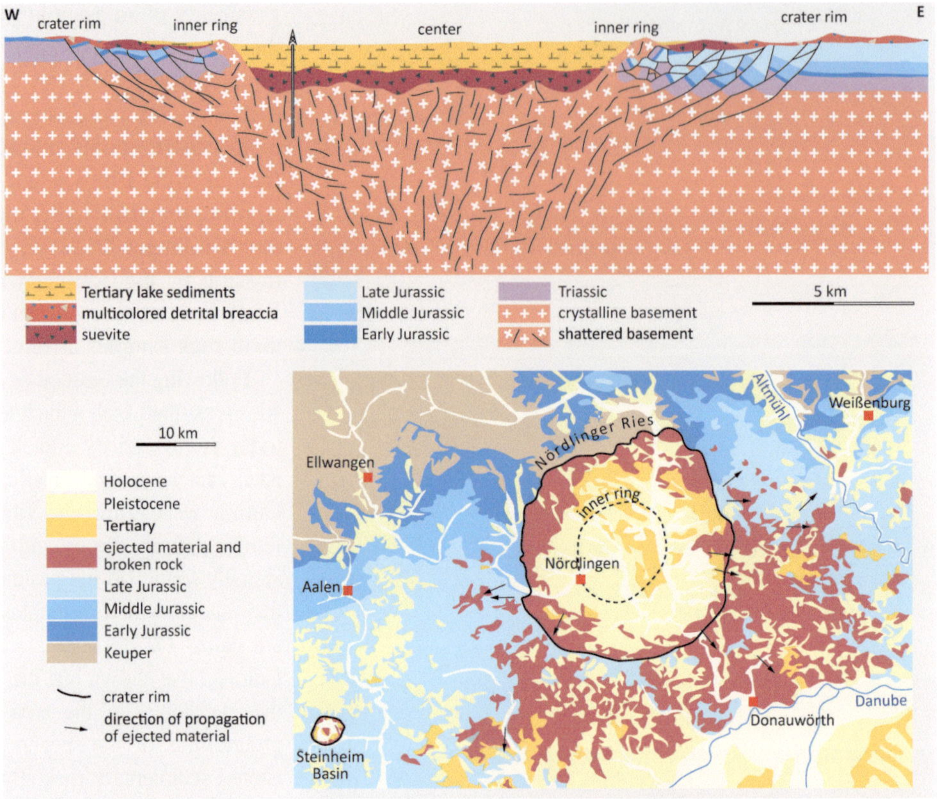

Fig. 15.3 Cross-section through the Nördlinger Ries and geological map of the Nördlinger Ries and Steinheim Basin area. Modified after the 1: 200,000 geological survey map, BGR (1981), Rothe (2005), and Pösges and Schieber (1994)

Fig. 15.4 Striations on the surface of the White Jurassic limestone plateau at the rim of the Nördlinger Ries crater exposed in a quarry, near Gundelsheim. The striations were formed immediately after the impact due to ejected material that sheared over the floor. [N 48°54′27″/E 10°49′52″]

Fig. 15.5 Moldavites from South Bohemia, Czech Republic. Collection of Stefan Meng (University of Germany) who kindly provided the photographs

and gradually with lake sediments over the following two million years. Even outside the Nördlinger Ries, lakes were formed by the damming of rivers. To the northeast of the Ries crater, precursors of the Altmuehl and the Rezat rivers were dammed to form the Rezat-Altmühl Lake. This lake temporarily extended up to the area of present-day Nuremberg (Fig. 15.9) and at that point was about twice the size of Lake Constance.

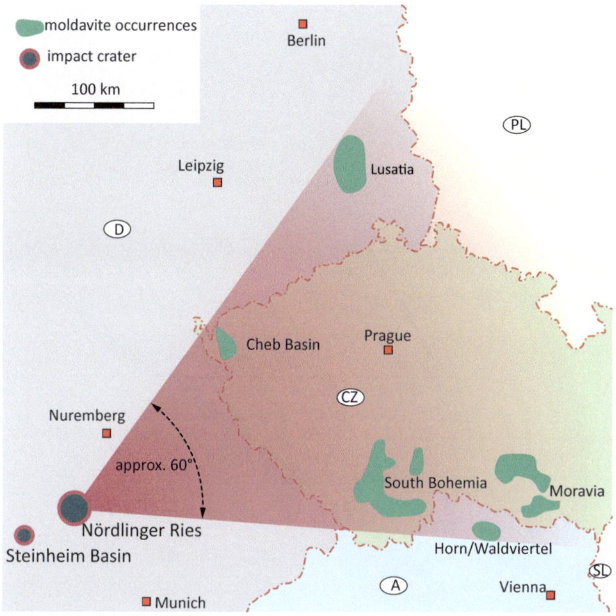

Fig. 15.6 Occurrence of moldavites from the Ries impact crater across central Middle Europe (southeast Germany, Czechia, Lower Austria). They are distributed in the form of an ENE oriented emission fan with an aperture angle of about 60°. Modified after Skála et al. (2009)

Fig. 15.7 Suevite in the Nördlinger Ries collected from the entrance to the village of Zipplingen about 10 km northwest of Nördlingen. [N 48°55′40″/E 10°24′34″]

Fig. 15.8 Conglomerate pebble fractured by the asteroid impact of the Nördlinger Ries. The photograph shows a specimen from the archive of Wolf Jürgen von Engelhardt (1910–2008), Tübingen, Germany

The **Ries Lake** was no ordinary lake. It was fed by several rivers but did not develop a major drainage system. As the water accumulated it became salty from solution processes in the subsurface. The rates of water evaporation and mineral dissolution were initially high due to the elevated temperatures of the impact site and the abundance of high-temperature minerals and glasses. The cooling of the Ries impact is a process that probably lasted for more than 1000 years. The lake developed an extremely species-poor fauna compared with today's salt or soda lakes in dry areas. However, some individual species proliferated. The lake is thought to have been initially almost 200 m deep and then gradually shallowed. The rather special depositional conditions formed dolomitic green algae bioherms (Fig. 15.10) and snail-rich carbonate sands on the crater rim and in the area of the inner ring wall. In the deeper waters, clays and organic-rich muds formed that resulted in oil shales. Only later, at the end of the sedimentary infilling of the basin, after about 2 Myr, did the living conditions return to a rich flora and fauna represented by thin brown coals. These contain fossil remains of various vertebrate species including turtles, small deer and different species of birds.

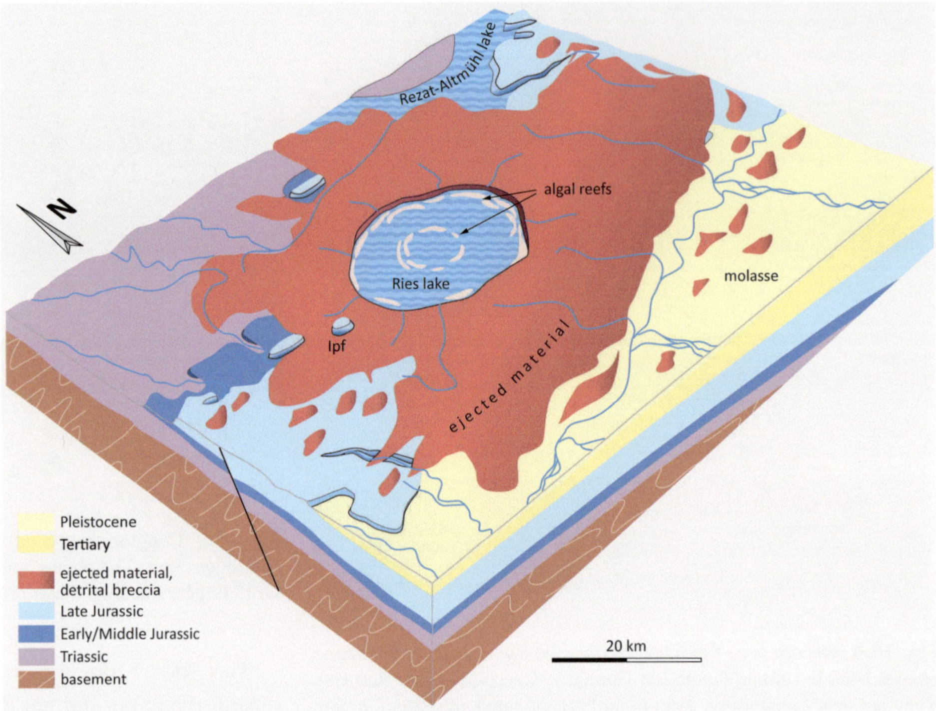

Fig. 15.9 The Ries and Rezat-Altmühl lakes formed because of the asteroid impact of the Nördlinger Ries. Modified after Eberle et al. (2010)

Fig. 15.10 Bioherm of green algae from the Tertiary Ries Lake located beneath the castle of Wallerstein, north of Nördlingen. [N 48° 53′21″/E 10°28′35″]

Fig. 15.11 Radiant limestone (*Strahlenkalk*, shatter cones) in the Steinheim Basin was formed during shock wave metamorphism as the asteroid impacted the White Jurassic limestone. Figure with the kind permission of the Steinheim meteor crater museum. [N 48°40′39″/E 10°04′07″]

The current crater shape of the Nördlinger Ries has been modified by erosion since the late Tertiary and above all during recent glacial periods. Originally, the Tertiary sediments were about 100 m higher than the current level of the Ries plain and a large part of the debris masses have since been removed.

The considerably smaller Steinheim Basin is also characterized by impact, shock-related radiant Jurassic limestones with shatter cones (Fig. 15.11). Debris breccia developed in the Steinheim Basin, but rocks formed by melting, such as the suevite, are missing. Following impact,

Fig. 15.12 Fossil-rich limestone of the Miocene sedimentary filling of the Steinheim Basin with numerous snail shells from the Pharion sand quarry, near Steinheim am Albuch. [N 48°41′12″/E 10°28′35″]

a lake also developed in this crater. It remained for about 1 Myr and accumulated fossil-rich carbonates (Fig. 15.12), together with the well known **Steinheimer snail sands** that were later formed by the redeposition of sediment. During late 2016, a fragment of meteorite fragment was discovered in a limestone block recovered from the Steinheim Basin (Buchner et al., 2017). In contrast to the meteorite of the Nördlinger Ries, the material was that of an iron meteorite which presumably orbited the Nördlinger Ries meteorite as a small moon and struck the Earth simultaneously.

Germany During the Glacial Periods 16

The Earth is currently experiencing a warm phase within a glacial period. The last glaciation phase, which occurred about 10,000–15,000 years ago, formed a thick ice sheet that covered north Germany and parts of the Alps. The northern area was affected by the large bodies of inland ice that extended down from the North Pole. Such ice advancements occurred several times across the area during the past 400,000 years and some glaciations reached as far as the edge of the German low mountains of the Rheinisches Schiefergebirge. The remains of glacial and periglacial sediments now characterize these landscapes. The evolution of the Baltic Sea and its present form was much linked to the melting of the ice sheets, whereby the various stages of development were accompanied by an isostatic readjustment of the Scandinavian continental shelf during glacial retreat.

Even though warm climatic conditions with relatively mild winters and temperate summers prevail at present, we should be aware that we are living in an interglacial period between ice ages. This period of glaciation began in the South Pole during the Eocene. In contrast, in the northern hemisphere, large glaciations have occurred over the past 2.6 Myr, albeit subject to regular fluctuations in the amount of ice coverage. Approximately one million years ago, cycles of colder periods were established. Each cycle lasted for about 100,000 years whereby the cold periods were separated by shorter warm phases of 10,000–20,000 years duration. As we are living in one of these **warm phases** today, the natural progression of events would predict we eventually reenter a new **cold period**. However, due to the high emissions of anthropogenic CO_2 released from the burning of fossil fuels (oil, natural gas, and coal) we have succeeded in disturbing the natural balance of events in an extremely short time interval in terms of the geological record. As a result, it has become rather difficult to predict how the climate will develop in the long-term future.

During the past three periods of glaciation, large areas of the polar region were buried beneath many kilometers of ice. The inland ice of North America extended in places down to the 40° latitude, whereas in Europe the ice exceeded the 50° latitude (Fig. 16.1). In relation to the long periods of the Earth's history (Chap. 3), the current glaciation that covers about 10% of the Earth's surface is rather unusual. Most of the time, the Earth has been free of ice with no glaciation of the poles or mountainous regions (Fig. 16.2). In general, the past global climate was much warmer than it is today, with different living conditions that were often rather sensitive to environmental changes.

The duration and intensity of the annual **solar irradiation** received by the Earth's surface is influenced by a complex interplay of (i) the circulation of ocean currents determined by the arrangement of the lithospheric plates, (ii) the topography and coverage of the continents such as the location of large mountain ranges, vegetation, and ice, and iii) the periodic fluctuations of the **Earth's orbit parameters** (Box 25: Fluctuations of the orbit parameters) involving the inclination of the Earth's axis (41,000 year cycle), the ellipticity of the Earth's orbit around the Sun (100,000 year cycle), and the shift of the seasons by the circular motion of the Earth's axis (25,800 year cycle).

30–40 million years ago, the Tasmanian Strait opened between Australia and Antarctica. This created a new circumpolar sea current around the Antarctic which isolated the continent from the heat exchange of the oceans and led to a cooling of the global climate. The closure of the Panama Strait about 2.7 million years ago also led to further profound oceanographic changes by affecting the circulation of the ocean currents. Since then, there have been strong periodic climatic fluctuations which, according to present knowledge, have contributed to the repeating phases of glaciation experienced at high latitudes. These periodic fluctuations are referred to as **Milanković cycles**, named after their discoverer (Box 25: Fluctuations of the orbit parameters). Initially, the climatic fluctuations that correlate with the closure of the Panama Strait essentially reflect changes in the

M. Meschede and L. N. Warr, *The Geology of Germany*, Regional Geology Reviews,
https://doi.org/10.1007/978-3-319-76102-2_16

Fig. 16.1 Maximum ice coverage of the northern hemisphere during the last Quaternary glaciation phase (light yellow) compared with the extent of today's ice (purple). Blue arrows indicate the assumed ice movement direction during glacial advance. The maximum ice extent was not attained at the same time in all locations: in the east of Eurasia, the maximum was reached about 90,000 years ago, but in the west, it was achieved only about 24,000–22,000 years ago. Modified after Ehlers and Gibbard (2007)

inclination of the Earth's axis. However, during the last 1 Myr, a 100,000-year cycle—caused by fluctuations in the Earth's elliptical orbit has dominated (Fig. 16.2).

If we travel to Greenland today and look at the barren, treeless landscape lying in front of the inland ice glaciers, an impression of what was happening in Central Europe just 12,000 years ago can be obtained. In northern Germany, the retreat of the inland ice left a substantial layer of till, glaciofluvial sand and varved clay with an average total thickness of about 100 m. In many places, large erratic boulders that were transported by the ice are found scattered across the glacial landscape. Today, almost the entire surface of Northern Germany is covered by glacial deposits or is characterized by related erosional features. The classical geomorphological features of a glacial landscape are best developed across the lake-ridden areas of Mecklenburg, Brandenburg, and Schleswig-Holstein, together with a number of prominent hill chains. The glacial sediments of these regions occur in a characteristic sequence known as the **glacial series** and are composed of ground moraine, end moraine, outwash plains, and glacial valley deposits (*Urstromtal*). In addition to Northern Germany, the Alpine region was also strongly affected by the glaciations. Almost the whole of the Alpine mountain chain and, in some cases, the foothills of the Alps were covered by large inland ice bodies. Typical features such as many U-shaped valleys and large ribbon-shaped lakes testify to intense glacial activity. Only the highest peaks of the Alps stood out as **Nunataks** above the Alpine glacial ice. The effects of past glaciations are also recognizable in the low mountain ranges of the Black Forest and the Bavarian Forest, one example being the Great Arber mountain and the Arber lake, which formed as an ancient glacial cirque.

16.1 Glacial and Interglacial Periods

The Nordic glaciations and those that affected the Alps have a variety of different names and in general are difficult to correlate. The first researchers who described the glacial deposits in the nineteenth and early twentieth centuries named the deposits according to the rivers that have marked

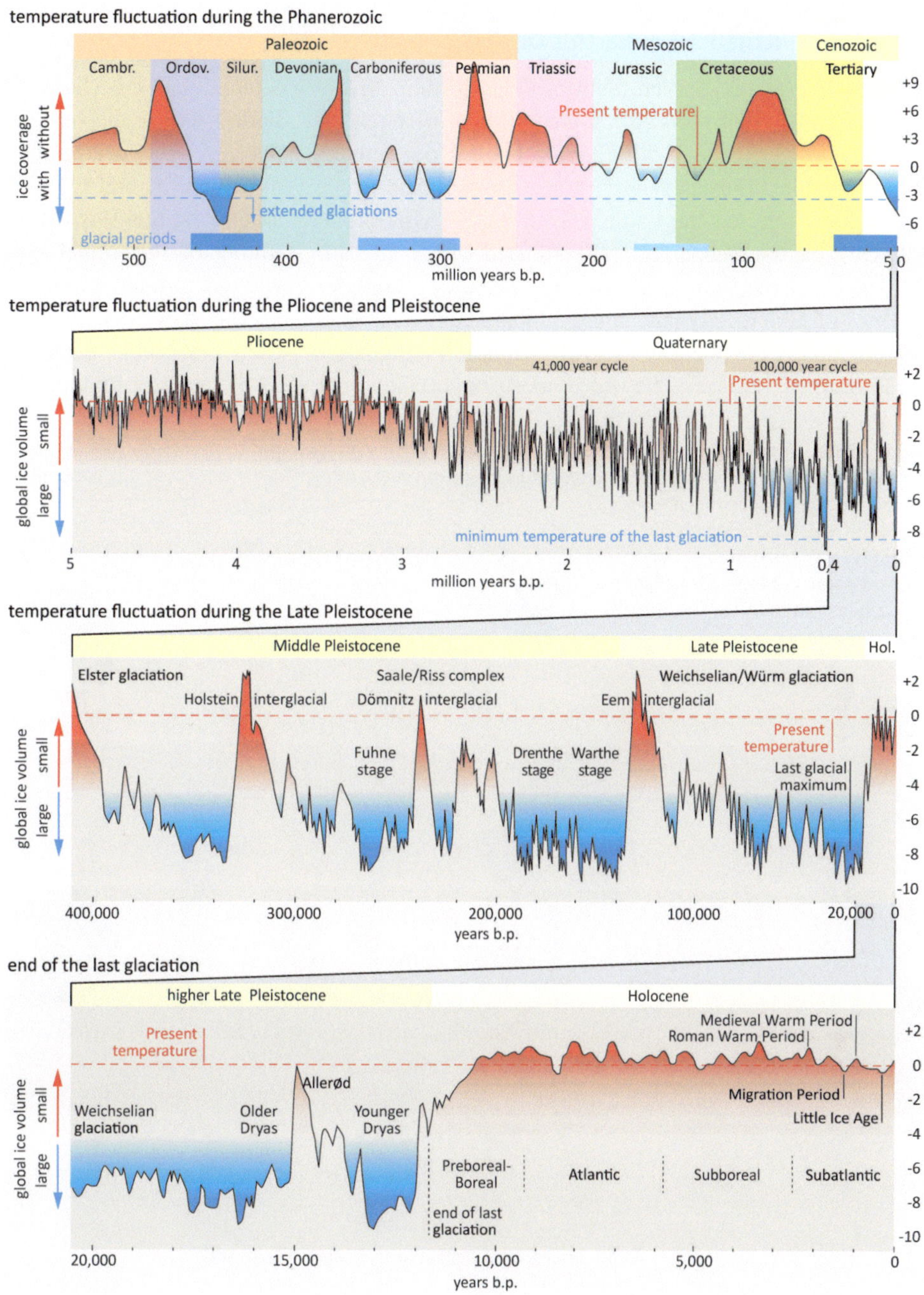

Fig. 16.2 Temperature changes and glaciation phases derived from the oxygen isotope ratios ($\delta^{18}O$) of fossils and from measurements on ice cores across different time periods. Compiled and modified from Veizer et al. (1999), Petit et al. (1999), Alley (2000), and Lisiecki and Raymo (2005)

the maximum extent of the individual ice advancements. For this reason, both the cold periods and ice movements in the Alpine region bear different names than in Northern Germany, which are also different from those used in the Netherlands or Poland. The Saale cold period (around 310,000–126,000 years ago), as reported across Northern Germany, was interrupted several times by warm phases. For this reason, it is now collectively referred to as the **Saale complex** (Fig. 16.6). The warm period between the early **Fuhne stage** and the later **Drenthe** and **Warthe stages** is termed the Dömnitz interglacial.

While the last two Alpine glacial periods of the **Würm** and **Riss** can be reliably correlated with the Northern German Saale and Vistula glaciations, the earlier Alpine **Mindel glaciation** remains difficult to correlate but approximates with the **Elster glacial period** in the north. The Saale/Riss complex is limited by the **warmer interglacials of** the older **Holstein** and the younger **Eem**. In the Alpine area, these warm phases are referred to as the Mindel/Riss and Riss/Würm interglacials. The Mindel glacial period (and the Elster glacial period) is in the southwest of Germany referred to as the **Hoßkirch glacial period**. The older **Cromer complex** is also called the

Box 25: Fluctuations in Earth's Orbital Parameters

In the 1920s, the Serbian mathematician Milutin Milanković (1879–1958) recognized a connection between the regular fluctuations in the **Earth's orbital parameters** and periodic climate changes. These astronomical cycles that have affected the Earth, known as the **Milanković cycles**, can be recognized and correlated with systematic changes in the sedimentary record. There is a complex series of periodically fluctuating orbital parameters, all of which affect the intensity of solar radiation impinging on the Earth's surface. The three most important cyclic orbital changes are (1) the **obliquity** of the Earth's axis (obliquity is the change in the inclination angle of the Earth's axis to the vertical of the orbital plane around the Sun), (2) the **precession** of the Earth's axis around a rotation axis, and (3) the **ellipticity** of the Earth's orbit around the Sun (Fig. 16.3). Also important is the **ecliptic** (tilting of the orbital plane in space), which fluctuates within a similar periodicity to the ellipticity.

The skewness of the Earth's axis is currently 23.4° from the orbital plane. It fluctuates between 22.1 and 24.5°, with a rhythm of 41,000 years. The lower the

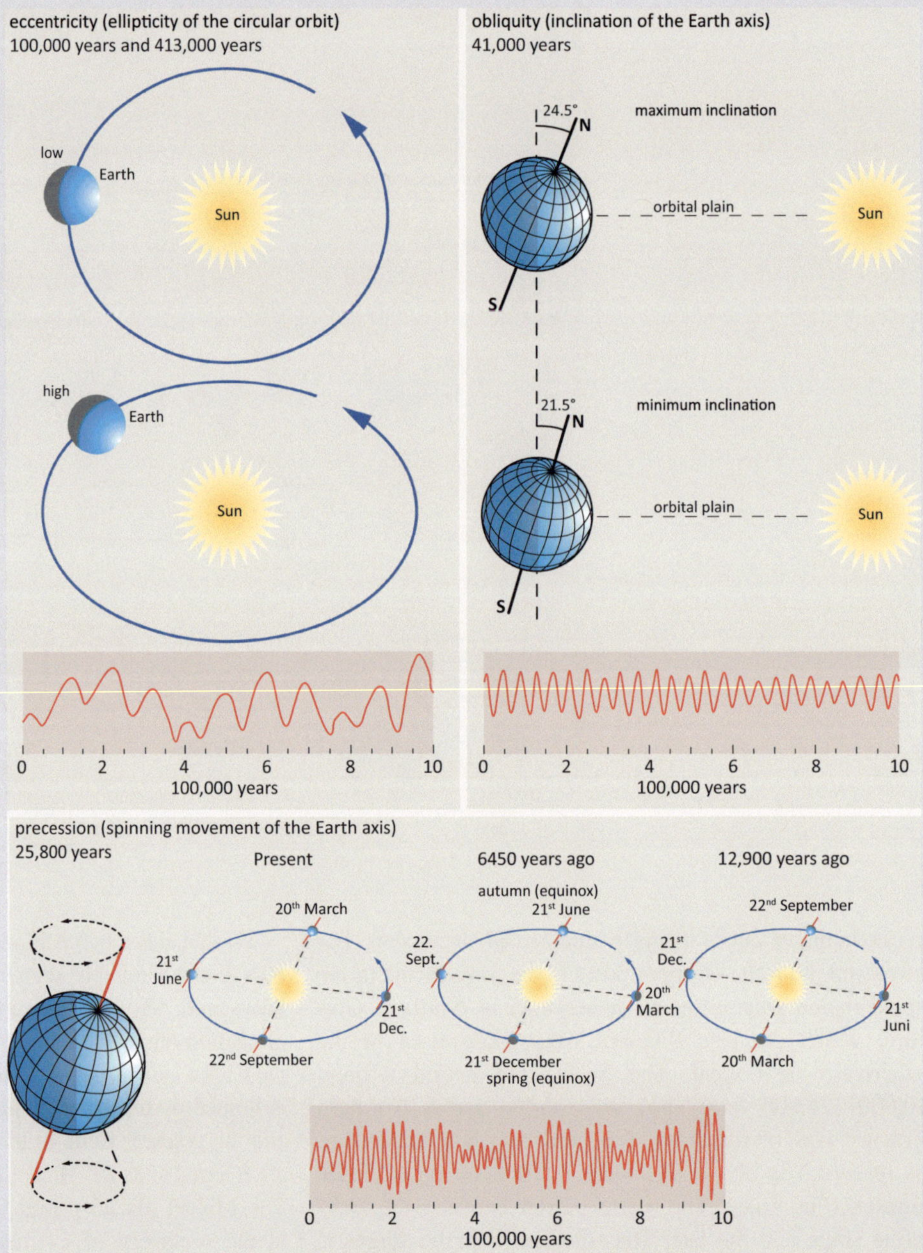

Fig. 16.3 Cyclic changes of the Earth's orbital parameters and fluctuations in the intensity of solar radiation: eccentricity, obliquity, and precession. Modified after Zachos et al. (2001)

skewness of the Earth's axis, the lower is the difference in solar radiation between the seasons. The ellipticity of the Earth's orbit around the Sun has the same effect. The more circular the orbital path, the more uniform the solar radiation. However, the ellipticity varies in rhythms of about 100,000 and 413,000 years so, sometimes, the factors intensify but other times they mutually reduce or cancel each other. The third parameter is precession. This refers to the gyro-motion of the Earth's axis around an axis perpendicular to the orbital plane. By this movement, which turns around its axis in the course of 25,800 years, the seasons on the Earth are continually shifting so that, after 12,900 years, winter in the northern hemisphere will be in June and summer in December. This effect can also enhance or mitigate the other two cyclic changes. If all the three factors act in the same direction, then the influence can be so strong that results in long-term climatic changes.

In Fig. 16.4, the fluctuations of the three most important Earth's orbital parameters are compared with the temperature curve of the last 400,000 years. It is clear that the recent glacial periods correlate with the ellipticity of the solar orbit. In this way, various correlations can also be drawn for other times. For example, the dominance of the 41,000-year rhythm in the temperature curve 2 to 1 million years ago. However, up to now, it is still unclear what triggers a glaciation period. The fluctuations of the Earth's orbital parameters certainly do their part, but they cannot be the only cause since there have been long periods in the Earth's history without large ice sheets (Fig. 16.2). However, the effects of the fluctuations can be amplified by different factors. Important roles are played both by the CO_2 content of the atmosphere and by plate tectonic activity that influenced the opening or closing of oceanic pathways. Other cyclic variables such as solar spots, the frequency of which has an average period of 11 years, are also likely to influence climatic change.

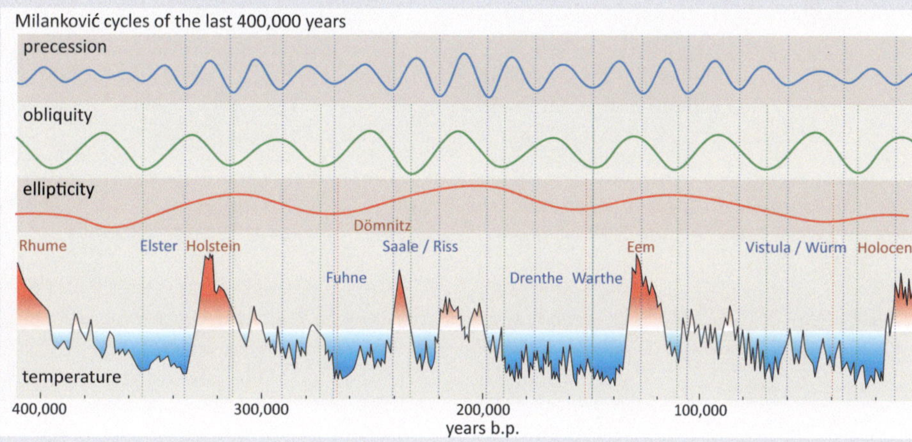

Fig. 16.4 Milanković cycles over the last 400,000 years compared to temperature variations obtained from the study of ice cores. Modified after Petit et al. (1999)

Cromer glacial period in the southwest, whereas the term **Günz glacial period** is used regionally in the Bavarian Alpine foothills.

The Elster glacial period, commonly known as the first glaciation at the beginning of the twentieth century, represents a relatively short period of glaciation between 390,000 and 330,000 years b.p. The Alpine Mindel glacial period in the Bavarian Alpine foothills has been placed much earlier in the past, but in the most recent stratigraphic classification (German stratigraphic commission 2016), it is now correlated with both the Hoßkirch glacial period of southwest of Germany and the northern Elster glacial period. The Ruhme interglacial event known in the north is not detectable in the Alpine area. The lack of correlation reflects different reactions to climatic changes between the maritime-influenced glaciated continental areas of the north and the southern high mountainous regions. In particular, the different sizes of the glaciers played a role, whereby the smaller mountain glaciers reacted more sensitively to environmental changes than the larger ice sheets did. In addition, the fact that there was no direct contact between the Nordic and Alpine ice masses results in different temporals and causal classifications. In order to avoid such problems of correlation, the cold and warm periods are now largely defined by the isotopic ratio ^{16}O to ^{18}O. The resultant **$\delta^{18}O$ curve** (Fig. 16.6) allows recognition of both cold and warm periods with high ^{18}O values indicating the cold periods and low ^{18}O values indicating the warm ones. These are used together with the numbered **Marine Isotope Stages** (MIS), which are based on well-dated marine sediments that provide an important additional stratigraphic framework. A relatively higher

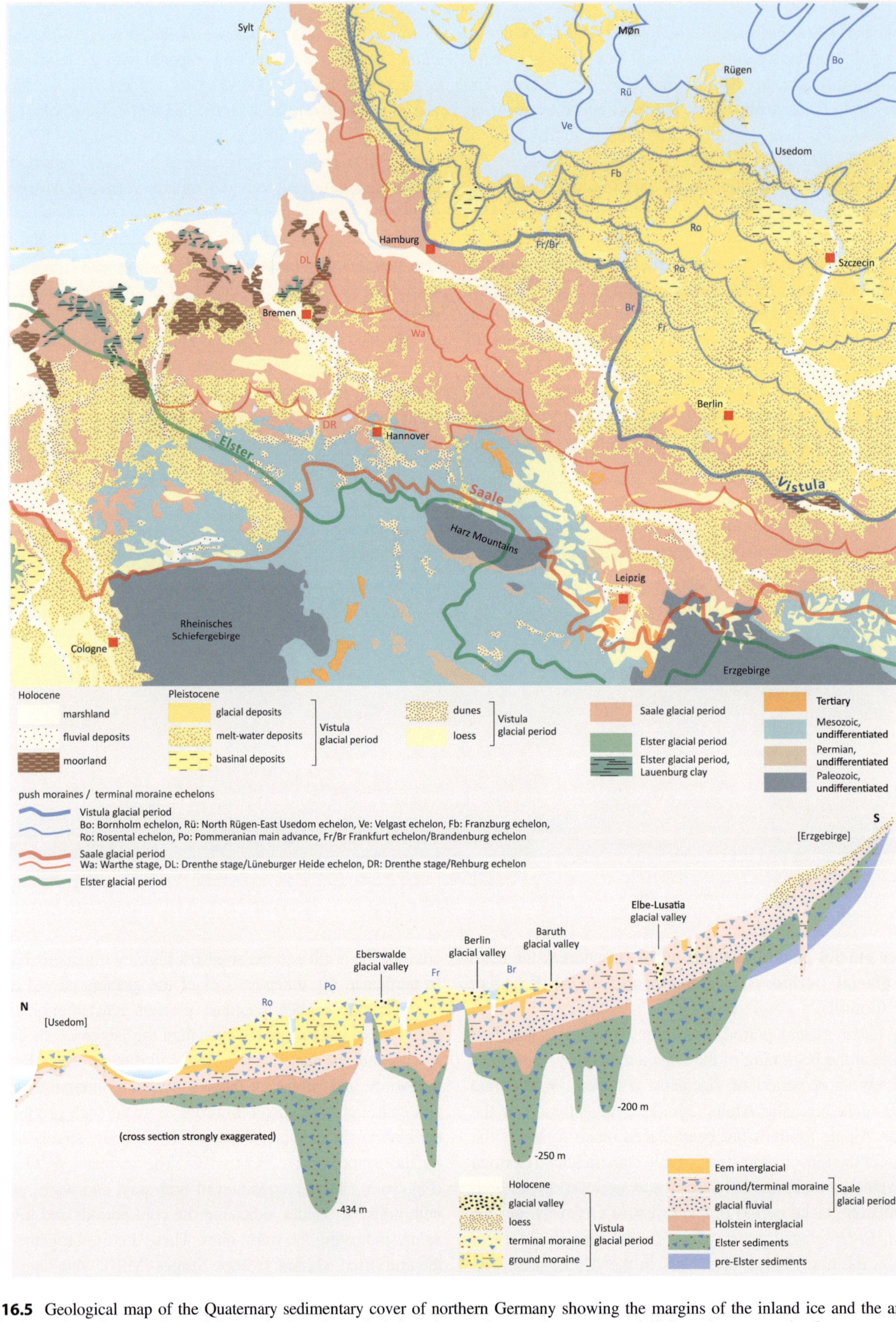

Fig. 16.5 Geological map of the Quaternary sedimentary cover of northern Germany showing the margins of the inland ice and the arrays of terminal moraines that was formed during the last three-glaciation phases. A strongly exaggerated N-S section across the Quaternary deposits through the eastern part of northern Germany shows the filling of deeply eroded valleys formed during the Elster glaciation phase. Map modified after BGR (1993) and the section modified after Eissmann and Litt (1994)

Fig. 16.6 Stratigraphic table showing glacial and interglacial periods since the Middle Pleistocene with a correlation of the nomenclature of Northern Germany with the Alpine region. H. Holocene. Modified after Cohen and Gibbard (2010), and Deutsche Stratigraphische Kommission (2016)

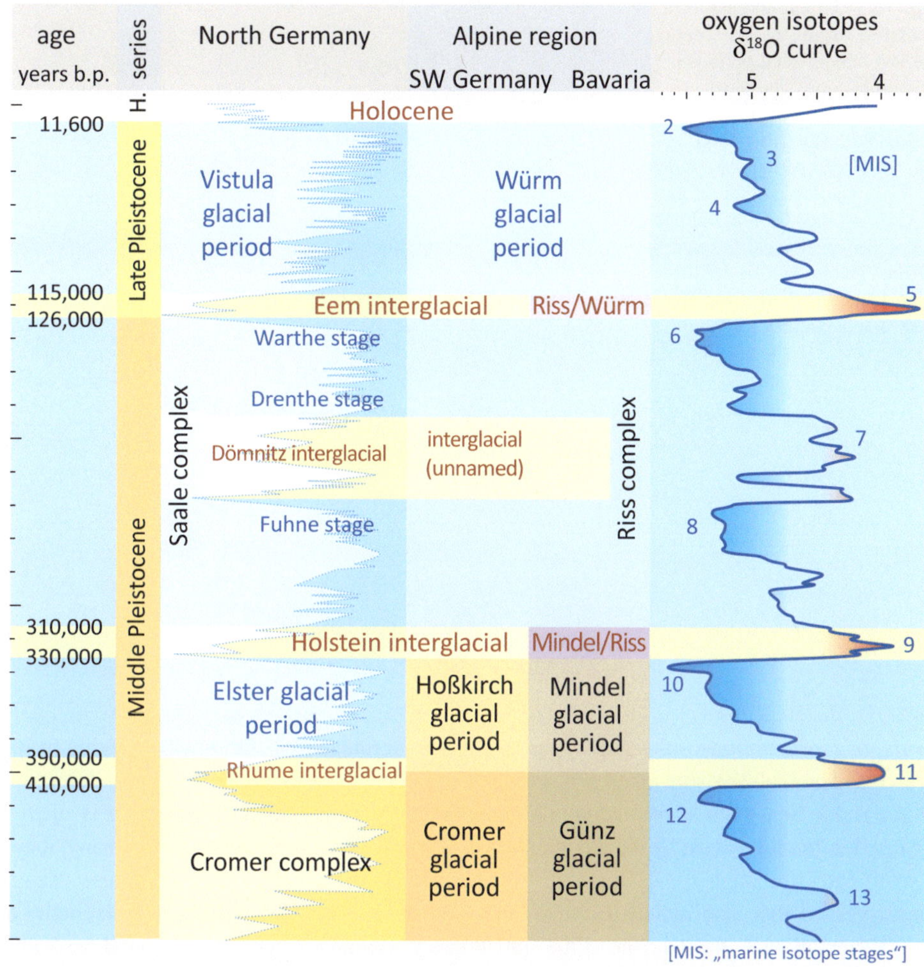

proportion of the (heavier) isotope ^{18}O in the sediment means that, at the time, water with predominantly lighter ^{16}O was preferentially bound into ice, thus indicating a cold period. Additionally, the lighter isotope ^{16}O tends to evaporate more easily and, as a result, it appears more commonly in ice formed by the accumulation of snowfall.

At its maximum coverage during the Elster glacial period, the inland ice of northern Germany extended to the edge of the German low mountain range (Figs. 16.5 and 16.7). Prior to the Elster glaciation, there were already strong climatic fluctuations with cold and warm phases (Fig. 16.2). However, it seems that the ice advances did not reach so far across northern Germany since they are no longer detectable. The Elster glaciation saw the ice advance as far south as Dresden and Erfurt. It also reached the northern flank of the Harz Mountains and, in North Rhine Westphalia, the northern edge of the Rheinisches Schiefergebirge. In these areas, debris was deposited from Scandinavia, together with flint from the Cretaceous chalk found in situ in locations such as Møn, Dover, and Rügen. The maximum extent of the ice can be determined by the occurrence of the numerous fragments of transported flint, the distribution of which is used to determine a so-called **flint line**. However, this line does not represent a uniform time boundary. In the middle and eastern part of Germany, as in the area around the Teutoburg Forest, it marks the extent of the ice during the Elster glacial period. However, to the west, it indicates the ice reach during the Saale glaciation.

Since the younger glacial deposits overprint and partially rework the older ones, it is often difficult to assign the sediments to a specific glaciation phase without using precise dating methods. The landscapes formed during the Vistula glacial period are referred to as **young moraine areas** (*Jungmoränengebiete*). In contrast, the landscapes of the Elster and Saale glacial periods are called **old moraine areas** (*Altmoränengebiete*).

Elster glacial period

The **Elster glacial period** (390,000–330,000 years b.p.) is the oldest glaciation recognized in northern and central Germany and was responsible for widespread deposition of glacial sediments across these areas. The advance of the ice southwestward to the limit shown in Figs. 16.5 and 16.7 led

Fig. 16.7 Extent and assumed thickness (m) of the ice sheet of the last three glacial periods in Central Europe. The older margins of the ice sheet (Elster- and Saale-, resp., Riss- and Mindel-glacial periods) are shown as lines. The topography of the map shows the shape of Europe with a sea-level stand lowered by 100 m. The margin of the ice sheet is modified after van Husen (1987) and Walter (2007)

to major reorientation of the large rivers and their drainage systems. Thus, all the streams that flowed into a predecessor of the Baltic Sea, such as the ancient Odra and Vistula, were diverted to the northwest by the advancing ice. The Elbe was also forced to change its watercourse, and hence formed a new riverbed that also included rivers further east. Prior to the Elster glacial period, the Elbe presumably drained through the present valley of the Odra into the predecessor of the Baltic Sea. However, this outflow was eventually blocked by the ice advance, damming the Elbe to form a lake in front of the glacial ice sheet. As the Elster glacier retreated, a new riverbed developed toward the North Sea, which was progressively widened and deepened during later cold periods. Today, the Elbe Valley forms one of the largest **glacial valleys** in northern Central Europe. In the up to 150-km-long Elbe lake that developed south of Dresden at that time, varved sediments were deposited from the melt-waters. The individual **varves** correspond to seasonal fluctuations in the sedimentary input, which can be used for the dating of the sediment in a similar way as tree rings.

Two ice advances are recognized to have occurred during the Elster glacial period. In the older Zwickau phase (older stadial), a maximum ice coverage was reached, whereas the subsequent Markranstaedt phase (younger stadial) remained 20–50 km north of it. No intra-Elster warm phase can be detected between these two stages of ice advancements. The characteristic features of the Elster glaciation are the deeply incised trenches developed in the subsurface of the northern German lowland (Fig. 16.8).

Occupying a stripe from Hamburg to Berlin, these trenches can reach several kilometers wide and over 100 kilometers long. Their depth is often more than 500 m whereas, in the northwest of Germany, they tend to be shallower with depths of up to 250 m. These deep trenches are rather unique and are known as **glacial trenches** or tunnel valleys. They have formed beneath the inland ice body as water at the base of the glacier that merged into large melt-water streams. Any surface melt-water rapidly descended to the base via clefts and crevices in the glacier, where it is placed under great pressure by the superposition of the ice. Such a pressure difference even caused the water to flow upward from the base toward the outer edge of the ice sheet. The enhanced water pressure had a strong erosional force capable of removing considerable amounts of unconsolidated subsoil sediment in a short time to form deep trenches. Sometimes, the trenches in northern Germany have very steep flanks and unlike the glacial **U-shaped valleys** in the Alps and the **fjords**, they were later filled with sediment derived from melt-water lakes and streams. Today, these sedimentary layers serve as important aquifers for drinking water supplies.

The deep glacial trenches, geographically oriented in the N-S to NE-SW directions (Fig. 16.8) mark the main drainage direction beneath the ice. These were possibly linked to the ice flow direction. However, this direction may also indicate zones of structural weakness in the subsurface rock such as fractures in the systems along which salt diapirs penetrated. Such structural weaknesses had an influence on the erosional forms and the thickness of Quaternary

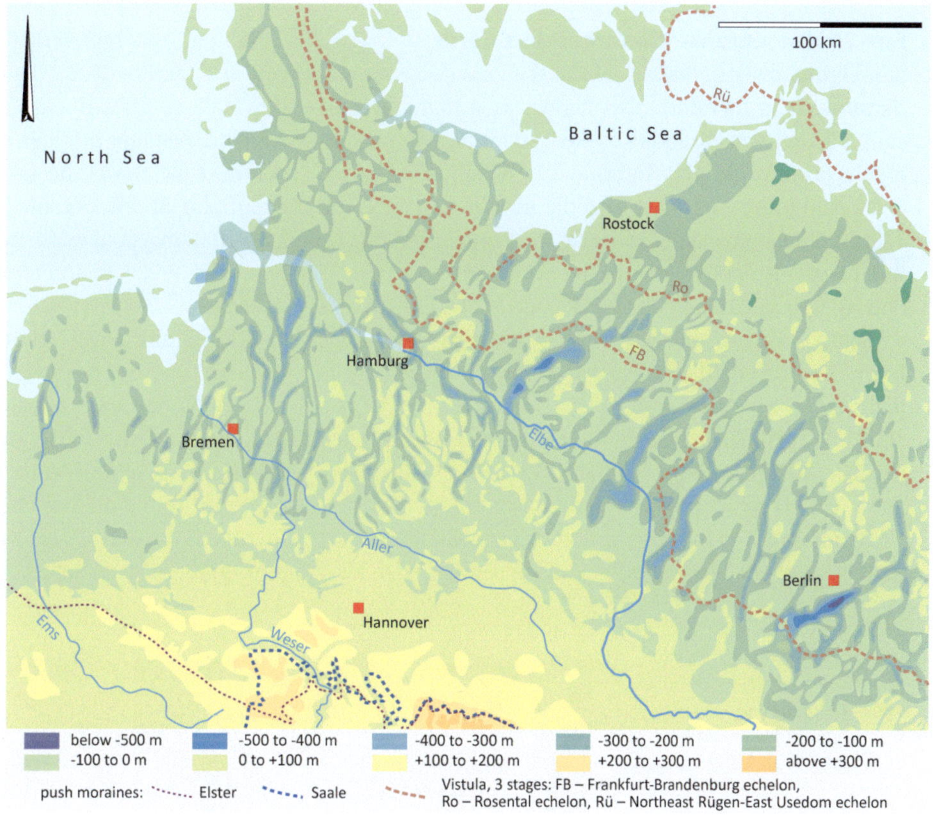

Fig. 16.8 Pleistocene trench systems formed during glacial periods and their depths in the subsurface of Northern Germany. Modified after Stackebrandt (2009)

sediments deposited. In the upper layers of the Elster deposits, the **Lauenburg clay** is particularly widespread. These deposits are formed in ice lakes at the edge of the glacier and reached more than 100 m thickness. The clay characteristically contains silt to fine sand, which distinguishes it from the marine Holstein clay that developed during the following warm period.

Holstein interglacial period

The strong glaciation of the Elster period was followed by the **Holstein warm period** characterized by similar climatic conditions as today. Following initial deposition of freshwater sediment, a maritime influence prevailed in the north with marine waters flooding large parts of Northern Germany. As evident from the deposition of marine sands and clays containing the mussel *cardium*, the transgression of the Holstein sea extended southward beyond Berlin. Such sediments occur on the west coast of Schleswig-Holstein, along the lower course of the Elbe, and in West Mecklenburg and Western Pomerania. Marine to brackish and limnic sediments, as well as river gravels and sand, are found further to the south. With its widespread distribution, the Holstein clay represents an important marker horizon for the stratigraphic classification of the Quaternary deposits.

As determining the age of Quaternary sediment is often problematic, easily recognizable marker horizons are of great stratigraphic value. The sedimentary deposits of the warm periods are typically dated using pollen and microfossil assemblages. However, as glacial deposits typically lack the organic remains useful for chronostratigraphic classification, other more recently developed dating methods are applied. One particularly useful technique is to measure the concentration of the beryllium isotope ^{10}Be (half-life period 1.387 Myr) produced by nuclear reaction of the ^{16}O isotope as cosmic radiation bombards grains of quartz. Under ideal conditions, this method dates the exposure time of a sediment or rock that is uncovered at the surface. Examples are glacier moraines or abraded rock faces that become exposed to cosmic radiation and are no longer covered by debris or ice. As ^{10}Be is formed only in the uppermost centimeters to decimeters of a sediment or rock, these isotopes begin to occur at the moment of exposure. As a result the time at which these materials were left exposed during the retreat of a glacier can be determined.

During the Holstein interglacial, thick diatomaceous deposits known as kieselguhr are formed in the Lüneburg Heath in the Unterlüss area and, to a lesser extent, near Klieken in Saxony (Box 26: Kieselguhr). Workable kieselguhr earth deposits exist not only from the warm periods of the Quaternary but also during earlier times such as those formed in the Tertiary at the Vogelsberg, near Altenschlirf.

Box 26: Kieselguhr—Diatomaceous Earth

Kieselguhr consists mainly of the shell residues of fossil **diatoms**. It is therefore also known as a diatomaceous earth. This typically fine-layered sedimentary rock is highly porous and very light, and is widely used in the chemical industry as a raw material for filters, paints, and additives. Diatoms are unicellular plant organisms (algae) that carry out photosynthesis. Its shell consists of amorphous, noncrystalline silicon dioxide (SiO_2), which is commonly known as quartz in its ordinary crystalline form. The size of the individual shells, which occurs in the most diverse forms (Fig. 16.9), lies between 10 and 100 µm. This means that about 10 billion shells and shell residues of diatoms are contained in a half-filled shot glass (ca. 10 ml) full of diatomaceous earth.

Diatoms are extremely prolific organisms since they reproduce every few hours by cell division. Under ideal conditions, one billion diatoms may form from one single cell in 1 month. These diatoms float in water and sink to the seafloor after dying. There they form thick layers, which depending on their organic content are different in color: at the top is the white diatomaceous earth with a maximum of 5% organic content, directly below is the gray diatomaceous earth with 10% organics and, at the base, the green diatomaceous earth with >30% organic content.

Kieselguhr has gained worldwide importance through its use as a carrier substance in dynamite. Alfred Nobel took advantage of the good damping properties of the kieselguhr for packaging the highly explosive glyceryl trinitrate (better known as nitroglycerin) in the diatom shell material and in the process invented dynamite. This was much safer to handle than pure glyceryl trinitrate.

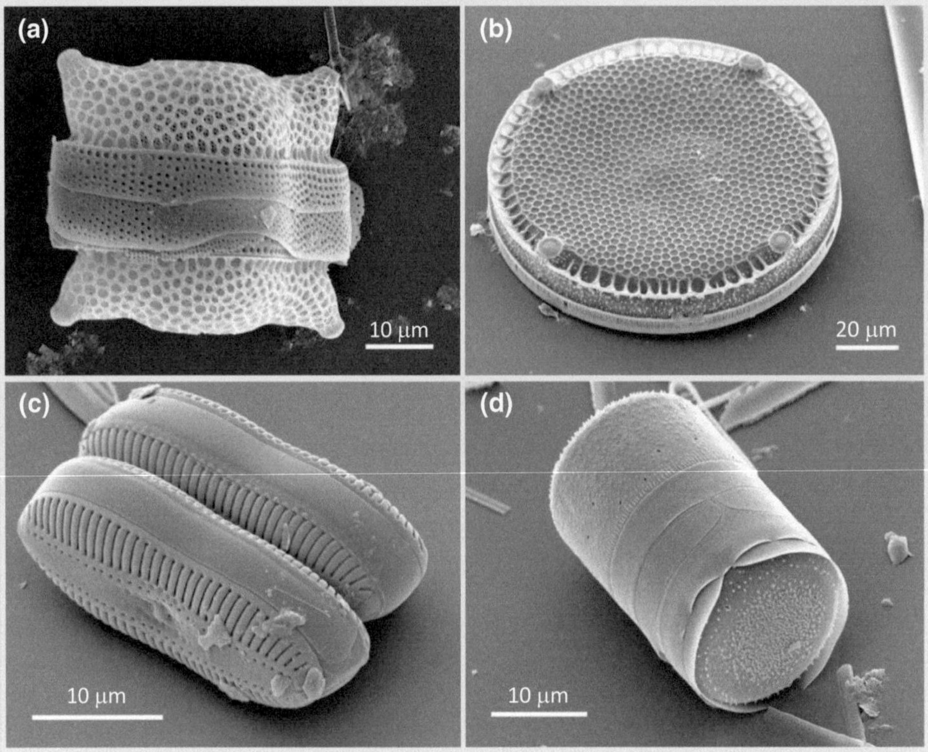

Fig. 16.9 Scanning electron microscope images of diatoms: **a** *Biddulphia reticulata*, **b** *Eupodiscus radiatus*, **c** *Diploneis sp.*, **d** *Melosira varians*. Photographs from Mary Ann Tiffany, San Diego State University; in: Bradbury (2004)

Saale complex

Numerous glacial features are also found scattered across Western Germany. The Münsterland Cretaceous Basin was at least once completely covered with ice during the Saale glacial period following the Holstein interglacial. The Teutoburg Forest and the Wiehen Hills were also overridden by the ice (Fig. 16.10). The ice margin reached the northern edge of the Rheinisches Schiefergebirge, where for a short time a large ice lake extended deep into the valleys of the Sauerland region. When the ice gradually melted at the end of the Saale glacial period, the melt-water rivers left behind sandy deposits in the Senne area. During the Saale period, two narrow and elongated elevations were formed composed of gravel and sand (Fig. 16.10): the Münsterländer Kiessandzug in the east and the Twente-Achterhoek trench in the west. The over 80-km long and up to 1-km-wide flat

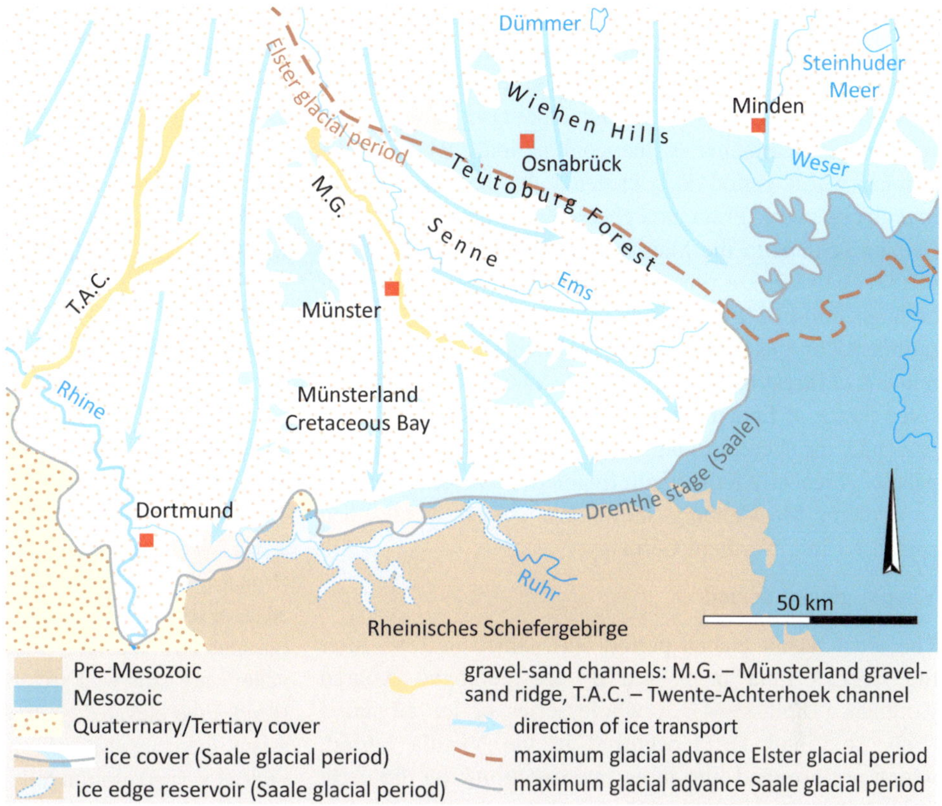

Fig. 16.10 Ice coverage of the Münsterland Cretaceous Basin during the Saale glacial period. The ice margin of the Elster glacial period is marked by a line. Modified after Liedtke (2007), and Thome (1990)

elevations either represent eskers (Sect. 16.2) or remnants of melt-water rivers that flowed along the edge of the active southward advancing glacier, which was limited to the east and west by dead ice. During the most recent Vistula glacial period, the inland ice did not reach these regions (Figs. 16.5 and 16.6).

The ice advances that formed the Saale complex did not extend so far southward as the Elster glaciation, except for a few places such as the Münsterland Cretaceous Basin where the ice advanced further than the Elster front (Fig. 16.10). The Saale complex consists of at least three and possibly even five different cold phases, each of which was not always associated with its own ice advance. According to the most recent stratigraphic classification based on the Marine Isotope Stages (MIS, see above), the older Fuhne stage (MIS 8) is separated from the two younger Drenthe and Warthe stages of the Saale glacial period (MIS 6) by the **Dömnitz interglacial** (MIS 7; identical in time to the Wacken and Schöningen warm periods; Fig. 16.6). The Drenthe ice advance was regionally the most extensive one of the Saale complex.

Remnants of the individual phases of ice advance can be determined from the different compositions of moraine material, for example, by variation in carbonate content or different debris characteristics. Sometimes, the various ice advances were separated by sandy intermediate layers derived from the melt-waters of the intervening warm phases, where varved clays were deposited that can be used for chronological dating. The youngest ice advance of the Saale period, the Warthe stage, reached the region between the Elbe and Weser rivers. At that time, the main drainage system for the melt-water passed through the Aller–Weser glacial valley. Farther east, the end moraine ridges of the Fläming and the Lausitzer Grenzwall were formed, which today still represent striking examples of the Warthe stage landscape.

Eemian interglacial

The Eemian interglacial period lasted just about 11,000 years. It followed the Saale glacial period and, according to climatic data obtained from glacial ice cores in Greenland, was characterized by stable climatic conditions with temperatures probably at least 2–5 °C above today's average. The warmer climate led to a sea-level rise, but there was no major transgression as during the Holstein interglacial that led to extensive flooding of Northern Germany. In contrast, the Eemian Sea covered the present mainland only in a few places. A clay deposit at Dornbusch on the island of Hiddensee, known as the **Cyprina Clay**, is not well stratigraphically constrained but is placed in the Eemian and taken to indicate a marine influence. During the eighteenth and nineteenth centuries, the clay was used as a source of raw material for the production of local faience, a particular type of glazed pottery.

The Eemian interglacial sediments are typically found on top of the moraine deposits of the Saale Complex

(Fig. 16.5). They are overlain by river gravels or aeolian deposits and in the northeast by ground moraines of the following Vistula glacial period. This horizon provides a good stratigraphic marker in the otherwise poorly dated surrounding sediments of the cold periods, whereby the Eemian warm period is considered the best constrained interglacial event. From an archeological point of view, the Eemian belongs to the Middle Paleolithic (about 200,000–40,000 years ago), a time when **Neanderthals** were settled in large parts of Europe and areas leading to Asia. The sediments of this warm period in Germany have yielded remains of **forest elephants** that were hunted with a wooden lance (**lance of Lehringen**; Fig. 2.3). This illustrated for the first time, the successful big-game hunting of the Stone Age people. A recently found tooth fragment of a forest elephant in northern Mecklenburg shows this animal species had spread far into Northern Germany.

Vistula glacial period

The most recent glacial period was named in 1899 after the river Vistula in Poland by the geologist Konrad Keilhack (1858–1944). A whole series of ice advances (Figs. 16.5 and 16.7) occurred in the Vistula period, which are grouped into three main sections: (i) the early glacial phase, (ii) the pleniglacial phase, and (iii) the late glacial phase. These sections are further subdivided into a series of stadial and interstadial stages. During these ice advances, none of them reached the maximum extent of the Elster and Saale glacial periods (Fig. 16.7). The glaciations were more limited to the northeastern region of Germany, where they covered about half of the areas of Schleswig-Holstein, Mecklenburg-Western Pomerania, and northern Brandenburg.

Importantly, the ice did not advance across the Elbe River but, during the **Brandenburg stage**, it did advance south of Berlin (Fig. 16.5). The main geomorphologic remnants of the Vistula glaciation are the end moraines that developed in the **Pomeranian stage**. The melting phase of the Vistula glacial period was interrupted several times by repeated ice advances, producing new glacial valleys that followed the ice margin. Impressive examples of glacial valleys are seen in the areas of Baruth, Berlin, and Eberswalde (Sect. 16.2; Fig. 16.17), where the rivers are today significantly smaller than they were during their time of development. Particularly important was the lower Elbe area, which drained all of the glacial valleys formed during the Vistula period before running into the ancient North Sea. Today, the Elbe flows along this classic **glacial valley** which, formed at the peak of the Vistula glacial period shortly before the end of the glaciation. The valley is characterized by melt-water deposits and aeolian sediment formed under periglacial conditions. The path of the glacial valley largely corresponds to the current course of the Elbe. Although the course of the valley already led to the North Sea during the Vistula glacial period, at that time the sea-level was about 120 m lower than today and the coastline was located about 600 km further northwest.

The sedimentary deposits of the Vistula glacial period (the young moraine areas) are much better preserved than the older Elster and Saale glacial periods (the old moraine areas), which were overridden and reworked by the younger ice advances. A number of stages can be recognized during the Vistula glacial period by a series of **end moraine ridges** of different ages, the so-called *Staffeln* (arrangement in echelons; Fig. 16.5). Due to its significant number of subdivisions, the Vistula glacial period is often referred to as the Vistula complex. This period of glaciation ended about 11.600 years ago with a relatively abrupt temperature rise of an average of 7 °C within a few decades (Fig. 16.2).

The Mecklenburgian and Pomeranian lake district, as well as the lake lands of Brandenburg and the Holstein Switzerland (*Holsteinische Schweiz*), are typical for the young moraine areas of the Vistula glacial period. In these areas, glacial series (ground moraine, end moraine, outwash plain, and glacial valley) are still well-preserved and thus easily recognizable. Besides the elevated geomorphology of the end moraines and eskers, there are also hollows in the relief caused by the melting of dead ice and erosion following the retreat of the glacier or by glaciofluvial erosion beneath the ice body. Many of these negative depressions are now filled with lakes. The periglacial areas not covered by ice were characterized by aeolian sediment that sometimes produced large inland dunes, generally, in the more flat-lying terrains. In front of the ice sheets, the sands and gravels of the melt-water rivers were deposited in large outwash plains (Sect. 16.2). The fine-grained silty particles were blown from these deposits and sedimented on the flanks of the large valleys, on top of older glacial sediments and on the slopes of the German low mountains that lay to the south. Only a few relics of these landscape forms were preserved in the flatter old moraine areas of the Saale and Elster glacial periods.

16.2 Deposits and Erosion Forms of the Glacial Periods

Some of the rocks found as erratic boulders in the glacial sediments are suitable for precisely determining their place of origin due to their specific properties and characteristics. These are referred to as **index rocks** and are used to reconstruct the paleo-ice transport directions at different times (Fig. 16.11). For example, the boulders found along the Baltic Sea shoreline originated largely from Northern Norway, Sweden, or Finland, and some from Bornholm and other areas of the Baltic Sea. Figure 16.11 gives an impression of the diverse types of rock that were carried southward by the ice.

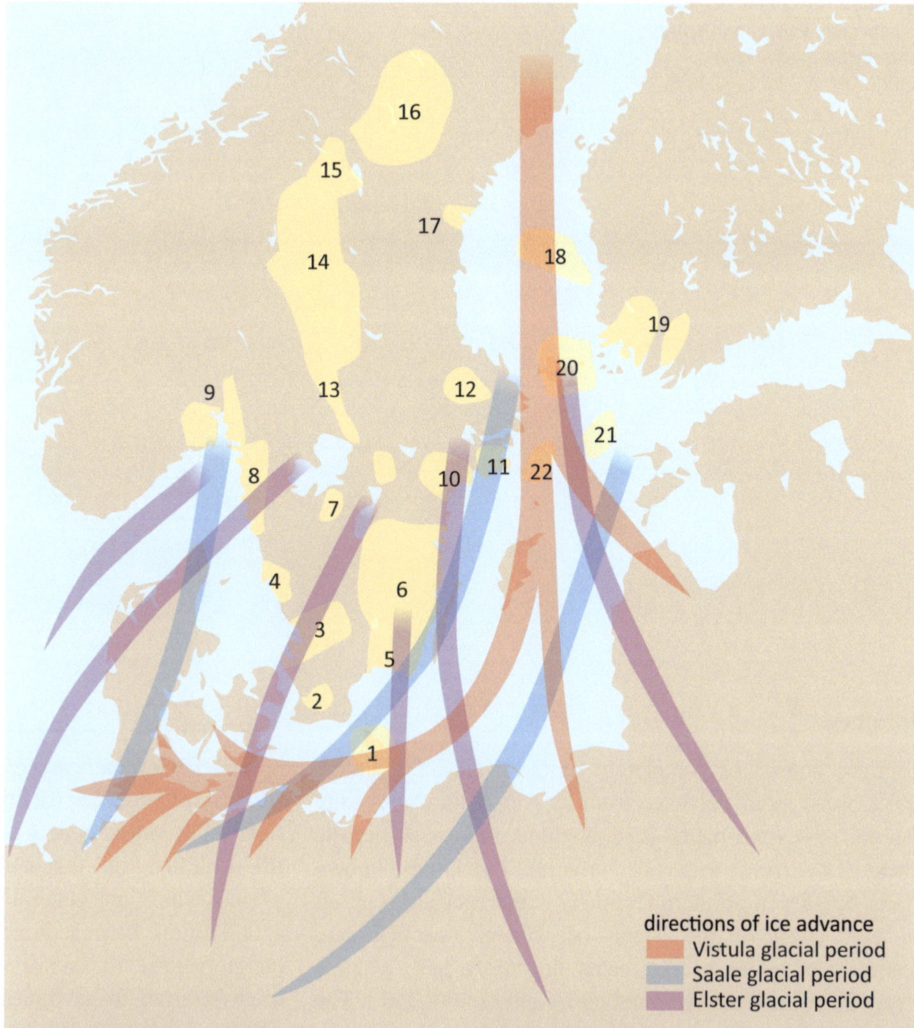

index erractic boulders

1 Bornholm- granite
2 Scania basalt
3 garnet amphibolite
4 Varberg charnockite
5 Karlshamn and Spinkamåla granite
6 Påskallavik porphyre, Järeda granite, Växjo granite, Småland granite
7 Kinne diabase
8 Bohuslän granite
9 rhombeporphyry, larvikite
10 Graverfors granite
11 Stockholm spotted quartzite
12 Uppsåla granite
13 Filipstad Granite
14 Dalarna rocks: Särna tinguaite (= phonolite), Garberg granite, Siljan granite
15 Rätan granite
16 Ragunda and Rävsund granite
17 Rödö granite and porphyry
18 Bottensee porphyry
19 SW-Finnish rocks (granite, porphyry)
20 Åland rapakiwi
21 red Baltic Sea quartz porphyry
22 brown Baltic Sea quartz porphyry

Fig. 16.11 Origin of various index rocks in the glacial deposits of northern Germany. Modified after Fraedrich (1996)

The Quaternary glaciation period represents the last geologically significant epoch in Central Europe. While thick glacial deposits were formed in Northern Germany and in the foothills of the Alps; erosion dominated the low mountainous areas and the Alps under completely different climatic conditions than at present. The landforms are today overprinted and often covered by soil and vegetation, so they are only partly visible. However, when traveling across these landscapes with a trained eye, it is possible to recognize many features that originated during the last glaciation period (Fig. 16.12).

In Northern Germany and in the foothills of the Alps, the soil is almost exclusively underlain by loose sediment. In most cases, these are glacial moraine deposits composed of gravel, sand, clay, or loess. Only in a few places are older and diagenetically lithified rocks exposed at the surface, mostly in areas uplifted to the surface by salt tectonics (salt diapirs or salt domes; Sect. 11.2.2) or along young fault zones.

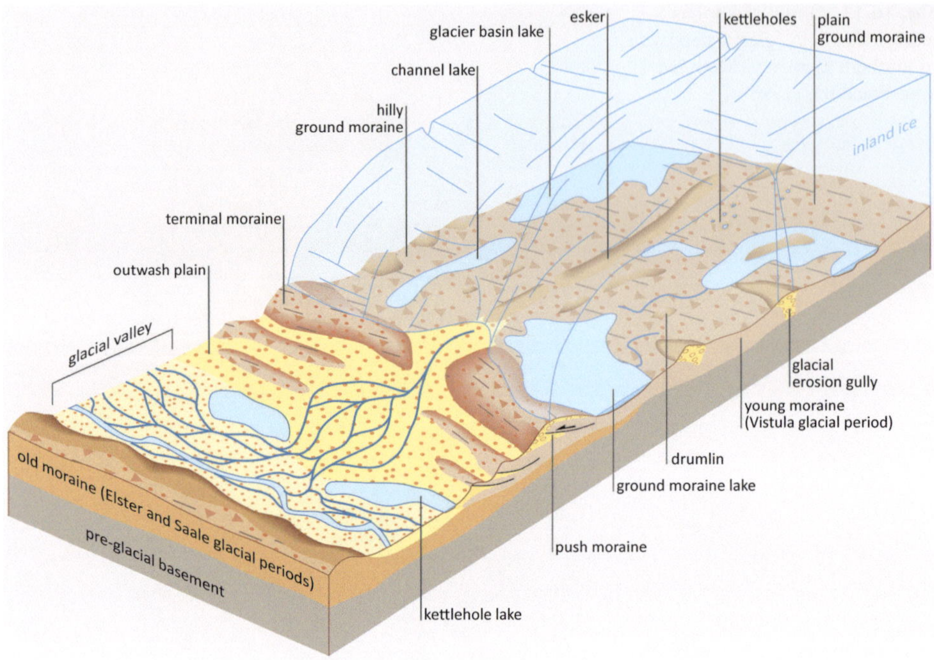

Fig. 16.12 Overview of different sedimentary and erosive features of a glacial series. Modified from various sources

Moraines

Moraines consist of completely unsorted clastic sediment with a broad grain-sized spectrum ranging from clayey silty material to several meter-sized boulders known as **erratic blocks** (lat. *errare*: to wander around). The largest known erratic boulder in northern Germany is the Buskam rock off the east coast of the island of Rügen, near Göhren. According to recent measurements, this erratic has a volume of ca. 206 m^3 and a corresponding weight of 500–550 t. The largest erratic boulder located inland was discovered at the beginning of the nineteenth century in the Rauen Mountains near Fürstenwalde/Spree. Known as the Great Markgrafenstein, it was estimated to have a weight of 700–750 t. Unfortunately, it was largely dismantled and part of it used for the construction of a large granite bowl almost 7 m in diameter, which is located today in front of the old museum in Berlin.

Ground moraines formed at the base of past glaciers are widespread throughout northern Germany and in the foothills of the Alps. These moraines, composed of clastic sediment that accumulated during melting of the ice, are common to all young and middle Quaternary glaciation periods. Along the edges of the glaciers, the moraine material piled up at the margins to form **lateral moraines**, while at the end of the glacier bow-shaped **end moraines** accumulated. Such end moraines can reach heights of more than 200 m. In northern Germany, they mark the course of the inland ice edge and, in the foothills of the Alps, they indicate the location of the glacier lobes during the stages of ice retreat as the most recent glaciations receded (Fig. 16.5).

In Germany, the moraine sediment deposited by a glacier is termed ***Geschiebemergel*** (marly till) or in its decalcified state ***Geschiebelehm*** (debris loam). However, more recently German scientific papers have begun to use the English term **till**. Lithified tills that originated during earlier glaciation events, which are generally older than the Quaternary, are called tillites. If the sedimentary conditions of brecciated and conglomeratic rocks are not clearly understood, they are often referred to as **diamictite**. An example is located in Thuringia, where dropstones in glaciomarine diamictites of Ordovician age are considered to represent an Early Paleozoic ice coverage (Sect. 9.2.6).

The origin of erratic boulders has long been a mystery in science. Although James Hutton (1726–1787) had proposed that they provide evidence of a glaciation across large parts of northern Europe, it took another 100 years for the **inland ice theory** to become established in the 1870s. A talk before the German Geological Society in Berlin by the Swedish geologist Otto Torell (1828–1900) played a major role in the acceptance of the ice theory when abrasion scratches in the Muschelkalk (Triassic) of Rüdersdorf (Sect. 11.4.2) were convincing attributed to the effects of glacial sliding.

When a glacier remains stable for an extended period, larger amounts of moraine material tend to accumulate at its end to form a **settled end moraine**. A well developed example is the Pomeranian main end moraine of northeastern Germany. However, if the glacier meets an obstacle it cannot override, it is compressed by the force of the pushing ice. This leads to the formation of **push moraines** with deformed and folded rocks that includes material from the obstacle interleaved with the glacial sediment. The resulting structures have similarities to the folds and thrusts that arise during mountain building processes. The construction of push moraines is considerably more complicated than that of settled end moraines. A good example of a push moraine can be found in the **Brohmer Mountains**, east of Neubrandenburg, overprinted at the end of the Vistula glacial period in a section of the Rosenthal end moraine train (Fig. 16.5) about 13,700 years ago. The range of hills had already formed earlier by ice advances and are composed of sediments of the Saale glacial period together with Tertiary strata piled up to heights of more than 100 meters.

Another well documented example of a push moraine is the **Muskau Arch** (*Muskauer Faltenbogen),* now accessible to the public in a transnational geopark, along the German border to Poland (Fig. 16.13). According to recent findings, the Muskau Arch was formed during the Elster glacial period about 350,000 years ago. Here, lignite-containing Tertiary layers have been visibly displaced by compression. The opencast brown coal mining (Sect. 13.1) along the outcrops of brown coal seams left a series of elongated lakes positioned around the Muskau Arch.

Good examples of end moraines belonging to the Saale complex can be found in the Göhrde–Drawehn hill range between Lüneburg Heath and Wendland, and in the form of the Schmiedeberg push moraine between the Elbe Valley and the Mulde Valley. End moraine step-like ridges are in fact widespread across northern Germany, where they are often found aligned together in a garland-like manner marking the stages of glacial retreat (Fig. 16.5).

The most impressive example of the deformation of rock layers caused by the force of an inland ice body is exposed in the chalky cliff line of **Jasmund on the island of Rügen**. The style of deformation is very similar to the well documented glaciotectonic structures on the Danish island of Møn. Both locations have steep cliff sections composed largely of white Late Cretaceous limestone. On Rügen Island, the limestone is referred to as **Rügen's writing chalk**. The chalk originated from the Early Maastrichtian, and therefore belongs to the youngest part of the Cretaceous period (Sect. 11.7.2).

The Cretaceous limestones of Rügen Island were folded and thrust, rather like roof tiles, into a large number of sheets as the ice advanced during the Vistula glacial period. In this case, the Vistula glacial sediments, which lie discordantly on the chalk (marly tills M_1 and M_2 with intermediate sediments I_1 and I_2), were overthrust by the Cretaceous limestone (Fig. 16.14). Since duplex-like overthrust complexes are discordantly overlain by an undeformed marly till of the youngest part of the Vistula glaciation (marly till M_3), the deformation was restricted to the earliest part of the glacial period. These structures are, therefore, formed after the last glacial maximum (LGM) about 20,000–25,000 years ago (Fig. 16.6). However, it is not yet possible to assign the compression with confidence to a specific ice advance. The **duplexing** of the Cretaceous chalk demonstrates well the enormous deformation forces that can arise during the motion of a glacier. Notable changes in the pressure direction of the glacial ice occurred on Rügen island, as evident from the orientation and alignment of the overthrust complexes (Fig. 16.14; Gehrmann et al., 2015). An older phase of ice transport directed toward the south to the southwest was followed by an ice sheet moving northwestward. More recent glacial movements resulted in repeated ice flow toward the south (not shown in Fig. 16.14).

Esker

Another characteristic feature of the glacial landscape is the **esker**, sometimes referred to as a serpent kame. These are elevated ridges comprised of sand and gravel materials that are deposited during ice retreat when rivers develop in melt-water tunnels under the glacier or when melt-waters drain from the glacier surface (Fig. 16.12). The ridges that develop may be several kilometers long and occasionally several tens of meters high. Once the glaciers thaw, they leave behind the elongated sedimentary ridge of gravel and sand that rise up clearly above the surrounding till deposits. A fine example of an esker is the 2-km-long ridge exposed at Rühlow, east of Neubrandenburg. The elongated hill range rises about 20 m above the surroundings and has a central depression (the ***Osauge***, eye of the esker) formed by a block of dead ice that was surrounded by the till of the esker (Fig. 16.15).

Drumlins

Occasionally streamlined hills of loose ground moraine accumulate under the ice to form **drumlins**. The hills are steep sided when facing the ice movement and more flat sided in the direction of ice flow. Morphologically they resemble the erosional structures of the roches moutonnées (see below), which are in contrast to the accretionary origin of drumlins. An impressive example is found at Brodowin, in the state of Brandenburg, where the Kleiner Rummelsberg is considered to have formed as a drumlin close to the main Pomeranian ice edge. Drumlins are more frequently encountered in the Bodensee region and in Upper Bavaria. Examples can be found at Friedrichshafen in the district of Raderach, a town that was built on a drumlin structure. Other drumlins are located in the surroundings of Weilheim, near Lake Starnberg, where the Eberfingen drumlin field is located.

Forms of glacial erosion

Glaciers have a direct effect on the overridden land surface and leave behind a variety of erosional structures. Scratched rock surfaces and (**glacial striations**) are commonly formed during motion of the overriding glacier. Rock material transported within the ice is abrasive and can act like sharp

Fig. 16.13 The end moraine ridge of the Muskau Arch formed during the Elster glacial period. Modified after Kupetz et al. (2004), and the 1:1.000.000 geological map of Germany, BGR (1993)

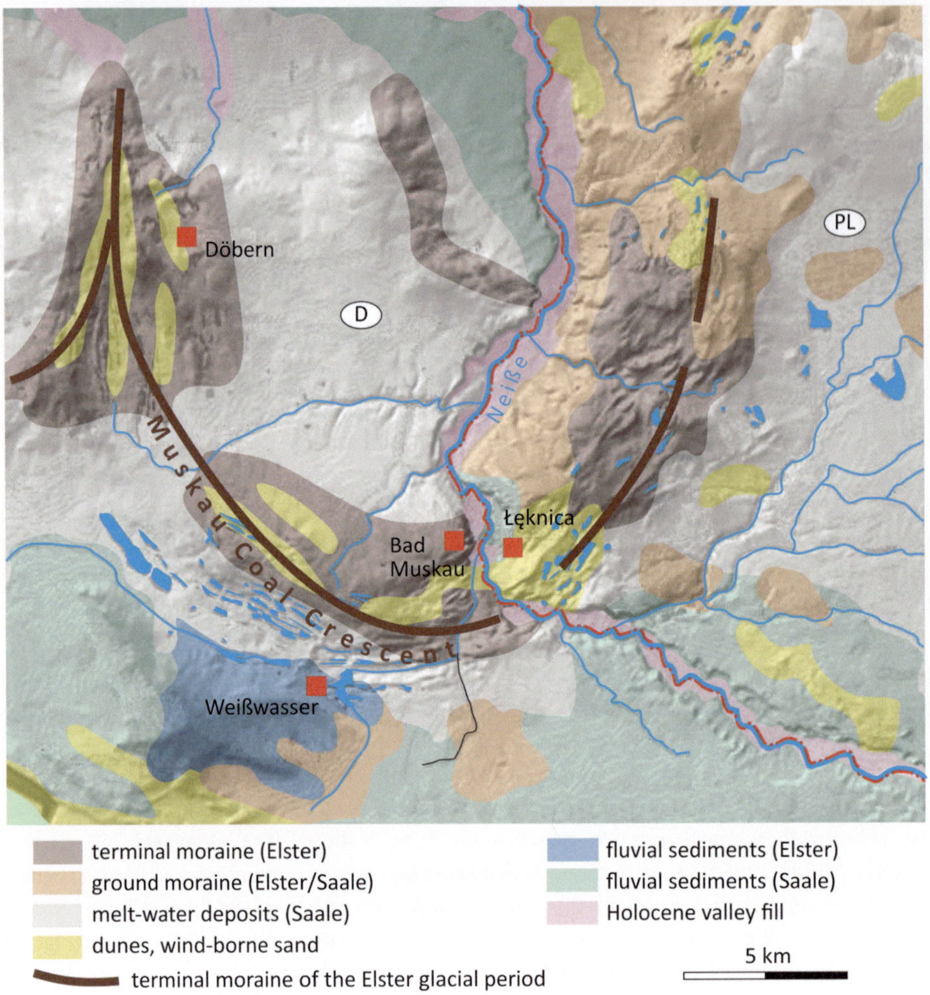

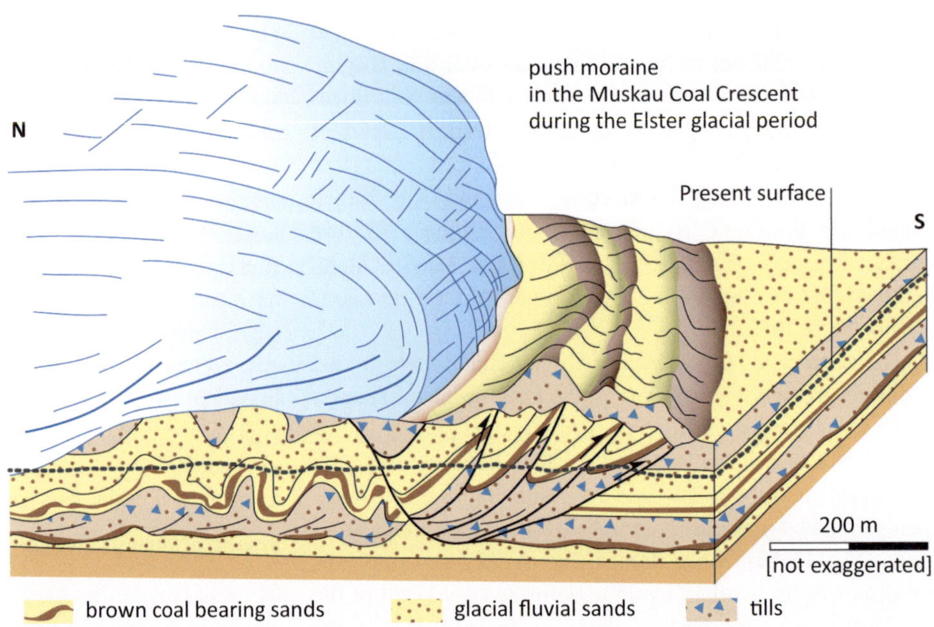

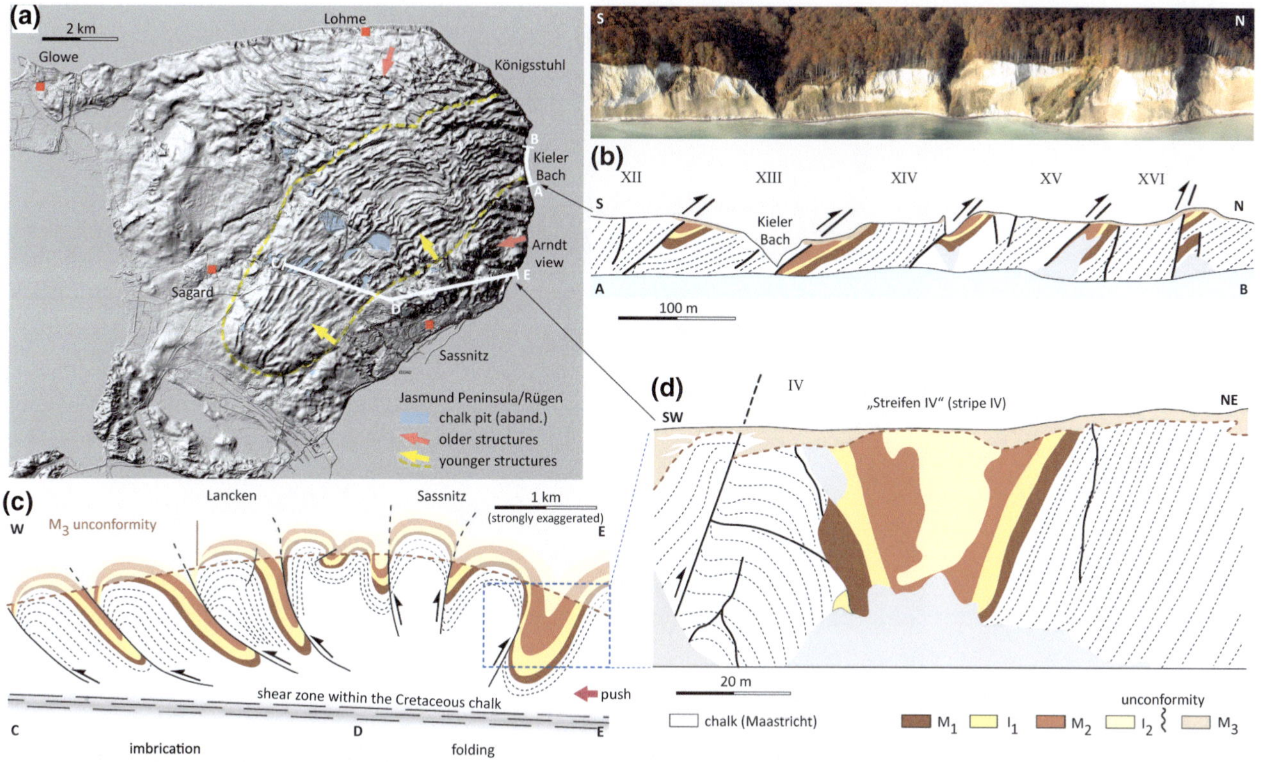

Fig. 16.14 Glacial tectonics on the Jasmund peninsula, Rügen. (**a**) morphologically visible structures on Jasmund viewed in a digital elevation model, based on data obtained using a laser scanning method (by determination of the last echoes, for further explanation, see Sect. 16.2, dunes, loess), (**b**) stacked Cretaceous sheets (named by earlier workers as *Streifen*, stripes) in the glacially deformed complexes of Jasmund, here stripe XII to XVI; (**c**) model showing the evolution of the glaciotectonic complex; (**d**) detailed view of the thrust zone of stripe IV. The digital elevation model [a] was kindly made available by the Mecklenburg-Vorpommern office of internal administration: © GeoBasis-DE/M-V 2013, stripe numbering [b] according to Steinich (1972); [d] modified after Steinich (1972), Herrig (1995), and Niedermeyer et al. (2010)

nails, forming scratch marks in the underlying bedrock. Examples of scratched surfaces are found at the formerly glaciated Feldberg in the Black Forest or at the Great Arber in the Bavarian Forest. In northern Germany, glacial striations and numerous scratched erratic boulders are found in the Hohburg Mountains, near Wurzen, or at the Petersberg, near Halle. In these areas, the mountain peaks emerged above the ice during the glaciation phase as **Nunataks**. Well-known are the previously mentioned glacial scratches on surfaces of the Muschelkalk limestone in Rüdersdorf, with which Torell gained acceptance for the theory of inland ice in 1875.

Larger scale erosive forms caused by the overriding glacier are the **roches moutonnée**. These rounded hills were abraded by the glacier but not fully leveled. They rose above the glacial plain with the characteristic shape of a low-angle flank on the side facing the approaching glacial ice and a higher angle steeper flank in the direction of the flowing ice. The steepness of the flank is caused by the extraction (detraction) of rock fragments by the glacier during freezing. However roches moutonnée are rare in northern Germany.

The gradual retreat of a glacier leaves behind various traces produced by the thawing process, features of which can still be observed in many places today. Glaciers do not move backward but retreat by melting when an insufficient supply of new ice is added at the rear. However, the thawing process is often not uniform. When smaller and even larger parts of the glacier are buried by debris, they remain protected from thawing as so-called **dead ice**. When the isolated ice blocks finally thaw, hollow depressions form where the ice was located. Today these hollows are often filled with water and form small circular or oval ponds in the landscape known as **kettle holes** (Fig. 16.16). They may also form larger dead ice lakes.

Outwash plains

In addition to the ground moraines that form directly from the melting glacier or its melt-waters and the erosive features caused by the ice movement, a variety of other landscape-shaping structures may develop in front of the glacial ice. These areas are subject to **periglacial conditions**, as found today around the glaciers in Greenland, Iceland, and Norway. At the terminus of the glacier, large flat areas of sand and gravel accumulate, together with the development of **dune structures** on the **outwash plains** or **sandurs** (Icelandic: sand). These sediments are deposited

Fig. 16.15 The esker of Rühlow, near Neubrandenburg (Mecklenburg-Vorpommern). Satellite data: Esri Digital Globe GeoEye, i-cubed, USDA, AEX, Getmapping, Aerogrid, IGN, IGP, swisstopo, and the GIS user). [N 53°33′50″/E 13°25′5″]

from the water emerging from the glacier and contain a high proportion of debris of sand and gravel-sized material. The extensive sandy soil areas around Brandenburg and in the Lüneburg Heath represent this type of glacial deposit. A striking example of a sandur district is found in the Zauche area of Brandenburg, southwest of Potsdam (Fig. 16.17). During the Brandenburg stage of the Vistula glacial period, a conical sandur developed, known as the Beelitz sandur. It was formed in front of the Brandenburg main ice advance and it is one of the largest ones in Brandenburg.

The broad river valleys in which the glacial melt-waters are collected and drained are called **glacial valleys** (*Urstromtal*; Fig. 16.12). In northern Germany, glacial valleys received the drainage waters from the continental areas lying in front of the inland ice as any outflow to the north was blocked by the glaciation. The present river systems located in the glacial valleys are generally too small to have formed such large many kilometer wide valley structures. These rivers often use only parts of the glacial valleys and, once the ice disappears, they form new courses to reach the sea in the shortest distance. The valleys were originally filled by large **braided river systems** similar to those observed today crossing the sandurs in Iceland or Greenland. Remains of these braided river systems are found in the Spreewald or the Oderbruch (Fig. 16.17), with a complex of old river beds aligned parallel. However, today the remaining riverbeds no longer alter their course. In an active braided river system, the riverbeds are actively displaced across the broad river valley, particularly during severe flooding. In the northern areas of glaciation, several parallel glacial valleys were formed, mainly running from east to west. These valleys testify to the glacial dynamics at the ice margins and the melt-water drainage as the inland ice thawed.

Dunes and loess

A periglacial area located in front of an ice sheet is characterized by strong and continually blowing winds. Fine

Fig. 16.16 Kettleholes formed by dead ice at a farmland west of Bergen, Rügen Island (view toward the north). [N 54°24′24″/E 13°23′12″]

particles from the deposits of the sandur and glacial valleys are separated out by these winds and transported to wind-protected sites in the surrounding region. In northern Germany, such processes led to the formation of large **inland dune** areas, the basic structures of which remain today, although no longer active. Large dune systems were formed close to the glaciers in particularly windy areas, where sandy hills accumulated with heights of more than 100 m. The largest preserved inland dune in Germany is located on the northern edge of the Elbe glacial valley in the neighborhood of Dömitz. Here, the preserved remains of the **Schmölen wandering dune** (Fig. 16.18) has an area of about 1.1 km^2 and a height of 30 m (43 m above sea-level). It formed as a longitudinal dune that ran parallel to the edge of the glacial valley during the last glaciation phase. It is considered to represent the largest inland dune in Europe and it is now a designated nature reserve.

An example of a dune field located in the eastern part of Saxonia in Upper Lusatia is shown in Fig. 16.19. It originated during the Vistula glacial period and was located upstream on the hillslopes of the glacial valley, directly in front of the glacier. Today, the glacial dunes are covered by

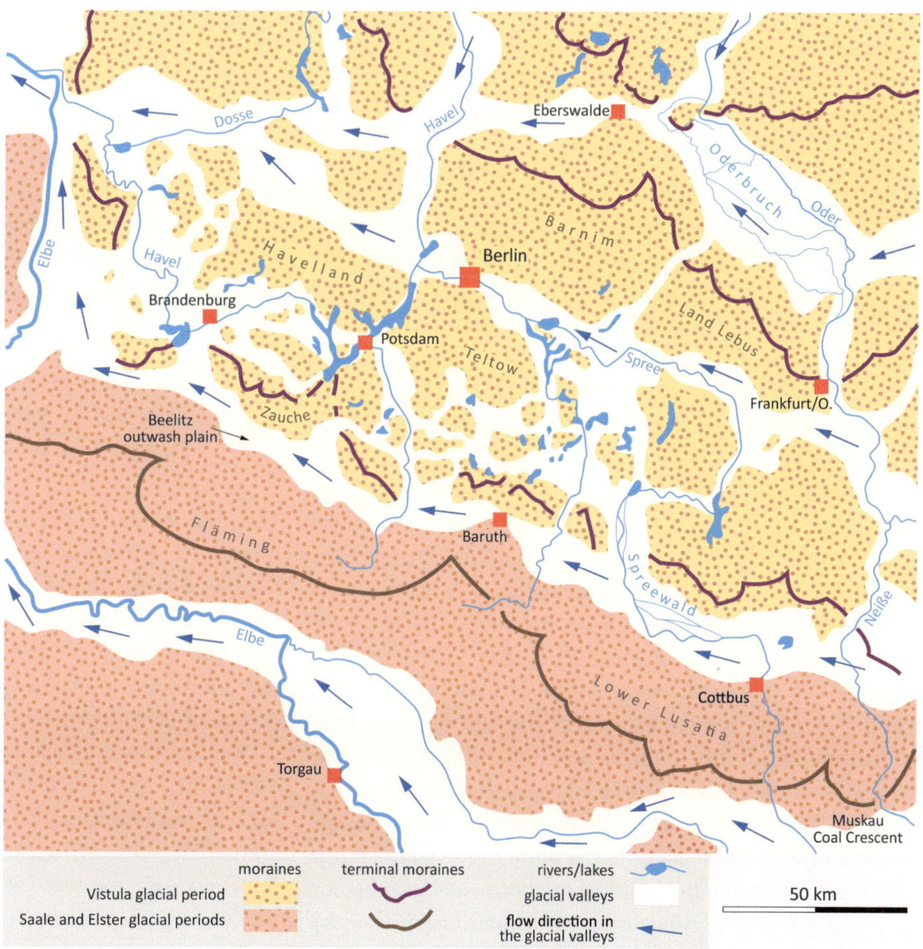

Fig. 16.17 Glacial valleys formed during the Vistula glacial period between Brandenburg and Frankfurt/Odra. Modified after the 1:100.000 geological map of Germany, BGR (1993)

Fig. 16.18 The inland dune of Schmölen, near Dömitz, along the Elbe. It is the largest preserved inland dune formed at the edge of the Elbe glacial valley during the Vistula glacial period. [N 53°07′26″/E 11°18′14″]

a dense pine forest that grows preferably on sandy soil. The dune structure is rather difficult to identify in aerial and satellite images but can be well visualized using a digital model created from elevation data measured with a special laser scanning method (Fig. 16.19). This method records the elevation data of a surface in three dimensions (length, width, and height) with an accuracy in the decimeter range. The distance measurements are carried out with a laser beam mounted on an aircraft. In addition to the distance to the surface of the vegetation (in forests, the tree surface), the actual distance to the ground (determination of the "last echo") can also be measured. After the appropriate filtering of the data, it is possible to remove the vegetation level to reveal the topography of the soil surface.

Figure 16.19 shows a series of **parabolic dunes** from which the wind direction at the time of their formation can be derived. The low-angle flanks of the dunes (on the left side) were formed on the windward side and thus faced the wind. The sand was blown up and over the crest of the dune and was deposited on the steeper lee side in the wind shadow (on the right side). The prevailing wind direction at the time of the formation of dunes was therefore from west to east.

A widespread fine-grained sediment characteristic of a periglacial area is windblown loess. It is composed of silt and clay blown from the vegetation-free moraines and sandur surfaces and deposited on wind-protected hillsides (windward side) or across basins when the transport force of the wind decreased. The accumulation of such sediment resulted in the formation of massive deposits of loess which, due to its carbonate content, produced very fertile soils popular for agricultural use. Loess deposits are often characterized by deeply cut narrow paths due to the angular, fragment-like fine particles that result in a high mechanical stability. As loess is not deposited in glacially covered areas, the distribution of this sediment in northern Central Europe reflects the location of the ice front at the time of the last ice advance. The so-called **loess girdle** extends across Europe and includes the Börde areas to the north of the German low mountains, where the thickest loess deposits are found (such as the Magdeburger Börde and the Hildesheimer Börde). Loess is also widespread in southern Germany. For example, on the slopes of the Kaiserstuhl, where thick loess deposits contribute to the quality of the Kaiserstuhl vineyards.

Permafrost soil

In the areas around the large inland glaciers, the annual average temperatures fell far below freezing point and created the necessary conditions for the formation of **permafrost soils**. These soils were permanently frozen throughout the year at a depth below a few decimetres. During the summer months, the permafrost soils are often thawed superficially to develop a number of characteristic structures. Such structures include **solifluction** (flow earth) features, **ice wedges**, and **stone rings**. Solifluction soils are widespread and are formed mainly on hill slopes. In some cases, the remnants of specific structures, such as **ice wedge pseudomorphosis**, occurred as the ice melted away. In favorable cases, preserved stone rings provide information on the climatic conditions during the glaciation phases. In exceptional cases, the experienced Quaternary geologist can even detect frost ground patterns across farmland based on the color differences imaged during aerial survey.

16.3 The Baltic Sea—A Relic from the Last Glaciation Period

The Baltic Sea is an extraordinarily young marginal or inland sea with an average water depth of about 50 m. The deepest waters are located at the Landsortief, to the west of the island of Gotland, where the water depth reaches 459 m. The Baltic Sea represents an epicontinental shallow sea that lies completely on continental crust. As a result, it does not represent an independent ocean in the plate tectonic sense.

The history of today's Baltic Sea began shortly before the end of the last glaciation period, about 15,000–16,000 years ago. Predecessors of the Baltic Sea, however, existed in a similar form already during earlier warm periods. The Baltic Sea that was formed during the Eemian interglacial had, unlike its present successor, two connections to the world's

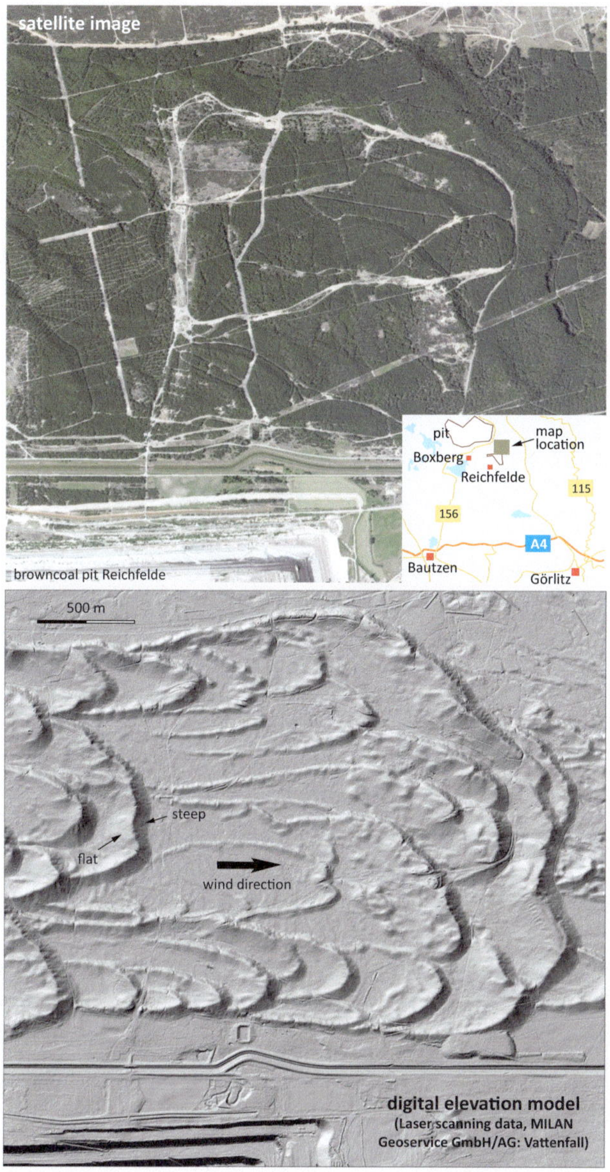

Fig. 16.19 **a** Parabolic dunes from the Vistula glacial period, today covered by woodland north of Reichwalde, near Görlitz (satellite data: Esri Digital Globe GeoEye, i-cubed, USDA, AEX, Getmapping, Aerogrid, IGN, IGP, swisstopo, and the GIS user) [N 53°33′50″/E 13° 25′05″]. **b** Digital elevation model of the parabolic based on elevation data obtained by laser scanning (determination of the last echo, for further explanation, see text; figure with kind permission by MILAN Geoservice GmbH/AG: Vattenfall)

oceans. One connection lay to the west and was similar to the today's Kattegat and Skagerrak. The other lay in the northeast and passed through the Gulf of Finland, the Karelian lakes (Ladoga and Onega) and the White Sea into the Barents Sea. Since the last glaciation, the history of the Baltic Sea can be divided into several stages, summarized in Fig. 16.20. Its development was strongly influenced by the thick inland ice body which, after weighting down the Scandinavian Block by pushing it deeper into the Earth's crust, had a global influence on eustatic sea-level fluctuations as the large ice mass melted. As the rebound movements following removal of the ice are not yet complete, the shape of the Baltic Sea can be expected to continue changing into the future.

Three interacting processes were important for the development of the Baltic Sea, each of which had a different impact on the Baltic regions and their coastlines: (i) **Eustatic sea-level fluctuations** had a global effect attributable to climatic change, such as rapid warming at the end of the last glaciation period. During the peak of the last glaciation (*last glacial maximum*, LGM: about 25,000–20,000 years b.p.), the global sea-level was about 100–120 m lower than at present. Since then, the sea-level has risen to today's stand as the large amounts of water contained in the ice bodies during the glaciation thawed and flowed back into the sea. This process is counterbalanced by (ii) the **isostatic adjustment** of the continental crust in northern Europe. This was caused by the extensive inland ice bodies, with thicknesses of over 3000 m reached in the center of the glaciers. Such a high load on the mainland over a period of several thousand years resulted in gradual subsidence of the Earth's crust into the mantle below. Vertical movements of this kind are possible because the asthenosphere below the lithospheric mantle reacts plastically to pressure and, due to partial melting, it can flow out laterally. The associated process of (iii) **tilting** is dealt further below after a brief consideration of ice-related isostatic rebound.

Presuming that the continental crust of northern Europe was in a state of thermomechanical equilibrium prior to glaciation, a load of 3000 m of ice would have led to a maximum subsidence in the order of 850 m (Fig. 16.21). A simple assumption is that the lithosphere of Scandinavia consisted of 30 km of continental crust and 70 km of lithospheric (solid) mantle. In its equilibrium state, a theoretical area of equal pressure (the compensating surface) at a depth of 200 km in the asthenosphere, would have resulted in a total pressure of 64,000 kg/cm^2 (i.e., equivalent to a column of 1 cm^2 stretching from the compensating surface to the height of the sea-level, weighing 64,000 kg). If a 3 km thick ice layer was added to this column, the pressure would increase to 64,275 kg/cm^2 (assuming the density (ρ) of ice is 0.918 g/cm^3). This difference would amount to an additional column length of 847 m to the asthenosphere (with a specific weight of ρ = 3.25 g/cm^3), which corresponds to the theoretical maximum for subsidence.

If the ice load was then removed by the melting of the ice body, the opposing forces would result in uplift that is stronger in the center and decreases toward the former edge of the ice body. At present, the highest uplift rates of 8 mm per year are measured on the west coast of the Gulf of Bothnia, located in the center of the former ice sheet (Fig. 16.21). In this area, former sandy beaches lie more than

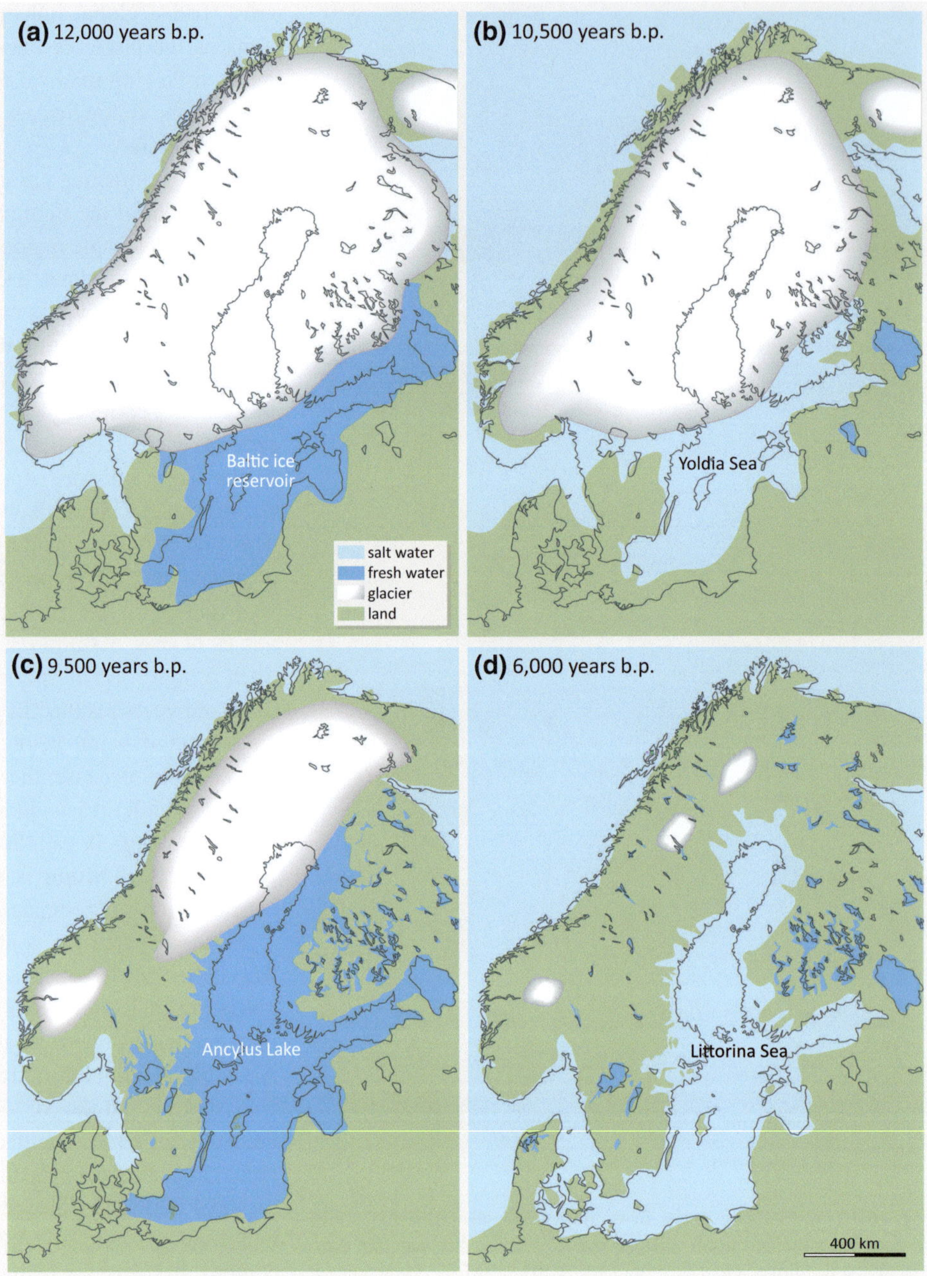

Fig. 16.20 Development stages of the Baltic Sea after the last glaciation period. Modified after Björk (1995)

300 m above the present sea-level. However, as the global sea-level rose strongly during the same period, the effective emergence in the center of the former ice sheet is considerably higher.

Since the volume beneath the uplifting crust has to be filled and the continental crust behaves like a rigid body, a tilting occurs in addition to isostatic uplift. The zero line of tilting runs through Jutland in Denmark and along the German and Polish coastline of the Baltic Sea, but over time it has shifted to the north as it followed the melting ice. While the southern part of Sweden is still undergoing slight uplift, the northern German plain and the coastal area of the Baltic Sea is continuously subsiding (Fig. 16.21). Although the uplift and subsidence movements are today not as rapid as they were immediately after the melting of the glacial ice that covered the area, the process is still not complete today.

At the end of the last glaciation period around 15,500 years ago a series of smaller ice lakes developed. These eventually merged to form a larger lake entity in the course of the gradually thawing inland ice. As a result, the **Baltic Ice Reservoir** (Fig. 16.20a) was formed fed by the melting glacier ice. This freshwater lake deposited a series of varved

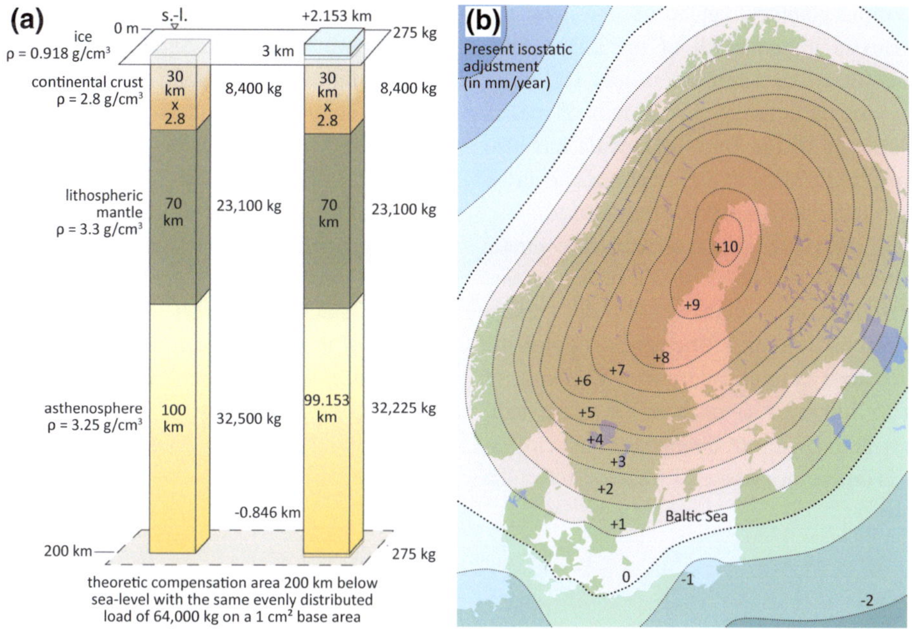

Fig. 16.21 Isostatic adjustment in Scandinavia. **a** It is assumed that the pressure per unit area (here given in kg/cm^2) is equal at a depth of 200 km along a hypothetical compensation plane in the asthenosphere. Caused by the mass of the ice, the lithosphere is pushed down until equilibrium is reestablished. The depth of downward motion does not correspond to the thickness of the ice, but to the weight equivalent of the asthenospheric material. By taking away the mass of the ice, the process is reversed. **b** The present rate of uplift and subsidence in Scandinavia and northeastern Central Europe. Modified after Harff and Meyer (2011)

clays that can be found in many places across the Baltic region.

The lake ended about 12,200 years ago with a sudden drop in the water level by ca. 25 m. It is assumed that in Southern Sweden, between the lakes of Vättern and Vänern at the so-called **Billing Gate**, a drainage path developed to the North Sea caused by the continuous degradation of the inland ice. From this time onward, there was an open connection to the North Sea, where glacial waters mixed with salt water. This is evident from species of bivalve mollusc that migrated into the lake. The mussel *Yoldia arctica* (now known as *Portlandia arctica*), which gives its name to the newly formed **Yoldia Sea** (Fig. 16.18b), inhabited the salt-water environments and can be detected in sediments of that time.

As the slow uplift of Scandinavia continued during pressure relief, the Yoldia Sea was blocked off from the North Sea after about 1500 years of connection. The water became once again fresh and the **Ancylus Lake** (named after the freshwater snail *Ancylus fluviatilis*) was formed (Fig. 16.20c). This lake continued to drain into the North Sea, but there was no longer any mixing with saline water. At this time, large areas of the present Baltic Sea coast were still part of the dry mainland.

About 9000 years ago, the effects of a global sea-level rise became more apparent. In particular, between 8800 and 8000 years before present, an extremely rapid rise in sea-level of ca. 2.5 cm/year occurred. For comparison, the average global sea-level rise today is just 0.32 cm/year. In fact, since the end of the last glaciation period until about 6000 years ago, the sea-level has increased on an average by about 1.3 cm per year. As a result, the flat mainland to the south was flooded and a new connection to the North Sea was established. At that time, the connection was not directly to the Kattegat, but lay further south through the Great Belt and Øresund to the Kattegat. With the introduction of saline water from the Atlantic Ocean, the saltwater snail *Littorina littorea* migrated into the Baltic Sea. This gave its name to the newly formed **Littorina Sea** (Fig. 16.20d).

The strong rise in sea-level lasted until about 6000 years ago and then slowed to a much lower average rate of about 0.17 cm/year. The slower rise in water level initiated erosional processes most noticeable on the coastline. The irregularly shaped coast, typical of glacial regions, was modified by the removal of prominent cliffs and the redistribution of eroded material to neighboring bays. Even today, these processes continue unabated. Within only a few years, recognizable changes resulted from the rearrangement of sediment. These rapid changes can be observed on the peninsula of New Bessin in the north of the island Hiddensee. The peninsula formed as a **spit** that is only 100 years old and continues to grow at a rate of several tens of meters per year. The northern end of the Darss peninsula has also formed from accumulating sediment and continues to grow continuously with a typical hook-shaped promontory.

This environment of deposition around Hiddensee contrasts to the retreating steep coastlines on the islands of Rügen and Usedom, and at the Stoltera, near Warnemünde. Cliff erosion with rates of **coastal retreat** averaging 30 cm/year supplies the abundant material for new coastal landforms. This rate of retreat is equivalent to 100 m in more than 300 years, indicating that the Rügen Island will not disappear as quickly as some assume. There is an obvious desire to want to protect the well known and popular cliffs of Rügen against erosion by installing wave breakers. Such action could perhaps slow the erosion a little but, in the long term, the beauty of the white cliff edges will anyway be lost. From the geological point of view, there is a little that can be done to stop the continuous erosional forces of nature that constantly break off material and transport it away by the sea. In the short term, however, if the process of erosion is interrupted, the oblique slopes will rapidly accumulate small fragments of rock debris and disappear under a dense cover of stabilizing rough vegetation and shrubs. A good example of this can be seen at Cape Arkona on Rügen Island, where a cliff section under the lighthouse was protected from further erosion by a wall. Here, the steep coast is already inhabited by bushes but continues to show its natural white chalk beauty, close by, in the unwalled sections of the cliff.

Appendix

Geological map of Germany and surrounding areas

simplified after the 1 : 1.000.000 geological map of the
Federal Republic of Germany, BGR, 1993

Quaternary:
- Holocene
- Late Pleistocene, deposits of the Vistula resp. Riss glaciation
- Middle and Early Pleistocene, deposits of older glacial periods

Tertiary:
- sedimentary rocks
- volcanics

- Cretaceous
- Jurassic
- Triassic

Permian:
- Zechstein
- Rotliegend
- Rotliegend volcanics

Carboniferous:
- Late Carboniferous
- Early Carboniferous

- Devonian
- Silurian
- Ordovician
- Cambrian
- Precambrian

- Pre-Mesozoic volcanics (Precambrian to Carboniferous)
- Variscan intrusive rocks (granite, granodiorite, gabbro)
- Cadomian intrusive rocks (granite, granodiorite, gabbro)
- Variscan metamorphic rocks
- Early Palezoic metamorphic rocks
- granulite

- Permian to Skythian, Austroalpine
- Palezoic rocks of the Graywacke Zone, Austroalpine

crystalline basement of the Tauern Window, Penninic:
- metamorphic rocks
- orthogneiss

- fault
- nappe thrust
- national border
- river

100 km

M. Meschede and L. N. Warr, *The Geology of Germany*, Regional Geology Reviews,
https://doi.org/10.1007/978-3-319-76102-2

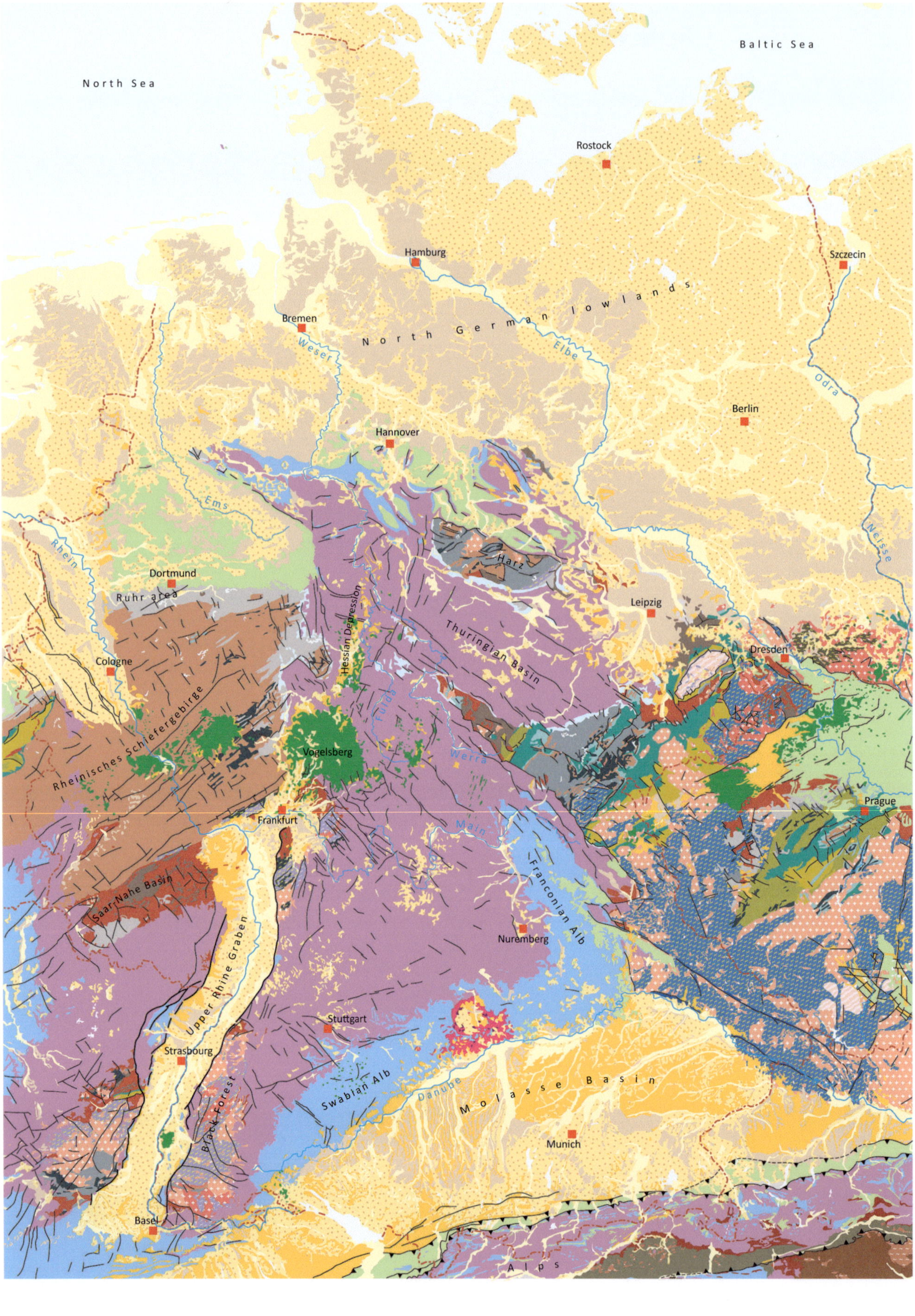
North Sea
Baltic Sea
Rostock
Hamburg
Szczecin
Bremen
North German lowlands
Weser
Elbe
Odra
Berlin
Hannover
Ems
Rhein
Neisse
Harz
Dortmund
Ruhr area
Leipzig
Hessian Depression
Thuringian Basin
Dresden
Cologne
Rheinisches Schiefergebirge
Fulda
Vogelsberg
Werra
Prague
Frankfurt
Main
Saar-Nahe Basin
Franconian Alb
Nuremberg
Upper Rhine Graben
Stuttgart
Strasbourg
Swabian Alb
Danube
Molasse Basin
Black Forest
Munich
Basel
Alps

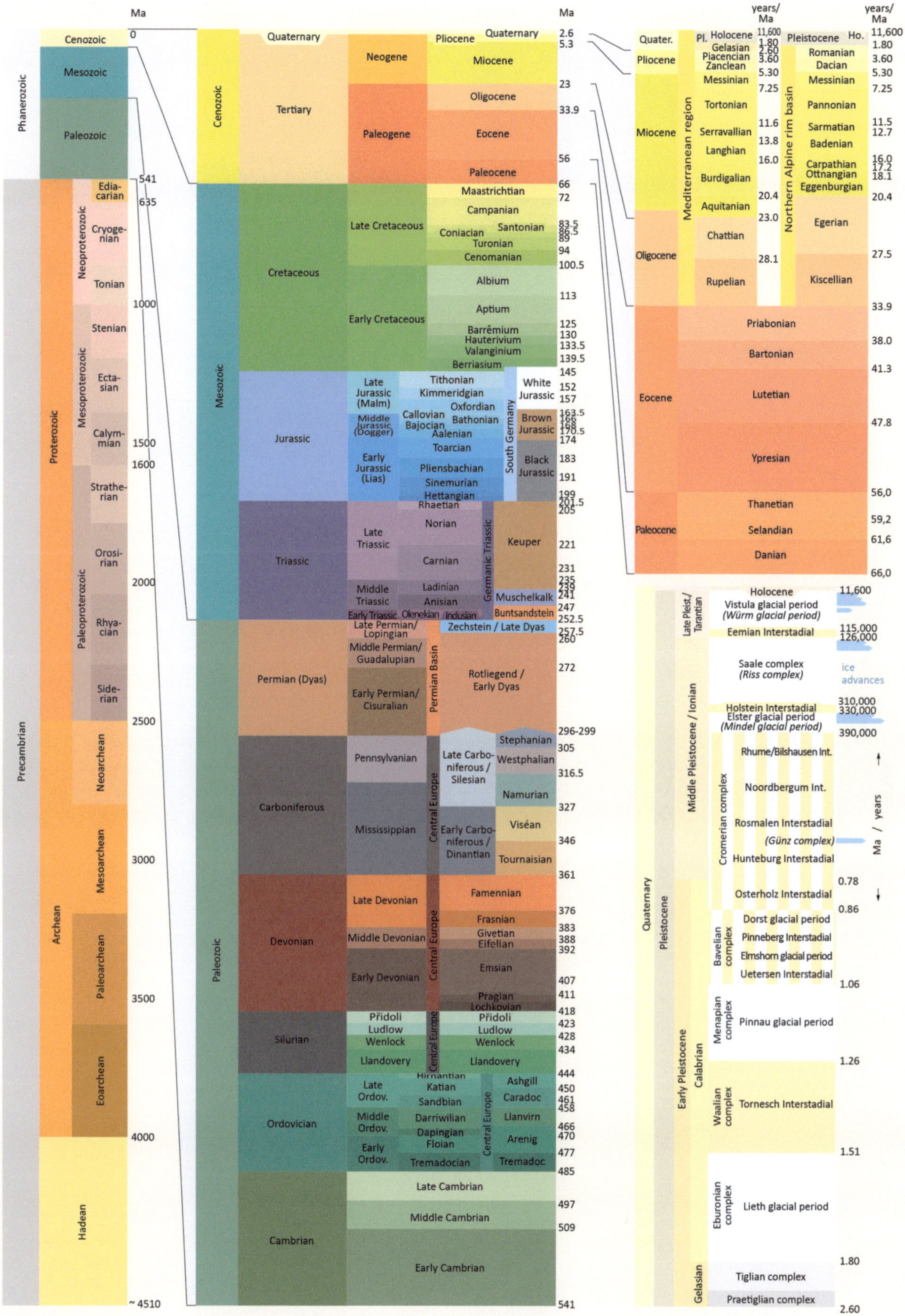
Ma
Cenozoic
Mesozoic
Paleozoic
Phanerozoic
Precambrian
Proterozoic
Archean
Hadean
Neoproterozoic
Mesoproterozoic
Paleoproterozoic
Ediacarian
Cryogenian
Tonian
Stenian
Ectasian
Calymmian
Statherian
Orosirian
Rhyacian
Siderian
Neoarchean
Mesoarchean
Paleoarchean
Eoarchean
0
541
635
1000
1500
1600
2000
2500
3000
3500
4000
~ 4510
Quaternary
Tertiary
Neogene
Paleogene
Pliocene
Miocene
Oligocene
Eocene
Paleocene
Cretaceous
Late Cretaceous
Early Cretaceous
Maastrichtian
Campanian
Santonian
Coniacian
Turonian
Cenomanian
Albium
Aptium
Barrêmium
Hauterivium
Valanginium
Berriasium
Jurassic
Late Jurassic (Malm)
Middle Jurassic (Dogger)
Early Jurassic (Lias)
Tithonian
Kimmeridgian
Oxfordian
Callovian
Bathonian
Bajocian
Aalenian
Toarcian
Pliensbachian
Sinemurian
Hettangian
White Jurassic
Brown Jurassic
Black Jurassic
South Germany
Triassic
Late Triassic
Middle Triassic
Early Triassic
Rhaetian
Norian
Carnian
Ladinian
Anisian
Olenekian
Indusian
Germanic Triassic
Keuper
Muschelkalk
Buntsandstein
Permian (Dyas)
Late Permian/ Lopingian
Middle Permian/ Guadalupian
Early Permian/ Cisuralian
Permian Basin
Zechstein / Late Dyas
Rotliegend / Early Dyas
Carboniferous
Pennsylvanian
Mississippian
Central Europe
Late Carboniferous / Silesian
Early Carboniferous / Dinantian
Stephanian
Westphalian
Namurian
Viséan
Tournaisian
Devonian
Late Devonian
Middle Devonian
Early Devonian
Famennian
Frasnian
Givetian
Eifelian
Emsian
Pragian
Lochkovian
Silurian
Přidoli
Ludlow
Wenlock
Llandovery
Ordovician
Late Ordov.
Middle Ordov.
Early Ordov.
Hirnantian
Katian
Sandbian
Darriwilian
Dapingian
Floian
Tremadocian
Ashgill
Caradoc
Llanvirn
Arenig
Tremadoc
Cambrian
Late Cambrian
Middle Cambrian
Early Cambrian
2.6
5.3
23
33.9
56
66
72
83.5
86.5
89
94
100.5
113
125
130
133.5
139.5
145
152
157
163.5
166
168
170.5
174
183
191
199
201.5
205
221
231
235
239
241
247
252.5
257.5
260
272
296-299
305
316.5
327
346
361
376
383
388
392
407
411
418
423
428
434
444
450
461
458
466
470
477
485
497
509
541
years/ Ma
Quater.
Pl.
Holocene
Gelasian
Piacencian
Zanclean
Messinian
Tortonian
Serravallian
Langhian
Burdigalian
Aquitanian
Chattian
Rupelian
Mediterranean region
11,600
1.80
2.60
3.60
5.30
7.25
11.6
13.8
16.0
20.4
23.0
28.1
Northern Alpine rim basin
Pleistocene
Ho.
Romanian
Dacian
Messinian
Pannonian
Sarmatian
Badenian
Carpathian
Ottnangian
Eggenburgian
Egerian
Kiscellian
11,600
1.80
3.60
5.30
7.25
11.5
12.7
16.0
17.2
18.1
20.4
27.5
33.9
Priabonian
Bartonian
Lutetian
Ypresian
Thanetian
Selandian
Danian
38.0
41.3
47.8
56,0
59,2
61,6
66,0
Quaternary
Pleistocene
Late Pleist./ Tarantian
Middle Pleistocene / Ionian
Early Pleistocene
Calabrian
Gelasian
Holocene
Vistula glacial period
(Würm glacial period)
Eemian Interstadial
Saale complex
(Riss complex)
Holstein Interstadial
Elster glacial period
(Mindel glacial period)
Cromerian complex
Rhume/Bilshausen Int.
Noordbergum Int.
Rosmalen Interstadial
(Günz complex)
Hunteburg Interstadial
Osterholz Interstadial
Bavelian complex
Dorst glacial period
Pinneberg Interstadial
Elmshorn glacial period
Uetersen Interstadial
Menapian complex
Pinnau glacial period
Waalian complex
Tornesch Interstadial
Eburonian complex
Lieth glacial period
Tiglian complex
Praetiglian complex
11,600
115,000
126,000
ice advances
310,000
330,000
390,000
Ma / years
0.78
0.86
1.06
1.26
1.51
1.80
2.60

References

Agricola G (1546) De natura fossilium libri, vol X. Basel

Agricola G (1556) De re metallica libri, vol XII. Basel

Ahlborn O, Richter-Bernburg G (1953) Exkursion zum Salzstock von Benthe (Hannover), mit Befahrung der Kaliwerke Ronnenberg und Hansa. Zt Dt Geol Ges 105:855–865

Ahrendt H, Clauer N, Hunziker JC, Weber K (1983) Migration of folding and metamorphism in the Rheinisches Schiefergebirge deduced from K–Ar and Rb–Sr age determinations. In: Martin H, Eder FW (eds) Intracontinental fold belts. Case studies in the Variscan belt of Europe and the Damara Belt in Namibia. Springer, Berlin, pp 323–338

Aigner T (1985) Storm depositional systems. Lect Notes Earth Sci 3:1–174

Aigner T, Bachmann GH (1992) Sequence-stratigraphic framework of the German Triassic. Sediment Geol 80:115–135

Aigner T (1999) Dynamische Stratigraphie des Oberen Muschelkalks am Beispiel Süddeutschlands. In: Hauschke N, Wilde V (eds) Trias, eine ganze andere Welt: Mitteleuropa im frühen Erdmittelalter. Pfeil-Verlag, München, pp 115–128

Alley RB (2000) The Younger Dryas cold interval as viewed from central Greenland. Quat Sci Rev 19:213–226

Alvarez LW, Alvarez W, Asaro F, Michel HV (1980) Extraterrestrial cause for the Cretaceous-Tertiary extinction. Science, Washington, pp 1095–1108

Anderle H-J, Franke W, Schwab M (1995) Metamorphic units (Northern Phyllite Zone – Stratigraphy). In: Dallmeyer RD, Franke W, Weber K (eds) Pre-Permian geology of Central and Eastern Europe. Springer, Berlin, pp 98–107

Baales M, Jöris O, Street M, Bittmann F, Weninger B, Wiethold J (2002) Impact of the Late Glacial Eruption of the Laacher See Volcano, Central Rhineland, Germany. Quat Res 58:273–288

Bachmann G, Beutler G, Hagdorn H, Hauschke N (1999) Stratigraphie der Germanischen Trias. In: Hauschke N, Wilde V (eds) Trias, eine ganze andere Welt: Mitteleuropa im frühen Erdmittelalter. Pfeil-Verlag, München, pp 81–104

Bachmann GH, Kozur H (2004) The Germanic Triassic: correlations with the international chronostratigraphic scale, numerical ages and Milankovitch cyclicity. Hallesches Jb Geowiss B 26:17–62

Bachmann GH (2010) Depositional cycles of the Southern Permian Basin. In: Doornenbal JC, Stevenson AG (eds) Petroleum geological atlas of the Southern Permian Basin Area. EAGE Publications b.v., Houten, 99 pp

Bachmann GH, Geluk MC, Warrington G, Becker-Roman A, Beutler G, Hagdorn H, Hounslow MW, Nitsch E, Röhling H-G, Simon T, Szulc A (2010) Triassic. In: Doornenbal JC, Stevenson AG (eds) Petroleum geological atlas of the Southern Permian Basin Area. EAGE Publications b.v, Houten, pp 149–173

Baier J (2009) Zur Herkunft und Bedeutung der Ries-Auswurfprodukte für den Impakt-Mechanismus. Jber Mitt oberrhein geol Ver, N F 91:9–29

Bankwitz P, Bankwitz E, Kramer W (1994) Early Paleozoic bimodal volcanism in the Vesser area, Thuringian Forest, eastern Germany. Zentralbl Geol Pal Teil I 1992:1113–1132

Bankwitz P, Schneider G, Kämpf H, Bankwitz E (2003) Structural characteristics of epicentral areas in Central Europe: study case Cheb Basin (Czech Republic). J Geodyn 35:5–32

Bartier D, Ledésert B, Clauer N, Meunier A, Liewig N, Morvan G, Addad A (2008) Hydrothermal alteration of the Soultz-sous-Forêts granite (hot fractures rock geothermal exchanger) into a tosudite and illite assemblage. Eur J Mineral 20:131–142

Baumann A, Grauert B, Mecklenburg S, Vinx R (1991) Isotopic age determinations of crystalline rocks of the Upper Harz Mountains, Germany. Geol Rdsch 80:669–690

Bayerisches Geologisches Landesamt (1996) Geologische Karte von Bayern 1:500.000 mit Erläuterungen. München, 329 pp

Bechtel A, Elliot WC, Oszczetpalski S (1996) Indirect age determination of Kupferschiefer-type mineralization in the Polish Basin by K/Ar dating of illite – preliminary results. Econ Geol 92:1310–1319

Becker A, Schwarz M, Schäfer A (2012) Lithostratigraphische Korrelation des Rotliegend im östlichen Saar-Nahe-Becken. Jber Mitt oberrhein geol Ver, NF 94:105–133

Behrmann J, Tanner D (1997) Carboniferous tectonics of the Variscan basement collage in eastern Bavaria and western Bohemia. Geol Rdsch 86(Suppl):S15–S27

Benton MJ (2005) When life nearly died: the greatest mass extinction of all time. Thames & Hudson, London, p 336

Berger H-J, Krentz O, Lapp M (2008) Granulitgebirge. In: Pälchen W, Walter H (eds) Geologie von Sachsen. Schweizerbart'sche Verlagsbuchhandlung, Stuttgart, pp 51–56

Berggren WA, Kent DV, Swisher CC, Aubry M-P (1995) A revised Cenozoic geochronology and chronostratigraphy. In: Berggren WA, Kent DV, Aubry M-P (eds) Geochronology, time scales and global stratigraphic correlations, vol 54. SEPM Special Publications, Darlington, pp S17–S28

Berner RA (1990) Atmospheric carbon dioxide levels over Phanerozoic times. Science 249(4975):1382–1386

Berner RA (1999) A new look at the long-term carbon cycle. GSA Today 9(11):2–6

Berner RA, Petsch ST, Lake JA, Beerling DJ, Popp BN, Lane RS, Laws EA, Westley MB, Cassar N, Woodward FI, Quick WP (2000) Isotope fractionation and atmospheric oxygen: implications for Phanerozoic O_2 evolution. Science 287(5458):1630–1633

Berner RA, Kothavala Z (2001) Geocarb III: a revised model of atmospheric CO_2 over Phanerozoic time. Am J Sci 301:182–204

M. Meschede and L. N. Warr, *The Geology of Germany*, Regional Geology Reviews,
https://doi.org/10.1007/978-3-319-76102-2

Bertrand M (1884) Rapports de structure des Alpes de Glaris et du bassin houiller du Nord. Bull Soc Géol France 12:318–330

BGR – Bundesanstalt für Geowissenschaften und Rohstoffe (1981) Geologische Übersichtskarte 1:200.000, Blatt Bayreuth

BGR – Bundesanstalt für Geowissenschaften und Rohstoffe (1993): Geologische Karte der Bundesrepublik Deutschland 1:1.000.000

Björk S (1995) A review of the history of the Baltic Sea 13.0–8.0 ka BP. Quart Int 27:19–40

Blakey RC (2003) Global Paleogeographic views on Earth history – Late Precambrian to Recent. Flagstaff, World Wide Web. http://deeptimemaps.com/map-lists/. Accessed July 2017

Blakey RC (2011) Paleogeographic and tectonic history of Europe. Flagstaff, World Wide Web. http://deeptimemaps.com/map-lists/. Last access July 2017

Blundell D, Freeman R, Mueller S (eds) (1992) A continent revealed. The European Geotraverse. Cambridge University Press, Cambridge

BNL (2004) Informationsblatt Wutachschlucht-Bezirksstelle für Naturschutz und Landschaftspflege Freiburg, 4 edn Freiburg

Bornemann O, Behlau J, Fischbeck R, Hammer J, Jaritz W, Keller S, Mingerzahn G, Schramm M (2008) Standortbeschreibung Gorleben Teil 3: Ergebnisse der über- und untertägigen Erkundung des Salinars. Geol Jb, C 73:211

Bouma AH (1962) Sedimentology of some Flysch deposits: a graphic approach to facies interpretation. Elsevier, Amsterdam, New York, 168 pp

Bousquet R, Schmid SM, Zeilinger G, Oberhänsli R, Rosenberg C, Molli G, Robert C, Wiederkehr M, Rossi P (2012) Tectonic framework of the Alps, 1st edn. Commission for the geological maps of the World, subcommission for magmatic and metamorphic maps, UNESCO (1:1.000.000)

Bowring SA, Erwin DH, Jin YG, Martin MW, Davidek K, Wang W (1998) U/Pb Zircon geochronology and tempo of the end-Permian mass extinction. Science 280:1039–1045

Bowring SA, Williams IS (1999) Priscoan (4.00–4.03 Ga) orthogneisses from northwestern Canada. Contrib Min Petrol 134:3–16

Bökenschmidt S (2006) Die Fossillagerstätte Korbacher Spalte – ihre Entstehung und Einordnung in den Zechstein Nord-Hessens. Diss. Philipps Univ. Marburg, 138 pp

Bradbury J (2004) Nature's nanotechnologists: unveiling the secrets of diatoms. PLoS Biol 2(10):e306. https://doi.org/10.1371/journal.pbio.0020306

Branco W, Fraas E (1905) Das kryptovulkanische Becken von Steinheim. Abh. königl. preuß. Akad. Wiss, Berlin, pp 1–64

Bräuer K, Kämpf H, Strauch G (2009) Earthquake swarms in non-volcanic regions: what fluids have to say. Geophys Res Lett 36(L17309):5

Bräuer K, Kämpf H, Niedermann S, Strauch G (2013) Indications for the existence of different magmatic reservoirs beneath the Eifel area (Germany): A multiisotope (C, N, He, Ne, Ar) approach. Chem Geol 356:193–208

Breitkreuz C, Kennedy A (1999) Magmatic flare-up at the Carboniferous/Permian boundary in the NE German Basin revealed by SHRIMP zircon ages. Tectonophysics 302:307–326

Breitkreuz C, Mock A (2004) Are laccolith complexes characteristic of transtensional basin systems? Examples from the Permo-Carboniferous of Central Europe. In: Breitkreuz C, Petford N (eds) Physical geology of high-level magmatic systems. Geol Soc Spec Publ 234:13–31

Breitkreuz C, Kennedy A, Geißler M, Ehling B-C, Kopp J, Muszynski A, Protas A, Stouge S (2007) Far Eastern Avalonia: its chronostratigraphic structure revealed by SHRIMP zircon ages from Upper Carboniferous to Lower Permian volcanic rocks (drill cores from Germany, Poland and Denmark. Geol Soc Am Spec Pap 423:173–190

Breitkreuz C, Ehling B, Sergeev S (2009) Chronological evolution of an intrusive/extrusive system: the Late Paleozoic Halle Volcanic Complex in the northeastern Saale Basin (Germany). Z Dt Ges Geowiss 160:173–190 (Stuttgart)

Brenchley PJ, Harper DAT (1998) Palaeoecology: ecosystems, environments and evolution. Chapman & Hall, London, 407 pp

Brink H-J (2013) Die Intrusion von Bramsche – ein Irrtum im invertierten Niedersächsischen Becken? ZDt Ges Geowiss 164:33–48

Buchner E, Schwarz WH, Schmieder M, Trieloff M (2010) Establishing a 14.6 ± 0.2 Ma age for the Nördlinger Ries impact (Germany) – a prime example for concordant isotopic ages from various dating materials. Meteorit Planet Sci 45:662–674

Buchner E, Hölzel M, Schmieder M, Rasser M, Fietzke J, Frische M, Kutterolf S (2017) A meteorite fragment trapped between positive and negative shatter cones in a limestone block stored at the Meteorkrater-Museum Steinheim, Germany. Meteoritics and Planetary Science. In: 80th annual meeting of the meteoritical society, Santa Fe, New Mexico, USA, vol 52, Iss. Suppl S1, 6014

Buggisch W (1991) The global Frasnian-Famennian »Kellwasser Event«. Geol Rdsch 80:49–72

Buness H, Felder M, Gabriel G, Harms F-J (2005) UNESCO-Welterbe Fossilienlagerstätte Grube Messel – Momentaufnahmen aus dem Eozän Vernissage, Reihe UNESCO-Welterbe, vol 21. Vernissage-Verlag, Heidelberg, 63 pp, 13 year

Burollet P, Dal Piaz GV, Franke W, Trümpy R (1992) Atlas map 1 – tectonics, south sheet. In: Blundell D, Freeman R, Mueller S (eds) A continent revealed. The European Geotraverse. Cambridge University Press, Cambridge, 73 + 275 pp, 14 maps

Büchel G, Krawczyk E (1986) Zur Genese der Dauner Maare im Vulkanfeld der Westeifel. Mainzer Geowiss Mitt 15:219–238

Chen F, Todt W, Hann H (2003) Zircon and garnet geochronology of eclogites from the Moldanubian Zone of the Black Forest, Germany. J Geol 111:207–222

Closs H (1938) Hebung, Spaltung, Vulkanismus. Geol Rdsch 30:401–527

Cloos H (1941) Bau und Tätigkeit von Tuffschloten. Untersuchungen an dem Schwäbischen Vulkan. Geol Rdsch 32:709–800

Cocks LRM, Torsvik TH (2006) European geography in a global context from the Vendian to the end of the Palaeozoic. In: Gee DG, Stephenson RA (eds) European lithosphere dynamics, vol 32. Geological Society, London, Memoirs, pp 83–95

Cohen KM, Gibbard PL (2010) Global chronostratigraphical correlation table for the last 2.7 million years. Subcommission on Quaternary Stratigraphy. International Commission on Stratigraphy, Cambridge

Condie K (1997) Plate tectonics and crustal evolution, 4. ed. Butterworth-Heinemann, Oxford, 282 pp

Cooper MA, Williams GD (1989) Inversion Tectonics. Geol Soc Spec Publ 44:375

Czamanske GK, Gurevich AB, Fedorenko V, Simonov O (1998) Demise of the Siberian plume: paleogeographic and paleotectonic reconstruction from the prevolcanic and volcanic records, North-Central Siberia. Int Geol Rev 40:95–115

Dahm T, Krüger F, Stammler K, Klinge K, Kind R, Wylegalla K, Grasso J-R (2007) The 2004 Mw 4.4 Rotenburg, Northern Germany, earthquake and its possible relationship with gas recovery. Bull Seismol Soc Am 97:691–704

Dallmeyer RD, Franke W, Weber K (eds) (1995) Pre-Permian geology of central and eastern Europe. 604 S. Springer, Heidelberg

DEKORP BASIN Research Group (1999) Deep crustal structure of the Northeast German Basin: new DEKORP-BASIN' 96 deep-profiling results. Geology 27:55–58

Dercourt J, Gaetani M et al (2000) Atlas Peri-Tethys, paleogeographical Maps. Commission for the Geological Map of the World, Explanatory Notes, Gauthiers-Villars, Paris

Deutsche Stratigraphische Kommission (Hrsg.; Koordination und Gestaltung: Menning M., Hendrich A) (2016) Stratigraphische Tabelle von Deutschland 2016 (STD 2016). Potsdam (Deutsches GeoForschungsZentrum). ISBN 978-3-9816597-7-1

Dézes P, Schmid SM, Ziegler PA (2004) Evolution of the European Cenozoic rift system, interaction of the Pyrenean and Alpine orogens with the foreland lithosphere. Tectonophysics 389:1–33

Di Vincenzo G, Skála R (2009) 40Ar-39Ar laser dating of tektites from the Cheb Basin (Czech Republic): Evidence for coevality with moldavites and influence of the dating standard on the age of the Ries impact. Ceochim Cosmochim Acta 73:493–513

Drozdzewski G, Wrede V (1994) Faltung und Bruchtektonik – Analyse der Tektonik im Subvariscikum. Fortschr Geol Rheinl Westf 38:7–187

Eberle J, Eitel B, Blümel W-D (2010) Deutschlands Süden – vom Erdmittelalter zur Gegenwart. Spektrum Akademischer Verlag, Heidelberg, p 192

Edel JB, Schulmann K, Rotstein Y (2007) The Variscan tectonic inheritance of the Upper Rhine Graben: evidence of reactivations in the Lias, Late Eocene-Oligocene to recent. Int J Earth Sci (Geol Rdsch) 96:305–325

Ehlers J, Gibbard PL (2007) The extent and chronology of Cenozoic global glaciation. Quat Int 6(20):164–165

Eichler B, Werneburg R (2010) Neufunde von Branchiosuriern aus dem Rotliegend von Freital bei Dresden im UV-Licht. Geologica Saxonica. J Central Europ Geol 56:137–157

Eidam J, Hammer J, Korich D, Bielicki K-H (1995) Characterization, distribution and genesis of amphibole-bearing Variscan granites in the Lusatian Granodiorite Massif (Northern margin of the Bohemian Massif. N Jb Miner Abh 168:259–281

Eisbacher GH, Linzer HG, Meier L, Polinski R (1990) A depth-extrapolated transect across the Northern Calcareous Alps of western Tirol. Eclogae Geol Helv 83:711–725

Eissmann L, Litt T (1994) Das Quartär Mitteldeutschlands. Ein Leitfaden und Exkursionsführer. Altenburger Naturwissenschaftliche Forschungen 7:1–458

Enderle U (1998) Signaturen in refraktionsseismischen Daten als Abbild geodynamischer Prozesse. Dissertation, Universität Karlsruhe, 221 pp

Engel W, Franke W, Grote C, Weber K, Ahrendt H, Eder FW (1983) Nappe tectonics in the southeastern part of the Rheinisches Schiefergebirge. In: Martin H, Eder FW (eds) Intracontinental fold belts. Springer, Berlin, pp 267–287

Eugster HP, Hardy LA (1975) Sedimentation in an ancient playa-lake complex – Wilkins peak member of green river formation of Wyoming. Geol Soc Am Bull 86:319–334

Finkenwirth A (1978) Die Braunkohle am Meißner. Der Aufschluß 28 (Sonderband):229–236

Flick H, Nesbor HD, Behnisch R (1990) Iron ore of the Lahn-Dill type formed by diagenetic seeping of pyroclastic sequences – a case study on the Schalstein section at Gänseberg (Weilburg). Geol Rdsch 79:401–415

Fraedrich W (1996) Spuren der Eiszeit, Landschaftsformen in Europa. Springer, Berlin, Heidelberg, p 184

Franke W (1984a) Late events in the tectonic history of the Saxothuringian Zone. In: Hutton DWH, Sanderson DJV (eds) Variscan tectonics of the North Atlantic region. Blackwell, Oxford, pp 33–45

Franke W (1984b) Variszischer Deckenbau im Raume der Münchberger Gneismasse, abgeleitet aus der Fazies, Deformation und Metamorphose im umgebenden Paläozoikum. Geotekt Forsch 68:1–253

Franke W (1989) Tectonostratigraphic units in the Variscan belt of central Europe. Geol Soc Am, Spec Pap 230:67–90

Franke W (1992) Phanerozoic structures and events in central Europe. In: Blundell D, Freeman R, Müller R (eds) A continent revealed. The European Geotraverse. Cambridge University Press, Cambridge, pp 164–179

Franke W (2000) The mid-European segment of the Variscides: tectonostratigraphic units, terrane boundaries and plate tectonic evolution. In: Franke W, Haak V, Oncken O, Tanner D (eds) Orogenic processes: quantification and modelling in the Variscan Belt. Geological Society, London, Special Publications 179:35–61

Franke W, Stein E (2000) Exhumation of high-grade rocks in the Saxo-Thuringian belt: geological constraints and geodynamic concepts. Geological Society, London, Special Publications 179:337–354

Franke W, Żelaźniewicz A (2000) The eastern termination of the Variscides: terrane correlation and kinematic evolution. In: Franke W, Haak V, Oncken O, Tanner D (eds) Orogenic processes: Quantification and modelling in the Variscan Belt, vol 179. Geological Society, London, Special Publications, pp 63–86

Franke W (2006) The Variscan Orogen in Central Europe: construction and collapse. In: Gee GD, Stephenson RA (eds) European Lithosphere Dynamics, Geological Society, London, Memoirs, 32:333–343

Franke W, Cocks LRM, Torsvik TH (2017) The Paleozoic Variscan oceans revisited. Gondwana Research, 48:257–284

Franzen JL, Gingerich PD, Habersetzer J, Hurum JH, von Koenigswald W, Smith BH (2009) Complete Primate Skeleton from the middle Eocene of Messel in Germany: morphology and paleobiology. PLoS One 4(5):e5723. https://doi.org/10.1371/journal.pone.0005723

Franzke H-J, Voigt T, von Eynatten H, Brix M, Burmester G (2004) Geometrie und Kinematik der Harznordrandstörung, erläutert an Profilen aus dem Gebiet von Blankenburg. Geowiss Mitt Thüringen 11:39–62

Franzke H-J (2009) Eckertalsperre/Eckergneis. Schautafel des Geopark Harz am Eckerstausee

Franzke H-J (2012) Exkursionspunkt 1: Liegender Grenzbereich der Ostharz-Decke im Ludetal 4. Workshop Harzgeologie, Roßla, 19.–20.10.2012. Hallesches Jahrbuch, Beiheft, vol 28, S 68–74

Friedl G, Finger F, Paquette JL, von Quadt A, McNaugthon NJ, Fletcher IR (2004) Pre-variscan geological events in the Austrian part of the Bohemian Massif deduced from U–Pb zircon ages. Int J Earth Sci 93:802–823

Frisch W (1979) Tectonic progradation and plate tectonic evolution of the Alps. Tectonophysics 60:121–139

Frisch W, Kuhlemann A, Dunkl I, Brügel A (1998) Palinspastic reconstruction and topographic evolution of the Eastern Alps during late Tertiary tectonic extrusion. Tectonophysics 297:1–15

Frisch W, Dunkl I, Kuhlemann J (2000) Post-collisional orogen-parallel large-scale extension in the Eastern Alps. Tectonophysics 327:239–265

Frisch W, Kuhlemann J, Dunkl I, Székely B (2001) The Dachstein paleosurface and the Augenstein Formation in the Northern Calcareous Alps – a mosaic stone in the geomorphological evolution of the Eastern Alps. Int J Earth Sci (Geol Rdsch) 90:500–518

Frisch W, Gawlick H-J (2003) The nappe structure of the central Northern Calcareous Alps and its disintegration during Miocene tectonic extrusion – a contribution to understanding the orogenic evolution of the Eastern Alps. Int J Earth Sci 92:712–727

Frisch W, Meschede M (2011) Plate tectonics – Continental drift and mountain building. Springer, Heidelberg, New York, 202 pp

Frisch W, Meschede M (2013) Plattentektonik und Gebirgsbildung, 5th edn. Wissenschaftliche Buchgesellschaft/Primus-Verlag, Darmstadt (196 pp, 1st edn, 2005)

Froitzheim N, Pleuger J, Roller S, Nagel T (2003) Exhumation of high- and ultrahigh-pressure metamorphic rocks by slab extraction. Geology 31:925–928

Fuhrmann R (2005) Die Bernsteinlagerstätte Bitterfeld, nur ein Höhepunkt des Vorkommens von Bernstein (Succinit) im Tertiär Mitteldeutschlands. Zt Dt Ges Geowiss 156:517–530

Gansslosser M (1996) Stoffbestand, Fazies und geodynamische Interpretation der Stieger Schichten (Ostharz-Decke). Braunschweiger Geol-Paläont Diss 18:1–162

Gast R (1988) Rifting im Rotliegenden Niedersachsens. Die Geowissenschaften 6:115–122

Gast RE, Dusar M, Breitkreuz C, Gaupp R, Schneider JW, Stemmerik L, Geluk MC, Geißler M, Kiersnowski H, Glennie KW, Kabel S, Jones NS (2010) Rotliegend. In: Doornenbal JC, Stevenson AG (eds) Petroleum geological atlas of the Southern Permian Basin Area. EAGE Publications b.v, Houten, pp 101–121

Gawlick H-J (2000) Paläogeographie der Ober-Trias Karbonatplattform in den Nördlichen Kalkalpen. Mitt Ges Geol Bergbaustud Österr 44:45–95

Gebauer D, Williams IS, Compston W, Grünenfelder M (1989) The development of the Central European continental crust since the Early Archean based on conventional and ion-micro-probe dating of up to 3.84 by old detrital zircons. Tectonophysics 157:81–96

Gehrmann A, Meschede M, Hüneke H, Rother H, Obst K (2015) Der glazitektonische Komplex von Jasmund (NE-Rügen): Geomorphologische Kartierung und Landformenanalyse auf der Grundlage von LiDAR-Daten, vol 79. Tagung der ARGE NDG, 26–29. Mai 2015, Güstrow

Geisler T, Vinx R, Martin-Gombojav N, Pidgeon RT (2005) Ion microprobe (SHRIMP) dating of detrital zircon grains from quartzites of the Eckergneiss Complex, Harz Mountains (Germany): Implications for the provenance and geological history. Int J Earth Sci 94:369–384

Geologischer Dienst Nordrhein-Westfalen (2003) Geotope in Nordrhein-Westfalen. Zeugnisse der Erdgeschichte, 2nd edn. Broschüre Geologischer Dienst Nordrhein-Westfalen, Krefeld

Gereke M, Luppold FW, Piecha M, Schindler E, Stoppel D (2014) Die Typlokalität der Kellwasser-Horizonte im Oberharz, Deutschland. Zt Dt Ges Geowiss 165:145–162

Geyer OF, Gwinner MP (2011) Geologie von Baden-Württemberg. In: Geyer M, Nitsch E, Simon T (eds) Geologie von Baden-Württemberg, 5 edn. Schweizerbart-Verlag, Stuttgart, 627 pp (1st edn 1964)

Glaser S, Lagally U, Loth G, Schmid H, Schwerd K (2008) Geotope in Oberbayern Erdwissenschaftliche Beiträge zum Naturschutz, vol 6. Bayerisches Landesamt für Umwelt, Augsburg, 192 pp

Glennie KW, Higham J, Stemmerik L (2003) Permian. In: Evans D, Graham C, Armour A, Bathurst P (eds) The millenium atlas: petroleum geology of the central and northern north Sea. The Geological Society, London, pp 91–103

Goette T (2004) Petrographische und geochemische Untersuchungen zu den postvariszischen Mineralisationen im devonischen Massenkalk des nordwestlichen Rechtsrheinischen Schiefergebirges unter besonderer Berücksichtigung der Kathodolumineszenz. Dissertation, Ruhr University Bochum, 205 pp

Goth K (1990) Der Messeler Ölschiefer – ein Algenlaminit Cour. Forsch.-Inst. Senckenberg, vol 131. Schweizerbart'sche Verlagsbuchhandlung Stuttgart, 143 pp

Goth K, Moschen R, Kober F, Daßinnies M, Nitsch E (2001) Übersichtskarte wichtiger Geotope des Freistaates Sachsen 1:400.000, 1st edn. Sächsisches Landesamt für Umwelt und Geologie, Dresden

Grafenstein U von, Böhmer P, Daut G, Mücke T von, Müller W, Prohaska J (1991) Zur Geologie Helgolands: Unterwasser-Kartierung des Gebietes südwestlich der Insel. Geol Jb A120:19–35

Grimm MC, Wielandt-Schuster U, Hottenrott M, Grimm KI, Radtke G (2011) Oberrheingraben. In: Grimm KI (ed) Stratigraphie von Deutschland IX, Tertiär, Teil 1: Oberrheingraben und benachbarte Tertiärgebiete. Schriftenreihe der Dt. Ges. Geowiss, pp 57–132

Grünthal G, Wahlström R (2003) An Mw based earthquake catalogue for central, northern and northwestern Europe using a hierarchy of magnitude conversions. J Seismol 7:507–531

Güldenpfenning M (1998) Zur geotektonischen Stellung unterkarbonischer Grauwacken und Vulkanite der Zone von Badenweiler-Lenzkirch (Südschwarzwald). Z Dt Geol Ges 149:213–232

Hagdorn H (1999) Seelilien und Crinoidenkalke des Muschelkalks. In: Hauschke N, Wilde V (eds) Trias, eine ganze andere Welt: Mitteleuropa im frühen Erdmittelalter. Pfeil-Verlag, München, pp 321–330

Hagdorn H (2004) Das Muschelkalkmuseum Ingelfingen. Edition Lattner, Heilbronn, 88 pp

Hagdorn H, Simon T (2005) Der Muschelkalk in der Stratigraphischen Tabelle von Deutschland 2002. Newslett Stratigr 41(1–3):143–158

Hallam A, Wignall PB (1999) Mass extinctions and sea-level changes. Earth Sci Rev 48:217–250

Hames WE, McHone JG, Renne P, Ruppel C (eds) (2003) The Central Atlantic Magmatic Province. Insights From Fragments of Pangea. Amer Geophys Union, Geophys Monograph Ser 136, 276 pp

Hammer J, Eidam J, Röber B, Ehling B (1999) Prävariszischer und variszischer granitoider Magmatismus am NE-Rand des Böhmischen Massivs – Geochemie und Petrogenese. Z Geol Wiss 27:401–405

Handy MR, Schmid SS, Bousquet R, Kissling E, Bernoulli D (2010) Reconciling plate-tectonic reconstructions of Alpine Tethys with the geological–geophysical record of spreading and subduction in the Alps. Earth Sci Rev 102:121–158

Hann HP, Chen F, Zedler H, Frisch W, Loeschke J (2003a) The Rand Granite in the southern Schwarzwald and its geodynamic significance in the Variscan belt of SW Germany. Int J Earth Sci 92:821–842

Hann HP, Chen F, Zedler H, Sawatzki G (2003b) Zircon ages and geochemistry of metavolcanic layers from the northern Badenweiler-Lenzkirch Zone (southern Schwarzwald, Germany). N Jb Geol u Paläont, Abh 230:451–469

Hann HP, Zedler H (2008) Variscan Internides: Black Forest (Schwarzwald). In: McCann T (ed) Precambrian and Paleozoic. The Geology of Central Europe, vol 1. The Geological Society, London, pp 606–611

Haq BU, Hardenbol J, Vail P (1987) Chronology of fluctuating sea levels since the Triassic. Science 235:1156–1167

Haq BU, Hardenbol J, Vail PR (1988) Mesozoic and Cenozoic chronostratigraphy and cycles of sea-level change. In: Wilgus CW, Hastings BS, Kendall CGSC, Posamentier HW, Ross CA, Van Wagoner JC (eds) Sea-level changes: an integrated approach, vol 42. SEPM Special Publications, Darlington, pp 71–108

Haq BU, Al-Qahtani AM (2005) Phanerozoic cycles of sea-level change on the Arabian platform. Geoarabia 10:127–160

Harff J, Meyer M (2011) Coastlines of the Baltic Sea – zones of competition between geological processes and a changing climate: Examples from the southern Baltic. In: Harff J, Björck S, Hoth P (eds) The Baltic sea basin. Springer, Berlin, Heidelberg, pp 149–164

Harms F-J, Aderhold G, Hoffmann I, Nix T, Rosenberg F (1999a) Erläuterungen zur Grube Messel bei Darmstadt, Südhessen. Kleine Senckenberg-Reihe Nr. 31 Schriftenreihe der Dt. Geol Ges 8:181–222

Harms F-J, Wallner H, Jacoby WR (1999b) Welterbe Grube Messel. 2 S., Geolog. Karte 1:25.000 mit Erläuterungen auf der Rückseite, Hess. Landesamt f. Bodenf. Forschungsinstitut und Naturmuseum Senckenberg, Wiesbaden

Hays JD, Pitman WC (1973) Lithospheric plate motion, sea-level changes and climatic and ecological consequences. Nature 246:18–22

Heidbach O, Tingay M, Barth A, Reinecker J, Kurfeß D, Müller B (2008) The world stress map database release. http://dx.doi.org/10.1594/GFZ.WSM.Rel2008. Accessed July 2017

Henningsen D, Katzung G (2006) Einführung in die Geologie Deutschlands, 7th edn. Spektrum Akad. Verlag, Heidelberg, p 234

Herrig E et al (1995) Die Kreide und das Pleistozän von Jasmund, Insel Rügen (Ostsee). In: Katzung G (eds) Geologie des südlichen Ostseeraumes – Umwelt und Untergrund, Exkursionsführer, vol 6. Terra Nostra, pp S91–S113

Hesselbo SP, Robinson SA, Surlyk F, Piasecki S (2002) Terrestrial and marine mass extinction at the Triassic-Jurassic boundary synchronized with major carbon-cycle perturbation: A link to initiation of massive volcanism? Geology 30:251–254

Hiete M, Berner U, Heunisch C, Röhling H-G (2006) A high-resolution inorganic geochemical profile across the Zechstein-Buntsandstein boundary in the North German basin. Zt Dt GesGeowiss 157:77–105

Hildebrand AR, Penfield GT, Kring DA, Pilkington M, Camargo ZA, Jacobsen SB, Boynton WV (1991) Chicxulub crater: a possible Cretaceous/Tertiary boundary impact crater on the Yucatan Peninsula, Mexico. Geology 19:867–871

Hofmann M, Linnemann U (2013) The Late Ordovician (Hirnantian) Lederschiefer Formation in Thuringia – evidences for a glaciomarine origin from the Berga Antiform (Saxo-Thuringian Zone). Geologica Saxonica J Centr Europ Geol 59:133–139

Hornung J, Böhme A, van der Lubbe T, Reich M, Richter A (2012) Vertebrate tracksites in the Obernkirchen Sandstone (late Berriasian, Early Cretaceous) of northwest Germany – their stratigraphical, palaeogeographical, palaeoecological, and historical context. Paläont Z 86:231–267

Huckriede H, Wemmer K, Ahrendt H (2004) Palaeogeography and tectonic structure of allochthonous units in the German part of the Rheno-Hercynian Belt (Central European Variscides. Int J Earth Sci 93:414–431

Hunter A, Easterbrook G (2004) The geological history of the British Isles. Open University, Milton Keynes, 143 pp

Hüneke H, Ruchholz K (2004) Olisthostromale Resedimentation im Harz: Resultat dextraler Strike-Slip-Bewegungen am passiven Kontinentalrand von Laurussia während des Tournaisium. Schriftenreihe der Deutschen Geologischen Gesellschaft 34:100. [Jahrestagung der DGG, Kurzfassungen, Leipzig 29.9–01.10.2004]

Ilgner J, Jeffries T, Faust D, Linnemann U, Ullrich B (2009) U/Pb-dating and geochemical characterization of the Brocken and the Ramberg Pluton, Harz Mountains, Germany. Geologica Saxonica 59:79–94

Illies JH (1965) Bauplan und Baugeschichte des Oberrheingebietes. Oberrhein Geol Abh 14:1–54

Jowett EC, Pearce GW, Rydzewski A (1987) A Mid-Triassic paleomagnetic age of the Kupferschiefer mineralization in Poland, based on a revised apparent polarwander path for Europe and Russia. J Geophys Res 92:581–598

Jöris O, Weninger B (2000) 14 C-Alterskalibration und die absolute Chronologie des Spätglazials. Archäologisches Korrespondenzblatt 30:461–471

Kaiser KF, Friedrich M, Miramont C, Kromer B, Sgier M, Schaub M, Boeren I, Remmele S, Talamo S, Guibal F, Sivan O (2010) Challenging process to make the Late-glacial tree-ring chronologies from Europe absolute – an inventory. Quat Sci Rev 36:78–90

Kalkowsky EL (1908) Oolith und Stromatolith im Norddeutschen Buntsandstein. Zt Dt Geol Ges 60:68–125

Kalt A (1991) Isotopengeologische Untersuchungen an Metabasiten des Schwarzwaldes und ihren Rahmengesteinen. Freiburger Geowiss Beitr, vol 3, 185 pp

Katzschmann L, Lepper J (1999) Naturwerksteine der Germanischen Trias. In: Hauschke N, Wilde V (eds) Trias, eine ganze andere Welt: Mitteleuropa im frühen Erdmittelalter. Pfeil-Verlag, München, pp 429–447

Katzung G (1961) Die Geröllführung des Lederschiefers (Ordovizium) an der SE-Flanke des Schwarzburger Sattels (Thüringen). Geologie 10:778–802

Katzung G (2001) The Caledonides at the southern margin of the East European Craton. N Jb Geol Paläont Abh 222:3–53

Kästner H, Seidel G, Wiefel H (1995) Regionalgeologische Stellung und Gliederung. In: Seidel G (ed) Geologie von Thüringen. Schweizerbart'sche Verlagsbuchhandlung, Stuttgart, 556 pp

Keiter M (2013) Erdbeben in Bielefeld 1612-Ein bemerkenswertes geologisches Ereignis vor 400 Jahren und seine Ursachen. Ber. Naturwiss. Verein für Bielefeld u. Umgegend 51: 16–31

Kemnitz H (2007) The Lausitz graywackes, Saxo-Thuringia, Germany – witness to the Cadomian orogeny. Geol Soc Am Spec Pap 423:97–141

Kemper E (1976) Geologischer Führer durch die Grafschaft Bentheim und die angrenzenden Gebiete mit einem Abriß der emsländischen Unterkreide. Das Bentheimer Land. Verlag Heimatverein der Grafschaft Bentheim e. V., Nordhorn-Bentheim, vol 64, 206 pp

Kennedy WJ, Kaplan U (1995) Parapuzosia (Parapuzosia) seppenradensis (Landois) und die Ammonitenfauna der Dülmener Schichten, Westfalen. Geologie und Paläontologie in Westfalen 33:1–127

Kirsch H, Kober B, Lippolt HJ (1988) Age of intrusion and rapid cooling of the Frankenstein gabbro (Odenwald, SW-Germany) evidenced by $^{40}Ar/^{39}Ar$ and single zircon $^{207}Pb/^{206}Pb$ measurements. Geol Rdsch 77:693–711

Klein T, Kiehm S, Siebel W, Shang CK, Rohrmüller J, Dörr W, Zulauf G (2008) Age and emplacement of late-Variscan granites of the western Bohemian Massif with main focus on the Hauzenberg granitoids (European Variscides, Germany). Lithos 102:478–507

Kley J, Voigt T (2008) Late Cretaceous intraplate thrusting in central Europe: effect of Africa-Iberia-Europe convergence, not Alpine collision. Geology 36:839–842

Kley J, Voigt T, Jähne F (2008) A case of distributed continental collision: late cretaceous intraplate shortening from central Europe to North Africa Am Geophys Union, Fall Meeting 2008. (T11 C-1897)

Klügel T, Ahrendt H, Oncken O, Käfer N, Schäfer F, Weiss B (1994) Alter und Herkunft der Sedimente und des Detritus der nördlichen Phyllit-Zone (Taunussüdrand). Z dt geol Ges 145:172–191

Klügel T (1997) Geometrie und Kinematik einer varistischen Plattengrenze – der Südrand des Rhenoherzynikums im Taunus. Geol Abh Hessen 101:1–215

Koch R, Sobott R, Lorenz HG (1999) Der Schaumkalk (Trias, Unterer Muschelkalk) am Naumbruger Dom als Baustein: Einfluss von Fazies und Diagenese auf die Gesteinsqualität. In: Hauschke N, Wilde V (eds) Trias, eine ganze andere Welt: Mitteleuropa im frühen Erdmittelalter. Pfeil-Verlag, München, pp 449–471

Kombrink H, Besly BM, Collinson JD, Den Hartog Jager DG, Drozdzewski G, Dusar M, Hoth P, Pagnier HJM, Stemmerik L, Waksmundzka MI, Wrede V (2010) Carboniferous. In: Doornenbal JC, Stevenson AG (eds) Petroleum geological atlas of the southern Permian Basin Area. EAGE Publications b.v, Houten, pp 81–99

Kopp J, Simon A, Göthel M (2006) Die Kupfer-Lagerstätte Spremberg-Graustein in Südbrandenburg. Brandenburg geowiss Beitr 13(1–2):117–132

Kossmat F (1927) Gliederung des varistischen Gebirgsbaues. Abh d Sächs Geol Landesamtes 1:1–39

Kowalski H (1978) Die Trilobitentrift bei Gees – Ein weltberühmter Fossilienfundort Eifel-Jahrb, vol 1979, pp 57–61

Kozur HW (1999) The correlation of the Germanic Buntsandstein and Muschelkalk with the Tethyan scale. In: Bachmann GH, Lerche I (eds) The Epicontinental Triassic. Zbl Geol Pal pp 701–725

König W (2008) Zeitliche und genetische Einordnung von tertiären Sedimentvorkommen im Mittelharz und im Harzvorland – Ein Beitrag zur Reliefentwicklung und zur Karstmorphogenese im Harz. Dissertation an der Martin-Luther-Universität Halle-Wittenberg, 229 pp

Königshof P, Nesbor H-D, Flick H (2010) Volcanism and reef development in the Devonian: a case study from the Lahn syncline, Rheinisches Schiefergebirge (Germany). Gondwana Res 17:264–280

Krauss M, Eidam J, Hammer J, Korich D (1992) Die cadomisch-variszische Entwicklung des Lausitzer Granodiorit-Komplexes. Zbl Geol Pal Teil I:71–85

Krawczyk CM, McCann T, Cocks LRM, England RW, McBride JH, Wybraniec S (2008) Caledonian tectonics. In: McCann T (Ed) Precambrian and Palaeozoic. The Geology of Central Europe, vol 1. The Geological Society, London, pp 303–381

Kroner U, Willner AP (1998) Time of formation and peak of Variscan HP-HT metamorphism of quartz-feldspar rocks in the central Erzgebirge, Saxony, Germany. Contrib Miner Petrol 132:1–20

Kroner U, Mansy J-L, Mazur S, Aleksandrowski P, Hann HP, Huckriede H, Laquement F, Lamarche J, Ledru P, Pharaoh TC, Zedler H, Zeh A, Zulauf G (2008) Variscan tectonics. In: McCann T (ed) Precambrian and Paleozoic. The Geology of Central Europe, vol 1. The Geological Society, London, pp 599–664

Kröchert J, Buchner E, Schmieder M, Maurer H, Walter M, Strasser A, Strasser M (2009) Effusiver melilithitischer Vulkanismus auf der Schwäbischen Alb – der Sternberg bei Gomadingen. Z Dt Ges Geowiss 160:315–323

Krüger JC, Romer RL, Kämpf H (2013) Late Cretaceous ultramafic lamprophyres and carbonatites from the Delitzsch complex, Germany. Chem Geol 353:140–150

Kuhlemann J, Frisch W, Dunkl I, Székely B (2001) Quantifying tectonic versus erosive denudation by the sediment budget: the Miocene core complexes of the Alps. Tectonophysics 330:1–23

Kuhlemann J, Kempf O (2002) Post-Eocene evolution of the North Alpine Foreland Basin and its response to Alpine tectonics. Sed Geol 152:45–78

Kuhlemann J, Frisch W, Meschede M (2009) Geologie von Korsika – Exkursionsführer. Sammlung geologischer Führer 99, Schweizerbart'sche Verlagsgesellschaft Stuttgart, 236 pp

Kupetz A, Kupetz M, Rascher J (2004) Der Muskauer Faltenbogen – ein geologisches Phänomen, Grundlage einer 130-jährigen standortgebundenen Wirtschaftsentwicklung und Geopark in Brandenburg, Sachsen und der Wojewodschaft Lebuser Land Ges f Geowiss, Berlin, 36 pp

Langenhain JH (2003) Plant resins: chemistry, evolution, ecology, and ethnobotany. Timber Press, Portland, Cambridge, p 612

Larson RL (1991) Geological consequences of superplumes. Geology 19:963–966

Laubscher H (1970) Grundsätzliches zur Tektonik des Rheingrabens. In: Illies JH, Müller S (eds) Graben Problems. Schweizerbart, Stuttgart, pp 79–87

Lawver LA, Dalziel IWD, Norton IO, Gahagan LM (2009) PLATES 2009 atlas of plate reconstructions (750 Ma to Present Day). © 2009, University of Texas Institute for Geophysics. http://www-udc.ig.utexas.edu/external/plates/recons.htm. Accessed July 2017

Le Maitre RW (2005) Igneous rocks: a classification and glossary of terms. Recommendations of the international union of geological sciences subcommission on the systematics of igneous rocks, 2nd edn. Cambridge University Press, Cambridge, 256 p

Lemcke K (1988) Geologie von Bayern. I. Teil: Das bayerische Alpenvorland vor der Eiszeit. Schweizerbart'sche Verlagsbuchhandlung, Stuttgart, 175 pp

Leydecker G (2011) Erdbebenkatalog für Deutschland mit Randgebieten für die Jahre 800 bis 2008. Geol Jb, E 59:1–198

Liedtke H (2007) Westfalen im Eiszeitalter. Landschaftsverband Westfalen-Lippe (LWL), Geogr. Komm. Westfalen, Klima/Umwelt. https://www.lwl.org/LWL/Kultur/Westfalen_Regional/Naturraum/Eiszeitalter. Last accessed July 2017

Linnemann U, Gehmlich M, Heuse Th, Schauer M (1999) Die Cadomiden und Varisziden im Thüringisch-Vogtländischen Schiefergebirge (Saxothuringisches Terrane). Beitr Geol Thüringen, NF 6:7–39 (Jena)

Linnemann U, Gehmlich M, Tichomirowa M, Buschmann B, Nasdala L, Jonas P, Lützner H, Bombach K (2000) From Cadomian subduction to Early Palaezoic rifting: the evolution of Saxo-Thuringia at the margin of Gondwana in the light of single zircon geochronology and basin development (Central European Variscides, Germany. Geol Soc Lond Spec Publ 179:131–153

Linnemann U, Romer RL, Pin C (2008) Precambrian. In: McCann T (ed) Precambrian and Paleozoic. The Geology of Central Europe, vol 1. The Geological Society, London, pp 21–101

Linzer HG, Ratschbacher L, Frisch W (1995) Transpressional collision structures in the upper crust: the fold-thrust belt of the Northern Calcareous Alps. Tectonophysics 242:41–61

Lisiecki LE, Raymo ME (2005) A Pliocene-Pleistocene stack of 57 globally distributed benthic δ18O records. Paleoceanography, 20, PA1003, https://doi.org/10.1029/2004pa001071

Lokhorst A (ed) (1998) The northwest European Gasatlas. Netherlands Institute of Applied Geoscience TNO, Haarlem

Lorenz V (2000) Formation of maar-diatreme volcanoes. In: Negendank JFW, Brüchmann C (eds) International Maar conference. Maar 2000. Terra Nostra, vol 6, pp 284–291

Martin-Gombojav U (2003) Petrographie und Petrogenese des Eckergneis-Komplexes, Harz. Dissertation, University of Hamburg, 108 pp

Martini E (1990) The Rhinegraben system, a connection between northern and southern seas in the European Tertiary. Veröffentlichungen Übersee-Museum, A 10: 83–98; Bremen

Martínez Catalán JR, Arenas R, Díaz García F, Gonzáles Cuadra P, Gómez-Barreiro J, Abati J, Castineiras P, Fernández-Suárez J, Sánchez Martínez S, Andonaegui P, Gonzáles Clavijo E, Díez Montes A, Rubio Pascual FJ, Valle Aguado B (2007) Space and time in the tectonic evolution of the northwestern Iberian Massif: Implications for the Variscan Belt. Geol Soc Amer, Mem, 200:403–423

Matte P (1986) Tectonics and plate tectonic models for the Varicsan belt of Europe. Tectonophysics 126:329–332

McCann T, Skompski S, Poty E, Dusar M, Vozárová A, Peryt T, Oplustil S, Stollhofen H, Schneider J, Wetzel A, Boulvain F, Dusar M, Török A, Haas J, Tait J, Körner F (2008) Permian. In: McCann T (ed) Precambrian and Palaeozoic. The geology of Central Europe, vol 1. The Geological Society, London, pp 599–664

McHone JG, Puffer J (2003) Flood basalt provinces of the Pangaean Atlantic Rift: Regional extent and environmental significance. In: LeTourneau PM, Olsen PE (eds) The Great Rift Valleys of Pangea in Eastern North America. Columbia University Press, New York, pp 141–154

Menning M, Deutsche Stratigraphische Kommission (2002) Eine geologische Zeitskala 2002. In: Deutsche Stratigraphische Kommission (eds) Stratigraphische Tabelle von Deutschland 2002

Menning M, Hendrich A (eds) (2005) Erläuterungen zur Stratigraphischen Tabelle von Deutschland 2005 (ESTD 2005). Newsl Stratigr 41(1/3):1–405 (19 Taf.; Berlin, Stuttgart)

Menning M, Schröder B, Plein E, Simon T, Lepper J, Röhling H-G, Heunisch C, Stapf K, Lützner H, Käding K-C, Paul J, Horn M, Hagdorn H, Beutler G, Nitsch (2011) Beschlüsse der Deutschen Stratigraphischen Kommission 1991–2010 zu Perm und Trias von Mitteleuropa. Z Dt Ges Geowiss 162:1–18

Menning M, Hendrich A, Deutsche Stratigraphische Kommission (2012) Eine geologische Zeitskala 2002. In: Deutsche Stratigraphische Kommission (eds) Stratigraphische Tabelle von Deutschland Kompakt 2012. Potsdam

Meschede M (1986) A method of discriminating between different types of mid-ocean ridge basalts and continental tholeiites with the Nb–Zr–Y diagram. Chem Geol 56:207–218

Mingram B (1998) The Erzgebirge, Germany – a subducted part of northern Gondwana: geochemical evidence for repetition of early Paleozoic metasedimentary sequences in metamorphic thrust units. Geol Mag 135:785–801

Missoni S, Gawlick H-J (2011) Jurassic mountain building and Mesozoic-Cenozoic geodynamic evolution of the Northern Calcareous Alps as proven in the Berchtesgaden Alps (Germany). Facies 57:137–186

Molzahn M, Anthes G, Reischmann T (1998) Single zircon Pb/Pb age geochronology and isotope systematics of the Rhenohercynian basement. Terra Nostra 98:67–68

Mönnig E (2005) Der Jura von Norddeutschland in der Stratigraphischen Tabelle von Deutschland 2002. Newsl Stratigr 41(1–3):253–261

Nickschick T, Kämpf H, Jahr T (2014) The "Triasscholle" near Greiz, Germany – a volcanic origin? Bull Volcanol 76(20S):806. https://doi.org/10.1007/s00445-014-0806-x

Niedermeyer R-O, Kanter L, Kenzler M, Panzig W-A, Krienke K, Ludwig AO, Schnick HH, Schütze K (2010) Die Insel Rügen (I) – Fazies, Stratigraphie, Lagerungsverhältnisse und geologisches Gefahrenpotenzial pleistozäner Sedimente der Steilküste Jasmund. In: Lampe R, Lorenz S (eds) Eiszeitlandschaften in Mecklenburg-Vorpommern, Exkursionsführer DEUQUA 2010, S51–S71

Nitsch E (2003) Wie kommt das Salz in den Keuper? Beitr GeolThür, N F 10: 75–110

Olsen JC (1987) Tectonic evolution of the North Sea region. In: Brooks J, Glennie KW (eds) Petroleum geology of Northwest Europe. Graham & Trotman, London, pp 389–403

Olsen PE, Kent DV, Sues H-D, Koeberl C, Huber H, Montanari A, Rainforth EC, Fowell SJ, Szajna MJ, Hartline BW (2002) Ascent of dinosaurs linked to an iridium anomaly at the Triassic-Jurassic boundary. Science 296:1305–1307

Oncken O (1997) Tranformation of a magmatic arc and an orogenic root during oblique collision and its consequences for the evolution of the European Variscides (Mid-German crystalline rise). Geol Rdsch 86:2–20

Oncken O, Franzke HJ, Dittmar U, Klügel T (1995) Rhenohercynian foldbelt: metamorphic units – structure. In: Dallmeyer RD, Franke W, Weber K (eds) Pre-Permian geology of Central and Eastern Europe. Springer, Berlin, pp 108–117

Oncken O, Plesch A, Weber J, Ricken W, Schrader S (2000) Passive margin detachment during arc-continent collision (Central European Variscides). Geol Soc Spec Paper 179:199–216

Oppel A (1856) Die Juraformen Englands, Frankreichs und des südwestlichen Deutschlands: Nach ihren einzelnen Gliedern eingeteilt u. verglichen. Ebner & Seubert, Stuttgart, 857 pp

O'Neil J, Carlson RW, Francis D, Stevenson RK (2008) Neodymium-142 Evidence for Hadean Mafic Crust. Science 321: 1828–1831

Park C, Schmincke H-U (2009) Apokalypse im Rheintal. Spektrum der Wissenschaft 2009:78–87

Patočka F, Kachlík V, Dostal J, Frána J (2003) Granitoid gneisses with relict orbicular metagranitoids from the varied group of Southern Bohemian Massif Moldanbicum: protolith derived from melting of Archean crust? J Czech Geol Soc 48:100–101

Paul J, Huckriede H (2004) Riffe, Gips und Erze: Zechstein zwischen Saalfeld und Neustadt/Orla. Schriftenreihe der Dt Ges Geowiss 35:75–91

Pearce JA, Cann JR (1973) Tectonic setting of basic volcanic rocks determined using trace element analysis. Earth Planet Sci Lett 19:290–300

Pearce JA (1983) Role of the sub-continental lithosphere in magma genesis at active continental margins. In: Hawkesworth CJ, Norry MJ (eds) Continental basalts and mantle Xenoliths, pp 230–250

Peryt TM, Geluk MC, Mathiesen A, Paul J, Smith K (2010) Zechstein. In: Doornenbal JC, Stevenson AG (eds) Petroleum geological atlas of the Southern Permian Basin Area. EAGE Publications b.v, Houten, pp 123–147

Petit JR, Jouzel J, Raynaud D, Barkov NI, Barnola J-M, Basile I, Bender M, Chappellaz J, Davis M, Delaygue G, Delmotte M, Kotlyakov VM, Legrand M, Lipenkov VY, Lorius C, Pépin L, Ritz C, Saltzman E, Stievenard M (1999) Climate and atmospheric history of the past 420,000 years from the Vostok ice core, Antarctica. Nature 399:429–436

Pfaff K, Romer RL, Markl G (2009) U-Pb ages of ferberite, chalcedony, agate, 'U-mica' and pitchblende: constraints on the mineralization history of the Schwarzwald ore district. Europ J Miner 21:817–836

Pfeiffer L, Suhr P (2008) Tertiärer Vulkanismus. In: Pälchen W, Walter H (eds) Geologie von Sachsen. Schweizerbart'sche Verlagsbuchhandlung, Stuttgart, pp 486–494

Pfiffner OA (2010) Geologie der Alpen, 2nd edn. UTB, Stuttgart, 359 pp

Pharaoh TC, Dusar M, Geluk MC, Kockel F, Krawczyk CM, Krzywiec P, Scheck-Wenderoth M, Thybo H, Vejbæk OV, Van Wees JD (2010) Tectonic evolution. In: Doornenbal JC, Stevenson AG (eds) Petroleum geological atlas of the Southern Permian Basin area. EAGE Publications b.v, Houten, pp 25–57

Plomerová J, Babuška V (2010) Long memory of mantle lithospheric fabric – European LAB constrained from seismic anisotropy. Lithos 120:131–143

Pohl WL (2005) Mineralische und Energie-Rohstoffe. Eine Einführung zur Entstehung und nachhaltigen Nutzung von Lagerstätten. In: Petrascheck W, Petrascheck WE (eds) Lagerstättenlehre. Schweizerbart'sche Verlagsbuchhandlung, Stuttgart (663 pp, 1st edn, 1950)

Pösges G, Schieber M (1994) Das Rieskrater-Museum Nördlingen. Museumsführer und Empfehlungen zur Gestaltung eines Aufenthalts im Ries. Pfeil-Verlag, München, p 112

Putis M, Gawlick H-J, Frisch W, Sulák M (2008) Cretaceous transformation from passive to active continental margin in the Western Carpathians as indicated by the sedimentary record in the Infratatric unit. Int J Earth Sci 97:799–819

Quenstedt FA (1843) Das Flözgebirge Württembergs. Mit besonderer Rücksicht auf den Jura. Laupp & Siebeck Verlag, Tübingen

Quenstedt FA (1846–1849): Petrefactenkunde Deutschlands. Erster Band: Die Cephalopoden. Fues, Leipzig, Abt. 1, 581 pp

RAG Anthrazit Ibbenbüren GmbH (2002) Die Ibbenbürener Anthrazit-Lagerstätte. http://www.dsk-anthrazit-ibbenbueren.de/bergwerk/lagerstaette.shtml. Accessed Jul 2017

Ratschbacher L, Frisch W, Linzer H-G, Merle O (1991) Lateral extrusion in the Eastern Alps, part II: structural analysis. Tectonics 10:257–271

Rauche H (1991) Spätvariszische Kompressionstektonik am Südwestrand der Elbezone. Z geol Wiss 19:463–475

Regionalverband Harz EV (2012) Kellwassertal – Das große Massenaussterben. Schautafel im Geopark Harz, Kellwassertal

Reinecker J, Schneider G (2002) Zur Neotektonik der Zollernalb: Der Hohenzollerngraben und die Albstadt-Erdbeben. Jahresber Mitt Oberrh Geol Ver 84:391–417

Reitner J (2011) Mass extinctions, phanerozoic. In: Encyclopedia of geobiology. Encyclopedia of Earth Sciences Series, pp 543–547

Richter CF (1935) An instrumental earthquake magnitude scale. Bull Seismol Soc Am 25:1–32

Richter F (1999) Elbsandsteingebirge – Landschaft im Meer geboren Geologische Wanderungen, vol 2. Sächsisches Druck- und Verlagshaus AG, Dresden, 32 pp

Richter U (2007) Dinosaurier-Freilichtmuseum und Naturdenkmal Saurierfährten Münchehagen. Weserdruckerei, Stolzenau, 89 pp

Ritter J (2004) Small-scale mantle plumes: imaging and geodynamic aspects. Report of the Eifel Plume Project

Roll A (1979) Versuch einer Volumenbilanz des Oberrheintalgrabens und seiner Schultern. Geologisches Jahrbuch A, 52:3–82

Rothe P (2005) Die Geologie Deutschlands. 48 Landschaften im Portrait. Wissenschaftliche Buchgesellschaft, Darmstadt, p 240

Röper M, Rothgänger M, Rothgänger K (2000) Die Plattenkalke von Schernfeld (Landkreis Eichstätt). Eichendorf Verlag, Eichendorf, 128 pp

Rössler R, Annacker V, Kretzschmar R, Mehlhorn S (2009) Auf Schatzsuche in Chemnitz – Wissenschaftliche Grabungen 2009. Veröff Museum Naturk Chemnitz 32:25–46

Rötzler J, Romer RL (2010) The Saxon Granulite Massif: a key area for the geodynamic evolution of Variscan central Europe. In: Linnemann U, Romer RL (eds) Pre-Mesozoic geology of Saxo-Thuringia – from the Cadomian active margin of the Variscan Orogen. Schweizerbart'sche Verlagsbuchhandlung, Stuttgart, pp 233–252

Ruchholz K (1983) Bretonischer Blattverschiebungsmechanismus im Harz und seine Bedeutung im Bereich des Rhenischen Troges. Thesen Koll., Zentralinst. Physik der Erde, Potsdam, pp 1–8

Ruchholz K (1988) Tiefenbrüche und präorogene Seitenverschiebung im Bau Mitteleuropas. Mitt Ges Geol Wiss 16:30

Ruchholz K (1989) Entwicklung und gerichtete Transformation von oberdevonisch-unterkarbonischer Flyschsedimentation und Olisthostromen im östlichen Rhenoherzynikum (Harz). Z Geol Wiss 17:581–588

Sach VJ (1997) Neue Vorkommen von Brockhorizonten in der Oberen Süßwassermolasse von Baden-Württemberg (Deutschland) – Zeugnisse der Rieskatastrophe im Mittelmiozän. N Jb Geol Paläont Abh 205(3):323–337

Sach VJ, Heizmann EPJ (2001) Stratigraphie und Säugetierfaunen der Brackwassermolasse in der Umgebung von Ulm (Südwestdeutschland). Stuttgarter Beitr Naturk, Ser B 310:95

Schäfer A, Korsch RJ (1998) Formation and sediment fill of the Saar-Nahe Basin (Permo-Carboniferous, Germany). Z Dt Geol Ges 149:233–269

Schäfer F, Kemnitz H, Romer RL, Oncken O (1998) The NW-Thuringian Vesser zone – a Variscan retro wedge. Acta Univ Carolinae Geologica 42:327–328

Schäfer A (2011) Tectonics and sedimentation in the continental strike-slip Saar-Nahe Basin (Carboniferous-Permian, West Germany. Z Dt Ges Geowiss 162:127–155

Scheck-Wenderoth M, Krzywiec P, Zühlke R, Maystrenko Y, Froitzheim N (2008) Permian to Cretaceous tectonics. In: McCann T (eds) Mesozoic and Cenozoic. The geology of Central Europe, vol 2. The Geological Society, London, pp 990–1030

Schleicher A, Warr LN, Kober B, Laverret E, Clauer N (2006) Episodic mineralization of hydrothermal illite in the Soultz-sous-Forêts granite (Upper Rhine Graben, France). Contrib Mineral Petrol 152:349–364

Schlüter T (1990) Baltic amber. In: Briggs DEG, Crowther PR (eds) Palaeobiology – a synthesis. Blackwell, Oxford, pp 294–297

Schmid SM, Pfiffner OA, Froitzheim N, Schönborn G, Kissling E (1996) Geophysical-geological transect and tectonic evolution of the Swiss-Italian Alps. Tectonics 15:1036–1064

Schmid SM, Fügenschuh B, Kissling E, Schuster R (2004) Tectonic map and overall architecture of the Alpine orogen. Eclog geol Helv 97:93–117

Schmid SM, Bernoulli D, Fügenschuh B, Matenco L, Schefer S, Schuster R, Tischler M, Ustaszewski K (2008) The Alpine-Carpathian-Dinaridic orogenic system: correlation and evolution of tectonic units. Swiss J Geosci 101:139–183

Schmidt-Kaler H (2003) Von der Frankenhöhe zum Fränkischen Seenland Wanderungen in die Erdgeschichte, vol 14. Pfeil-Verlag, München, 128 pp

Schmincke H-U (2008) Volcanism of the East and West Eifel. In: Litt T, Schmincke H-U, Frechen M, Schlüchter Q, McCann T (eds) Mesozoic and Cenozoic. The geology of central Europe, vol 2. The Geological Society, London, pp 1318–1333

Schmincke H-U (2009) Vulkane der Eifel – Aufbau, Entstehung und heutige Bedeutung. Spektrum Akad. Verlag, Heidelberg, p 160

Schmincke H-U, Mertz DF, Kämpf H, Bräuer K, Hinzen K-G, Ritter J, Walter TR, Wenzel F (2011) Öffentliche Erklärung zu „Aktive Magmatische Prozesse im Untergrund der Eifel?“ Eifel-Symposium, Dt. Vulkanolog. Ges. (DVG), Mendig, 15. Januar 2011

Schmincke H-U (2013) Vulkanismus, 4. ed Wissenschaftliche Buchgesellschaft/Primus-Verlag, Darmstadt (264 pp, 1st edn, 2000)

Schreiber U (2006) Die Flucht der Ameisen: eine geokalyptische Vision. Shayol Verlag, Berlin, p. 350

Schreiner A (1977) Tektonische Bewegungen und Entstehung der Landschaft im Tertiär und Quartär. Erläuterungen Geol. Karte Freiburg i.Br. und Umgebung 1:50.000

Schubert G, Turcotte DL, Olson P (2001) Mantle convection in the earth and planets. Cambridge University Press, Cambridge, 941 pp

Schwerdtner G, Anger H, Störr M (2007) Die Kaolinlagerstätten des Kemmlitzer Reviers Bergbaumonographie; Bergbau in Sachsen, vol 13, 118 pp

Sebastian U (2013) Die Geologie des Erzgebirges, 1st edn. Springer, Berlin, 268 pp

Seidel G, Kastner H, Wiefel H, Heuse H, Heuse T, Huckriede H, Katzschmann L, Wucher K (2002) Geologische Übersichtskarte von Thüringen 1:200000, 1st edn. Thüringer Landesanstalt für Umwelt und Geologie, Jena

Seifert W, Kämpf H, Wasternack J (2000) Compositional variation in apatite, phlogopite and other accessory minerals of the ultramafic Delitzsch complex, Germany: implication for cooling history of carbonatites. Lithos 53:81–100

Seilacher A (1999) Fossillagerstätten im Muschelkalk. In: Hauschke N, Wilde V (eds) Trias, eine ganze andere Welt: Mitteleuropa im frühen Erdmittelalter. Pfeil-Verlag, München, pp 299–308

Sengör AMC (1984) The Cimmeridge Orogenic System and the Tectonics of Eurasia. J Geol Soc London 161:583–592

Sepkoski JJ Jr (1996) Patterns of Phanerozoic mass extinction: a perspective from global data bases. In: Walliser OH (ed) Global events and event stratigraphy in the Phanerozoic. Springer, Berlin, pp 35–51

Sereno PC (1999) The evolution of dinosaurs. Science 284:2137–2147

Seton M, Gaina C, Müller RD, Heine C (2009) Mid-Cretaceous seafloor spreading pulse: Fact or fiction? Geology 37(8):687–690

Shoemaker EM, Chao ECT (1961) New evidence for the impact origin of the Ries, Basin, Bavaria, Germany. J Geophys Res 66:3371–3378

Skála R, Strnad L, McCammon C, Čada M (2009) Moldavites from the Cheb Basin, Czech Republic. Ceochim Cosmochim Acta 73:1145–1179

Sole RV, Newman M (2002) Extinctions and biodiversity in the fossil record. In: Timmerman P (ed) The Earth system: biological and ecological dimensions of global environment change. Encyclopedia of global enviromental change, vol 2. John Wiley & Sons, New York, pp 297–391

Sommer M, Katzung G (2004) Abgrenzung, Gliederung und strukturelle Entwicklung des Neoproterozoikums im Schwarzburger Antiklinorium (Thüringisch-Fränkisch-Vogtländisches Schiefergebirge, Saxothuringische Zone). Z Geol Wiss Berlin 32:191–226

Stackebrandt W (2009) Subglacial channels of Northern Germany – a brief review. Z Dt Ges Geowiss 160:203–210

Stadler G, Teichmüller R (1971) Zusammenfassender Überblick über die Entwicklung des Bramscher Massivs und des Niedersächsischen Tektogens. Fortschr Geol Rheinld u Westf 18:547–564

Stedingk K (2012) Geologie und Erzlagerstätten im Oberharz. Exkursf und Veröfftl D Ges Geowiss 247:9–81

Steinich G (1972) Endogene Tektonik in den Unter-Maastricht-Vorkommen auf Jasmund (Rügen). Geologie 71/72:207

Stets J, Schäfer A (2002) Depositional environments in the lower Devonian Siliciclastics of the Rhenohercynian Basin (Rheinisches Schiefergebirge, W-Germany). Contrib Sediment Geol 22, 78 pp

Stille H (1925) Rheinische Gebirgsbildung im Kristianiagebiete und in Westdeutschland. Abh Preuß Geol Landesant, NF 95:110–132

Stüwe K (2000) Einführung in die Geodynamik der Lithosphäre – Quantitative Behandlung geowissenschaftlicher Probleme. Springer, Berlin, 405 pp

Suess E (1883) Das Antlitz der Erde, vol 2. Temsky, Prag & Wien, Freytag, Leipzig, 703 pp

Sues H-D, Munk W (1996) A remarkable assemblage of terrestrial tetrapods from the Zechstein (Upper Permian: Tatarian) near Korbach (northwestern Hesse). Paläont Zeitschrift 70:213–223

Szulc J (2000) Middle Triassic evolution of the northern Peri-Tethys area as influenced by early opening of the Tethys Ocean. Ann Soc Geol Poloniae 70:1–48

Tackley PJ (2008) Geodynamics: layer cake or plum pudding? Nat Geosci 1:157–158

Tait JA, Bachtadse V, Franke W, Soffel HC (1997) Geodynamic evolution of the European Variscan fold belt: palaeomagnetic and geological constraints. Geol Rdsch 86:585–598

Teipel U (2003) Obervendischer und Unterordovizischer Magmatismus im Bayerischen Wald – Geochronologische (SHRIMP), geochemische und isotopengeochemische Untersuchungen an Metamagmatiten aus dem Westteil des Böhmischen Massivs. Münchner Geol Hefte, Reihe A, Allg. Geologie 33, 98 pp

Teipel U, Eichhorn R, Loth G, Rohrmüller J, Höll R, Kennedy A (2004) U–Pb SHRIMP and Nd isotopic data from the western Bohemian Massif (Bayerischer Wald, Germany): implications for upper Vendian and lower Ordovician magmatism. Int J Earth Sci 93:782–801

Thome K (1990) Inlandeisvorstöße in das Ruhrgebiet (nebst der Entwicklung einer spätglazialen Rheinrinne). In: Schirmer W (ed) Rheingeschichte zwischen Mosel und Maas. Deutsche Quartärvereinigung, Hannover, pp 273–292

Thürach H (1912): Erläuterungen zur Geologischen Spezialkarte des Großherzogtums Baden, Blätter Karlsruhe und Daxlanden Nr. 50 und 51 (Unveränderter Nachdruck als Geol. Kt. 1:25000 Baden-Württemberg, Bl. 6915 Wörth am Rhein und 6916 Karlsruhe-Nord, Stuttgart, 1985

Timmermann M (2008) Palaeozoic magmatism. In: McCann T (ed) Precambrian and Palaeozoic. The geology of central Europe, vol 1. The Geological Society, London, pp 665–748

Tollmann A (1977) Die Zentralalpen. Geologie von Österreich, vol 1. Deuticke, Wien (XVI + 766 pp)

Torsvik TH, Carlos D, Mosar J, Cocks LRM, Malme TNM (2002) Global reconstructions and North Atlantic paleogeography 440 Ma to recent. In: Eide EA (ed) BATLAS – Mid Norway plate reconstructions atlas with global and Atlantic perspectives. Geol Surv, Norway, pp 18–39

Torsvik TH, van der Voo R, Doubrovine P, Burke K, Steinberger B, Ashwal LD, Trønnes R, Webb SJ, Bull AL (2014) Deep mantle structure as a reference frame for movements in and on the earth. Proc Nat Acad Sci USA (PNAS) 111:8735–8740

Trusheim F (1960) Mechanism of salt migration in northern Germany. Am Ass Petrol Geol, Bull 44:1519–1540

Trusheim F (1971) Zur Bildung der Salzlager im Rotliegenden und Mesozoicum Mitteleuropas. Geol Jahrb, vol Beih 112, pp 1–51

Van Husen D (1987) Die Ostalpen und ihr Vorland in der letzten Eiszeit (Würm) 1:500.000. Geol Bundesanstalt Österreich, Wien

Veizer J, Ala D, Azmy K, Bruckschen P, Buhl D, Bruhn F, Carden GAF, Diener A, Ebneth S, Godderis Y, Jasper T, Korte C, Pawellek F, Podlaha O, Strauss H (1999) 87Sr/86Sr, δ13 C and δ18O evolution of Phanerozoic seawater. Chem Geol 161:59–88

Vejbæk OV, Andersen C, Dusar M, Herngreen GFW, Krabbe H, Leszczyñski K, Lott GK, Mutterlose J, van der Molen AS (2010) Cretaceous. In: Doornenbal JC, Stevenson AG (eds) Petroleum

Vinken R (1988) The Northwest European Tertiary Basin: results of the international geological correlation programme, Project No. 124. Geol Jb, A 100:508 (Hannover)

Vinx R (2011) Gesteinsbestimmung im Gelände, 3rd edn. Spektrum Akad, Verlag, 480 pp

Voigt T (2009) Die Lausitz-Riesengebirgs-Antiklinalzone als kreidezeitliche Inversionsstruktur: Geologische Hinweise aus den umgebenden Kreidebecken. Z Geol Wiss 37:15–39

Voigt T, Wiese F, von Eynatten H, Franzke H-J, Gaupp R (2006) Facies evolution of syntectonic Upper Cretaceous deposits in the Subhercynian Cretaceous basin and adjoining areas (Germany). Z Dt Ges Geowiss 157:203–244

Von Buch L (1822) Über Dolomit als Gebirgsart. Abhandl. der Berl. Akad., Berlin Von Buch L (1839) Über den Jura in Deutschland: eine in der Königlichen Akademie der Wissenschaften am 23. Februar 1837 gelesene Abhandlung. Dümmler, Berlin

Von Eynatten H, Voigt T, Meier A, Franzke H-J, Gaupp R (2008) Provenance of Cretaceous clastics in the Subhercynian Basin: constraints to exhumation of the Harz Mountains and timing of inversion tectonics in Central Europe. Int J Earth Sci 97:1315–1330

Wagner G (1960) Einführung in die Erd- und Landschaftsgeschichte mit besonderer Berücksichtigung Süddeutschlands, 3 edn. Rau, Öhringen, 694 pp

Wagreich M (2003) Die Etnwicklung des Gosaubeckens während der Kreidezeit. In: Weidinger JT, Lobitzer H, Spitzbart I (eds) Beiträge zur Geologie des Salzkammerguts. Gmundner Geo-Studien, pp 21–28

Walliser OH, Alberti H (1983) Flysch, olistostromes and nappes in the Harz Mountain. In: Martin H, Eder FW (eds) Intracontinental fold belt. Springer, Berlin, pp 145–170

Walter H, Schneider JW (2008) Perm – Rotliegend. In: Pälchen W, Walter H (eds) Geologie von Sachsen. Schweizerbart'sche Verlagsbuchhandlung, Stuttgart, pp 223–257

Walter R (2007) Geologie von Mitteleuropa, 7th edn. Schweizerbart'sche Verlagsbuchhandlung, Stuttgart, 511 pp

Walther HW (1986) Federal Republic of Germany. In: Dunning FW, Evans AM (eds) Central Europe. Mineral Deposits of Europe, vol 3. Institute of Mining and Metallurgy and Mining Society, London, pp 194–198

Warr LN (2012) The Variscan Orogeny: the welding of Pangaea. In: Woodcock N, Strachan R. (eds) Geological history of Britain and Ireland, 2nd edn. Wiley-Blackwell, pp 275–298

Wegener A (1912) Die Entstehung der Kontinente. Geol Rdsch 3:276–292

Wegener A (1915) Die Entstehung der Kontinente und Ozeane. Sammlg. Vieweg, Braunschweig, 94 pp

Wegener A (1929) Die Entstehung der Kontinente und Ozeane, 4. ed Vieweg, Braunschweig, 231 pp

Weller H (1991) Facies and development of the Devonian (Givetian/Frasnian) Elbingerode complex in the Harz area/Germany. Facies 25:1–50

Weller H (2008) Elbingeröder Komplex. In: Deutsche Stratigraphische Kommission (eds) Devon. Stratigraphie von Deutschland, vol VIII. Schriftenr Dt Ges Geowiss, 52, pp 525–531

Weller H (2012) Eine neue Geologische Karte des Elbingeröder Komplexes/Harz – GK EK 2011. Hallesches Jb, Beih 28:57–59

Wellnhofer P (1983) Das Land der Isar im geologischen Blockbild. In: Frank, Jung, Werner, W, Geologie rechts und links der Isar. In: Plesse M-L (ed) Die Isar. Ein Lebenslauf. München, 24 pp

Wendt JI, Kröner A, Fiala J, Todt W (1993) Evidence from zircon dating for existence of approximately 2.1 Ga old crystalline basement in southern Bohemia, Czech Republic. Geol Rdsch 82:42–50

Werner AG (1786) Kurze Klassifikation und Beschreibung der verschiedenen Gebirgsarten. Abh Böhm Ges Wiss 2:272–297

Wernicke B (1985) Uniform-sense normal simple shear of the continental lithosphere. Can J Earth Sci 22:108–125

Wilde SA, Valley JW, Peck WH, Graham CM (2001) Evidence from detrital zircons for the existence of continental crust and oceans on the Earth 4.4 Gyr ago. Nature 409: 175–178

Wunderlich H-G (1968) Endogene Dynamik Einführung in die Geologie, vol 2. BI Hochschultaschenbücher, 341, 197 pp

Zachos J, Pagani M, Sloan L, Thomas E, Billups K (2001) Trends, rhythms, and aberrations in global climate 65 Ma to present. Science 292:686–693

Zeh A (2005) Exkursionsführer in das Kyffhäuser Kristallin und Ruhlaer Kristallin, 44 pp

Zeh A, Wunderlich J (2003) Mitteldeutsche Kristallinzone. In: Seidel G (ed) Geologie von Thüringen. Schweizerbart'sche Verlagsbuchhandlung, Stuttgart, pp 24–51

Ziegler PA (1990) Geological Atlas of Western and Central Europe, 2nd edn. Shell Internationale Petroleum Maatschappij B.V. Geol Soc Publ House, Bath, 239 pp

Ziegler PA (1992) European Cenozoic rift system. Tectonophysics 208:91–111

Ziegler PA (1994) Cenozoic rift system of Western and Central Europe: an overview. Geol Mijnbow 73:99–127

Ziegler PA, Cloetingh S, van Wees J-D (1995) Dynamics of intra-plate compressional deformation: The Alpine foreland and other examples. Tectonophysics 252:7–59

Ziegler PA, Dèzes P (2006) Crustal evolution of Western and Central Europe. Mem Geol Soc Lond 32:43–56

Zirngast M (1991) Die Entwicklungsgeschichte des Salzstocks Gorleben – Ergebnisse einer strukturgeologischen Bearbeitung. Geol Jb, A 132:3–31

Zittel KA (1884) Handbuch der Paläontologie. Abt. 1, vol 2. R. Oldenbourg, München, Leipzig, 893 pp

Zolitschka B, Negendank JFW, Lottermoser BG (1995) Sedimentological proof and dating of the early Holocene volcanic eruption of Ulmener Maar (Vulkaneifel, Germany. Geol Rdsch 84:213–219

Index

M. Meschede and L. N. Warr, *The Geology of Germany*, Regional Geology Reviews,
https://doi.org/10.1007/978-3-319-76102-2

MIX
Papier aus verantwortungsvollen Quellen
Paper from responsible sources
FSC® C105338

If you have any concerns about our products,
you can contact us on
ProductSafety@springernature.com

In case Publisher is established outside the EU,
the EU authorized representative is:
Springer Nature Customer Service Center GmbH
Europaplatz 3, 69115 Heidelberg, Germany

Printed by Libri Plureos GmbH
in Hamburg, Germany